C. L. Marlatt, Edwin Willits, U. S. Department of Agriculture

INSECT LIFE

C. L. Marlatt, Edwin Willits, U. S. Department of Agriculture

INSECT LIFE

ISBN/EAN: 9783741124891

Manufactured in Europe, USA, Canada, Australia, Japa

Cover: Foto ©Klaus-Uwe Gerhardt /pixelio.de

Manufactured and distributed by brebook publishing software (www.brebook.com)

C. L. Marlatt, Edwin Willits, U. S. Department of Agriculture

INSECT LIFE

U. S. DEPARTMENT OF AGRICULTURE.
DIVISION OF ENTOMOLOGY.
PERIODICAL BULLETIN. JULY, 1889.
Vol. II. No. 1.

INSECT LIFE.

DEVOTED TO THE ECONOMY AND LIFE-HABITS OF INSECTS, ESPECIALLY IN THEIR RELATIONS TO AGRICULTURE, AND EDITED BY THE ENTOMOLOGIST AND HIS ASSISTANTS.

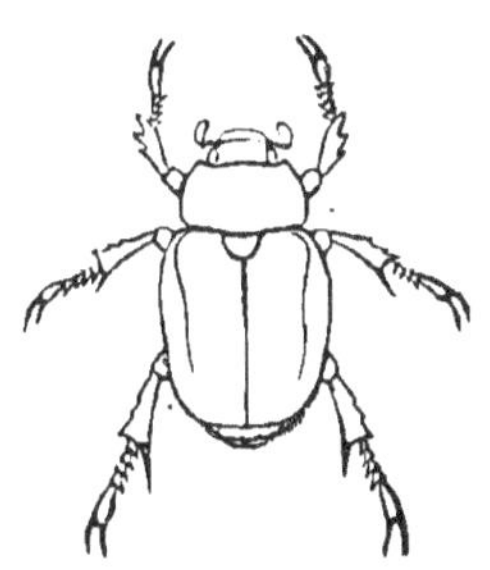

WASHINGTON:
GOVERNMENT PRINTING OFFICE.
1889.

CONTENTS.

SPECIAL NOTES.

With this number we commence the second volume of INSECT LIFE. The last number was somewhat delayed by the preparation of the extensive indices, which, however, we feel will greatly increase the value of volume I. Largely through the kindness of the authorities of the Government Printing Office we were able to print the numbers during the past year more regularly and promptly than we had anticipated, and we hope to continue this regularity through the coming volume. As stated in the salutatory to the first volume, however, the force of the Division of Entomology is so actively engaged during the larger part of the year with field work and experimentation that some lack of promptness in publication can not but ensue. The publication of the bulletin met with even more favor than we had hoped at the start, and almost no adverse comments have reached our eye. The only criticism which we have noticed was published in the review column of the *Atlantic Monthly*, in which slight exception was taken to the idea of the publication of a magazine by the Government, which by its free distribution would compete on unfairly advantageous terms with private enterprises. We have no comment to make except that the various branches of the Government are constantly publishing bulletins, many of which differ but slightly in character from this, so that if the title-page were only slightly changed, INSECT LIFE would escape all criticism of this kind. We trust that the interest of our readers will continue, and that the tendency which INSECT LIFE has so far shown, to increase the correspondence, and therefore the range of benefit of the Division of Entomology, will also continue.

South African Insects.—That indefatigable worker, Miss Eleanor A. Ormerod, has just brought out privately a little work entitled "Notes and Descriptions of a Few Injurious Farm and Fruit Insects of South Africa." The book is published by Simpkin, Marshall & Co., of London, and the price is 2*s.* 6*d.* The descriptions and identifications of the insects are by Mr. Oliver E. Janson, and the species are figured in nearly all instances. Many items of interest strike us in glancing through the pages, and while many of the species seem to be vicarious with our own in the damage which they do, but one (barring scale insects) seems to be identical with any American injurious species. This is the Diamond-back Moth (*Plutella cruciferarum*), which damages cabbage in the East Province. The Fluted Scale (*Icerya purchasi*) of course occupies a considerable space, while the Flat Scale (*Lecanium hesperidum*) is also mentioned.

Among the vicarious pests may be mentioned the Orange Fly (*Ceratitis citriperda*), which damages oranges in the same way as does the Morelos Orange Fly (*Trypeta ludens*—see INSECT LIFE, August, 1888, page 45); the Orange Butterfly (*Papilio demoleus*), which works upon the foliage of the Orange in the same way that the Orange Dog (*Papilio cresphontes*) works in Florida; the Bean Seed-weevil (*Bruchus subarmatus?*), which damages beans just as does the Bean Weevil (*Bruchus obsoletus*) in America; a large Cantharid (*Mylabris oculata*), which injures beans and peas in a similar manner to the damage done by Meloids in the West; and the Cetoniid (*Rhabdotis semipunctata*), which injures figs and peaches just as does the "Fig Eater" (*Allorhina nitida*) in our Southern States.

Professor Forbes' Correction.—We are glad to make room in this number for an article received from Professor Forbes which corrects a statement in the article on the Plum Curculio in the Annual Report of the Department for 1888. Our information in the Annual was derived from a newspaper report which we supposed reliable.

A Phase of Buffalo Gnat Injury.—A report by Mr. Marlatt on a trip made in April, published in the present number, is comparatively interesting as indicating an unexpected result of certain operations by the Army engineers. We have already published Mr. G. A. Frierson's letter concerning this peculiar case (see INSECT LIFE, April, 1889, Vol. I, page 313), and in the light of Mr. Marlatt's observations our opinion there published is confirmed. It is a hard case, and the only remedy can come through Congress in the shape of an item in the river and harbor appropriation bill.

Bibliography of American Economic Entomology.—The first part of this long-delayed work is now being printed, and the second part will probably be in the printer's hands by the close of the year. The preparation of this work has been in the hands of Mr. Samuel Henshaw, of Boston, for the past two years, and the first part, just now being printed, comprises the writings of B. D. Walsh and C. V. Riley.

Bulletin on Root-knot Disease in Florida.—This bulletin, mentioned in our Special Notes in the last number of INSECT LIFE, has been delayed for the plates, which the printer could not have executed until after July 1. We hope soon, however, to have it ready for distribution.

ARSENICAL POISONS FOR THE PLUM AND PEACH CURCULIO.

By S. A. FORBES

The following report of results of my recent experimental work on the common peach curculio is intended to correct and complete a reporter's summary of remarks made in August, 1888, at a meeting of the Central Illinois Horticultural Society, at Champaign, as republished in the last Report of the U. S. Entomologist, page 75. The experiments there alluded to were not generalized by me, but were described as merely preliminary to a much more elaborate series which I have since carried through.

The object of these experiments has been to ascertain some details of the food and feeding habits of the curculio and to test its sensibility to arsenical poisons when distributed on the trees which the insect frequents. In the case of the peach it was important also to find what amount of these poisons the leaves might receive without marked injury.

FEEDING EXPERIMENTS.

June 15, 1888, plum curculios confined with plum leaves. June 16, one observed making a deep, sharp, oblong excavation in the midrib; similar work on other midribs, petioles, and stems. Beetles also seen gnawing the surfaces of the leaves, especially the fresher terminal ones. Leaves removed and green plums substituted. June 19, plums peppered with holes, some containing eggs, others not. July 2, fresh lot of beetles imprisoned with both leaves and green plums. The next day both had been eaten, the plums perhaps the more freely.

Several examples taken April 14, 1889, before peach trees were in bloom, were proven by dissection to have last fed on dead vegetation, as shown by the absence of chlorophyl and the presence of some of the fungi of decomposition. Curculios confined April 19, with both dead and living peach leaves, fed only on the latter, not having touched the dead leaves at the end of three days. Peach blossoms being placed in the cage, with fresh leaves also, April 22, both were freely eaten at once, the blossoms being, however, evidently preferred. Both calyx and corolla were perforated with small round holes, and eaten away from the edge.

Three specimens taken in southern Illinois were dissected April 23, and found to contain vegetable tissues, chiefly of leaves (as shown by the fragments of spiral vessels), without fungi and with more or less chloryphyl. Vegetable hairs and peculiar pollen grains, not those of fruit blossoms, were also recognized.

Thinking it possible that the curculio might feed on flowers somewhat indiscriminately, we put a number under a bell glass with roses in full bloom. The next day, May 19, the petals were much eaten, and two days later calyx and peduncles had likewise been attacked. The rose leaves were nct injured. When rose blooms and peach leaves together were offered the imprisoned beetles, they fed freely on both.

Again, May 23, curculios were confined with both bush honeysuckles and snowballs in blossom. The next day the honeysuckle blossoms were eaten, and on the second day those of the snowball also. On the other hand, beetles shut up with peach leaves and peony flowers ate the peach at once, as usual, but refused the peony entirely, not having eaten it at all after ten days.

INSECTICIDE EXPERIMENTS.

My first experiments with insecticides for the curculio alone were made July 6, 1888. Two lots were placed under glass, with leaves and green fruit of the plum, the food of one being sprayed with Paris green, 1 pound to 50 gallons of water, and the other not. The first beetle died in the poisoned lot July 9, and the next day all were dead, the check lot continuing without loss. July 28 a similar experiment was made with Paris green, 1 pound to 100 gallons, applied until the leaves began to drip. The poisoned beetles commenced to die the next day, and five of the six were dead on the 31st. In the check lot of six, on the other hand, only one was dead.

An experiment begun with 1 pound to 200 gallons was unavoidably suspended in two days, before results were reached.

Next, April 19, 1889, a lot of curculios, greatly exhausted by long confinement in transit, were divided into five lots—the first, of twenty-four, a check; the second and third of twelve each, the fourth of nine, and the fifth of twelve. The food of the second lot was treated with Paris green mixed with water at the rate of 1 pound to 100 gallons; that of the third, with a pound to 200 gallons; the fourth, a pound to 300, and the fifth, a pound to 500 gallons.

The previous hardships of the check lot caused many of them to die, most of them having been insensible, in fact, when first released; but the effects of the poisons were nevertheless evident, as shown by the subjoined table:

Paris-green experiment No. 1, April 19, 1889.

Died.	Check lot. Number used, 24.	1 lb. to 100 gals. Number used, 12.	1 lb. to 200 gals. Number used, 12.	1 lb. to 300 gals. Number used, 9.	1 lb. to 500 gals. Number used, 12.
April 22	....	3	3	1	2
23	2	2	....	2	3
24	1	2	3	2	1
25	....	2	4	2	3
26	3	1	2	....	1
27	2	2	....	1	1
29	2	....	....	1	1
Total	10	12	12	9	12

May 4 this experiment was repeated with a fresher lot of beetles, with more marked results, curculios commencing to die two days after treatment in all the poisoned lots but one, all of one lot being dead in nine days, and in ten days all of every poisoned lot but a single beetle. In the check lot, meanwhile, only one had died.

Paris-green experiment No. 2, May 4, 1889.

Died.	Check lot. Number used, 12.	1 lb. to 100 gals. Number used, 12.	1 lb. to 200 gals. Number used, 12.	1 lb. to 300 gals. Number used, 22.	1 lb. to 500 gals. Number used, 22.
May 6	....	3	2	....	1
7	....	1	....	4	2
8	....	1	2	2	1
9	....	2	3	3	3
10	1	3	1	4	4
11	....	1	....	6	4
13	....	....	4	2	5
14	....	1	....	1	1
Total	1	12	12	22	21

In both the above experiments, as also in the following, peach leaves were used as food, and these were sprayed but once.

All strengths of the poison mixture here killed the beetles feeding on it, the difference being seen in the rapidity with which they took effect. In four days from poisoning the ratios killed were 42 per cent. in lot two, 33 per cent. in lot three, 27 per cent. in lot four, and 18 per cent. in lot five.

Finally, May 17, a still more extensive experiment was begun with London purple, three hundred and forty-seven curculios being divided into five lots as before, their treatment differing from that of the foregoing only in the substitution of London purple for Paris green. The results were rendered, however, somewhat less satisfactory by the lateness of the season, which probably accounts for the number of deaths in the check. Other parallel observations led to the conclusion that spent adults, doubtless the earliest to emerge, were already beginning to die spontaneously. The experiment was continued for eight days, when all the curculios of the first lot were dead, and nearly all of the other poisoned lots, a fourth of the check having also perished.

London purple experiment, May 17, 1889.

Died.	Check lot. Number used, 47.	1 lb. to 100 gals. Number used, 100.	1 lb. to 200 gals. Number used, 100.	1 lb. to 300 gals. Number used, 50.	1 lb. to 5.0 gals. Number used, 50.
May 19	....	35	37	16	12
20	....	18	19	4	6
21	1	18	10	2	4
22	....	10	11	9	10
23	5	5	7	7	8
24	4	6	5	3	5
Total	10	92	89	41	45

EFFECT ON THE FOLIAGE.

It is well known to fruit-growers that the leaves of the peach are much more sensitive to the scorching effect of the arsenical poisons than those of the apple or plum, and it is important to know just how strong a mixture of the common arsenical insecticides that tree will bear under favorable, and also under unfavorable, conditions. My experiments on this point are incomplete, but they are given here for what they are worth:

First. Two branches of a peach tree were sprayed May 18 with London purple mixtures, a pound to 100 and a pound to 200 gallons, respectively. A week later no noticeable difference could be made out between the condition of the two branches, the tips of the leaves in both being somewhat deadened and dry. May 20 identical applications were made, with no apparent effect on the foliage by May 22. Heavy rains followed, and no further observations were made.

June 6 two other branches were sprayed as before. A heavy rain followed June 8, and more upon the 9th. On the 10th the effects of the poison were somewhat apparent on both branches, reddish discolorations occurring where the fluid had gathered in drops and also along the margins of the younger leaves. Further rains occurred on the 16th and 17th. On the 18th the discolored spots had increased in size, those on the branch sprayed with the stronger solution being somewhat larger and more numerous. No leaves had fallen, but those worst affected were easily detached, and doubtless would have fallen eventually. This loosening of the leaves was evidently due, not to damage to the petiole, but to premature ripening of the leaf,* consequent on the chemical injury to the blade. June 8 two other branches were sprayed

* Ascertained by studying sections of the petiole.

as before, substituting Paris green for London purple in both mixtures. Light rain followed the same day, and more on the 9th. On the 10th a scorching of the leaves was somewhat evident, a little more so where the stronger mixture was used, while on the 18th the condition of the foliage was practically the same as on those branches treated with London purple—if anything, a little less severely injured. There was also a barely perceptible difference in favor of the weaker mixture. Supposing that all the worst injured leaves were rendered practically useless to the tree, the loss of foliage would probably amount to 4 or 5 per cent.

There can certainly be no further question of the liability of the curculio to poisoning by very moderate amounts of either London purple or Paris green while feeding on the leaves and fruit of peach or plum; but much additional experiment is needed to test the possibility of preventing serious injury to these fruits by this means. The pupal hibernation and late appearance of a considerable percentage of the curculios make it possible that sprayings must be several times repeated, and perhaps carried further into the season than is consistent with safety; and the limit of tolerance of these poisons by the peach under ordinarily trying circumstances has not been clearly ascertained. Further, the observations above reported on the food plants of the curculio make it likely that, in nature, a smaller proportion of the food of these beetles comes from the peach or plum than has hitherto seemed probable, and that poisons there applied would kill less certainly. It seems worth while to make the attempt to attract the adult to flowering plants in the orchard other than the peach, with the hope of poisoning it there (especially late in the season) without using these dangerous insecticides on fruits afterwards to be eaten.

REPORT OF A TRIP TO INVESTIGATE BUFFALO GNATS.

By C. L. MARLATT, *Assistant.*

WASHINGTON, D. C., *April* 22, 1889.

SIR: In accordance with your letter of instruction of April 5, 1889, I proceeded to Frierson's Mill, La., and studied, as far as the conditions would permit, the relation of the raft of logs in Bayou Pierre to the injurious abundance of the gnats in that immediate locality. Examination was also made to determine the feasibility of removing the raft to prevent the further breeding of the gnats thereon.

I wish here to express my thanks to Mr. G. A. Frierson and brothers for their kind hospitality, and for the efficient aid rendered by them in the investigation of the raft and bayou.

Respectfully,

C. L. MARLATT.

Prof. C. V. RILEY,
U. S. Entomologist, Washington, D. C.

As you had surmised would be the case, the Buffalo gnats had already disappeared when I arrived at Frierson's Mill. A few Turkey Gnats were observed about horses and cattle, but it was evidently somewhat early for this species to be about abundantly.

The severity of the attacks of the Buffalo Gnat the present season was plainly indicated by the general emaciated condition of the cattle and mules—the effect also of the repeated application of oils on the latter being shown on many of them by the loss of large patches of hair. The remains of smudge fires were frequently seen in the vicinity of the negro houses and through the woods. In addition to these visible indication of the *Simulium* attacks was the unvarying testimony to that effect of the planters and negroes questioned, all of whom ascribed the abundance of the gnats to the presence of the raft, and manifested no little anxiety to have the Government take measures to prevent the yearly recurrence of this pest.

As shown in the letter from Mr. G. A. Frierson, and also by my own inquiries, the planters have, from their extended experience with the gnats, learned how to prevent loss of stock, by the use of train oil to which a small amount of sulphur is commonly added for the work animals; and smudge fires for cattle, sheep, etc. But the annoyance during the six weeks of the spring from the immense swarms of gnats, practically stopping field work, and also preventing the stock from feeding, can not be avoided.

As shown later the raft was formed in 1872–'73. The gnats were not especially troublesome, however, previous to the spring of 1885, since which time they have appeared in increasing numbers every year. They seem to have extended the present season 5 to 10 miles out from the bayou, swarming in greater numbers on cleared and particularly on meadow land.

As indicating the abundance and probable source of the gnats the present year, the report of several planters living near the raft is here recorded, viz, that the water in the neighborhood of the logs in the time of the greatest abundance of the gnats seemed to be in ebullition from the great numbers of flies constantly popping to the surface.

A heavy rain on the day of my arrival (April 13) prevented an immediate examination of the raft and bayou, and, unfortunately for my work, the rain continued with increased violence during the night and part of the day following. The bayou became much swollen, rising, in fact, nearly up to the high-water mark of the spring (February and March) floods, and 6 to 8 feet above the level of the few weeks previous, during which the Buffalo Gnats had been abundant. By this means much of the raft, and especially that portion likely to bear evidences of the gnats, either as eggs, larvæ, or cocoons, was covered with water; and as the raft consisted of large logs tightly wedged together, it was impossible to remove them for examination with the means at hand, except in a few instances. The floating portion of the raft was not likely to contain cocoons in any quantity, and larvæ were not found on these logs, although they afforded excellent breeding places in the numerous whirls of water caused by the rapid current of the stream impinging against them.

Careful and continued search on April 15 over 2 or 3 miles of the lower portion of the raft, near Lake Cannisnia, resulted in the finding of a few isolated cocoons on logs which were partially upright, and thus projected several feet into the water. Logs so placed, and possible of removal for examination, were not commonly met with. Nearly all of the floating logs extended lengthwise on the surface of the water, being submerged but a few inches, and hence did not afford suitable conditions for the cocoons, and if larvæ of the buffalo gnat were on the logs their small size prevented their discovery.

On the day following (April 16) that portion of the raft near Red Bluff was examined, and here again were found excellent breeding places for Simulium larvæ, viz, a swift current striking against the logs and rubbish of the raft forming innumerable whirls and eddies, and somewhat better success attended our search here. On submerged branches, twigs, etc., which projected several feet below the surface of the water and which were evidently raised with the floating lower portion of the raft, were found large numbers of cocoons (some few of which contained pupæ) and larvæ. A few cocoons and larvæ were also found attached to water plants growing from the logs. These specimens were found only where the current set strongly against the raft, this causing the riffles known to be necessary for the larval and pupal existence of Simulium species.

The larvæ and pupæ found proved to be largely if not altogether those of the Turkey Gnat (*S. meridionale* Riley). Many of the larvæ did not exceed 1^{mm} in length, were almost hyaline and apparently but recently hatched (?); others were full grown, and spinning cocoons. These larvæ were found attached to the smaller branches and twigs which were in nearly every instance already thickly crowded with cocoons. It is probable from the association of the Turkey Gnat larvæ with most of the deserted cocoons that the latter had contained the earlier appearing gnats of this species. Some few of the cocoons may have been those of the Buffalo Gnat, as also some of the minuter larvæ, but this could not be satisfactorily determined.

The height of the water prevented any satisfactory examination of the trees and shrubs growing near the bayou, but wherever possible branches or vines so situated and extending into the water were drawn out and examined. No evidence of gnats, however, was found.

Mr. G. A. Frierson has promised to look for cocoons here as soon as practicable. The reported appearance of the gnats coming to the surface in such places in quantity as well as about the raft would indicate that the larvæ had during the spring flood attached their cocoons to such submerged trees and branches. Examination will also be made at low water for further evidences of the gnats on the lower and at present inaccessible portion of the raft, which is more likely to bear cocoons in quantity than the floating material at high water.*

The relation of Bayou Pierre to the Red River is such, as shown in the report of Captain Bergland contained in the Annual Report of the Chief of Engineers, U. S. Army, Part II, 1885, Appendix U, pages 1487–1493, that in times of high water three-fourths of the discharge of the Red River is through this bayou, and in times of low water but one-fourth. Before the formation of the raft, this very great augmentation of the bayou in high water had no ill effect, but now the water, checked by the raft, floods every spring much of the adjacent low-land, thus furnishing additional foothold for larvæ, and possibly also driving the adults in larger swarms to the higher land.

During the summer months the water is confined by moderately high clean banks and is free from drift, except where such material is held by the raft. This would indicate that the raft is largely responsible for the abundance of the gnats in that locality.

The smaller streams in the neighborhood dry up in the course of the summer, and hence could not breed gnats; however, a number of the principal ones were carefully searched for cocoons or larvæ without the discovery of any evidence of them.

The Buffalo Gnat was reported to be quite abundant on the Sabine River; and it also occurs in less numbers on the Red and Washita Rivers.

The raft in Bayou Pierre originated in the attempt of the United States Government (in 1872–'73) to close Tone's Bayou, which connects Bayou Pierre with the Red River, and to confine the water of Red River to its own channel. A large raft which was being removed from above Shreveport was run into Tone's Bayou and the attempt made to retain it there by means of a boom. This raft and also a second one formed later were entirely swept away by floods and carried into Bayou Pierre, where they are at present lodged. As described by Captain Bergland (l. c.) the raft "extends 5.3 miles above and 2.8 miles below Red Bluff at the mouth of Wallace Lake. The upper portion is fragmentary, of recent formation and loose structure, occupying in the aggregate one-fifth of the area of the water surface. That below is nearly continuous and gradually becomes denser until at its lower end it becomes solidified." This lower portion of the raft has now become almost entirely solidified by the massing of the logs and the accumulation of débris, and trees and shrubs are now growing upon it.

---

*Mr. Frierson subsequently collected and forwarded to the Department a considerable quantity of material—cocoons, larvæ, etc., from this place, concerning which we quote briefly from his letter of May 3 as follows:

"The water has fallen about 2 feet below its level when the gnats were hatching out. * * * The current is very swift, * * * and I found that every overhanging tree, logs sticking out of the water, and the millions of roots on the bank were literally plastered over with the cocoons for the distance of 2 feet above and below the water."

In the estimates made by the Government Engineers for the clearing of Bayou Pierre, the principal item has been the cleaning out of this lower raft. The removal of this portion of the raft is not now necessary, however, as the water has made for itself a new channel through Bennett's Bayou on the west. This natural change in the course of the stream, and the slow but constant breaking up of the remaining and less stable portions of the raft, will make the clearing of Bayou Pierre at the present time comparatively inexpensive. The raft, even if left to take its own course, would in time go out of itself, and if the work of loosening that portion above Red Bluff should be undertaken in time of high water, the bayou might be freed of logs with little difficulty. Taking the estimates in the report above cited as a basis, $25,000 would probably cover the expense of removing such portions of the raft as is now necessary. The final disposition of the material of the raft would occasion some difficulty. It could, however, be directed into Lake Cannisnia and secured there in still water, beyond the reach of the current of the Bayou Pierre, which crosses this lake. If this were done the gnats would not breed on the raft, and in a short time, by the accumulation of sediment and growth of plants, it would become entirely solidified, as is now the case in its lower portion in Bayou Pierre.

The utility of the stream as a water-way and the reclaiming of much valuable land which would result from such improvement, while having no direct bearing on the question at issue, may still be mentioned as an additional reason for removing the obstruction of logs, if this is thought not to be warranted by the presence of the gnats alone.

NOTES ON NOISES MADE BY LEPIDOPTERA.

By Henry Edwards.

The article by Mr. A. H. Swinton on "Stridulation in *Vanessa antiopa*," published in the last number of "Insect Life," Vol. 1, p. 307, has directed my attention to the subject, and I venture to add a few notes on this interesting phase of entomological study. It is not alone among the *Vanessas* that *antiopa* has the power of making a sound, for many years ago in England, when I began to collect butterflies and moths, I observed that the beautiful *Vanessa io*, the favorite of every young entomologist, gave out a slight rasping sound when many specimens were flying together, or when a male was in hot pursuit after the opposite sex. But the sound was very slight and could only be distinguished when "all around was still," and when there was no conflicting influence to deaden the insect's expression of love. The projecting vein which is shown in Mr. Swinton's cut is also quite apparent in *V. io*, and probably is a character of the whole of the genus. Still more remarkable is the noise produced by various species of the Nymphalid genus *Ageronia*, to which attention was first called by the late Charles Darwin in his "Naturalist's Voyage Round the World." This was his famous expedition in H. M. S. *Beagle*, which enabled him to contribute so largely to our knowledge of the fauna of the various countries visited. During his stay in Brazil he paid considerable attention to entomology, and his notes upon the singular habit of *Ageronia* are worth transcribing in full. He says:

I was much surprised at the habits of *Papilio feronia* (*Ageronia feronia* of later authors). The butterfly is not uncommon, and generally frequents the orange groves. Although a high flier, yet it very frequently alights on the trunks of trees. On these occasions its head is invariably placed downwards, and its wings are expanded in a horizontal plane, instead of being folded vertically, as is commonly the case. This is the only butterfly which I have ever seen that uses its legs for running. Not being aware of this fact, the insect, more than once, as I cautiously approached with my forceps, shuffled on one side just as the instrument was on the point of closing, and thus escaped. But a far more singular fact is the power which this species possesses of making a noise. Several times when a pair, probably male and female, were chasing each other in an irregular course, they passed within a few yards of me, and I distinctly heard a clicking noise, similar to that produced by a toothed wheel passing under a spring catch. The noise was continued at short intervals, and could be distinguished at about 20 yards distance. I am certain there is no error in the observation. (Nat. Voyage, Appleton's edition, p. 33.)

As a boy, I had read this interesting note by the great naturalist, and in the last months of the year 1866 I had, during a stay of four weeks in Panama, the opportunity of observing for myself this curious butterfly habit. The species *Ager. feronia*, *A. ferentina*, and *A. amphinome*, and more especially the two former, are particularly common in the forests around the city of the Isthmus, and it is not possible to walk a mile through them without meeting with many examples. The sound made by the first-named species is like that of the next, and somewhat recalls the noise produced by a boy's imitation of the old watchman's rattle. It is a decided "click," "click," very often repeated, and can be, as Mr. Darwin says, distinctly heard at the distance of 20 yards. Indeed, I should be disposed to extend this to at least 40 or 50 yards on a clear day, and when no wind could carry the sound away. The noise of *A. amphinome* is a heavier and more grating sound, and the two species can be readily distinguished without being seen. The trees on which they are accustomed to sit are species of *Cassia* or *Mimosa*, and their gray color, closely resembling that of the bark, renders them rather difficult to be seen when at rest.

I once went into the forest some time after sundown to see if they remained at night upon the trunks of the trees, as moths do in the daytime, but I could not find a single specimen, although many trees on which I had noticed them during the day were carefully examined. In a foot-note to page 33 of Mr. Darwin's narrative, he quotes Mr. Edw. Doubleday as having described before the Entomological Society, March

3, 1845, "a peculiar structure in the wings of this butterfly which seems to be the means of its making its noise." He says:

It is remarkable for having a sort of drum at the base of the forewings, between the costal nervure and the subcostal. These two nervures, moreover, have a peculiar screw-like diaphragm or vessel in the interior.

Darwin also alludes to a statement in Langsdorff's travels (1803–'07) that "a butterfly called *Februa hoffmanseggi* makes a noise when flying away like a rattle." This name probably refers to *Ager. ferentina*, which was called *Ager. februa* by Hübner.

In addition to this genus, I have observed the power of stridulation in two other butterflies, viz, in those of the genus *Prepona*, also natives of Tropical America, and in *Charaxes sempronius* of Australia. The noise of *Prepona* is only made as it takes wing from the trunks of the trees, on which it is also fond of resting, and is not repeated during its flight) It is therefore most probably in this case used as a defense against birds or other enemies. The *Charaxes* as it alights upon a bunch of the beautiful and sweet-scented flowers of *Bursaria spinosa* closes its wings with a grating sound not unlike that of the *Prepona*, and repeats the same as it is disturbed from its resting place. In butterflies it would appear that the noises are all caused, as Mr. Swinton suggests, by the rubbing of one vein of the upper wing against a corresponding vein in the lower wing, and probably they are all produced by slight modifications of the same structure, and it would appear that the power of stridulation is confined to the *Nymphalid* group, in which, as will be readily seen, a large development of the veins of the wing, particularly towards their bases, occurs.

There is very considerable difference in the sounds produced by the moths, that of one species having been likened by the older authors to "the voice of anguish, the moaning of a child, the signal of grief." This description applies to the well-known European *Sphinx* (*Acherontia*) *atropos*, familiarly known as the "Death's Head Moth," which gives out a very singular and plaintive cry, not unlike that (though in a greater degree) produced by a captive beetle of the *Geotrupid* or *Coprid* group when pressed between the finger and thumb. The noise of the great *atropos* has caused it to be regarded with superstitious terror, and this added to the grotesquely horrid mark of the skull and eye-sockets upon the thorax has made it in the districts in which it abounds an object of awe and terror. It is somewhat strange that, in this age of entomological research, the means by which the sound is produced by this species is yet unknown, comparative anatomists being considerably at variance in their opinions on the subject. Some observers have stated that the larva of this insect has also the power of emitting a sort of squeaking noise.

In our own country, if any one has ever noticed a large swarm of the pretty little moth, so injurious to our grape-vines (*Alypia octomaculata*), about a bush of flowers, he will have been conscious, if his ears were attuned to the finer harmonies of nature, of a slight breezy sensation rather than a sound, but one quite appreciable by a clear hearing. If the moths are driven away, the sound ceases, and there is no doubt but that it had been produced by the males in paying court to their mates, and probably by rubbing the antennæ at their tips across the costal nervure, which will be seen to be considerably thickened about its middle, just where the apex of the antenna would reach it. This thickening of the costa is much more apparent in an allied species, *Alyp. lorquinii*, than in our common form. With the latter I was enabled some few years ago, while walking across the Public Garden in Boston, to notice the peculiarity I have spoken of. The insect was in the greatest possible abundance upon a small bush of a plant of the Composite family, the name of which I do not know, not less, I should think, than from two hundred and fifty to three hundred specimens being about the single shrub. I distinctly heard the slight humming noise to which I have alluded, and am quite confident that it did not proceed from the vibration of the wings.

A more remarkable instance of stridulation, and certainly the most striking that has come under my notice, I was fortunate enough to witness during my residence in Australia. I was collecting insects in the Plenty Ranges, about 20 miles from Melbourne, and in the burning heat of mid-day had sat down to rest and pin my captures under the shade of a thick acacia tree. I was astonished and almost startled at a peculiar sound apparently very near me, which was unlike anything I had ever heard, and which I at first thought was the voice of some unfamiliar bird. I listened intently, looking in the direction of the noise, but could see nothing. I took up my net and walked up the opening in the woods, the sound still continuing, and greatly exciting my curiosity. It was very loud and distinct and not unlike "whiz, whiz," repeated by the mouth with the teeth closed. I had proceeded about thirty yards when the noise suddenly stopped. I sat down and waited, thinking that I should again hear it and be able to trace it to its source. I was not disappointed, for in a few minutes it again appeared, and this time quite close to me. I looked very carefully and in an opening, buzzing about with a swaying lateral motion, were two or three insects, which at first sight I took to be some species of Hymenoptera. I gave a sweep with my net and made a capture which was soon safe within my collecting bottle. My heart beat violently, as I found that I had taken a lovely black and orange moth, such as I had never before seen. I was alone, and had no one to whom I could communicate my pleasure, but I clearly understood Mr. Wallace's feeling upon his first capture of *Ornithoptera crœsus*, which he so graphically describes in his "Malay Archipelago," and I felt as if I should have gloried in making those primeval woods echo with my shouts.

Three more of the beautiful little creatures soon found their way to my collecting box, and the records of that day's excitement still remain with me in a treasured corner of my collection. The whole of my specimens are males, and it was not until some years after that I became acquainted with the other sex of this singular moth. It belongs, as does *Alypia*, to the family *Zygænidæ*, as we at present understand that very incongruous group, and the generic name is *Hecatesia*, my species being *H. fenestrata*. The structure by which the insect is enabled to produce the singular and striking sound is the thickening of the costal membrane about the apical third, behind which, and nearer to the center of the wing, is a rather broad vitreous space extending almost to the median nerve, this space being transversely ribbed, as are the bundles of eggs in some species of *Orthoptera*. The antennæ are thickened at the tips into a sort of prolonged club, pointed at the extreme end, and with the under side of the terminal joints horny and devoid of cilia. These, striking as they would do in flight at the will of the insect against the transverse muscles of the transparent space, cause the whizzing and characteristic sound which so attracted me, and which is doubtless intended as a call of love to the individual of the weaker sex, who sits enthroned in the branches listening with delight to the noisy homage of her many lovers.

Another species of this most curious group is found in the southeastern part of the province of Victoria, and was called by the late Adam White *H. thyridion*. I took several examples of this in the summer of 1856 at Westernport, the females, differing in this respect to the other species, being much more common than the opposite sex. In this the clear space is much smaller than in *H. fenestrata*, the sound produced by it being weaker and more closely resembling the buzzing of a bumble-bee. A third species of the genus, *H. exultans*, from Western Australia, is figured by Boisduval in Trans. Linn. Soc., London, 1877, and a fourth is described and figured as a native of Mexico by Mr. H. H. Druce, in the Biol. Centr Amer., but of the habits of this last mentioned nothing as yet is known.

A LETTER ON ICERYA PURCHASI.

The following letter was written June 10, 1889, by Hon. Edwin Willits, Assistant Secretary of Agriculture, to Hon. Ellwood Cooper, President of the California State Board of Horticulture, in response to a letter from Mr. Cooper transmitting certain resolutions of the fruit-growers of California. It is here published as a good summary of the past work of the Division of Entomology relative to this pest, and as a statement of the present condition of affairs:

DEPARTMENT OF AGRICULTURE, *Washington, D. C.*

HON. ELLWOOD COOPER,

President State Board of Horticulture, San Francisco, Cal.

I have the honor to acknowledge the receipt of your letter of May 20, transmitting the petition of the fruit-growers of California in convention assembled, to the effect that this Department send a qualified agent to Australia to collect and export to this country the parasites of the Fluted Scale (*Icerya purchasi*). Your petition is timely, and I abundantly realize the importance of the action which you suggest. In reply let me recite briefly the steps which have been taken during the past three years by this Department in regard to this great pest of the California fruit-growers, in order to place clearly before you the present condition of affairs.

As a result of numerous petitions from your State, in the spring of 1886 a competent agent of the Division of Entomology was appointed and was located at Los Angeles with instructions to carry out a certain line of experimentation which was mapped out for him by the Entomologist, Professor Riley. Later in the season another agent was sent to the same spot and the results of their combined work were published in the Annual Report of this Department for 1886, in an extended article by Professor Riley, which detailed thoroughly the life history of the pest and contained authoritative recommendations concerning remedies. Some of the washes recommended in this report were proven by careful experimentation to be perfectly efficacious and quite within the means of the most indigent fruit-grower.

Early in the spring of 1887 Professor Riley visited California in person and investigated the sections of the State in which the *Icerya* occurs, and in an address before your State Board at Riverside summarized his conclusions. Among other points brought out in this address was the suggestion that it would be very desirable to introduce its natural enemies and parasites from Australia. He expressed his regret that he would be unable to send one of his agents for the reason that Congress had limited the field of his investigation to the United States, but said that California, or even Los Angeles County, could well afford to appropriate the funds for the sending of an expert to Australia to devote some months to the study of the parasites there and to their artificial introduction into California.

During the summer of 1887 the two agents previously mentioned—Messrs. D. W. Coquillett and Albert Koebele—were continued in their work upon *Icerya*, and the Division at Washington was engaged in an industrious correspondence with entomologists in South Africa, Australia, and New Zealand, with a view of ascertaining facts bearing upon the natural habitat of this species and upon its natural enemies in these countries. The results of the additional experiments by the agents were published in the Annual Report of the Department for 1887. Those reached by Mr. Coquillett concerned chiefly the matter of treating trees with gases, while those attained by Mr. Koebele related entirely to washes. Meantime it had been found by correspondence that at least one important parasite existed in Australia, and strong efforts were made by the Department and also by the California delegation in Congress to secure a specific appropriation for the purpose of studying and importing this parasite. These efforts, as you well know, failed, as did also the equally strong effort on the part of this Department to have the clause in the appropriation bill, restricting the payment of traveling expenses to expenses within the United States, removed from the bill. The Department was thus rendered by Congress apparently powerless in the matter, but, fortunately, by a happy chance, which however will not occur again, we were able to send an agent after all through the courtesy of the Department of State. Congress had appropriated a large sum to enable this Government to exhibit at the Melbourne exposition, and the Secretary of State and the chief of the commission, Mr. McCoppin, of California, were kind enough to set aside a sufficient sum for this purpose, and Mr. Koebele went to Australia in August and accomplished the results with which you are already familiar.

During the winter of 1888-'89 strong efforts were again made by this Department to secure the removal of the restricting clause concerning foreign travel with the idea that, should Mr. Koebele's results warrant further importation of parasites, we would desire to send him or another agent again during 1889; in fact, to take just the action which you have petitioned us to undertake. This effort was apparently successful, and, as the Entomologist understood, the appropriation clause passed Congress in this modified form. On my assumption of my present office, in discussing this matter with the acting entomologist, I was put in possession of these facts, but was surprised to find, upon examination of the appropriation bill, that, in some way which I can not at this time explain, the restricting clause had been again inserted after it had been considered certain that it would be removed. The result is that the Department now finds itself in the same condition in which it was last year, and the only hope of Government help in this matter will rest in securing independent legislation the coming winter. The Department will urge strongly either the passage of an independent resolution or the addition of a clause to the appropriation bill which will set aside enough funds for this purpose, and we hope for your earnest co-operation in this direction.

Your Board should pass further resolutions and place them in the possession of the Senators and members of Congress from your State, urging such legislation, and in this way some action may possibly be brought about.

I have entered into this matter at some length in order to place strongly before you the fact that the Department has in no way been blind to the importance of the subject and that the interests of California have not suffered at its hands, as well as to show you definitely the impossibility of taking such action as you suggest at the present time, and to indicate, moreover, that efforts to obviate this state of affairs have been by no means wanting.

Meantime, however, I may express myself as strongly of the opinion that it will not do for California fruit-growers to tamely await Government aid in the way of the importation of parasites. I have seen myself that the *Icerya* can be overcome by persistent toil, and am quite inclined to indorse the sentiments expressed by Professor Riley upon page 164 of the December number of INSECT LIFE, a copy of which is sent you by accompanying mail. I would also call your attention to Professor Riley's latest article upon this insect, which you will find in the Annual Report for 1888, a copy of which has doubtless already reached your office.

Yours, respectfully,

EDWIN WILLITS,
Assistant Secretary.

EXTRACTS FROM CORRESPONDENCE.

American Insecticides in India.

A copy of your valuable periodical INSECT LIFE, Vol. I, No. 9, has to-day been sent to me. On page 293 you remark as follows:

"It has for some time seemed to us that the scale insects of the coffee plant which do so much damage in Ceylon and other parts of India could be successfully treated with the remedies which we have found in this country so valuable against the scale insects of the orange, namely, the kerosene soap emulsions, and we hope soon to bring this before the attention of the British Government."

You are probably not aware that kerosene emulsion has already been tried on Green Coffee-scale (*Lecanium viride*) in South India, and that so far as the experiments went it was found to be successful. Arrangements are being made for further experiments, and it is confidently hoped that this insecticide, with which Dr. Riley's name is so honorably associated in America, will prove of equal service in India. An account of what has been done in the matter of the introduction of kerosene emulsion and other American insecticides into India will appear in my forthcoming report, which has been in type for some months, and which will probably be published before this reaches you. A copy of the complete report, which deals with the whole investigation of Indian economic entomology, undertaken by the trustees of this museum, will be forwarded to you as soon as it appears.—[E. C. Cotes, Indian Museum, Calcutta, India, May 22, 1889.

Sciapteron robiniæ in Cottonwood in Washington Territory.

By to-day's mail we send you what appears to be the borer that destroys the Cottonwood and Balm trees of the West. While holding the creature on the blade of a saw, the pretty winged bug that you will find in the box shed off the dry skin, which you will also find with it. I took the creature directly from a hole in a Cottonwood tree which had apparently been bored by a borer. Please give us all the information about it that you can, its habits and the way to kill it, for publication in our paper.—[Legh R. Freeman, editor Washington Farmer, North Yakima, Wash. Ty., March 10, 1889.

REPLY.—Your letter of the 10th with specimen just received. The insect which you send is one of the Western Clear-winged Moths and is known as *Sciapteron robiniæ*. It breeds in Locust and White Poplar in Nevada and has been found in Cottonwood in California. It is a near relative to the common Peach-tree Borer of the East and belongs to a group of moths the larvæ of which all bore into the stems of trees and plants. It is probably neither sufficiently abundant nor destructive with you to occasion a demand for a remedy.—[May 18, 1889.]

A Fodder Worm in the South.

Mr. W. H. Peel, of this place, has called my attention to a worm which during the winter for three years has infested the stacks of dry corn blades, here universally called "fodder" and the main representative of hay in this country. The grown worm (I have seen but one) is over an inch long, a uniform brown, without hair, almost translucent, has full complement of feet for crawling rapidly, something like the Tortricidæ, but does not roll the dry leaves nor make a web till the chrysalid condition. Very abundant it seems and destructive—a new pest to the farmers of this region: yet as the fly has been coming out some two weeks I could get only a few, which are sent in a small box to-day. They come to light, but with others, and I refrain from catching them for fear of getting them mixed. According to Mr. Peel the worms are active for months, webbing up about the 1st of March and coming out the last of the same month, three to four weeks.—[Lawrence C. Johnson, Waterford, Miss., May 4, 1889.

REPLY.—Your letter of the 4th instant, inclosing specimens of an insect which attacks the stalks of dry corn, from the place of Mr. W. H. Peel, of Waterford, Miss., has been received. The specimens are very interesting, and belong to a species of Pyralid known as *Helia æmula*. The larva of this species has previously been found feeding upon the dry leaves of various plants in the woods, and also upon a number of fodder plants during the winter. The remedy will depend altogether upon the particular method in which the fodder corn is stored. Will you kindly request Mr. Peel to write us a full account of the way in which this insect works, and the manner in which he stores his fodder during the winter, and we will then advise him as to remedies. If he can send other specimens we shall be glad to get them.—[May 15, 1889.]

SECOND LETTER.—Your favor of the 15th instant received. Much obliged for your prompt information about *Helia æmula*. I found some dry clover hay once in process of destruction by a worm similar to this one, but on that occasion failed to get a fly, and had no one to watch them. I can tell you now all that is known of this specimen in Mississippi. As I wrote before, no one seems ever to have noticed its ravages until three years ago. The fodder in question consists of the blades stripped from standing corn (maize) as the fashion is at the South, and dried in the field in the sun. When dry or nearly so it is taken up and tied by a withe of its own leaves into bundles of about two pounds' weight. These when considered cured are carted up to points selected and stacked, with the butts within next the stack-pole, the ends without. A little of the ends take the weather as in any fodder that is stacked, and becomes worthless. This item is mentioned because it is the only part of the bundle not attacked by the insect. Externally, therefore, the stack seems perfectly sound and safe, when within it may be a mass of fragments and dung. The manner of eating the blades you may see in the bits put in the box sent you. They eat it pretty much all except the central vein; especially at the binds, where most compact, they eat all, running from that towards the ends. But a moldy or spoiled spot they never touch. The stack of fodder I saw had been put up about the last of August, 1888, and as remarked appeared perfectly sound till opened about the 1st of April. I am told the larvæ were then numerous, but they had already begun to web up. This is about all I can tell you; I never saw the egg.—[Lawrence C. Johnson, Bolivar, Tenn., May 19, 1889.

REPLY.—Your letter of the 19th, from Bolivar, Tenn., has just come. Thank you very much for the additional information relative to the habits of *Helia æmula*. I should imagine from what you write that the value of the fodder stacks is so slight that altogether the most satisfactory remedy will be to burn those which are infested with this insect. It strikes me that in this way and at slight expense the numbers of this pest can be greatly reduced.

The worm which you found in dry clover was probably a different thing, and I have no doubt that it was the common Clover-hay Worm (*Asopia costalis*), which you will find figured and described on pages 102 to 107 of Professor Riley's Sixth Report on the Insects of Missouri.—[May 23, 1889.]

Colonel Pearson's Method of fighting Rose Beetles.

I kill Rose-bugs by *smashing* them. I know of no insecticide which is also an insecticide for the Rose-bug—that is, which will kill the bugs and yet not injure the plant. Pyrethrum will intoxicate or stupefy them. They will fall from their perch and after a time recover and fly again. I have been experimenting for the past two weeks with all the poisons procurable in the drug shops, and without desired results. In dealing with Rose-bugs in my vineyards I send my men along the trellis early, from 6 to 10 a. m. They strike the vines with paddles; the bugs fall on the ground, and then they *smash* them with the paddles. The vines are trained upon a single wire, and the ground is made smooth and clean beneath, so that when the bugs fall they are at our mercy. This job must be done every morning until the bugs leave the vines for other foods. They are now on my strawberries and roses by myriads. Even if we could find something medicinal to kill the bugs, it would be of no use during such an invasion as we have had for the past three years in Vineland. Kill one and four more come to attend the corpse. They migrate and travel onward like the Army-worm. They must be fought by killing them as fast as they come. I have by this constant work for two or three weeks saved most of my vines, and I am now searching for something which will be offensive to them and drive them away from the plants they infest. Carbolated lime is the best I have found thus far.—[Alex. W. Pearson, Vineland, N. J., June 15, 1889.

Lyctus sp. in Bamboo.

I send you by mail to-day three bugs that are eating up a bamboo work-basket from Japan that I bought in Chinatown, San Francisco, Cal., a year ago last April. I have given it a thorough heating with flat-irons, which did not kill the pests, and then I gave the basket as thorough a bath of benzine, and that has not destroyed them. * * * The basket is being perforated with round holes, under which I find little dust piles. The dust I send with the bugs.—[Mrs. N. W. C. Holt, Winchester, Mass., June 20, 1889.

REPLY.—I beg to acknowledge the receipt of your letter of June 20. The insect found in your bamboo work-basket is not unknown as an enemy to bamboo imported from China and Japan. It is a species of a genus of wood-boring beetles called by entomologists Lyctus. You need not fear the spread of this insect, as they feed on nothing but bamboo. Keep up your benzine treatment and you will kill the insects. * * *—[June 25, 1889.]

The Texas Cattle-tick.

Will you please give me the history of the Texas Cattle-tick (*Ixodes bovis*) or refer me to the literature on the subject? They are a terrible pest here.—[M. Francis, D. V. M., College Station, Texas, June 17, 1889.

REPLY.—I beg to acknowledge the receipt of your letter of the 17th instant, requesting information concerning the Texas Cattle-tick (*Ixodes bovis*). This species was described by Professor Riley in a special report of this Department (Report of Commissioner of Agriculture on Diseases of Cattle in the United States, 1871, p. 118, foot-note). It is a reddish, coriaceous, flattened species, body oblong oval, contracted just behind the middle, and the whole insect is from one-quarter to one-half an inch in length. It occurs from the Northern States to Nicaragua, and lives not only on cattle but even on the rattlesnake, the iguana, and on small mammals. It no doubt attaches itself to almost any animal that brushes against it in going through the grass. The species is mentioned in a treatise on the external parasites of domestic animals, by A. E. Verrill, in the report of the Connecticut Board of Agriculture for 1870, page 46. It is found in our Northern States, but is, however, most abundant in the Southwest, Missouri to Texas, and has been taken in large numbers by Mr. J. McNeil on horned cattle on the west coast of Nicaragua.

As to remedies, the kerosene emulsion has been recommended for lice on cattle in Bulletin 5 of the Iowa Agricultural Experimental Station, May, 1889, page 185. This would no doubt be the best and most practical remedy for the Cattle-tick also, and is indorsed by Dr. Cooper Curtice, of the Bureau of Animal Industry of this Department, who recommends that the emulsion be made with soap according to the formula originally proposed by this Division. The emulsion should be applied in an 8 per cent. solution with a force pump, using the Riley or Cyclone nozzle and a few feet of hose. It thus easily penetrates the hair of the animal, and at that strength can not injure stock.—[June 24, 1889.]

The Boll Worm in Texas.

I take the liberty to report to you the condition of affairs in regard to the Boll Worm (*Heliothis armigera*) and its yearly destruction of cotton, with the view of asking your opinion and advice for my own and the public benefit. I live in one of the northern counties, where cotton is the principal crop. We raise what is known as the Moon cotton, one inch and a quarter staple. This county loses yearly from the ravages of Boll Worms and moth from $300,000 to $400,000 on cotton alone, the moth, in my opinion, doing nine-tenths of the damage. The first crop of the caterpillars appeared in the corn near the 20th of May. On examination the 1st of June four-fifths had left the corn to transform to pupæ, but I found caterpillars up to the 10th of June, though scarce. In order to destroy them the planters generally put lamps in the field in the month of May, and expect to continue their use until October. The lamps are similar to those described in the Agricultural Report for 1880, page 239. The field crop of corn is now in silk and tassel.

Usually from the 1st to the 10th of August the Boll Worm moth leaves the corn and adopts the cotton as its home. This brood does immense damage, the moth laying her eggs in the squares in the blooms and in young bolls from the size of a garden pea to a partridge egg in preference to any other place. She pierces them as if done by a needle or pin, and in a few days they drop from the plant. Some farmers, not knowing what insect does this, have given them the name of sharpshooters, and it is yet a mooted question with us. By the time the cotton puts on a new crop of squares and blooms the moth is ready for it again, and if the weather is moist and warm it thus keeps on until frost; but should a drought prevail, with hot, drying winds, the eggs will not hatch, and this puts an end to them for that year, with the exception of a few scattering ones. Thus a dry and hot July and August is always a heavy crop year on the heavy, black, waxy prairie lands. Now I wish to know whether we have adopted the best course for the destruction of the Boll Worm. Is there any other course that has been successful in destroying them? Any advice or suggestions that you may choose to give us will be thankfully received. * * * —[William Somerville, Bagwell, Red River County, Tex., June 17, 1889.

REPLY.—I beg to acknowledge the receipt of yours of June 17 in reference to the damage done by the Boll Worm in your State. I can best answer your question by sending you a copy of the Fourth Report of the U. S. Entomological Commission, published in 1885, and which you do not seem to have seen. You will find the Boll Worm treated on pages 355 to 384. The destruction of the moths by trapping is not a satisfactory remedy, for experiments have proven with other species that the great majority of the insects so captured are either males, or females which have already laid their eggs. The first business of the female moth after issuing seems to be to lay her eggs, so that very few of them are caught in this way. The result is that other remedies are of much greater avail. The suggestion regarding the worming of corn while the first brood of worms is at work is a most excellent one, and the use of the arsenical poisons as indicated upon page 381 also affords a good remedy. The suggestion upon page 380, that in localities where no corn is grown over a considerable space it will pay to grow small patches here and there as traps for the early worms, is also a good one. It will be unnecessary to elaborate further, as the information is all contained in condensed form in this report.—[June 22, 1889.]

A cosmopolitan Flour Pest.

We send you herewith specimens of insects which are breeding in our flour mill. They seem to breed under basement floors and come up and fly away on warm days. There seems to be a difference of opinion as to what they are, and as there are no entomologists in this section we would be pleased to have your opinion and whether or not they will be likely to become a pest. They do not seem to work in wheat bins, but rather in flour dust in dark places. They breed all winter and spring and are now very numerous. We have tried several remedies, but Persian insect powder is the only thing that killed them.—[McPherson & Stevens, Sprague, Wash. Ty., May 18, 1889.

REPLY.—Your letter of May 18 with accompanying specimens has been received. The beetle which occurs in your flour mill is *Philetus bifasciatus*, a cosmopolitan species which feeds everywhere in flour and farinaceous products. Inasmuch as you find that Persian insect powder kills them readily we would advise you to use it very thoroughly and to hold them in complete subjection, for otherwise they will doubtless become quite a pest with you.—[May 27, 1889.]

Mites on a Neck-tie.

I send you in a tin box a neck-tie covered with Acari which a gentleman sends me from San Francisco. He says the tie has lain in a drawer and has been worn at intervals. He first noticed the "foreign substance" two weeks ago and thought it sand until he detected motion in the particles. What mite is it? How can garments be best treated to get rid of it?—[E. J. Wickson, Berkeley, Cal., May 25, 1889.

REPLY.—Yours of the 25th ultimo and mites duly received. We can not distinguish between the specimens found on the neck-tie and the common Cheese Mite (*Tyroglyphus siro*), and there must have been something very peculiar about those neck-ties or else the gentleman who sent the specimens must have been a bachelor and have kept his crackers and cheese in the same drawer with his clothes. The same mite, as you know, is found in flour of all kinds and milk. Sulphur is the best remedy. Either fumigate with burning sulphur or sprinkle with flowers of sulphur mixed in water.—[June 1, 1889.]

The Potato Beetle in the South.

The Potato Beetles herewith should have been sent you some weeks ago. They are from Madison Station, Madison County, Miss., the beetles occurring in several potato fields at and within a mile of the station. This is the first year I have seen them in Mississippi. If they have been here at an earlier date you may, perhaps, know it. I send them as a note of the spread of the beetle so far south.—[Dr. D. L. Phares, Agricultural College, Mississippi, May 11, 1889.

REPLY.—I beg to acknowledge the receipt of yours of the 11th instant, with accompanying specimens of the Colorado Potato Beetle (*Doryphora 10-lineata*). I believe that this is the first time they have been noticed so far south in your longitude. I will make a note of this matter for INSECT LIFE.—[May 18, 1889.]

Swarming of Urania boisduvalii in South America.

I take the liberty of mailing to you two specimens of butterfly captured at Colon, Republic of Colombia, March 18 and 25, 1889. When within a few hours of that port these insects were seen flying from the mainland in a northerly direction across the bay. This migration continued daily from the date of arrival, March 18, for nearly a week. When the flight began I could not ascertain. Its duration daily was from just before sunrise until sunset; it was protracted, however, until late at night on three evenings near and at full of the moon. The point which attracted my attention was the vast number of the insects. The air was actually full of them. It resembled an unremitting shower of forest leaves in autumn. I could learn nothing of its family history from the residents, but it is doubtless familiar to you. The excavations in each specimen were beautifully done by the Red Ant (*Formica rufa?*) in spite of the suspension of the tray in which the butterflies were placed from the ceiling by one string, and the saturation of said string with turpentine and castor oil.—[Dr. S. A. Davis, 107 West 47th street, New York City, May 9, 1889.

REPLY.—Your letter of May 9 transmitting specimens of a "butterfly" captured at Colon, United States of Colombia, has been received. The insect sent is not a butterfly but a moth, and is known as *Urania boisduvalii*. It bears, however, a striking resemblance to some of the large swallow-tailed butterflies of the genus Papilio. Your note concerning the abundance of this insect is very interesting.—[May 20, 1889.]

Letter on the proposed "American Entomologists' Union."

* * * I see in the March (1889) number of INSECT LIFE you ask for ideas concerning the proposed Society of Economic Entomologists. I do not think my views on the subject are worth much, but such as they are, they are as follows: I should like to see an organization founded, with members in every State in the Union (and I do not see why not also in Canada and Mexico), with the headquarters at the Department of Agriculture at Washington. Such a society to be called, perhaps, the "American Entomologists' Union," and to appoint a secretary in every State at least, and in the case of big States, like Texas and California, two or more; these to collect all the information they can relative to insects, especially from an economic point of view, and forward each one a report, at stated intervals, to Washington. These reports to be preserved and examined by a committee appointed, and the essence of them printed in INSECT LIFE or as a special bulletin. This I think would (1) bring economic entomologists in touch with one another; (2) enable them to benefit from one another's discoveries; (3) and especially the facts thus collected might be seen often to have a significance which would be totally lost were they to remain isolated among their discoverers; (4) although apparently adding to the work of the Department of Agriculture it would really diminish it, as you would have only the secretaries' reports to deal with, and it would be their duty to receive and collate reports of others within the boundaries of their own States.—[Theo. D. A. Cockerell, West Cliff, Custer County, Colo., May 12, 1889.

STEPS TOWARDS A REVISION OF CHAMBERS' INDEX,* WITH NOTES AND DESCRIPTIONS OF NEW SPECIES.

By Lord Walsingham.

[*Continued from page 291 of Vol. I.*]

LITHOCOLLETIS Z.

In revising the index to the genus *Lithocolletis*, one group of six supposed species has given me more trouble than the others. These are: *ulmella* Chamb., *modesta* F. & B., *conglomeratella* Z., *bicolorella* Chamb., *quercivorella* Chamb., and *obtusiloba* F. & B.

The first two are described as mining the upper side of elm leaves. The food-plant of the third is not known; and the three last are upper-side miners on the leaves of species of oak.

Zeller, in describing his *conglomeratella*, mentions two varieties of that species, differing chiefly in the extension of the white line along the dorsal margin of the fore-wings, and Chambers uses this character to distinguish his *bicolorella* from *ulmella*, with which he had at first placed it. He further says that *bicolorella* has *two* costal streaks, while *ulmella* has *three;* but in describing *quercivorella*, also with *three* costal streaks, he says the third streak is a mere spot before the cilia. In short, it is doubtful whether there are sufficient differences between the six descriptions to jastify the separation of any one of these species from the others on the ground of coior or markings. The evidence I have to rely upon in forming a conjecture (for it can scarcely be more than a conjecture) as to their distinctness is as follows:

(1) An authenticated specimen of *modesta* F. & B. from Boll's collection.

(2) A specimen received from Miss Murtfeldt, regarded by her as *ulmella* Chamb.

(*a*) A figure of a specimen in the collection of the American Entomological Society at Philadelphia, probably received from Chambers.

(*b*) A figure of a second specimen in the collection of the Peabody Academy of Sciences at Salem, Mass., received from Chambers under the above name, and presumably equal to his type.

(3) A specimen of *conglomeratella* referred to by Zeller in his description of that species as the second of the varieties from which his description was taken.

(4) Two specimens, unnamed, received from Miss Murtfeldt, bred from mines on the upper side of the leaves of white oak.

(5) An authenticated specimen of *obtusiloba* F. & B. from Boll's collection.

It is most improbable that the elm and oak feeders should be the same, although Miss Murtfeldt's specimen of the supposed *ulmella* is scarcely distinguishable from those bred from oak, and Boll's specimen of *modesta* actually bred from elm is still less so. We may at once admit that there are at least two distinct but very closely allied species, one on elm, the other on oak, but I think there can be no doubt whatever that *ulmella* and *modesta* are the same. The name *ulmella* takes precedence for the elm-feeder. I fear that some years ago in naming specimens for some of my American correspondents I may have been guilty of some confusion as to this species, having been misled by seeing specimens of *bicolorella* distributed by Chambers under the above name. We now come to the far more difficult identification of the oak-feeding species.

Zeller's specimen of *conglomeratella* is labelled "Dallas, Tex., Boll." This differs from the other specimens here referred to only in its somewhat duller color, but it is not in good condition, although the markings are easily visible. It agrees pre-

* Index to the describe l Tineina of the United States and Canada. V. T. Chambers. Bull. U. S. Geol. and Geog. Surv., IV (1), 1878.

cisely with the figure of the specimen in the collection of the American Entomological Society, but Chambers admits having mixed his specimens of *bicolorella* with *ulmella*, and this figure probably represents the oak-feeder. Zeller's descriptions of the three forms, which he regarded (probably with good reason) as varieties of one species, are extremely clear and precise. The first is an admirable description of my specimen of *obtusiloba* F. & B., and the third is an equally good one of the specimens received from Miss Murtfeldt. I have no doubt whatever that these are varieties of one species feeding on various oaks. There are no sufficient differences to distinguish *conglomeratella* Z. from these, or from *bicolorella* Chamb., which would certainly be included under Zeller's descriptions. I think it will be safe to regard three of the four names as applying to one and the same variable insect, for which the name *conglomeratella* takes precedence. The specimens mentioned as received from Miss Murtfeldt were bred from the upper side of leaves of *white* oak, but this would certainly not distinguish them from *quercirorella* or *bicolorella*, both upper-side mines, the one bred from *Q. bicolor*, the other from *Q. obtusiloba*. The main differences upon which Chambers seems to rely in separating these two species are as follows:

Bicolorella.	*Quercirorella.*
Fore wings yellowish saffron, dorsal stripe extending to cilia. Oblique dorsal streak absent. Two costal streaks, followed by small dots. Hind tarsi white.	Fore wings reddish orange, dorsal stripe extending beyond middle of dorsal margin. Oblique dorsal streak present: three costal streaks, the third a mere spot. Hind tarsi annulate with black.

In all other respects the two descriptions are approximately the same. The darker ground-color and spotted hind tarsi of *quercirorella* may perhaps be relied upon to distinguish this species from its allies. The synonymy of these species should therefore stand thus:

(1) *Ulmella* Chamb. = *modesta* F. & B.

(2) *Conglomeratella* Z. = *bicolorella* Chamb. = *obtusiloba* F. & B.

(3) *Quercirorella* Chamb.

Note.—Chambers, in distributing specimens to his various correspondents, frequently appears to have attached a wrong name to them. This he admits in more than one instance in his writings. The utmost caution is required before accepting a specimen in any collection as a co-type of any one of his species. Dr. Hagen's notes of Frey's examination of specimens in the Cambridge Museum (Papilio, IV, 151–3) show that in some cases the professor failed to recognize specimens that he must certainly have seen before. This may be partially accounted for by the condition of the specimens, but where Clemens' species are referred to it must be remembered that these were determined by Chambers, who had not seen Clemens' types at Philadelphia and who may have wrongly identified them in some cases.

Lithocolletis tubiferella Clem.

It may be worth while to mention that when I saw Dr. Clemens' type of this species in the collection of the American Entomological Society, Philadelphia, in 1871, I made a note, "Hind wings gone; very unlike a *Lithocolletis*." It is perhaps doubtful whether Chambers was rightly acquainted with the species. The larva supposed by him to belong to it (Can. Ent., III, 165–6) was proved to be Coleopterous (Can. Ent., IV, 123–4), and he does not mention the true larva, so far as I am aware, in any of his writings. He compares the perfect insect with his *bifasciella* (unknown to me), and says of the former that the tuft is white, and it has no costal and no dorsal streaks behind the fascia, and the apex is not dusted. Chambers described his *bifasciella* from a single bred female, and if the subapical markings were not conspicuous it is possible that Clemens may have omitted to mention them. In Dr. Hagen's paper (Papilio, IV, 152) mention is made of specimens (one good) of *tubiferella* Chamb. from Kentucky in the Cambridge Museum, and a comparison of these with the remains of Clemens' type at Philadelphia would decide the point; but for the present I should not be justified in attempting to correct their synonymy, and scarcely in suggesting that either of them may be identical with *lebertella* F. & B., which must be at least a nearly allied species.

Lithocolletis basistrigella Clem. = intermedia F. & B.

I have authenticated specimens of *basistrigella* Clem., compared with the type in the collection of the American Entomological Society of Philadelphia, and also of *intermedia* F. & B., from the Zeller collection, received from Frey, and I am able to say positively that these two species are the same. I have met with it also in Mendocino and Siskiyou Counties, Cal., Rouge River, in Oregon, and have received it from Miss Murtfeldt from Missouri.

Lithocolletis rileyella Chamb. = tenuistrigata F. & B.

I received from Miss Murtfeldt, in December, 1878, a *Lithocolletis* labeled "Tentiform mine on under side leaf of red oak." This specimen agrees precisely with Chambers' description of *L. rileyella*, and is obviously that species. It is undistinguishable from *tenuistrigata* F. & B., of which I have specimens and mines.

Lithocolletis quercibella Chamb. = subaureola F. & B.

I was at first disposed to think that *quercibella* could only be regarded as a synonym of *argentifimbriella*. Chambers writes that it resembles closely his *fuscocostella*, which I have shown to belong to that species; but after a careful study of his description by the side of a specimen of *subaureola* F. & B. I find that this is applicable in all particulars to that species, although the first, *quercibella*, is described as glistening snowy-white, with the apical third pale golden, and the other as pale golden-brown, with white markings. Chambers describes the subcostal streaks as pale golden. Frey and Boll regard this as corresponding with the ground color of the wing, and mention the straight, rather broad basal streak as being white, whereas Chambers regards white as the real ground color. With a specimen before one it is easy to see that the two descriptions are both accurate and precise in every detail.

Lithocolletis clemensella Chamb.

Another species that must be nearly allied to these is *clemensella*. I am induced to regard this species as distinct, owing to its feeding on *Acer saccharinum*, and by Chambers' remark that "the hinder marginal line at the base of the dorsal cilia reaches to, but does not pass around, the apical spot." I find this peculiarity well marked in a figure of the species taken from a specimen in Professor Fernald's collection, and I know of no allied species in which the same thing occurs. This insect is omitted from the Index, although it is given in the List of Food-plants of Tineina (Bull. U. S. G. G. Surv., IV, 109, 1878).

Lithocolletis argentifimbriella Clem.

= *Argyromiges quercialbella* Fitch.
= *Lithocolletis longestriata* F. & B.
= *Lithocolletis fuscocostella* Chamb.

In the Canadian Entomologist (Vol. III, 57) Chambers suggests that *argentifimbriella* Clem. may be the same species as *quercialbella* Fitch, but he appears to have never fully satisfied himself that this was the case owing to the differences between the descriptions of the larvæ. On page 182 of the same volume he points out that whereas Fitch describes the larva of *quercialbella* as being "flat," no known flat larva of this genus makes a tentiform mine, or an oval cocoon, such as Fitch describes. The larva of Clemens' species is cylindrical, and as Fitch's description is *not comparative* it is presumable that the word "flat" was not used in the sense in which Clemens and Chambers use it for larvæ of this genus, as distinguishing them from the cylindrical form.

Frey and Boll (Stett. Ent. Zeit., XXXIV, 209) themselves suggest the possibility that their *longestriata* may be the same as *argentifimbriella* Clem., and their description is so clear that, taking into consideration the similar larval habits, I think there can be no doubt that this is so.

In the Cincinnati Quarterly Journal of Science (II, 229), Chambers professes an acquaintance with *argentifimbriella* Clem. and confirms its identity with *longestriata* F. & B. (although he subsequently treats them as separate species in his index), but he fails to recognize his own *fuscocostella*, described shortly before that date, as falling under the same description. Chambers does not mention ever having taken or bred *argentifimbriella*, but there is a single specimen from Kentucky in his collection, now in the Museum of Comparative Zoology, Cambridge, Mass., about which Dr. Hagen writes (Papilio, IV, 151): "*Argentifimbriella* Chb., I, Ky. (very bad condition; perhaps, ? *longestriata* Frey)." It was probably owing to the condition of his specimen that Chambers failed to see that his description of *fuscocostella* corresponded with it. I have a specimen of the latter species from Dr. Riley, from Washington, D. C., and a specimen of *argentifimbriella* compared with Clemens' type in the collection of the American Entomological Society at Philadelphia. They are evidently the same.

It is somewhat doubtful whether this insect was first publicly named by Clemens or Fitch. Fitch's d scription was publish d in the annual report of the New York State Agricultural Society, issued as Vol. XVIII of the Transactions of that society, professedly for the year 1858. The title-page is dated "Albany, 18 9." The letter of presentation from Mr. B. P. Johnson to the Hon. D. W. C. Littlejohn, headed "In assembly, April 7, 1859," evidently antedates the real publication, for on page 585 is a letter from his excellency Joseph A. Wright, American minister at Berlin, dated "Berlin, May 11, 1859." In my copy is pasted the following letter:

"STATE OF NEW YORK, AGRICULTURAL ROOMS,
"*Albany, May* 19, 1860.

"SIR: Will your lordship be pleased to accept for your library the eighteenth volume of the Transactions of the New York State Agricultural Society for the year.

"I am, most respectfully, your very obedient servant.

"B. P. JOHNSON,
"*Corresponding Secretary.*

"Lord WALSINGHAM,
"*President Royal Agricultural Society of England.*"

The wording of this letter seems to show that this volume of the Transactions was not actually distributed until the year 1860, especially as the first three figures of the date "18 0" are *printed* (not written) on the paper. Now, the date of Clemens' paper in the Proceedings of the Academy of Natural Science, Philadelphia, is November, 1859, and if Vol. XI, in which it appeared, was issued before the agricultural volume, Clemens' name must take precedence.

Leaving my American friends who have access to the required information to correct me if I am wrong, I propose in the revised Index to give precedence to *argentifimbriella* Clem. over *quercialbella* Fitch.

GENERAL NOTES.

TWO LOCAL OUTBREAKS OF LOCUSTS.

Two locust occurrences worthy of note have come to our notice this season, one in Utah and another in Louisiana.

Under date of April 29, Mr. James B. Darton, of Nephi City, Utah, wrote the Secretary of Agriculture that millions of grasshoppers were at that time hatching out on the borders of the grain fields in the vicinity of Nephi City. At our request and to save time Mr. Bruner, our agent at Lincoln, Nebr., took up the correspondence and wrote us May 17 that he had received from Mr. Darton eight or ten specimens of the locust. These, however, from having been treated like botanical specimens, and evidently put through a press, could not be specifically determined. They were the young of Melanoplus, but might belong to any one of five species. A second lot, which was requested to be forwarded alive in a tin box, was reported on June 5 by Mr. Bruner, but still left us much in the dark as to the exact species doing the injury. The first lot seemed to be composed of at least three species, *M. bivittatus*, *M. spretus*, and *M. femur-rubrum* or *M. devastator*; but the other sending, consisting of a quarter pint of decaying pupæ, were nearly all *Camnula pellucida*, and just what other species were with them can not be said. In this outbreak several species were evidently united in the work of devastation. For several years back various causes have been working together to produce the injurious numbers appearing this year, but no great damage is to be looked for at the present in this region.

In Louisiana the species which occurred was *Melanoplus cinereus*, regarding which the Hon. T. J. Bird, Commissioner of the State Bureau of Agriculture, at Baton Rouge, wrote us June 8, mailing specimens. The damage done was slight and consisted in the leaves of young cotton plants being eaten. This is a local non migratory species, all of which, though liable to multiply to such an extent as to cause some little alarm, seldom really do any appreciable damage. Probably the best method of treatment is by the use of the bran-arsenic mash, concerning which several paragraphs will be found in the Annual Report of the Department for 1885, pages 300 and 301.

TENT CATERPILLAR IN ARKANSAS.

Mr. J. W. Bland, of War Eagle Mills, Benton County, Ark., has sent us a specimen of the moth of the American Tent-caterpillar (*Clisiocampa americana*) with its eggs, which he found the moth in the act of depositing on a peach limb on the 8th of June. We place this on record as giving an idea of the time of egg laying of this species in that part of the country. These eggs were for the second brood, which it is not unlikely may be followed by a third in Arkansas. Our correspondent writes us that this insect is very destructive to fruit trees in his county.

THE THISTLE CATERPILLAR IN WASHINGTON TERRITORY.

Mr. E. O. Schwägerl, of Naomi, Kitsap County, Wash. Ty., sent to us the middle of June specimens of the larvæ of the common Thistle Butterfly (*Pyrameis cardui*) infesting thistles and nettles there and which he has not been able to find on any other plants. This is a common butterfly, which is known to feed on the thistle the world over, and helps much in keeping this noxious plant in check in thistle infested localities. Our correspondent writes us that 90 per cent. of the thistles around Seattle are infested. The larvæ attack first the head or young shoots, eating out the flower buds, and then work down inside the stems, thus effectually destroying the seed crop. Birds do not eat these larvæ on account of their short, sharp spines.

THE CECROPIA SILK-WORM AGAIN.

In INSECT LIFE, for November (page 155), was mentioned the great abundance of the Cecropia near Calaway, Nebr. As we wished to obtain some of these cocoons, Miss Brown was written to and at the same time cautioned not to take any old cocoons, as the abundance which she referred to might be due to the accumulations of many years. In her reply she says:

A little boy collected me about half a bushel, but when I assorted them I found that about half of them were poor. A good many were stung and filled up with small grubs of some other insect, and others were last year's cocoons. I suppose you know that there is not much timber here, excepting where it has been set out and planted, and it is only on the cultivated box-elders, and then only in certain localities, that the cecropia silk-worm is found in numerous quantities.

Under date of December 16, 1888, M. Natalis Rondot writes us:

You notice, in No. 5 of INSECT LIFE, the remarkable abundance of Cecropia in one of the counties of the State of Nebraska, Miss Clara E. Brown having asked if the cocoons had any commercial value. To this question you replied that on the account of the difficulty in reeling the filament of the cocoon it could hardly be used industrially. This is true; but we may well ask if these cocoons may not be used for spinning into *schappe* (spun silk) or for articles of fantaisie. I do not know whether these cocoons have been studied from this point of view; in France at least no serious trial has been made of them, though I have had some samples of them combed as a matter of pure curiosity. The first question to study is that of the quantity of these cocoons. Miss Brown has, perhaps, personal reasons for complaining of the damage done by these wild worms; but it is possible that in reality the product in cocoons would be very light. It would be important to know how many of these cocoons could be obtained; for, in order to make a proper test, it would be necessary to have several pounds. This Cecropia is little known to us, and I find in fact that we even have no specimens of it. It would be interesting to have some, at least some of the cocoons, such as are found attached to trees, and some of the moths. In examining my notes I find that I saw, some years ago, cocoons and moths of certain species, one of which was very probably the *Platysamia cecropia*, while the others were of one or two species very similar to it. Were they hybrids of the Cecropia? I do not know. As the Cecropia is abundant in the United States you ought to know whether it is of a unique species or whether there are others allied to it.

Early in January we were able to send M. Rondot a few live cocoons of the Cecropia, and in transmitting them gave the following reply to his questions:

The species which is the most common in the United States is the *Attacus cecropia* of Linné. There are two species in this country very closely allied to it and by some held to be simple varieties of the Cecropia; they are the *Columbia* and the *Gloveri*. It is possible that it is to one or both of these that you refer as being mentioned in your notes.

It is hoped, from the live specimens sent M. Rondot, and a similar quantity sent to M. Quajat, at Padua, that these scientists may raise a sufficient crop of Cecropia cocoons to satisfy themselves of their value for the production of schappe. In this connection it may be added that Mr. L. G. Wilson, of Parsons, Dak. (statistical correspondent of the Department), informs us, under date of December 18, that wild cocoons are found in large quantities in his neighborhood, and that he wishes to send specimens of them to the Paris Exhibition. He has been requested to forward specimens to this Department.—[Philip Walker.]

SPRAYING FOR THE ELM LEAF-BEETLE.

Prof. John B. Smith, in *Garden and Forest* for June 19, gives an account of his experiments in spraying large elm trees on the Rutgers College campus. He used a Seneca Falls force-pump, mounted on a tank holding 40 gallons and provided with a 50-foot hose. The end of the hose is attached to a 10-foot pole, and by means of a light ladder 20 feet in length the foliage of the largest trees, some of which are over 50 feet in height, can be reached. Professor Smith finds that the addition of a small quantity of kerosene emulsion to the mixture of London purple and water is of use in enabling the spray to penetrate the pubescence on the under side of the leaves and to spread wherever it touches instead of collecting in drops and falling. He recommends the addition of a pint of kerosene emulsion to 20 gallons water containing one-fifth of a pound of London purple, and states that this amount of the mixture is sufficient for one of the largest trees.

THE DINGY CUT-WORM (AGROTIS SUBGOTHICA Haw.).

Late in May, 1886, Mr. Henry Nobes, a fruit-grower in the vicinity of La Fayette, Ind., called our attention to the fact that some insect, unknown to him, was destroying the ripening fruit in his strawberry field, large berries being wholly or for the most part devoured. A visit to the field soon revealed the depredator to be this cut-worm, which occurred in great numbers under the straw mulch. Worms were not only caught in the act of eating the berries, but many were found gorged with the fruit, the red color distinctly showing through the skin of the culprits. In places where the mulch had been removed they did not appear to trouble the fruit, except to a very limited extent.—F. M. Webster.

THE EUROPEAN WHITE GRUB.

We do not know which to wonder at the most, the industry of the woman or the numbers in which the White Grub (larva of the European *Melolontha vulgaris*) must have occurred in the soil, in the statement made by M. Reiset and quoted in "*La Nature*" for the 18th of May, where it is stated that in a field of about one hectare (2.471 acres) a single woman collected 759 pounds (344 kilograms) of these White Grubs or Cock Chafer larvæ in 15 days. The actual number of grubs was estimated at 180,000.

A WHEAT PEST IN CYPRUS.

Mr. A. E. Shipley, of Cambridge, England, has just published a preliminary report on the species of Tineina which injures wheat crops in Cyprus (Bulletin of Miscellaneous Information, Royal Gardens, Kew, No. 30, June, 1889, pages 133–135). This insect is *Œcophora temperatella*, a species which occurs at Beyrout and Libya, and is widely distributed throughout Palestine. The damage is done by the larva in mining the leaves and stems of the wheat. Many thousands of bushels of grain are lost through its work. The information which Mr. Shipley has received has so far been very fragmentary.

THE ENTOMOLOGICAL SOCIETY OF WASHINGTON.

June 27, 1889.—Mr. G. W. J. Angell, of New York City, was elected a corresponding member of the society.

Dr. Marx read a note giving the record of the numbers (216,000,060) of May beetles collected and destroyed in Tuchel, Pomerania. Mr. W. H. Ashmead read a descriptive paper entitled "An Anomalous Chalcid," in which he erected a new genus and species (*Hoplocrepis albiclavis*), for a Chalcid collected by the late Dr. R. S. Turner, at Fort George, Fla. The paper was discussed by Messrs. Howard and Schwarz.

Mr. L. O. Howard called the attention of the society to some enlarged figures of the mouth parts of *Periplaneta orientalis* in Miall and Denny's work on the Cockroach, in which no indication is given of a *digitus* proceeding from near the tip of the *lacinia* corresponding to the one occurring in *P. americana* described by him at a recent meeting of the society. Mr. Howard then briefly reviewed Miss Ormerod's recent book on South African Insects, and concluded his contributions by reading Hy. Edwards's paper, prepared for INSECT LIFE and published in the present number, on Noises made by Lepidoptera. This very interesting paper called forth a considerable discussion by various members relative to the noises of Lepidoptera and other insects.

Mr. E. A. Schwarz presented a paper entitled Myrmecophilous Insects and a catalogue of Myrmecophilous Coleoptera, exhibiting specimens of the Coleoptera treated. The paper was a very valuable contribution to our knowledge of the insect parasites and messmates of ants, and was discussed by Dr. Marx, Mr. Ashmead, and others.

C. L. MARLATT,
Acting Recording Secretary.

o

U. S. DEPARTMENT OF AGRICULTURE.

DIVISION OF ENTOMOLOGY.

PERIODICAL BULLETIN. AUGUST, 1889.

Vol. II. No. 2.

INSECT LIFE.

DEVOTED TO THE ECONOMY AND LIFE-HABITS OF INSECTS, ESPECIALLY IN THEIR RELATIONS TO AGRICULTURE, AND EDITED BY THE ENTOMOLOGIST AND HIS ASSISTANTS.

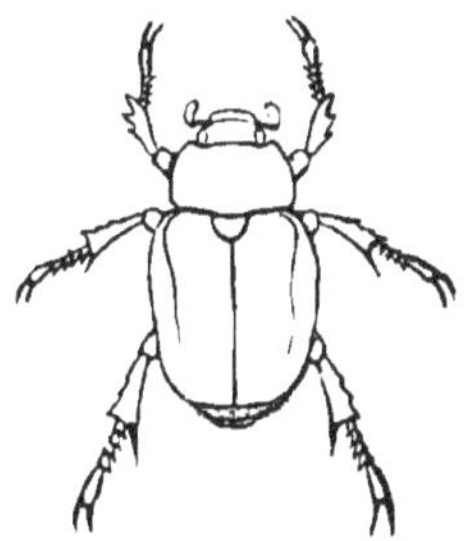

[PUBLISHED BY AUTHORITY OF THE SECRETARY OF AGRICULTURE.]

WASHINGTON:
GOVERNMENT PRINTING OFFICE.
1889.

CONTENTS.

SPECIAL NOTES.

The Grain Louse.—The common Grain Aphis (*Siphonophora avenæ*) has quite outdone itself this season. Appearing in enormous numbers in parts of Illinois, Indiana, Michigan, Wisconsin, and Ohio, it remained in the fields much later than usual, and it was not until nearly time for wheat harvest that its natural enemies had sufficiently increased to destroy it. Toward the end the parasites and predaceous insects were present in startling numbers and we have been able to rear many new ones, as well as to recognize at least two of Fitch's species. The insect enemies which we have so far found comprise eight species of hymenopterous parasites, one dipterous parasite, three species of Syrphid flies, two Chrysopas, and a number of Coccinellids.

The Grain Louse itself is a difficult insect to fight, and it is most fortunate that it is usually killed off by its enemies before appreciable damage is done. Its operations this year have doubtless caused some shrinkage of the crops, the amount of which can not be estimated at the present time.

The proposed Economic Entomologists' Union.—At about the time when this number of INSECT LIFE is being mailed an earnest discussion as to the advisability of such an association as we proposed in our January number will be going on at Toronto. Mr. James Fletcher, Dominion Entomologist of Canada and president of the Entomological Club of the American Association for the Advancement of Science, impressed with the great desirability for such an association and encouraged by favorable comments from a number of prominent workers, has issued a call for a preliminary meeting at the Toronto meeting of the American Association for the Advancement of Science.

We earnestly hope that an organization will be effected, for we feel sure that it would result in great benefit to the members and to the country at large.

Statistics of Loss from Insects.—As a contribution to the interesting study of the damage done by insects, computed in dollars and cents, we publish in this number a careful summary of the damage done by

Cotton Worms in Texas during 1887, compiled from the first annual report of the commissioner of agriculture of Texas, by Mr. B. W. Snow, assistant statistician to this department. This summary had been promised us by Mr. Dodge, but as he was called away Mr. Snow has kindly prepared it for our use.

Professor Cook's Bulletin on the Grain Louse.—Prof. A. J. Cook has just published, as Bulletin No. 50 of the Michigan Experiment Station, a short account of the Grain Plant-louse, giving a brief summary of the known facts concerning this insect. The bulletin is preliminary in its character and no remedies are suggested.

East Indian Rhynchota.—We have just received from Mr. E. T. Atkinson a continuation of his valuable papers upon this subject. The present installment comprises some nine y pages and includes descriptions of species numbered 295 to 443.

AGGREGATE DAMAGE FROM COTTON WORMS IN TEXAS, CROP OF 1887.

By B. W. Snow, *Assistant Statistician.*

The commissioner of agriculture of Texas, in his first annual report, presents a statement of the aggregate cotton crop of that State for 1887 by counties. In many parts of the State the season was an unfavorable one for this crop, drought and worms very much reducing the yield per acre. An estimate of the damage done by worms is presented for each county, ranging from nothing in many counties to a loss of 50 per cent. of the crop in others of large production, and an even heavier loss in some counties where the crop is of little importance and insecticides are not made use of. For the whole State the amount of damage done averaged about 21 per cent. of the crop.

According to this return the total number of bales gathered was 1,125,499, while had there been total exemption from insect damage the farmers of Texas, according to this authority, would have gathered a crop of 1,422,948 bales. This would make the aggregate loss from worms equal to 297,449 bales. The value per bale of the crop which was made at the place of production averaged slightly over $40. Presuming that an increase of less than half a million bales in the aggregate crop would have made but little difference in price, the actual money loss to the farmers of Texas in one year from the Cotton Worm alone was $11,897,960.

It is not claimed that these figures are absolutely accurate, but they are undoubtedly approximately correct, and will give some idea of the enormous tribute levied upon American agriculture by injurious insects. In that year Texas produced but 21 per cent. of the cotton crop of the country, and the Cotton Caterpillar and Boll Worm were active in all sections of the cotton belt. The injury elsewhere may not have been so heavy, but it would swell the aggregate loss in one crop to startling proportions.

The following statement has been prepared from the data presented in the report quoted from, and shows by counties the actual crop gathered, with the aggregate product which would have been picked had there been no loss from worms. In a number of counties damage from worms is not mentioned, and it is presumed that no loss occurred.

Counties.	Bales.	Loss from insects.	Product without loss.	Counties.	Bales.	Loss from insects.	Product without loss.
		Per cent.	*Bales.*			*Per cent.*	*Bales.*
Anderson	11, 818	20	14, 773	Hardin	94	$33\frac{1}{3}$	141
Angelina	2, 629	....	2, 629	Harris	2, 781	23	3, 612
Atascosa	348	....	348	Harrison	15, 556	10	17, 281
Austin	17, 378	12	19, 748	Haskell	16	29	23
Bandera	144	....	144	Hays	3, 253	$4\frac{5}{10}$	3, 406
Bastrop	13, 274	18	16, 167	Henderson	8, 773	17	10, 570
Baylor	9	....	9	Hidalgo	146	$89\frac{9}{10}$	1, 446
Bee	121	....	121	Hill	13, 188	58	31, 400
Bell	21, 481	20	26, 851	Hood	1, 082	87	8, 323
Bexar	1, 268	34	1, 921	Hopkins	14, 230	20	17, 788
Blanco	890	25	1, 187	Houston	10, 716	10	11, 907
Bosque	3, 618	36	5, 653	Hunt	29, 701	12	33, 751
Bowie	6, 679	25	8, 905	Jack	1, 088	5	1, 145
Brazoria	6, 344	25	8, 459	Jackson	741	25	988
Brazos	14, 229	15	16, 740	Jasper	1, 099	20	1, 374
Brown	2, 374	39	3, 892	Jefferson	87	25	116
Burleson	10, 489	....	10, 489	Johnson	11, 489	33	17, 148
Burnet	1, 849	15	2, 175	Jones	366	32	538
Caldwell	8, 669	8	9, 423	Karnes	967	44	1, 727
Calhoun	2	50	4	Kaufman	21, 236	....	21, 236
Callahan	745	25	993	Kendall	419	$\frac{4}{100}$	421
Cameron	390	25	520	Kerr	256	40	427
Camp	4, 356	20	5, 445	Knox	1	....	1
Cass	13, 546	....	13, 546	Lamar	29, 252	38	47, 181
Chambers	29	25	39	Lampasas	783	20	979
Cherokee	13, 137	10	14, 597	Lavaca	15, 246	20	19, 058
Childress	2	....	2	Lee	8, 126	11	9, 130
Clay	533	10	592	Leon	9, 443	10	10, 492
Coleman	469	48	902	Liberty	1, 693	20	2, 116
Collin	33, 112	17	39, 894	Limestone	13, 020	25	17, 360
Colorado	20, 526	15	24, 148	Live Oak	33	5	35
Comal	2, 315	20	2, 894	Llano	761	....	761
Comanche	4, 894	44	8, 739	Madison	4, 252	8	4, 622
Concho	19	45	35	Marion	6, 165	5	6, 489
Cooke	11, 109	45	20, 198	Mason	725	11	815
Coryell	6, 161	5	6, 485	Matagorda	3, 123	26	4, 220
Dallas	27, 796	....	27, 796	McCulloch	167	25	223
Delta	8, 514	24	11, 203	McLennan	16, 823	40	28, 038
Denton	13, 288	30	18, 983	McMullen	4	....	4
De Witt	7, 565	25	10, 087	Medina	178	5	187
Duval	52	50	104	Milam	14, 773	23	19, 186
Eastland	2, 456	32	3, 612	Mills	369	63	997
Edwards	13	20	17	Montague	7, 548	14	8, 777
Ellis	40, 735	19	50, 290	Montgomery	5, 315	10	5, 906
Erath	5, 375	44	9, 598	Morris	3, 702	....	3, 702
Falls	9, 750	15	11, 471	Nacogdoches	9, 468	5	9, 966
Fannin	38, 296	10	42, 551	Navarro	11, 730	....	11, 730
Fayette	35, 187	13	40, 445	Newton	1, 036	13	1, 191
Fisher	62	1	63	Nolan	3	....	3
Fort Bend	10, 139	25	13, 519	Nueces	3	....	3
Franklin	3, 897	21	4, 933	Orange	142	34	215
Freestone	6, 202	34	9, 397	Palo Pinto	803	34	1, 217
Frio	215	5	226	Panola	12, 658	5	13, 324
Galveston	38	....	38	Parker	4, 786	31	6, 936
Gillespie	1, 454	25	1, 939	Polk	3, 214	8	3, 493
Goliad	2, 806	32	4, 126	Rains	3, 795	20	4, 744
Gonzales	10, 382	15	12, 214	Red River	22, 512	25	30, 016
Grayson	24, 904	29	35, 076	Refugio	62	44	111
Greer	311	48	598	Robertson	18, 963	25	25, 284
Gregg	4, 854	10	5, 393	Rockwall	6, 665	50	13, 330
Grimes	16, 563	12	18, 822	Runnels	52	90	520
Guadalupe	9, 376	20	11, 720	Rusk	15, 967	20	19, 959
Hamilton	1, 940	55	4, 311	Sabine	2, 917	12	3, 315
Hardeman	10	50	20	San Augustine	4, 156	11	4, 670

Counties.	Bales.	Loss from insects.	Product without loss.
		Per cent.	*Bales.*
San Jacinto........	5, 342	15	6, 285
San Patricio	160	50	320
San Saba	708	8	770
Shackelford	145	....	145
Shelby.............	11, 415	....	11, 415
Smith..............	16, 589	20	20, 736
Somervell	498	95	9, 960
Stephens	1, 044	10	1, 160
Tarrant	9, 781	26	13, 217
Taylor.....	209	5	220
Throckmorton.....	7	....	7
Titus..............	5, 844	15	6, 875
Tom Green	21	....	21
Travis.............	18, 664	8	20, 287
Trinity	3, 759	2	3, 836
Tyler	2, 788	36	4, 356
Upshur	8, 212	22	10, 528
Uvalde	22	....	22
Val Verde	5	....	5
Van Zandt.........	10, 482	20	13, 103
Victoria	3, 710	33	5, 537
Walker.	6, 726	19	8, 304
Waller.............	7, 823	15	9, 203
Washington	30, 644	13	35, 223
Wharton	8, 875	18	10, 823
Wichita	39	55	87
Wilbarger	32	17	39
Williamson	11, 391	15	13, 401
Wilson	3, 793	32	5, 578
Wise.	5, 495	38	8, 863
Wood..............	8, 881	30	12, 687
Young.............	391	$16\frac{86}{100}$	469
Miscellaneous	27. 150	....	27, 150
Total	1, 125, 499	.. ..	1, 422, 948

A NEWLY-IMPORTED ELM INSECT.

By L. O. Howard.

Our first knowledge of this insect in this country was gained in 1884, when Mr. Charles Fremd, of Rye, Westchester County, N. Y., wrote Professor Riley, under date of June 22, as follows:

> My elm trees in the nursery are troubled this year with a red-looking mealy bug. Thousands of them are between the cracks of the bark, and are destroying the vitality of the trees. I have made one application of kerosene emulsion, but I presume not strong enough. I will go over them again with a stronger emulsion. * * *

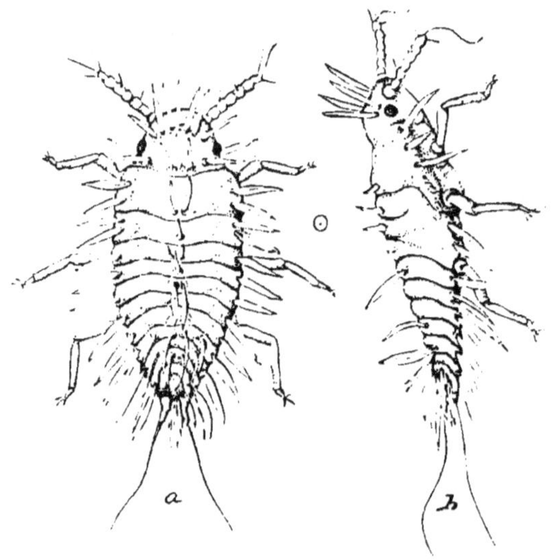

Fig. 1.—Gossyparia ulmi: *a*, young larva from above; *b*, young larva from side—greatly enlarged (original).

Professor Riley was in Europe at the time, and we therefore wrote Mr. Fremd for specimens, which he promptly sent, June 30. All of them had been saturated with kerosene emulsion, however, and were not in fit condition for study. It was plainly to be seen that they were new to the Coccid fauna of the United States, and our impression then was that they belonged near the genus *Eriococcus*.

The following month Mr. Fremd sent other specimens, all old females, and offered as a surmise as to the cause of their occurrence on his place the suggestion that they were very similar to bark-lice which he had noticed four or five years previously on some Chinese azaleas which he had procured from a New Jersey nursery, and which ultimately died, perhaps from the effects of the remedies applied for the Coccids.

This information unfortunately put us on the wrong track, and, supposing that it might be a new Chinese insect, we allowed other more important matters to intervene.

In June, 1887, this insect was sent to the Department again by Mr. John G. Jack, who found it at Cambridge, Mass., on the bark of *Ulmus fulva* (Slippery Elm). In Professor Riley's absence we wrote Mr. Jack the facts which had come to our notice, and that the species was undetermined in the collection of the National Museum and the Department of Agriculture, and advised him to send specimens to Professor Comstock, who was studying the group critically. A month later Mr. Jack wrote that he had followed our advice and that Professor Comstock reported that the species was undetermined, that it had been in his collection for some time, and that the previous winter he had found that it occurred abundantly on some elm trees in New York City.

In the summer of 1888 Mr. Jack sent other full-grown specimens, and the same summer it was found upon several elms in the grounds of the Department of Agriculture, at Washington, by Mr. W. B. Alwood. In the fall of 1888 we found it also upon *Ulmus americana* in two localities in the streets of Washington. Up to this date only old females had been found, and these presented much the appearance of *Eriococcus azaleæ* Comst.,* except that the white, somewhat ribbed excretion is not continuous over the back, but is abundant around the sides, curling up over the back and leaving the central portion brown and bare.

April 29, 1889, Mr. Jack sent to the Department some bits of bark and small limbs carrying non-impregnated females, male cocoons, and just-issued males, and, as Professor Riley was again unfortunately absent, this time as representative of the Department to the Paris Exposition, we undertook some further study of the species from Mr. Jack's material, and from that found in Washington had careful drawings made, and had little difficulty in determining that the insect was identical with the European *Gossyparia ulmi* Geoffroy, described by Signoret in the *Annales de la Société Entomologique de France* for 1875, page 21, and which occurs commonly upon *Ulmus campestris* in Europe. According to Signoret, *alni* Modier, *farinosus* De Geer, *spurius* Modier, and *lanigera* Gmélin are synonyms of this species. The specific name of the first-mentioned synonym would indicate that the species also occurs upon *Alnus*, and indeed Signoret states that he has collected it in the Bois de Boulogne on Alder.

---

* This was probably the scale which Mr. Fremd noticed upon his Chinese Azaleas and which he confounded with his Elm Coccids.

Signoret describes the newly hatched larva, the adult female before and after impregnation, and the immature male. Concerning the latter stage he writes:

We have collected a large number of active male nymphs, but no complete males. As with the preceding genus [*Nidularia*], when one disturbs these insects during their state of metamorphosis, they are apt to run away. This is what M. Lichtenstein has noticed with *Dactylopius vitis*, which he has pointed out as having an active nymph; but, according to us, it is to avoid danger, and under natural conditions the nymphs do not leave the sort of sac which serves them as a cradle [*berceau*].

In this conclusion Signoret has been at fault. The true pupa is not active, and from the nature of its sheathed limbs can not be active. The form which Signoret describes and calls the "nymph" casts off the pupa skin while yet in the cocoon and issues with its wings as yet unfolded and represented simply by pads, as shown in Fig. 3. It remains in this condition for some time (several days ?), runs freely about, with great activity, as we have seen, and, according to Mr. Jack's observations, even copulates with the female before its wings expand. It was in this condition that Signoret always found it. Others issue later with expanded wings and of the appearance shown in Fig. 5*c*, possessing long anal filaments. No casting of the skin has been observed between the two stages, but one may have taken place, and the form with the wing pads should be considered a pseudimago comparable with the form so-called in the Ephemerids.

Signoret's descriptions of the different stages are sufficiently accurate, and we may simply give a brief résumé of the appearance, adding a fuller description of the adult male.

The newly hatched larva is of an elongated oval form, narrower behind, of a clear yellow color, each segment with a strong lateral spine and the front border of the body with six spines. The genito-anal ring has six hairs, around which is later formed a secretion which renders them invisible. There is a double row of spines down the middle of the back; the antennæ are six-jointed, the first three joints longest, the fourth and fifth shortest. (See Fig. 1.)

The adult female before impregnation is of a similar shape, but the terminal lobes of the abdomen are more developed. Each segment is covered with spiny spinnerets secreting wax. The antennæ are six jointed, second and third longest, fourth and fifth shortest. There is an elongated protuberance each side of the antennæ. The legs are short and slender, with the tibia shorter than the tarsus. The genito-anal ring has 8 hairs. (See Fig. 5*a*.)

The full-grown male larva has seven-jointed antennæ, joint 7 longest, the rest equal. After impregnation the female becomes more round, fixes herself, the secretion becomes much more abundant on the sides, making at first lammellæ, which afterwards unite into a continuous cushion. The back becomes smooth and the segmentation is plainly visible. The dorsum is plane transversely, but curved longitudinally. Particularly after the birth of the young, the female becomes well separated from the waxy cushion and is easily removed from it (even jarring will accomplish the removal), leaving the noticeable empty white cup with its fringed edges. (See Fig. 2.)

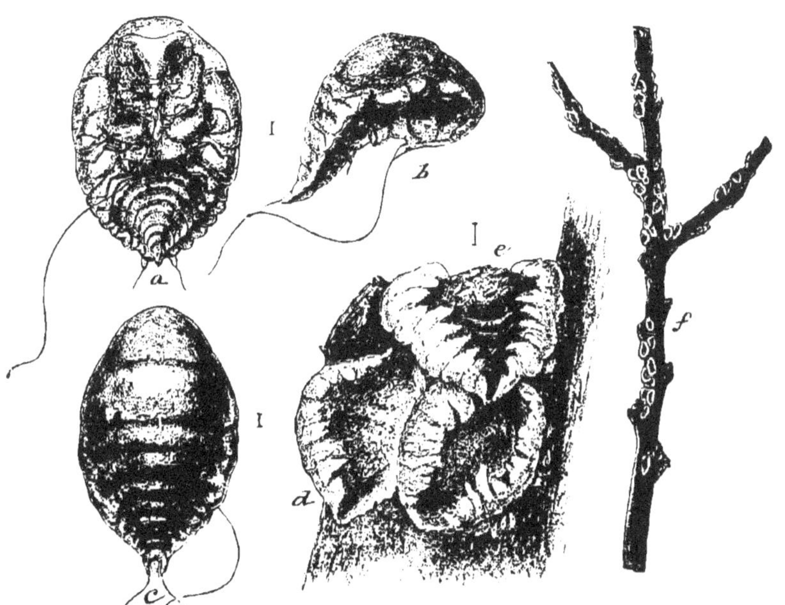

Fig. 2.—GOSSYPARIA ULMI: *a*, adult female from below; *b*, adult female from side; *c*, adult female from above—all greatly enlarged; *d*, empty waxy cushion; *e*, females in natural position—enlarged; *f*, shrivelled females—natural size (original).

The male presents a puzzle, and neither Mr. Jack's observations nor our own have solved it. The active form with wing pads issued some days before fully-fledged males were noticed. Specimens under observation in Washington were observed to copulate in this condition. The antennæ are ten-jointed, the joints well separated; the wings are represented by pads of varying length. The poisers appear rather thick and fleshy, but lack the terminal hook. The abdomen is very stout, suboval, considerably broader than the thorax, and when seen from above covers coxæ, trochanters, and bases of the femora. Its segments are not well marked. (See Fig. 3.)

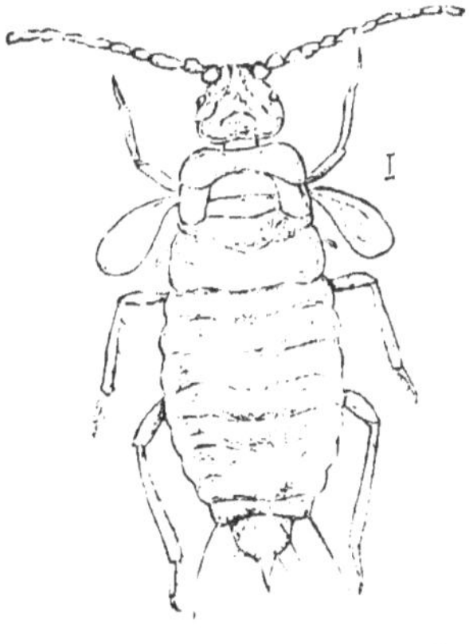

Fig. 3.—GOSSYPARIA ULMI: Imperfect male—greatly enlarged (original).

A few days after this form makes its appearance the cocoons begin to give out the perfect males, which issue with wings fully expanded. (See Fig.

5 *c.*) There really seems to have been a molt between this pseudimago and the perfect male, for in no other way can we account for the difference in form. The antennæ possess the same number of joints (ten) of about the same relative proportion, although joints 3 and 4 are longer, but the incisures are rather better marked. The poisers are lighter in color and less fleshy in appearance, and the curved hook is plainly visible at tip. The abdomen is rather longer, much more slender, and tapers gradually from base to tip. Its segments are well incised and plainly separable from above. It does not cover the hind coxæ and trochanters. The tibiæ are longer in proportion to their tarsi. The anal segment gives off two waxy filaments as long as the entire body. These filaments were not noticed in the pseudimago.

The cocoon itself is rather close though thin, flattened oval, and pure white, about 2^{mm} long by 1^{mm} wide, and is composed of rather coarse wax fibers. (See Fig. 4.)

Fig. 4.—GOSSYPARIA ULMI: Cocoon of male, showing anal filaments and edges of wings extruding—greatly enlarged (original).

According to one season's observations, therefore, this peculiar pseudimaginal form issues under perfectly natural conditions several days before the true imago; it is active and copulates. We have not observed it develop into a true imago. We have seen the true imago, however, issue from the cocoon, fully fledged, several days later. Why it ever issues as a pseudimago we do not know. That this is common is shown by the observations of Signoret, who never saw the fully-fledged male. We are not certain whether the copulation of the pseudimago with the female is a perfect one or is abortive and prompted by premature instinct, although the intromittent organ of this form is apparently complete and unsheathed.

From Mr. Jack's notes and our own observations at Washington we are able to give the round of the insect's life in general terms. The young lice are apparently born viviparously as with the Mealy Bugs, and issue from their living mothers in late June and early July and scatter actively over the tree, the majority of them with *Ulmus fulva* in which the twigs are pubescent or bristly, settling temporarily upon the leaves, mainly upon the upper surface in the angles of the midrib and principal veins, but also upon the under surface. With *Ulmus racemosa*, however, the twigs being smooth, large numbers settle about the buds and on the surface of the twig, many others also occurring on the leaves. With *Ulmus montana*, which is the species upon which we have principally studied them, they settle very abundantly upon the under sides of the leaves along the midrib and preferably just at the forkings of the veins. We have never found them settled upon the upper surface of the leaves, nor, in this stage, upon the twigs.

In August the lice desert the leaves and new twigs and return to the larger branches and trunk where they soon settle themselves in crevices of the bark. At this time they secrete a great deal of honey-dew which attracts ants and other insects, and gives off curiously enough a pungent odor which Mr. Jack states is noticeable where large numbers of the coccids are at work, but which we have not noticed at Washington, probably on account of the comparative scarcity of the lice.

This settling into the crevices of the trunk and limbs is purely for hibernation and is not a permanent fixture, as when Mr. Jack took some branches into the house in December they became quite active, moved about the limbs and escaped to different parts of the room.

As warm weather comes in the spring they begin moving once more, the females cast their last skin and the males form their cocoons. The adult males issue about May 1, and while still in the pseudimago state, were observed both in Cambridge and at Washington in many cases to copulate with the females. The fully developed males are seen in abundance a few days later; the great majority of the late ones issuing from their cocoons with the wings fully expanded and the anal filaments complete. Indeed the long filaments protrude from the cocoon and by laying hold of them the insect can be pulled out. It issues naturally backwards as do the males of other Coccidæ.

Soon after copulation the females fix themselves permanently and the males disappear. This occurs the latter part of May. The females at this time are attached mainly to the trunk and larger limbs. From this stage (the impregnated female) the secretion of honey dew is more pronounced than from the young females described in an earlier paragraph. It is given off in minute drops, which, according to Mr. Jack, are plainly visible while falling in the bright sunlight. The trunk, branches, and lower leaves are blackened, and many ants, wasps, and flies, as well as some beetles, are attracted.

The young lice begin to hatch in from three to four weeks after impregnation, and thus the life round is completed.

Mr. Jack's original specimens were found upon *Ulmus fulva* in the Arnold Arboretum near Boston, and he afterwards found the species quite widely distributed in the vicinity of Boston, occurring upon *U. americana* and *U. racemosa* as well as upon the European species, *U. montana* and *U. campestris*. He found it more common on the American species than upon the European, and more abundantly upon *U. fulva* than upon *U. americana*. Upon the latter species he found that the Coccids preferably left the coarse bark of the trunk and ascended to the higher parts of the tree.

In Washington specimens have been found upon the Department grounds in considerable numbers only upon one of the varieties of the European *Ulmus montana* (probably var. *rubra*), only occasional specimens being found upon *U. campestris* and the American species growing side by side with *U. montana*. *U. fulva*, which is so badly infested at Boston, is apparently untouched in Washington. In other parts of the city the Coccids have been found in several instances upon the trunks of the large *U. americana*, but these trees are too tall to mount readily to ascertain the numbers on the limbs. On the infested *U. montana* at the Department the old females cluster thickly along the under sides of the lower limbs, and through July the young are scattered over the leaves feeding vigorously and growing rapidly. Were we considering this question of the varieties attacked from the Washington trees only we could very plausibly account for the occurrence of the species so abundantly upon *montana* and not on *campestris* for the reason that the leaves of *campestris* are completely skeletonized every summer by the larvæ of the imported Elm-leaf Beetle, while the leaves of *montana* are only partly eaten, thus giving the young Coccids abundant opportunity to develop on the latter and none at all on the former species; but unfortunately the facts from Cambridge obviate this simple conclusion.

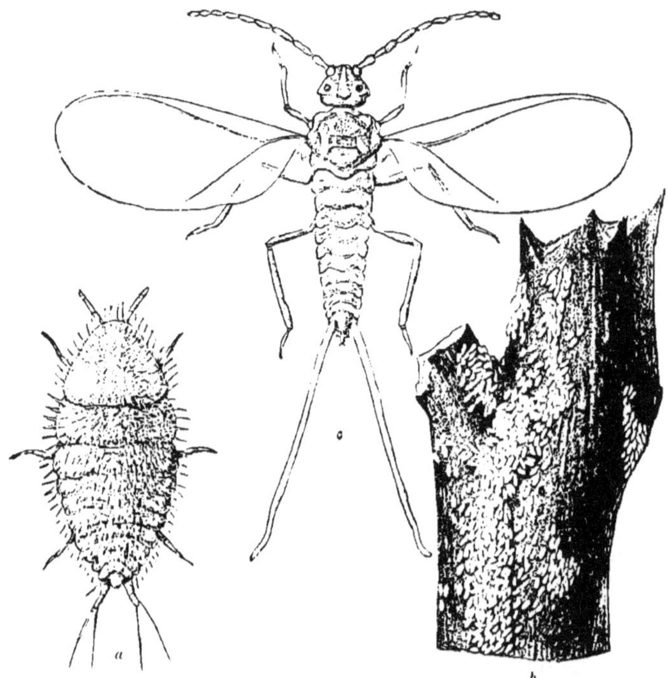

Fig. 5.—GOSSYPARIA ULMI: *a*, female before impregnation—greatly enlarged; *b*, male cocoons in natural position on limb—natural size; *c*, perfect male—greatly enlarged (original).

Upon ascertaining definitely during May the identity of the species with the European *Gossyparia ulmi* it immediately occurred to us as a matter of course that it was quite natural that the insect should be abundantly found in the two localities of Boston and Washington in the arboretums in which European elms were largely growing; but there was still their earlier occurrence at Rye, N. Y., to be explained. We therefore wrote to Mr. Fremd, June 26, to ascertain whether there were any European elms in his vicinity and whether the insects had increased, and received promptly the following very satisfactory reply:

I am just in receipt of yours, and will answer at once. At the time I wrote to you, in 1884, regarding the elm louse, I had several hundred of European elms in the nursery, and there are also quite a number of large trees, etc., in a number of lawns about Rye. The louse has disappeared from our trees altogether, how I don't know. * * *

The probable reason for the disappearance of the insects with Mr. Fremd was his use of the kerosene emulsion in 1884, as he wrote us under date of June 22, 1884, that he had used a weak emulsion and was about to try a stronger one. This leads us to Mr. Jack's statement that whale-oil soap with kerosene was successfully tried against the old scales on the trunks and larger limbs in the Arnold arboretum, but those upon the smaller limbs escaped. He did not know the strength of the solution.

This finding of *Gossyparia ulmi* upon American elms and upon European elms in this country was quite to be expected, and the only wonder is that it has not been found and recognized before. The species of Coccidæ have already extremely wide ranges, and every season still further extends them. Of our admitted North American Coccid fauna twenty-three species are of European origin (one more doubtfully so), three are from Australia and New Zealand, while sixty-nine are either truly North American or their original home is unknown. As several of these are found only on hot-house plants, they are certainly not North American. Several others are found on both native and imported plants and there are no data upon which to decide upon their proper faunal position. The fact that the *Gossyparia* prefers American elms at Cambridge is by no means without precedent in the group, and as another instance it may be mentioned that the beautiful oak-scale *Asterodiaspis quercicola* (Bouché), recognized by Comstock in 1880 upon foreign oaks on the Department of Agriculture grounds, is at the present time to be found almost solely upon American oaks in the same grove.

Since the completion of this article Professor Comstock has written us that he had himself recently decided that this insect is the European *Gossyparia ulmi*, and states that last winter he found it abundant upon elms in Saxony. He also states that it has been sent him by Mr. Henry Edwards from New York City, and by Dr. Lintner from Marlborough, N. Y. Mr. Edwards informs us by letter that his New York specimens were obtained from English elms of three years' growth.

SOME MICHIGAN NOTES RECORDED.

By Tyler Townsend.

The few notes here incorporated are selected and rewritten from an account of injurious insect appearances in the vicinity of Constantine, St. Joseph County, Mich., prepared by the writer three years ago (1886), and which it is not now thought advisable to publish in its original condition. The majority is omitted, only a few points being brought out which are considered of sufficient interest to be worthy of record.

Passing the Hymenoptera with the remark that the Raspberry Saw-fly (*Selandria rubi*) did some yearly injury from 1881 to 1886, we find in the Lepidoptera a number of species to be noticed. Of the two Cabbage Butterflies (*Pieris oleracea* and *rapæ*), it is worthy of note that the native species was (up to 1886) usually the more abundant, both species, however, being quite injurious every year. Scudder records *P. rapæ* as reaching this part of Michigan in 1877, on the authority of A. J. Cook and E. W. Allis. Thus for ten years at least the native butterfly has held its own against the foreign one, as it seems to have done for a shorter period of time in Colorado (see Insect Life, I, p. 382).

The Peach-tree and Currant Borers (*Ægeria exitiosa* and *tipuliformis*) are prominent, the first, aided perhaps by the hard winters, having exterminated the peach crop in this neighborhood. For several years up to 1881 a fine crop of this fruit was realized, and that year there was a splendid yield. In 1882 the yield was very small, many trees having died. Since then the trees were especially infested with this borer, which had previously been gaining steadily in its injuries for several years, and many trees had died every year, while none yielded fruit, until in 1886, in this immediate vicinity at least, hardly a live peach tree was to be found.

The Orange-striped Oak-worm (*Anisota senatoria*) was very abundant from 1879 as far back as 1874, stripping red oaks especially of their foliage to an alarming extent. It gradually became less injurious each year until it almost disappeared. With the exception of a few isolated larvæ seen in 1886 and some a year or two before, there had been none noticed for several years back. Accounts this year (1889) indicate that it has again made its appearance.

The Boll Worm (*Heliothis armigera*) came under notice only once during a period of twelve years. This was in 1881, when the worms were frequently met with in ears of green corn. The Army Worm (*Leucania unipuncta*) also appeared here in 1881, being in good force and entirely destroying many fields of grain, especially oats.

In 1886 the moths of a species of *Agrotis* (probably *subgothica*) were found in great numbers about houses, being especially numerous and active every evening during the latter part of May, the whole of June, and the first part of July, swarming on the upper-story windows of houses.

In the Diptera several species never known to be injurious occurred at times in some abundance. A very sleek-looking, black, pubescent fly (*Laphria canis* Will., determined for me by Dr. Williston) appeared in very large numbers in May, 1886. They covered the grass as well as raspberry and currant bushes, and were to be seen on almost everything, yet it could not be ascertained that they did any injury. The species passes its larval state in the ground, probably feeding on the roots of plants or other vegetable substances, while in the perfect state, together with other members of its genus, it is rapacious. Some members of the family are even predaceous in their larval state, devouring the larvæ of beetles found in grassy places (Williston). In two local lists of Diptera, one of Montreal and the other of Philadelphia, this species is not included, but it was described by Dr. Williston from two specimens, ♂ and ♀, taken in Connecticut, June 25. Three other Diptera were observed in considerable numbers on currant bushes in 1882, on May 9 and later in the same month. They are *Bibio femoratus* Wied., ♂ and ♀, a smaller undetermined species of *Bibio*, and *Scatophaga stercoraria* Linn. The first of these is given the locality "Atlantic States," in Osten-Sacken's list, and in the local lists just mentioned is recorded from Philadelphia, but not from Montreal; the last species occurs in both local lists. These three species appeared in more moderate numbers at the time than did *Laphria canis* in 1886, but were still quite numerous. They doubtless occur in smaller numbers every year, but were not noticed as particularly abundant after 1882.

Of the Coleoptera, one of the May beetles, *Lachnosterna prunina*, rather rare in collections, though locally abundant as will be seen, occurred in good numbers in 1886 on raspberry, blackberry, oak, and apple, in the evening, and there is good reason to believe that it has been numerous in previous years. It first appeared May 2. On May 22, at 11 o'clock in the evening, 82 specimens were beaten from raspberry bushes in the course of a half hour. It would seem that where there were so many of the beetles on the leaves they would be apt to cause some damage, yet the leaves had not been eaten. The beetles were abundant only on bushes in grass or sod, those kept clean of grass and weeds yielding very few specimens in proportion. In the larval state this species is, as are its congeners, destructive to the roots of grass. Numbers of the beetles were found every fine evening buzzing about in the grass in various places and finally flying away, these being no doubt individuals which had but recently emerged from the pupa state. This is in explanation of their being found in abundance only on the bushes that were in grassy places.

In the Hemiptera, Brood XXII of the Periodical Cicada may be recorded for this locality in 1885. Several other insects in this order may be noticed. Prominent among them is the Grape-vine Leaf-hopper (*Erythroneura vitis*) which was very abundant in all its stages during the first part of September, 1886, on the leaves of the grape. It caused considerable injury by puncturing, and thus disfiguring the leaves. The perfect insects that were noticed here did not have the transverse reddish bands nearly so broad as generally represented in the figures of them, but very narrow, while all the rest of the insect is of a pale yellow.

The Grain Plant-louse (*Siphonophora avenæ*) occurs some years on wheat and oats, but has never done particular damage. However, this year (1889), reports from the vicinity of Constantine, and the local papers, state that it has appeared in large numbers.

The Maple Scale (*Pulvinaria innumerabilis*) was very abundant on the maples in 1884, being conspicuous and causing some alarm. It however disappeared without particular injury.

A greenish-yellow or grayish plant-bug (*Euschistus variolarius*) was found in some numbers in July, 1886, on red raspberries. Quite a number of the berries were noticed on the bushes, each one having a specimen of this bug upon it, which from appearances seemed to have been engaged in the nefarious practice of piercing the berry and sucking its juices. One of these individuals was a nymph. This species is very common at present, and it would not take much increase to make it abundant, in which case some of our small fruits might sustain a slight amount of injury, though nothing probably that would be appreciable.

In the Orthoptera, many species of Acrididæ are common. The Red-legged Locust (*Caloptenus femur-rubrum*) was very abundant in August and September, 1886, in clover-stubble, meadows and pastures, and along roadsides everywhere; yet they were not particularly injurious. Specimens were taken *in coitu* from September 3 to October 12 on fences along the roads in the country. The first winged specimens were noticed this year on August 9. The Lesser Locust (*Caloptenus atlanis*) occurs occasionally with the preceding. This species was taken *in coitu* from September 13 to October 13. Other species occurring with these are *Caloptenus bivittatus* and *C. differentialis*, which are usually numerous. These two species were taken with *C. atlanis*, August 9, early in the morning, on hollyhock seed-cups beginning to turn yellow, which they had evidently been eating, as holes were found in their outer coverings.

PRELIMINARY NOTE UPON CHIONOBAS (ŒNEIS) MACOUNII, Edw.

By James Fletcher, Ottawa, Can.

In the *Canadian Entomologist* (XVII, p. 74, 1885) Mr. W. H. Edwards describes the male of *Chionobas macounii* from about a dozen specimens discovered June 28, 1884, by Prof. John Macoun, the Canadian Government Botanist, at Nepigon on the Canadian Pacific Railway at the northern extremity of Lake Superior. In the last week of June, 1885, the same collector took a male and two females at a far distant locality, Morley, in the district of Alberta, N. W. T., lying at the eastern base of the Rocky Mountains. Up to the present time these are the only known stations for this handsome species, which, in some respects, is the most remarkable and distinct species of the whole genus. In size and general appearance it approaches nearest to *C. californica*, but the sexual bar of *androconia*, such a conspicuous feature in the males of Chionobas, is entirely wanting in the present species. The average expanse of the wing is, ♂ 55–65^{mm}, ♀ 65–70^{mm}. In the Annual Report of the Entomological Society of Ontario, 1888, page 85, is an account of an expedition I had the pleasure of making with Mr. S. H. Scudder to Nepigon in the beginning of July, 1888, for the purpose of getting eggs so as to obtain a knowledge of the earlier stages. Although local, the species was found to be comparatively abundant and about 250 eggs were secured. To reduce as much as possible the chance of failure in breeding these were distributed to about twenty different entomologists in various parts of America and Europe. The eggs hatched in three weeks, and notwithstanding that the larvæ ate readily of all grasses and sedges offered them there was great mortality amongst the growing caterpillars, and the only specimens I know of which were carried safely through the winter were those sent to Mr. C. E. Holmgren, in Sweden, and three which I had myself at Ottawa. These hatched July 27, 1888, passed first molt August 17, grew very little before winter, and hibernated in the second stage. They were left out-of doors upon a living plant of *Carex pedunculata* and rested exposed upon the leaves, where they finished feeding without any protection and without spinning any silk.

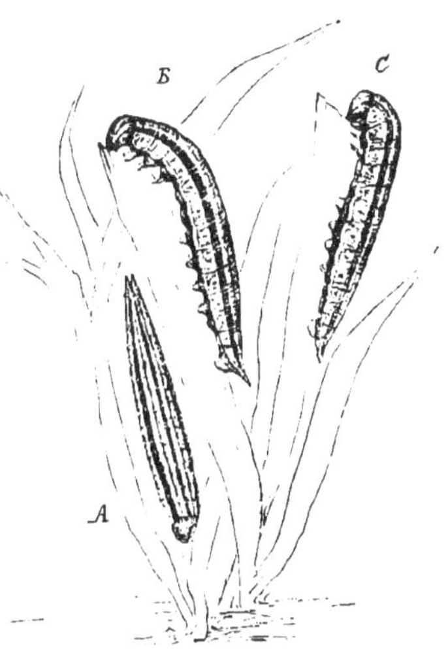

Fig. 6.—Œneis macounii: Full-grown larva; *A*, from above; *B*, from side, beginning of cut; *C*, from side, end of cut—natural size (original).

The cold during the first part of the winter was very severe, the mercury frequently dropping to 20° below zero (Fahr.), and this, too, without any snow upon the ground. During February, 1889, however, much snow fell, and they were covered by 4 feet of snow until the middle of March. When the spring opened three larvæ revived, but only one would feed; this passed its second molt on April 15, the third on June 13, and the fourth on July 6. In Mr. Scudder's Butterflies of New England (pp. 1775–1777), appear descriptions of the male, the female, and the first three stages of the larva. What I take to be the mature larva is figured life size above* (eighteen days after fourth molt). The general colour is grayish-brown, striped with black and pale lines. As with many other grass-feeders, this caterpillar furnishes a good instance of protective mimicry. It is extremely sluggish in its habits, generally feeding very early in the morning, and then resting for several hours, head downwards, at the base of the tuft of sedge, when the colour, shape, and longitudinal stripes give an exact resemblance to the dead leaves and scales always found at the base of these plants. The distinct dorsal and lateral stripes divide the body into widths equal to the leaves, and the faint subdorsal and stigmatal lines indicate the midribs, whilst many small black dots around these lines not a little resemble the minute parasitic fungi which so often discolour the leaves of grasses.

EXTRACTS FROM CORRESPONDENCE.

Pieris rapæ in California.

In INSECT LIFE, just received, I notice a note upon *Pieris rapæ*. In May, 1883, I captured in this place one male of that species (identified by George D. Hulst), since when I have never seen another specimen, although collecting butterflies every year, and usually extensively. That sample I have yet in my cabinet.

P. protodice is abundant here, but no great damage is done by it.—[W. G. Wright, San Bernardino, Cal., July 13, 1889.

Poisonous Spiders.

I send to-day in glass tube a specimen of *Latrodectus verecundus*, or "poison spider." It is believed to occasionally bite people, with serious effect. I have myself known two people (one of them a lady) who were bitten, presumably by this species of spider, while in privies, and both persons were seriously ill for weeks. I presume that the interest in this subject is about over; if not, I can interview the doctor who attended the lady and the gentleman bitten and send you the results of the inquiry. Personally I know that this spider frequents such places as old buildings and privies, and it is my custom always to brush out with leafy twig all dark places before running any risk.—[W. G. Wright, San Bernardino, Cal., July 13, 1889.

* This figure was drawn by Miss Sullivan from photographs and notes brought by Mr. Fletcher to Washington on a recent visit.—L. O. H.

A Spider-bite Contribution.

As my brother receives INSECT LIFE, in which I have found very many interesting things, I see that you are taking up the question of the bite of spiders and that observations are desired. Here is one which unfortunately does not accord in its results with those which have already been published on this subject.

In 1858, if I recollect rightly, being in Silao de la Victoria, near Guanajuato, they brought to me a little girl who had been bitten by one of those enormous spiders, quite common there, and which Mr. Leon Becker has named *Metriopelma breyeri*. The oblong tumified border was about 3 lines high, of a livid, violaceous color, filled with a serosity which I was not able to examine, not wishing to puncture the very thin epidermis. The center of the tumor was concave, and filled with hard pus. Eight days after the accident there was a little pain but there were no general symptoms. Unfortunately I was unable to follow the case, so that the observation remains incomplete, but I think that they would have brought the child back to me if there had been any serious consequences. It is impossible for me to recall the treatment which I employed. Since that time I have never had occasion to see any one bitten by *Metriopelma, Theridion*, or *Scolopendra*.—[Dr. Alfred Dugés, Guanajuato, Mexico, May 28, 1889.

Blackbirds *vs.* Boll-worms.

On page 351, INSECT LIFE, after comments on Blackbirds and the Boll-worm, it is remarked, "This is an interesting experience, but was the evidence sufficient," etc. To my mind it was, to encourage me that I had a friend in the blackbird, and that he was destroying boll-worms by the thousand. The facts are these:

My field of corn was in full roasting-ear, and the blackbirds were swarming in it. My hired man came to my library and told me we must get some boys with guns to shoot blackbirds, or they would ruin our corn. He added, "The neighbors are all in their corn-fields shooting to drive away the blackbirds." I told him to wait until I had time to see what the blackbirds were doing. On entering the field there were enough blackbirds in sight to have ruined the field of corn in a short time. I spent an hour or more in the field of 24 acres, and did not find an ear that showed the birds were eating the corn. The birds would light on the ears, and spend but a short time there, and pass to another ear. I noted ear after ear that I had seen a bird on, and I always waited until the bird had finished his work on it. I found on every such ear the marks of the boll-worm. They were developed enough to have commenced eating the grains. There were the evidences that the worm had been there, and I saw the blackbirds there, and making passes as if picking out the worms, and after the bird had left the ear I could find no worm. The birds seemed to be busy hunting and eating this destructive and disgusting pest. I left the field pleased and grateful to the blackbirds. I told my hired man he need not waste any time or powder on the birds. They were welcome to hunt worms, and could take what corn they wanted to make a variety. Now, this is not sufficient to show that blackbirds are in the habit of feeding on the boll-worms, I know, but it satisfied me that the birds were destroying thousands of them for me. The season was dry, the meadows were short, and the grass dried on the hillsides overlooking my bottom fields. The conditions were these corn in full roasting-ear, the earth dry, and the weather hot. The corn at husking time was not injured by birds more than usual, which is so light as to be almost inappreciable. I hope I may have opportunity this season to make further observations, and that the good work of the blackbirds may be established by many witnesses.—[L. N. Bonham, Columbus, Ohio, June 7, 1889.

Further on American Insecticides in India.

I have to thank you for No. 9 of your valuable publication, INSECT LIFE, containing my remarks upon insect pests and your foot-note to the same. With reference to my note about the Lecanium found upon Mango trees, I have since heard from Mr.

Douglas, who originally identified it as *L. acuminatum* of Signoret, that upon closer examination he considers it to be a distinct species. At his request I have accordingly described it as a new species in the April (No. 299) *Entomologists' Monthly Magazine* under the name of *Lecanium mangiferæ*. Mr. Douglas appends a note to this article in which he mentions that the specimens received from Demerara should also be referred to this species.

From small experiments with kerosene soap emulsions I feel sure that your proposed remedy would successfully exterminate the scale-bug so destructive to our coffee plants. But there are many serious difficulties in the way of its application on a sufficiently large scale. Some of these difficulties I note below for your consideration. The large size of plantations, varying from 200 to 1,000 acres, which, at the average rate of 1,500 trees per acre, gives from 30,000 to 1,500,000 individual trees to be treated on a single plantation. These plantations are situated on steep hill-sides, intersected only by narrow and rough foot-paths; consequently the liquid and apparatus would have to be transported entirely by hand labor. Unless this treatment were simultaneously undertaken by every planter, the infection would be continually re-imported. And even if united action could be made compulsory it would still be impossible to disinfect the indigenous trees and plants which at present act as reservoirs of the pest. I fear that the expenditure necessary to meet all these difficulties would be quite prohibitive. But if you still consider otherwise, and would kindly give me an idea of the probable cost of apparatus (or refer me to a manufacturer of the special nozzles and force-pumps used in this work), I would estimate the cost of the treatment and lay the plan before our Planters' Association.—[E. Ernest Green, Eton, Punduloya, Ceylon, June 1, 1889.

REPLY.— * * * The fact that the crop is grown upon hill-sides and that the field is only intersected by narrow foot-paths would render one of the knapsack pumps the only one which could be used for this purpose. European manufacturers have placed upon the market a number of desirable knapsack pumps, some of them holding several gallons, and all of them fitted with some modification of the Riley nozzle, which insures a fine spray and an economical distribution of the liquid. Knowing so little about the value of the crop and the amount of damage which the scale insects really cause, I can not pass judgment upon the advisability of the introduction of this remedy extensively, but I should surely say that it would pay to import one of the Vermorel pumps complete and make some careful experiments by its use with a good emulsion. * * * [July 3, 1889.]

A new Quince Enemy.

I inclose herewith a match-box containing Quince leaves infested with insects. The Quince tree is in a garden among pears, peaches, plums, pomegranates, figs, grapes, apples, etc. This is the second year that the Quince has been infested, and to such an extent as to check its growth and render it unfruitful, but I can discern the insects on no other tree. I should be glad to know the name of the pest and how to destroy it.—[W. Jennings, Thomasville, Ga., June 24, 1889.

REPLY.—Your letter of June 24 and the accompanying specimens of the insects found upon the leaves of your Quince tree have been received. The insect is one which has no distinctive common name. It feeds upon a variety of plants and is usually called, when found upon any particular one, by the name of the plant; as, when found upon hawthorn, it is called the "Hawthorn Tingis," when found upon butternut it is called the "Butternut Tingis." Its scientific name is *Corythuca arcuata*. It has not previously been recorded upon quince so far as I know, and this habit will enable it to do considerable damage when occurring in great numbers. If you will spray your trees with a dilute emulsion of kerosene and soap you will be able to destroy the insects which are now present, and if you will burn the rubbish under the tree in the fall instead of making a mulch around the base you will probably lessen the appearance next season. * * *—[June 28, 1889.]

New Food-plant and Enemy of Icerya.

* * * For the first time I have found the Icerya infesting a Conifer—the Cedar of Lebanon (*Cedrus libani*). The tree is growing in a yard in this city, and is infested with large numbers of the Icerya in all stages. In Professor Riley's report for 1886 no mention is made of this insect having been found infesting any Conifer in California, although Mr. Maskell records having found it on pines, firs, and cypress in New Zealand.

I have also to record a new insect enemy of the Icerya. Mr. J. W. Wolfskill and Mr. Alexander Craw, of this city, both of whom are close observers of the habits of insects, inform me that they saw a long, slender, pale brownish beetle—the *Telephorus consors* of Le Conte—feeding upon the eggs of the Icerya, having first torn open the cottony covering of the eggs. I have bred this beetle from a larva found under a stone near the margin of a small stream of water, but have not been able as yet to ascertain what the larva feeds upon. I confined one of them in a box with a cutworm, the larva of *Tæniocampa rufula* Grote, but the Telephorid larva did not attack it, and finally died. Is it possible that this beetle has learned to feed upon the eggs of the Icerya from having seen the larvæ of the Australian Lady-bird do so?—[D. W. Coquillett, Los Angeles, Cal., May 29, 1889.

The Red-legged Flea-beetle Again.

In regard to the Red-legged Flea-beetle, of which we wrote you last spring, stating that they were doing considerable damage (a reply having been received from you), will say from one year's experience that they are not so damaging as was at first supposed. The beetle does not migrate, as was first supposed, but remains on or near the ground that has been recently cleared of timber. We used a solution of Paris green on our infested trees last spring, and later in the season, finding that they did not disturb the trees of any account outside of their original haunts, we did nothing further, but waited for later developments. Early in the season the trees presented a dead appearance, but later they threw out a number of side branches, and by cutting out this spring the main branches, which are dead, and otherwise trimming the trees, they look about as well as ever, but have been thrown back one year and will be later in bearing in consequence. These same insects are noticeable where they were found last year, but not in such large numbers. They are damaging trees now, but principally on ground just cleared up.—[Stover and Stover, Edgemont, Md., April 23, 1889.

The Tarnished Plant-bug on Pear and Apple.

I inclose you in package and send by to-day's mail sample of pear-tree foliage injured by what I take to be the Tarnished Plant-bug, also samples of bug. These insects have been working on the pear and apple trees ever since foliage started, and over more than half of this (McPherson) county have destroyed from one-fourth to one-third of the pear bloom and a smaller proportion of the apple. They appear to do the most damage to the tender terminal buds toward the top of the tree. The bugs are in larger quantities the present season, and, while I have observed them almost every year, this is the first time they have created such marked damage. If I am wrong in the determination of the insects let me know.—[W. Knaus, McPherson, Kans., April 20, 1889.

REPLY.—I beg to acknowledge the receipt of your letter of the 20th instant, together with specimens of an insect which is damaging the foliage of pears and apples in your vicinity. This insect is, as you suppose, the Tarnished Plant-bug, which, as you may know, has been ascertained to be synonymous with the European *Lygus pratensis* Linn., the names *lineolaris* and *oblineatus* falling before the old Linnæan title. You are of course familiar with the habits of this bug as published in Riley's Second Missouri Report, pages 113 and 114, and in Forbes' report as State entomologist of Illinois for 1883, and in Professor Riley's report as Entomologist to this Department for 1884, pages 312 to 315. Kerosene emulsion will be the most effective remedy against it. * * * —[April 24, 1889.]

Walshia amorphelia and the Loco Weed.

By to-day's mail I send you a small tin box containing a piece of the Loco Weed or Crazy Plant. You will observe that there are worms or grubs in the roots and stems. From observations made by myself and a fellow stock-grower we are led to believe it possible that the worms, eaten by stock, produce the craziness and sometimes death, instead of the plant, as is generally supposed. Upon opening animals we always find many worms. An insect lays the egg upon the plant, and the worm, when hatched, descends into the root. The insect is longish and bronze winged. We desire information as to whether our theory be a plausible one or no. If we are right in our conclusions, we hope to find some remedy. Anything you may be able to suggest or knowledge you may be pleased to impart will be very gratifying to us.—[Thomas J. Quillian, Birmingham, Huerfano County, Colo., April 9, 1889.

REPLY.—I beg to acknowledge the receipt of yours of the 9th instant, together with the box containing a piece of the Loco Weed, supposed to be infested by grubs in the roots and stems. On arrival at Washington the work done by the grubs was evident, but not a specimen of the grub itself was to be found. However, we have received what is probably the same thing on several occasions from your State, and the sender has always been under the same impression, that the worms were the cause of the peculiar effect upon live stock. The maggots are harmless larvæ of a little moth known as *Walshia amorphella*, which occurs also in other allied plants, boring into the roots and stems. It has long since been decided that the peculiar effect of Loco Weed upon stock is due to some peculiar virtue of the plant itself, which I believe can not be ascertained by chemical analysis. *Post-mortem* examinations of diseased cattle and chemical examinations of the plant itself have been made by Dr. L. E. Sayer, dean of the department of pharmacy of the Kansas State University, from whom you might be able to ascertain something of value regarding treatment. In an article published in 1887 in the *Drug Record* concerning a *post-mortem*, he shows that the disease was one of the mucous and serous membranes, and recommends the following treatment:

"Pul. ext. belladonna .... grs. x.
Corrosive sublimate .... gr. j. to gr. jss.
Licorice .... ℥j.
Glycerine .... q. s.

"Mix. Make a thin paste and give a tablespoonful. The belladonna and mercury may be increased according to the severity of the symptoms. Opium, combined with belladonna, might be advantageous at the beginning of the disease. Mild and non-irritating articles of food only should be given, such as oil-cake, etc."—[April 19, 1889.]

STEPS TOWARDS A REVISION OF CHAMBERS' INDEX,* WITH NOTES AND DESCRIPTIONS OF NEW SPECIES.

BY LORD WALSINGHAM.

[*Continued from page 26 of Vol. II.*]

Lithocolletis fragilella F. & B.

The introduction of the name *trifasciella* Hw. into the North American lists rests first on the authority of Frey and Boll, who regarded specimens bred by them from *Lonicera sempervirens* as a form of this species. This was subsequently confirmed by Chambers, who, however, confused the species with his *mariaella* bred from a nearly allied plant—*Symphoricarpa*. I subsequently pointed out that *mariaella* was quite distinct from *trifasciella*, but confirmed the occurrence of *trifasciella* in America on the authority of a specimen, received from Dr. Riley, bred "from leaves of honey-suckle." I am now in a position to make further corrections. Frey and Boll in their last paper (Stett. Ent. Zeit., XXXIX, 270-271), described *fragilella* from larvæ feeding on leaves of *Lonicera albida*, and specimens of this are now before me, together with a leaf mined by the larvæ. Notwithstanding the remarks of these authors that this species is not nearly allied to any European form, I find it is so close to *trifasciella* Hw. as to be almost undistinguishable from it. It differs from that species precisely in the same peculiarity as Frey and Boll pointed out to distinguish their supposed variety from the European form, viz, in the different markings towards the apex of the wing including one extra small, white, costal streak. I have little doubt that this species is the one originally regarded by them as a variety of *trifasciella*. On again referring to the specimen received from Dr. Riley I find it to be the same as *fragilella* F. & B.; the close affinity of this species with *trifasciella* may be sufficient excuse for my previous error, as at that time I was unacquainted with Frey & Boll's species. Under these circumstances *trifasciella* must be erased from the American lists. The most noticeable characters by which *fragilella* may be distinguished from it are, first, the presence of an extra small, whitish, costal streak, beyond the interrupted third fascia, and secondly the absence of a subcostal shade of dark fuscous scaling, which in *trifasciella* commences at the base of the wing and reaches to the first fascia. In *fragilella* this fascia is densely dark-margined on the inner-side but in no one of the five specimens now before me does the dark dusting reach to the base of the wing.

Lithocolletis consimilella F. & B. and affinis F. & B.

Frey and Boll described *Consimilella* in 1873, bred from mixed mines, and in 1876 *affinis* from a red-fruited *Lonicera*. I have authentic specimens of both these from Boll's collection; *consimilella* from Zeller's cabinet, and *affinis* from Monsr. Ragonot, named by Boll, and although there is a slight difference in their size, they are scarcely distinguishable from each other. In *affinis*, the smaller of the two species, the frontal tuft is of a darker and more reddish-saffron, and the whole costal portion of the third fascia is decidedly more triangular and more conspicuous than in *consimilella*, in which it is confined to a very narrow line, scarcely wider than the black marginal dusting which precedes it. Moreover, at the base of the cilia, below the apex, there is no trace in *affinis* of the dusting of dark scales which is to be seen in *consimilella*, and the whole insect is also distinguished by a somewhat brighter and more glistening appearance, both of the ground-color and also of the silvery markings. The larva of *consimilella* being at present unknown, I hope to promote its discovery by pointing out these distinguishing differences.

---

* Index to the described Tineina of the United States and Canada. V. T. Chambers. Bull. U. S. Geol. and Geog. Surv., IV (1), 1878.

Lithocolletis lucetiella Clem.

= *ænigmatella* F. & B.

I find in Zeller's collection a specimen of *ænigmatella* F. & B., received from Boll, which agrees with a specimen compared with Clemens' type of *lucetiella* in the collection of the Entomological Society at Philadelphia. I am therefore able to say that these two names are synonyms for one species, so distinct in appearance from any other known *Lithocolletis* that confusion is rendered impossible.

Lithocolletis celtifoliella Chamb.

= *nonfasciella* Chamb.
= *celtisella* Chamb.
= *pusillifoliella* F. & B.

From actual date of publication *nonfasciella* would take precedence, but both the name and the description being founded on peculiarities which only exist in worn specimens, it falls under Strickland's Rule XI: "A name whose meaning is glaringly false may be changed." Chambers himself (Bull. U. S. G. G. Surv., IV., 155) says of *nonfasciella*, "This must be dropped from the list; there is no such species. It was described from varieties and old specimens of *L. celtisella* Chamb." The name *nonfasciella* must consequently be treated as a synonym. Chambers's description of *celtifoliella* differs from that of *celtisella* especially in having a third fascia, but this appears to be very near the apex of the wing, and frequently somewhat obliterated by the dark dusting. Since Chambers has admitted that he was somewhat confused in the first instance by the apparently different habits of the larvæ, I think we may conclude that his two species, *celtifoliella* and *celtisella*, come fairly within the range of varieties noticed by Frey and Boll. In the Stett. Ent. Zeit., XXXIX, 274–5, Frey and Boll admit that their *pusillifoliella* is the same as *celtisella* Chamb., although in the notes by Professor Frey, published by Dr. Hagan (Papilio IV, 152) we find "*celtisella* Chb. 15 Ky. (new to me)." They confirm Chambers' observations as to the peculiarity of the larva mining both sides of the leaf, and remark upon the extreme variability of the perfect insect, some specimens of which might easily be regarded as belonging to a distinct form.

In the absence of further proof to the contrary I should regard *celtisella* Chamb. and *pusillifoliella* F. & B. as synonyms of *celtifoliella* Chamb.

Lithocolletis morrisella Fitch.

= *texanella* Z.

Fitch, in describing his *Argyromiges morrisella*, remarks that it differs from *A. pseudacaciella* Fitch (= *robiniella* Clem.), in that "the inner half of the fore wings is black, slightly tinged posteriorly with golden yellow, and interrupted at equal distances by three white spots or short bands narrowing towards their inner ends, and between each of these is a less distinct white spot or cloud. Forward of the anterior white spot the color is more pure and coal-black, forming an oblong square spot occupying the inner half of the base of the wing, which spot is bordered along its inner side by a slender white stripe placed upon the middle of the wing at its base, its hind end uniting with the inner end of the anterior white spot."

Now, with the exception of the intermediate white spots or clouds, which are not recognizable in Zeller's figure, the differences described are precisely those which separate *texanella* Z. from *robiniella* Clem. The dark dorsal margin is particularly noticeable in Zeller's figure and specimens (his type is now before me), and the slightest abrasion of scales between the white dorsal streaks produces the effect of an indistinct intermediate cloud. I am unable to resist the conclusion that Dr. Fitch had before him the three closely allied species which have since been found to feed respectively upon *Robinia, Amorpha, and Amphicarpæa,* and are best known under the names of *robiniella* Clem., *amorphæella* Chamb., and *texanella* Z. There can be no doubt as to the precedence in nomenclature as between *morrisella* and *texanella,* if my theory is correct, the name *morrisella* having been published many years before Zeller's paper.

Lithocolletis uhlerella, Fitch.

= *amorphæella,* Chamb.
= *amorphæ,* F. & B.

Fitch's description of *Argyromiges uhlerella,* although brief, applies with sufficient precision to the *Amorpha*-mining *Lithocolletis,* described by Chambers as *amorphæella,* and by Frey and Boll as *amorphæ.* Fitch states that "it resembles *pseudacaciella* (= *robiniella* Clem.), but it is throughout of paler colors, its forewings being golden-grey" (rather than "uniform brilliant golden") and "the black dot on the tip of the wings is replaced by a short black stripe thrice as long as wide." This precisely describes the differences that separate *amorphæella* from *robiniella,* and we may at once give precedence to Fitch's name *uhlerella* for this species.

Lithocolletis ostensackenella, Fitch.

= *ornatella,* Chamb.

Another species of which the description is clear and absolutely unmistakable is *Argyromiges ostensackenella,* Fitch. Specimens of *ornatella,* Chamb., are now before me, and I can see no reason to doubt that this was the species from which Dr. Fitch wrote his description, although I have not had an opportunity of seeing his type.

Lithocolletis gemmea, F. & B.

When describing this species Frey and Boll were doubtful whether it were distinct from *Parectopa robiniella* Clem., not having properly recognized the latter species at that time, and Chambers asserts positively (Cin. Qr. Jr. Sc. I, 209–10) that *L. gemmea* F. & B. = *Parectopa robiniella* Clem. I am at a loss to understand how he could have made such a mistake. I have a specimen of the insect from the Zeller collection collected by Boll which agrees precisely with the description of *gemmea* and is so labeled. It would be utterly impossible to apply to it the description of *Parectopa robiniella,* which does not possess a transverse fascia and is of a totally different color. I observe that Chambers subsequently discovered his mistake and recanted (Can. Ent. XI, of 144–5).

L. gemmea is a true Lithocolletis and apparently a good and distinct species.

Lithocolletis ostryæfoliella, Clem.

= *mirifica,* F. & B.

Chambers suggests (Cin. Qr. Jr. Sc., I, 202) that *mirifica* may be the same as *ostryæfoliella.* I am inclined to agree with him.

Lithocolletis tritæniella, Chamb.

= *consimilella,* F. & B.

On the same page Chambers expresses his opinion that Frey and Boll have redescribed *tritæniella* under the name *consimilella.* I have a figure of a specimen of *tritæniella,* named by Chambers himself, and presented by him to the Peabody Academy of Sciences, Salem, Mass., and an authenticated specimen of *consimilella* from the Zeller collection. There is, I think, no doubt that these two names apply to the same species.

Lithocolletis guttifinitella, Clem

Chambers (Can. Ent., III, 111) describes *æsculisella* as a variety of *guttifinitella,* but notices that the larva differs decidedly from that of the type. It seems impossible to believe that the same species mines leaves of *Rhus toxicodendron*, one of the *Anacardiaceæ*, and also those of *Æsculus glabra* belonging to the *Sapindaceæ*. He then proceeds to describe another species, *coryliella*, also very nearly allied to *guttifinitella* but feeding on *Corylus americana*, and his variety *ostryaella* mining *Ostrya virginica* is said to bear the same relationship to *coryliella* as *æsculisella* bears to *guttifinitella*. It is more possible to conceive that this is only a variety, since the two food plants belong to the same family. He gives a table showing the differences between the larvæ of these four species, or varieties, which he finds to be constant and striking. It would seem perhaps to be a somewhat arbitrary proceeding to raise to specific value an insect described as an undistinguishable variety. I shall content myself with drawing special attention to these two descriptions of supposed varieties in the hope that at some future time those who have the opportunity of breeding the species will clear up the doubts that certainly exist in my mind about them.

Lithocolletis atomariella, Z.

Zeller placed *atomariella* in his cabinet between *pastorella* Z. and *populifoliella* Tr., and the differences, although slight, are sufficient to separate it from both.

Lithocolletis salicifoliella Chamb.

This species is also very closely allied to, but distinct from, *pastorella* Z. and *populifoliella* Tr. It is in all probability identical with the larva described under the same name by Clemens.

Lithocolletis ambrosiella Chamb.

A group of species allied to this typical form has been described by Chambers and Frey and Boll. These include *ignota* F. & B., *heleanthirorella* Chamb., *bostonica* F. & B., *elephantopodella* F. & B., *amœna* F. & B., *actinomeridis* F. & B., and *nobilissima* F. & B. (the latter can only be treated as an MS. name, no detailed description having been published), all feeding upon various *Compositæ*. The name *ambrosiaella* was corrected to *ambrosiella* by F. & B. (Stett. Ent. Zeit., XXXIX, 267). *L. ignota* F. & B. seems to be the same as *heleanthirorella* Chamb., as suggested by Chambers—*ignota* takes precedence.

I have not sufficient material at hand to determine whether the other species should, or should not, be retained as distinct. For the purpose of the revised index and until more evidence is forthcoming to identify them, they must certainly be respected.

(*To be continued.*)

GENERAL NOTES.

HONORS TO AMERICAN ENTOMOLOGY.

Professor Riley, chief of this Division, has just been elected an honorary fellow of the Entomological Society of London. Dr. Riley is the third American who has received this honor, the others being Dr. H. A. Hagen of Cambridge, who was elected in 1863, and Dr. A. S. Packard, elected in 1884. The Transactions for 1888 show that there are only ten living honorary fellows.

Professor Riley has also been created chevalier of the Legion of Honor by the French Government. This action had no reference to his official connection with the Exposition, but was taken on account of his researches in applied Entomology, particularly with reference to their value to French agriculture. This latter honor has been offered to Professor Riley before, but he has previously declined it on the supposition that an officer of this Government is not allowed to accept such decorations. His acceptance at the present time is conditional, of course, on the permission of this Government.—L. O. H.

A NEW EAST INDIAN GENUS OF COCCIDÆ.

Mr. E. T. Atkinson, of Calcutta, has just published, in the Journal of the Asiatic Society of Bengal (VOL. lviii, Part ii, No. 1, 1889), descriptions and figures of a new genus of Bark-lice found at Mungphu, in Sikkim, on *Quercus incarna*, *Castania india*, and *C. tribuloides.* The insect resembles *Pulvinaria* except that its larvæ have distinct anal tubercles. It is a Hemicoccid resembling the Lecanids in general appearance. The secretion is abundant and close during the larval state. In the second stage it becomes more waxy so as to approach, in appearance, the genus Orthesia, and the mass of wax on the leaves is more like detached or attached plates than threads.

CANNIBALISM WITH LADY-BIRDS.

Mr. J. W. Slater, in *Science Gossip* for July, 1889, states that he has seen the larvæ of *Coccinella dispar* attack the pupæ of its own species and destroy them. He has witnessed such instances of cannibalism not merely in a glass box in which he had placed some larvæ and pupæ, but on a row of currant bushes where Aphids were swarming. He fears that the Coccinellids are deliberate and habitual cannibals, and that this practice seriously interferes with the multiplication of the species and limits their usefulness as plant-louse destroyers. He has never observed the adults engaged in this reprehensible habit.

DAMAGE BY THE PEAR MIDGE.

Rev. E. N. Bloomfield, of Hastings, England, reports in the July number of the *Entomologist's Monthly Magazine* that considerable damage was done to Pears this spring in his vicinity by this insect (*Diplosis pyrivora*, Riley).

ICERYA PURCHASI NOT IN FLORIDA.

The several recent scares concerning the supposed appearance of the Fluted Scale of California in Florida appear, upon the best information which we have been able to secure, to have been founded upon errors in determination. In two instances the common Mealy Bug (*Dactylopius citri*) was the insect mistaken for Icerya, and in one case the insect causing the scare was the Florida Wax-scale (*Ceroplastes floridensis*).

A NEW STATE BOARD OF HORTICULTURE.

The legislative assembly of the State of Oregon passed last February an act to create a State board of horticulture and to appropriate money therefor. The board has been appointed and consists of one commissioner from each of five districts and one from the State at large. It has published two bulletins in circular form—No. 1, dated April 10, and No. 2, dated June 1—which deal with entomological matters. We notice from these circulars that the arsenical mixtures must be used in greater dilution than in the East. This point had already been brought out by California experiments. The Oregon people have found that one pound of London purple to 150 gallons of water will burn the foliage of apple.

THE ARMY WORM IN INDIANA.

The Army Worm has appeared this spring in several localities in the State of Indiana, and an account recently received, the latter part of June, from Mr. A. E. Mogle, of Kewanna, indicated that so much damage was being done in Fulton and other counties that our Mr. Webster was directed to visit the spot. He reached Kewanna July 3 and found that the worms had entirely disappeared. He visited the principal field infested, which was a 25-acre rye field, and found the crop a total loss. The field was on boggy land and was growing very rank, and there seemed no doubt but that this was where the insect originated. No attempt was made to save this field, but all energy was spent to prevent the worm from migrating to others by ditching and flooding the ditches. Cattle were also driven back and forth to trample upon the worms. Very few healthy pupæ were found, but many Tachinid puparia.

DOINGS OF AGROTIS CUPIDISSIMA.

In the early spring of this year and just as the buds upon grape-vines had expanded there appeared numerous examples of half grown larvæ of what afterwards yielded the *Noctua* (*Agrotis*) *cupidissima* Grote. These larvæ were in immense numbers, causing the loss of the first vintage in some vines, while in others the vines were kept alive only by the breaking forth of latent buds. This condition of things occurred over wide-spread areas in different parts of the State as far apart as Napa Valley and Tulare.

Visitations of this kind of caterpillar had not been observed before and it was chronicled as a new pest of the grape-vine. I visited the afflicted district of Napa County and found some few larvæ of *Plusia californica* and also some Mamestra-like larvæ likewise feeding upon the vines. I received many letters and consignments of worms; the persons sending always asked for remedies. As Agrotis, Plusia, and Mamestra larvæ do not ordinarily select the grape-vine as food, I concluded there must be an unusual cause. I think the cause to be this: The rain fall of this season was much prolonged; the weeds grew rank, feeding and harboring an unusual number of Noctuid larvæ, and when plowing became practicable the worms were already of large size; the plowing destroyed their food plants so that the larvæ had no choice but to fall upon the grape-vines or perish, but they proved themselves equal to the change of pabulum.

The remedy under like conditions should be earlier plowing, but if cultivation is retarded by late rains then plowing should be deferred still later to allow the broods of caterpillars to pupate.—J. J. Rivers.

THE DISAPPEARANCE OF ICERYA IN NEW ZEALAND.

Mr. R. Allan Wight, who seems to have kept track of the Icerya in New Zealand better than any one else, in a recent letter gives the following interesting facts concerning the disappearance of this pest, apropos to a recently published statement of Mr. Maskell's to the effect that Icerya was present in Grafton Road Valley six years ago:

The Iceryæ, were not only in millions in Grafton Road Valley, as he describes *six years ago*, but such was the case *fourteen months* ago. Yes, and also at Takapuna, Ponsonby, New Market, Waikomiti, Wairoa South, and several other places, where Mr. Maskell never saw them at all. These beetles have sprung up suddenly, and the work they have done is positively incredible. In March, 1888, I passed through Auckland to go to Whangarei, in the north, to advise people on the Icerya question (it had broken out there), and I found the pest white on everything in and around the city and for 20 miles in several districts. In February, 1889, I was again in Auckland and lo, it was gone! I found some, of course, but only "here and there a one." Did I not do well, then, to advise Mr. Koebele to go to Napier, where there was still a retreating host of the enemy? Yes; and, believe me, if you can only succeed in keeping these beetles from your birds they will clear the Icerya as the sun melts the snow from the mountain. Last March I visited the Wairoa South, where I saw the last of Icerya hanging to the *Acacia undulata* twigs, with ova sacs torn and empty, and I saw thousands upon thousands of the little *C. Nova Zealandia* in imago pupa and larva form, but mostly in the two first stages. My daughter, who lives there and who inherits her father's love of nature, undertook to watch them for me, and she now reports that the Coleoptera are all gone out of sight, and no more Iceryæ are as yet to be seen.

A PECULIARITY OF CERTAIN CADDIS-FLIES.

Mr. K. Flach, in the *Wiener Entomologische Zeitung* for June 25, mentions the fact that among the species of the genera *Aderces*, *Astatopteryx* and especially *Neuglenes*, specimens occur provided with wings and large black eyes, while others are found in which these organs are rudimentary or entirely wanting. Several explanations of this peculiarity have been advanced. Gillmeister and Erichson considered the forms as distinct species. Matthews considered those provided with eyes as females and the blind ones as males. Reitter insisted that, in conformity with all known analogous cases, the blind ones are the females and those with eyes the males. Flach's investigations have, however, proven without a doubt the rather surprising fact that sexes occur in both forms indicating the existence of alternating generations, the blind form being stationary while those provided with eyes and wings are migratory. He found *Neuglenes apterus* at different times in decaying poplars, without being able to explain how it was possible for them to get to such situations on account of their feebleness and awkwardness and the dryness of the air. The distribution of the species, however, over the whole of Europe points with certainty to a greater agility than the blind and wingless form could possibly have. In the same way Flach had not been able to explain the wide distribution in the East of the blind *Ptiliolum oedipus* until the mystery was solved by the discovery of a female with well-developed eyes and wings among seventy specimens of the degraded form from the Caucasus. He concludes that as forms with eyes as a rule appear to be much scarcer than the blind ones it would be a very interesting investigation to endeavor to decide to what particular conditions of their mode of life the change is due (light or dryness), or, have such changes taken place at cyclic intervals?

CATERPILLARS STOPPING TRAINS.

Under this caption we printed in No. 1, Vol. I, page 30, an occurrence in South Carolina, which turned out on investigation to be a great exaggeration.

On June 29 of the present year we received a letter from Mr. Stark Webster, of Mattawamkeag, Me., inclosing a clipping from the *Upper River News* of May 25, detailing a very similar circumstance. Mr. Webster also stated that in the Northern Penobscot region the same worm defoliated most of the orchards and all of the poplars, leaving them as bare as in mid-winter. He also noted that many of the cocoons spun in the latter part of June contained a large white maggot. A subsequent letter, dated July 6, was accompanied by specimens in which it was seen that the insect they contained was the Tent Caterpillar of the Forest (*Clisiocampa sylvatica*), and Mr. Webster wrote further that they seemed to prefer Poplar, and also fed upon Oak and Cherry, and after all these are stripped they attack the Elm, Gray Birch, Willow, Rock Maple, and some other trees.

In the first volume of the *American Entomologist*, page 210, the occurrence of this same species upon a railroad track in great numbers was recorded.

The newspaper clipping which Mr. Webster sent is here reprinted with its head-lines, although for the sake of brevity we do not use the same display.

The grand march of the caterpillars.—They blockade a train on the Canadian Pacific.—Freight locomotives and railroad men powerless.—Mosquitoes join in the raid and do bloody work.—Additional motive power and sand effect their release.

The first freight train run in connection with the Bangor and Piscataquis over the Canadian Pacific met with a novel and what at one time threatened to be a serious as well as a laughable mishap Sunday. Our managing editor was in it. At a point a few miles from Sebois, on the Canadian road, the Messrs. Pierce Brothers, of Milo, had collected 1,500 ship knees, and Superintendent Van Zile sent down a big engine and eleven flats to draw them up to Brownville crossing.

They were loaded, and the return trip of 15 miles was begun, which occupied ten hours. When the train had proceeded a few miles, and when it was on a short grade, it was brought to a standstill by an army of small, gray caterpillars, greasing the track and driving-wheels to such an extent as to almost entirely suspend friction between the rails and the driving-wheels. In some places they were half an inch thick, and the army stretched out 11 miles.

The night previous, as the time-keeper, who had about 20 miles to cover, was working homeward on his jigger, or railroad velocipede, he encountered the advance guard, and for half a mile pushed his machine along the rail by hand.

Section men undertook to sweep them off with alder bushes, but the slight touch of the twigs would crush them and lubricate the rails, and the mass formed like dough upon the driving-wheels.

The train in going down passed through these and others, but the big collection came during the forenoon and while the knees were being loaded. Of course, sand was used, but it did not avail much, and Superintendent Van Zile was wired, and he ordered out another locomotive from Sebois.

On her arrival there began a series of charges at that grade, which now had been liberally sprinkled with sand, but the animal life was so thick that various attempts were unsuccessful, and it was not until late at night and the sun had gone down that the creeping things desisted in their march.

With these there had come clouds of mosquitoes, and they very materially aided the other insects by pitching most vigorously into the men, seemingly drawing blood from all nationalities alike, and the sight of a sweating, swearing railroad laborer, frantically brandishing alder boughs over his head with one hand, while with the other he scraped caterpillars, was laughable in the extreme.

The matter has at once engaged the attention of Superintendent Van Zile, who is trying to find out from the encyclopedia how long the march of these Maine hosts continues, and it is quite likely that the road alongside this section will be ditched and flooded with running water. Nothing like it was ever known hereabouts before, but then sunlight was never before let into the wilds of Maine as the Canadian road has let it in, and there may be unknown difficulties to come consequent upon it.

LOCUSTS IN ALGERIA.

The French Government has lately been seriously occupied with the question of Locust ravages in Algeria, while the Algerians have been doing the best they know how to defend themselves against the plague. That they are yet unfamiliar with some of our American methods is shown by the following abstract of a communication from Constantine, Algeria, dated June 14, to the Paris *Petit Journal* of June 19.

The Algerians levied a tax of 4,000,000 francs to carry on the war against these Locusts, but unfortunately this subsidy was only available at the time when the Locusts, having passed their last stage of development, die after laying their eggs and stocking the country for another year.

The Algerians had offered pay for the collecting of Locust eggs. The price given was small (75 centimes per decaliter), but the 14,000,000 decaliters which were collected and destroyed were but a fraction of what remained.

The hatching of the remaining eggs, being retarded by violent rains, did not take place before the end of April last. As soon as the first hatching occurred vigorous measures of defense were taken by beating the ground with branches of trees in leaf.

When the Locusts have hatched in such large quantities that the force of men at hand is not sufficient to destroy them immediately after hatching, this beating is no longer employed. The *Melhafa* must then be used. This consists of a cloth 5 by 2 meters, which is set on end perpendicularly upon the ground, and folded at an obtuse angle; the Locusts are then driven into this cloth, which is then folded over them, when they are crushed, thrown into pits, and covered with quick-lime.

A last means of defense, the Cypriote machine (of which we have no description) is employed when the two former methods fail. Locusts which escape from this machine have been flying in such compact masses as to obscure the sunlight, generally flying before the wind.

All able-bodied men of any nationality, from the ages of 18 to 55, have been pressed into service. Even the army of Algeria, including the troops in Alger and Oran, were sent to the hatching points. The Algerians submit to this requisition willingly and without complaint.—C. V. R.

THE NEW CATTLE-FLY OR HORN FLY.

Many notes have appeared in the papers during the last summer and the present summer concerning a new pest which is worrying cattle in Pennsylvania, New Jersey, Delaware, Maryland, and northern Virginia. It is a small fly half the size of a house fly, which settles in great numbers around the base of the horns and on other portions of the body where it can not be reached by either the tail or the head of the animal. It sucks a moderate amount of blood, reduces the condition of the cattle, and lessens the yield of milk by from one-third to one-half. It has been named by Dr. Williston *Hæmatobia cornicola*. We are investigating its Virginia and Maryland occurrences, and have succeeded in tracing its life history. We find that the fly lays its eggs, usually at night, in freshly dropped cow dung, and that for the development from the egg through the maggot stage to the perfect fly a space of only twelve days is necessary. This rapidity of reproduction accounts for the wonderful numbers in which these flies appear, and it follows with reasonable certainty that thoroughly liming the dung in places where the cattle preferably stand at night will kill off many larvæ and greatly lessen the numbers of the flies.

On large stock farms little else can be done, but applications may be made to milch cows and valuable animals which will keep the flies away. The applications may be (1) fish-oil and pine tar with a little sulphur added; (2) tobacco dust, when the skin is not broken; (3) tallow and a small amount of carbolic acid. The latter application will also have a healing effect where sores have formed.

We expect to publish a full and illustrated account of this insect at the close of the season.

U. S. DEPARTMENT OF AGRICULTURE.

DIVISION OF ENTOMOLOGY.

PERIODICAL BULLETIN. SEPTEMBER, 1889.

Vol. II. No. 3.

INSECT LIFE.

DEVOTED TO THE ECONOMY AND LIFE-HABITS OF INSECTS, ESPECIALLY IN THEIR RELATIONS TO AGRICULTURE, AND EDITED BY THE ENTOMOLOGIST AND HIS ASSISTANTS.

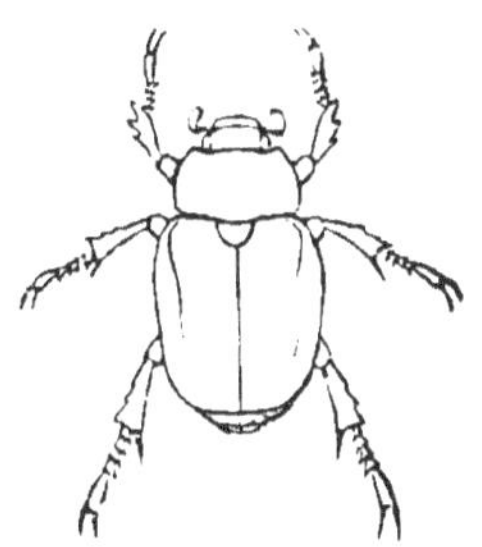

[PUBLISHED BY AUTHORITY OF THE SECRETARY OF AGRICULTURE.]

WASHINGTON:
GOVERNMENT PRINTING OFFICE.
1889.

CONTENTS.

II

SPECIAL NOTES.

Insect Pests in East India.—We have just received through the kindness of Mr. E. C. Cotes, of the Indian Museum, Calcutta, a very interesting paper, entitled "Notes on Indian Insect Pests," which forms No. 1 of Vol. I of the "Indian Museum Notes," published by the trustees of the museum and by the authority of the revenue and agricultural department of the Government of India. This publication is to take the place of "Notes on Economic Entomology," of which two numbers have appeared. The present number is divided into three parts; the first part contains "Notes on Rhynchota," by Mr. E. T. Atkinson, and includes short articles upon the Rice Sapper (*Leptocorisa acuta*), an insect which considerably injures the autumn rice by settling upon it when it is milky and sucking out the juice, leaving the husk dry; as many as 6 to 10 of the insects have been seen upon a single ear; the Chora-poka (probably *Carbula biguttata*), an insect which appears in vast numbers when the sesamum crop is gathered and stacked on the threshing floor and eats out the kernel of the seed, leaving only the husk; the Green Bug (*Nezara viridula*), which occurs upon potato halms; also several species of *Capsidæ*, *Jassidæ*, *Aphidæ*, and *Coccidæ*. A new species of *Cerataphis* and a new species of *Pemphigus* are mentioned as feeding upon Cinchona. The second part is by Mr. L. de Nicéville, and treats of a Butterfly injurious to Rice and the Ceylon cardamom pest. The butterfly is *Saustus gremius*, and the larvæ feed upon the leaves of rice. The cardamom pest is *Lamphides elpis*, the larva of which bores circular holes into the capsules and destroys the contents. The damage done by this latter pest is sometimes as great as 80 to 90 per cent. to young plantations. Between from 5 to 10 per cent. of the fruit capsules are perforated.

In the third part Mr. E. C. Cotes gives us further notes on the Wheat and Rice Weevil, on the Sugar-cane Borer-moth (*Chilo saccharalis*), the Sorghum borer (species not determined), a caterpillar injurious to tea, cut-worms, a moth injuring a cultivated timber tree known as *Cedrela toona*, Clothes moths, *Hispa ænescens* injuring rice, a species of *Tomicus* which bores in the Makai tree (*Shorea assamica*), a bamboo borer, the Leather Beetle (*Dermestes vulpinus*), which is mentioned as damaging silk-worm cocoons, further notes on insecticides, short notes on miscellaneous insect pests, and extracts from correspondence.

Among the short notes on miscellaneous insects we may mention as of especial interest the damage done by *Heliothis armigera* to the poppy crop in Patna and Arrah, the occurrence of a bag worm upon tea bushes, the damage done to the castor-oil plant by the larva of a noctuid moth known as *Achæa melicerte*, the damage done to jute crops by caterpillars, the *Spilarctia suffusa*, the injury by *Tinea lucidella* to the horns of hollow horned ruminants, damage to the leaf covering of opium balls by *Lasioderma testaceum*, a species which also injures manilla and Indian cheroots. Many other insect notes of considerable interest occur and many of them are accompanied by both their Indian names and particulars of the plants which they infest. The paper is illustrated by four very good plates reproduced by a photo-etching process.

The Lesser Migratory Locust.—Since the destructive year 1883, this insect has not done much damage in the interesting region of southern New Hampshire, which we wrote up at some length in the Annual Report of this Department for that year, but the present season has brought another outbreak, and in July we sent Mr. Marlatt, of this Division, into the field to look into the condition of affairs, to advise with the farmers concerning remedies, and to collect facts relating to the years intervening between the present date and 1883. We publish in this number his report of his short investigation, and this account will bring the history of locust damage in that locality down to the present time.

New Injury by the Leather Beetle.—Mr. F. M. Jones, of Wilmington, has called our attention to the damage done by this insect in many of the large establishments of that city to goat-skins used in the preparation of morocco leather. Mr. Jones has prepared a short article at our request, which we publish in this number.

The Official Association of Economic Entomologists.—We print under the head of general notes the constitution of this new organization, together with the lists of officers and charter members. The next meeting will soon be held, and we would urge all economic entomologists to read the constitution carefully, and, if they feel themselves in sympathy with the Association, to send their credentials and names to the secretary, Prof. J. B. Smith, at New Brunswick, N. J. That this association will have a successful future and that it will accomplish the results anticipated can hardly be doubted. The greatest enthusiasm was exhibited at the meeting, and every letter received carried with it the expression of warm approval.

DERMESTES VULPINUS IN GOAT-SKINS.

By Frank M. Jones, *Wilmington, Del.*

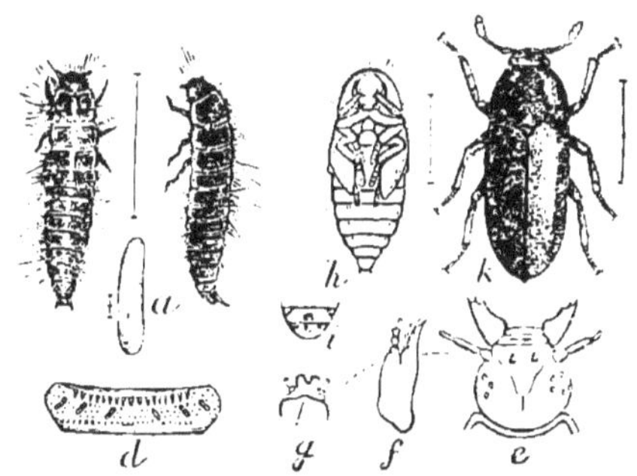

Fig. 7.—*Dermestes vulpinus*: *a*, egg; *b*, *c*, larva, lateral and dorsal view; *h*, pupa, ventral view; *k*, beetle—enlarged; *d*, dorsal view of one of the middle joints of larva denuded to show spines and tubercles; *i*, ventral view of tip of abdomen in ♂ beetle; *e*, head of larva; *f*, left maxilla of same, with palpus; *g*, labium of same, with palpi—enlarged. (After Riley.)

Mr. James Fletcher, in his address before the Entomological Society of Ontario, in October last, divided injurious insects into three classes—first, second, and third class pests—"according to the amount of injury they are answerable for"; and the insect under consideration, the leather beetle, *Dermestes vulpinus*, belongs to the second of these classes; for, while it is always to be found, throughout the summer months, in the baled goat-skins stored in the ware-rooms of the importers and morocco manufacturers in various parts of the country, it is only occasionally that it occurs in sufficient numbers to do any great amount of injury. The larvæ are usually most abundant upon the hair side of the skins, but an examination of skins which have been damaged by them proves that they often commence their attack on the flesh side. When they occur in large numbers, and when no attempt is made to check their ravages, the skins are quickly eaten into holes, rendering them almost worthless. The pupa is not inclosed in any cocoon, but lies loosely in the hair or in a fold in the skin; and it is a common sight to see larvæ of various ages, pupæ, and the perfect insects inhabiting the same skin.

Skins which are naturally of a greasy nature, such as the Kassan (from Russia) and the Angora skins, appear to be most liable to attack; and heavily-salted skins, such as the Mochas (Arabian), are comparatively free from the pest; but even the poison-cured skins are not entirely exempt. Tampico (Mexican) skins are sometimes very badly damaged by this insect, which must now be very widely distributed; for whether the skins come from Russia or Cape Town, Turkey or Mexico, Arabia or South America, the same species of insects is found in them all.

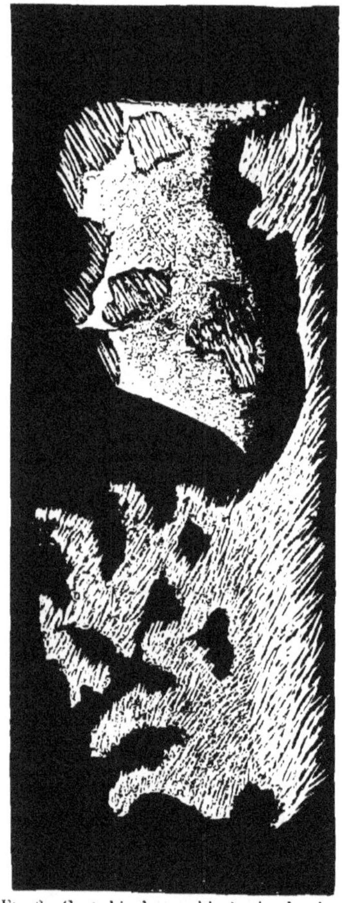

FIG. 8.—Goat-skin damaged by leather-beetle—nat. size. (Original.)

It is said that fifteen or twenty years ago this insect was much more injurious than now; but this is probably due to the fact that, the demand being much greater, the skins are used up much faster, and the insects do not have time to multiply to any great extent. The only method employed to destroy them is to beat or shake each skin separately and crush the insects which fall to the floor; but where there are thousands of skins this is a tedious process, and is probably only a temporary check, as many insects are undoubtedly left in the skins. Placing the bales in a close compartment and killing the insects by means of vapor of bisulphide of carbon, or by burning sulphur, has been proposed; but the practical value of these methods has not been tested.

THE JAPANESE PEACH FRUIT-WORM.

In the August (1888) number of Insect Life we published some correspondence between the Rev. W. J. Holland, who was then serving as naturalist to the U. S. *Eclipse* expedition, and the United States minister to Japan and the Commissioner of Agriculture, relative to the ravages of a worm which damages the peach crop of Japan. Those who read this correspondence will recollect that we suggested through Commissioner Colman that the matter be referred to Prof. C. Sasaki, of the Agricultural and Dendrological College at Tokio, and that Professor

Sasaki be directed to make a full report concerning this insect. It seems that this suggestion was adopted, and that Professor Sasaki was instructed by Count Okuma, the Japanese minister for foreign affairs, to prepare the report, which he did with his customary care. The report was submitted to the Secretary of Agriculture through the United States minister to Japan and the Secretary of State during July. Meantime we sent to Professor Sasaki for specimens of the insect, which have not yet arrived, but upon their receipt we shall reproduce some of his figures and give the insect a definite name, and shall publish his somewhat elaborate account in full. Meantime, however, the matter is of so much interest to the fruit-growers of the Pacific coast that we submit a short abstract.

The moth, according to Professor Sasaki (and judging from his figures he is correct) is a species of *Carpocapsa* very closely allied to our Codling Moth, and hence it is called by Professor Sasaki "a new Codling Moth injurious to the Peach." The peach crop is very large in Japan, and during some seasons more than 90 per cent. of the fruit is injured by this insect. Not infrequently more than one larva are found in a single peach. No means have been heretofore suggested for the protection of the crop. Professor Sasaki's studies were begun in April 1, 1888, and concluded in May, 1889. The moth appears twice in the year, viz, in June and in August, although certain individuals of the first brood are delayed until July and others of the second brood until September. They hide in the day-time and at twilight fly about the trees. The eggs are deposited singly on the apex of the fruit or along the suture passing from the apex toward the base. Usually one or two, but sometimes more, eggs are deposited in a single fruit. The eggs are spherical in form, measuring one-half millimeter in diameter. They are yellow in color. They hatch in a few days, and the larva molts four times. Upon first hatching it crawls actively about in search of a suitable spot at which to enter the fruit; it then gnaws its way in, turns its head towards the opening and closes it with silk, sometimes pushing its excrement outside. It then burrows to the stone and makes a large excavation around it. Occasionally a larva will leave one peach and enter another.

The fruit is continuously infested from June until September, those containing larvæ ripening early and dropping off. Infested fruit may be recognized in the following ways:

(1) It becomes soft and may be crushed by a slight pressure on account of the central excavation.

(2) It has usually a small cluster of yellowish-brown excrement on its surface.

(3) It bears irregular patches of a greyish-yellow or reddish-blue color.

The larva attains its full growth in from three to four weeks after hatching; it then leaves the fruit and falls to the ground, if the fruit has not already fallen.

The larva enters the ground to a depth of 1 or 2 inches, where it makes an oval cocoon of light gray silk. The cocoon is very strong and elastic. The larva of the first brood remains within this cocoon about a week and then changes to pupa, while the larva of the second brood remains within the cocoon in the larval state through the winter and changes to pupa in the month of May.

Professor Sasaki makes but one suggestion as to remedies, and that is to gather the fallen fruit every day and to dispose of it in such a way as to destroy the larva. We have already written him that he will unquestionably find a good remedy in the application of arsenical poisons for the first brood.

A REPORT ON THE LESSER MIGRATORY LOCUST.

By C. L. Marlatt, *Assistant.*

The following account of the recurrence in injurious numbers of the Lesser Locust (*Melanopus atlanis*) the present season in the Merrimac Valley near Franklin, N. H., may be considered as supplementary to the extended article in the report of the United States Entomologist for 1883, in which a full record of the earlier occurrence of this species in northern New England (1743–1883) is given; its life-history and habits, natural enemies, and means against it.

As stated in the article cited, Professor Riley visited the infested region in person in 1882 and 1883, and with the aid of some of his assistants introduced and explained to the farmers some of the machines for collecting and destroying the locusts successfully used against the closely allied but more destructive Rocky Mountain species.

The value of these appliances was immediately recognized by the intelligent farmers of the Merrimac Valley, and numbers of them were constructed after the pattern of the one described on p. 176 of the report for 1883 and figured Pl. VII, 1; and with the incentive of a bounty of $1 per bushel, granted by the State, they were used with such effect against the locusts in the two years following (1884 and 1885) that no serious injury has, previous to the present season, been occasioned by them since 1880.

To illustrate the success which attended their use, the statement of Mr. George B. Mathews may be given, viz, that no less than 500 bushels were caught at the Webster place in 1884, a much less number in 1885, since which time they have occurred in but small numbers.

A letter to the entomologist from Mr. E. A. Fellows, July 3, 1889, quoted below, again called attention to a serious outbreak of locusts in the Merrimac Valley, near Franklin, N. H., and but a few miles above the region unusually infested in 1882 and 1883, and seemed to warrant the investigation recorded in this article.

To the ENTOMOLOGIST:

DEAR SIR: My farm this season is infested with grasshoppers, the hay, oat, and rye, and part of vegetable crop, being nearly a complete failure. I find on many of the grasshoppers a small parasite or egg of a deep orange-red color, clinging to different parts of the locust's body, being mostly on and under the wings. What I would like to know is, whether this parasite is likely to check the increase of the locusts another season, as it don't pay for me to plant crops to be devoured by these ravenous locusts. I have caught some sixty bushels from one piece of oats, containing 3½ acres, but am satisfied I can never exterminate them that way. They were quite bad last season, but not to be compared to this.

Respectfully,

E. A. FELLOWS.

FRANKLIN, N. H., *July* 3, 1889.

Mr. Fellows's communication is interesting not only because it records the abundance of the Locust Mite (*Trombidium*), previously found here in but limited numbers, but as still further emphasizing the peculiar local habit of *Atlanis* in this region, noted in the report already cited.

Mr. Fellows's farm, which was visited July 11, is situated in an "intervale" or small valley of about 300 acres, shut in by high hills, and thus separated from similar intervales above and below.

In these small intervales the locusts find a permanent home, only occasionally assuming the migratory tendency; and under favorable circumstances, especially if left unchecked, they after a year or two become suddenly numerous enough to do great injury, while at the same time in the similar valleys above and below their numbers may be significant only of future increase.

This state of things is well illustrated on the Fellows farm the present year. These locusts, always present in small numbers, had last year become quite abundant, and as no measures were taken against them, they this spring appeared in destructive hordes. The grasses suffered most. Timothy, red-top, chess, and clover were reduced to mere innutritive stalks; both blades and the heads of the oats were eaten; all garden vegetables were attacked. Squashes, melons, and corn were only eaten when very young. The tassel of the latter, however, is also eaten by the locusts.

At the time of examination the locusts were generally winged, and while still quite thick in the oats had scattered somewhat over adjoining meadow-land, and were especially abundant near the river, which had perhaps, by forming a barrier to their half-migratory movements going on at this time, caused them to collect there. A small percentage (5 to 10) were *in coitu;* but none were found ovipositing, although in the dissection of a large number of females one or two were found with empty ovaries, indicating that oviposition had already begun.

Examination of the ground, and, as observed by Mr. Fellows, the first appearance of the young locusts in the spring agree in indicating that the eggs are deposited more particularly in certain sandy knolls in the interval, and perhaps to a certain extent on the lower portion of the bordering hill-sides.

If this be the case, the destruction of the eggs by harrowing or plowing in the fall, or of the young locusts in the spring either by plowing them under or by the use of trapping or kerosene machines, should be comparatively easy.

The parasite mentioned by Mr. Fellows, the young of the locust mite (*Trombidium locustarum* Riley), was very common, but on the authority of Mr. Fellows was becoming rapidly less abundant. He stated that during the active operations with the hopper-doser the "catch" was markedly colored by them, and that he had observed this spring on his land unusual numbers of a red spider-like mite, which, from his description, was undoubtedly the adult of the locust mite.

A considerable variation in the percentage of infested locusts in different parts of the intervale was noted, and this holds also for the parasites mentioned below.

On the oat-field, fully 95 per cent. of the locusts bore from one to fifty mites, while of those near the river less than 50 per cent. were infested, a fact easily explained perhaps by the greater activity of the non-infested locusts.

Large numbers of dead locusts, mostly hollowed out and reduced to mere shells, were observed over the infested tract on the ground or clinging to grass or oat stems. Some of the fresher specimens contained Dipterous larvæ (*Tachina* and *Sarcophoga*), and examinations of living locusts taken from the oat field showed that about 5 per cent. were thus parasitized, each parasitized locust containing from one to four maggots.

A slightly larger percentage proved to be infested with hair worms (*Mermis*). The abundance of these parasitic enemies the present year would indicate a very considerable reduction in the next year's crop of locusts; but this should not form an excuse for neglecting any direct measures that can be employed against the eggs this fall, or early work against the young should they appear in numbers next spring.

Mr. Fellows's operations against the locusts, which were confined to the use of the collecting pan mentioned above after the locusts had become mostly winged and the damage largely accomplished, while unsatisfactory to himself would have doubtless been much more effective if undertaken earlier, or if measures had been taken against the early stages.

In all seventy-two bushels were caught and buried in a trench during a period of about two weeks in the latter part of June and the first of July. Of these, sixty bushels were taken from the three and a half acres of oats into which the locusts migrated from adjacent fields during this time.

In place of the kerosene and water or kerosene emulsion ordinarily used in these pans, Mr. Fellows employed a strong soap-suds, which assisted in retaining the locusts in the pans.

Locusts were reported to be moderately abundant above Franklin, at Hill, and also below, near North Boscawen, at the Webster place. On the farm of Mr. Wright, near Hill, they had practically destroyed several acres of grass and were at the time of examination working in the oats. Mr. Wright stated that the locust had not been previously very abundant there since 1884 and 1885, when a number of bushels had been caught.

A number of farms in the neighborhood of the Webster place were also examined, and the farms of Mr. Gordon Burleigh and Mr. Benjamin Hancock were found to be somewhat thickly stocked with locusts, and the grass had been considerably injured.

Mr. Geo. B. Mathews, of the same place, a very intelligent farmer, assured me that the locusts could be easily controlled, and that he was not troubled at all except as they drifted onto his land from the farms adjoining. He had used the "hopperdoser" with good success in 1884 and 1885, and since then, by carefully noticing the breeding ground of the locusts and plowing the young under in the spring, he had succeeded in reducing their numbers to a minimum, with very little loss to himself. He was of the opinion that an officer empowered to compel the plowing of the infested fields at the proper time, with perhaps a compensation to the farmer for the crop turned under, would be the only practical solution of the locust trouble.

While investigating the locusts about Franklin, reports came to me of the serious depredations of this pest on the Connecticut River, near Bellows Falls, Vt., and at the direction of the Secretary of Agriculture this locality was visited and the following data collected.

The occurrence of the locusts here is especially noteworthy because it illustrates most pointedly the local habits of *Atlanis* already described.

In answer to inquiries made at various points from Hanover to Bellows Falls, I was informed by various parties and particularly by Professor Whicher, Director of the New Hampshire Experiment Station, at Hanover, that the locusts were not known to be abundant elsewhere on the river.

The infested area proved to be an intervale of about 500 acres extent, similar to those of the Merrimac Valley, and contained the farms of Mr. Marvin W. Davis, member of the State Board of Agriculture, and of Mr. R. H. Blair. Both of these gentlemen were seen, and to the former, on whose farm the locusts were especially abundant, I am indebted in part for the following facts:

The locusts were first noticed in this valley some fifteen years since, when they ruined the tobacco crop by eating the leaves of the young plants full of holes. Their attacks at that time and afterwards were so severe that the growth of this crop was abandoned. The locusts have increased from year to year, and the present has witnessed them more abundantly than ever before; the fences and roads being reported as black with them as they moved from the hatching-grounds to other fields.

Grasses and oats, young corn and garden vegetables, even the onions, were eaten.

The statements of these farmers, confirmed by my own observations, show that the eggs are deposited, in great part, in a sandy-clay knoll thinly clothed with grass and of but few acres area, from which the locusts migrate to all parts of the valley.

No effort has been made here to control the locusts except an ineffectual attempt to use a large flock of turkies for this purpose, but it would seem, in view of the limited area in which eggs are placed, to be a comparatively easy matter to keep them in subjection by the use of the measures already given.

The Locust Mite, Dipterous larvæ, and Hair-worms were found to infest the locust here in somewhat less numbers than at Franklin.

THE IMPORTED AUSTRALIAN LADY-BIRD.

Vedolia cardinalis.

By D. W. COQUILLETT, *Los Angeles, Cal.*

In his annual report for the year 1888, published in the report of this Department for that year, Professor Riley has given an account of "The Importation of Parasites and Predaceous Insects from Australia," containing an account of the importation by the Department of certain kinds of insects which naturally prey upon the Fluted or Cottony-cushion Scale (*Icerya purchasi*, Maskell). At the time of writing the above report only a few specimens of the black and red Lady-bird had been received, so that very little could be said in regard to its habits and early stages. As I have now carefully worked them out, I give herewith a brief account of them, in accordance with directions from the Division of Entomology.

EARLY STAGES.

EGG.—Elongate-ovate, or rarely elongate ellipsoidal, its width never more than one-half its length; very rough, or scabrous; deep orange-red; length, one-half millimeter.

LARVA (*first stage*).—Dark orange-red; first segment with two small black warts placed subdorsally, and with two long whitish bristles on each side; segments two to eleven each, with three dark-brown warts each side—those on segments two and three situated in the subdorsal, supra-stigmatal, and stigmatal regions, while those on the remaining segments are situated in the dorsal, supra-stigmatal, and stigmatal regions; each of those in the stigmatal region bears two long whitish bristles, while each of the others bears a single shorter whitish bristle, those on the eleventh segment the longest; head about five-sixths as wide as the first segment and slightly darker, its sides blackish; six thoracic legs orange-red, the tibiæ darker; last segment furnished with a retractile proleg.

Second stage.—Same as in the first, with these exceptions: Head about three-fifths as wide as the first segment; this segment bears two additional bristles near each corner, and two others in front of the middle; second and third segments each with an additional but much smaller wart in front of those in the stigmatal region, each bearing a single short bristle; bristles, except those in the stigmatal region, black, the warts in this region reddish, and larger than the others.

Third stage.—Same as in the second, except that the head is proportionately narrower, being only about one-half as wide as the first segment.

Fourth stage.—Same as in the third, except that the warts in the subdorsal and supra-stigmatal region on either side of the third, and usually of the second segment, are connected by a black spot, and the body finally becomes covered over with a light gray powder; length when fully grown, about 6 millimeters (Fig. 9).

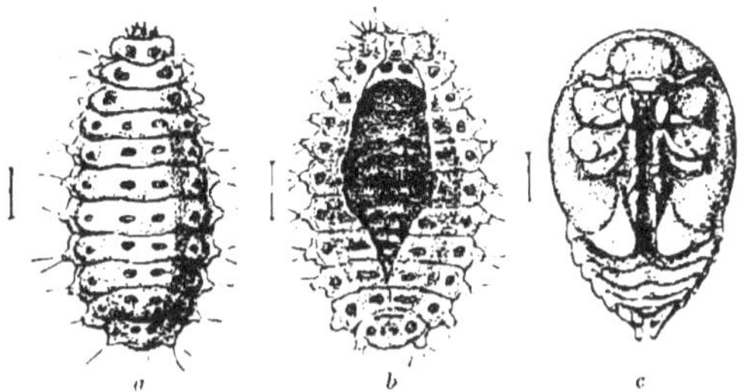

FIG. 9.—*Vedolia cardinalis*: *a*, Full-grown larva; *b*, pupa, dorsal view, enclosed in last larval skin; *c*, pupa, naked, ventral view—all enlarged. (Original.)

PUPA.—Partially inclosed in the old larval skin, which is of a whitish color, marked with black dots, which indicate the position of the warts on the larva as described above; this skin is rent from near the front edge of the first segment to the middle of the eighth; the exposed part is mottled light and brownish red, the first segment marked with two dorsal black dots, or the entire dorsum of this segment, and also that of the second and third segments, black; abdomen with a polished-black interrupted dorsal line; length, 4 millimeters (Fig. 9).

The following table exhibits the length of time passed by these Lady-birds in their different stages:

Egg laid.	Egg hatched.	First molt.	Second molt.	Third molt.	Pupated.	Beetle issued.
Apr. 20	Apr. 26	May 3	May 5	(?)	May 14	May 21
Apr. 23	Apr. 29	May 3	May 7	May 15	May 19	May 26
	Apr. 27	May 3	May 5	May 11	May 19	May 26
	May 6	May 11	May 14	May 19	May 29	June 5
		May 11	May 13	May 17	May 23	May 31
			May 9	May 12	May 20	May 26
			May 17	May 22	May 31	June 5
				May 10	May 17	May 25
				May 11	May 19	May 27
				May 12	May 19	May 26
					Apr. 25	May 4
					Dec. 5	Dec. 18

Averages: *Egg*, six days. *Larva*, nearly twenty-two days (*i. e.*, first stage, five and a half days; second stage, two and three-fifths days; third stage, five and one-sixth days; fourth stage, seven and five-ninths days). *Pupa*, seven and three-fourths days. Egg to beetle, a little over thirty-five days.

Three of the beetles which issued from the pupa May 4 were kept in a breeding cage in a sunny window of my office and supplied with an abundance of food; one of them died on the 20th of May, another on the 26th, and the third died on the 5th of June. It is probable, therefore, that in the open air in summer the beetles live about four weeks after issuing from the pupa, so that their existence from the time the egg is laid until the adult which originated from it dies a natural death covers a period of about two months. During the colder portion of the year, however, this period is doubtless extended considerably beyond this limit, as will be seen by reference to the above table; for instance, the larva that pupated December 5 was changed to a beetle thirteen days later, whereas the one that pupated May 31 produced the beetle five days later.

HABITS AND NATURAL HISTORY.

The eggs are usually thrust beneath the Iceryas, but are sometimes attached to the cottony egg-masses; they are placed on one of their sides, sometimes singly but usually in pairs or in groups of three or more. In hatching, the egg-shell is rent nearly the entire length along its upper side, and after the young larva has issued the shell becomes of a whitish color, and retains nearly its original form. The recently laid egg is more slender and of a deeper red color than the egg of the Icerya.

The young larvæ usually burrow into the egg-masses from below and feed upon the eggs; later they attack the Iceryas of all sizes, usually making the attack on the under side of the abdomen. The young larva is easily distinguished from the young Iceryas by lacking the long black antennæ so conspicuous in the latter. When about to cast its skin the larva attaches the posterior end of its body to some object, and at the proper moment breaks away the whole anterior end of the old skin and crawls out of the opening thus made.

When about to pupate the larva attaches the posterior end of its body to the bark or leaf of the tree and suspends itself head downward. It remains in this position about three days, when the skin along its back splits open, exposing a portion of the pupa to view. When the beetle is fully formed the old pupa-skin partially breaks away, showing the beetle to be of a pale reddish color. It remains in this situation about two days longer, when the beetle issues clad in its normal colors of black and red, as shown in the figure (Fig. 10). Coition occurs shortly afterward. In fact I have frequently seen the males standing by and waiting for the females to issue, even going so far as to tear away the old pupa-skin and uniting with the female while she is still soft and helpless. Egg laying begins the next day, and is continued during nearly the entire life of the beetle. One that I kept in a breeding-cage and supplied with an abundance of food, deposited 42 eggs in eight days. The total number deposited by one female will probably average from 150 to 200 eggs.

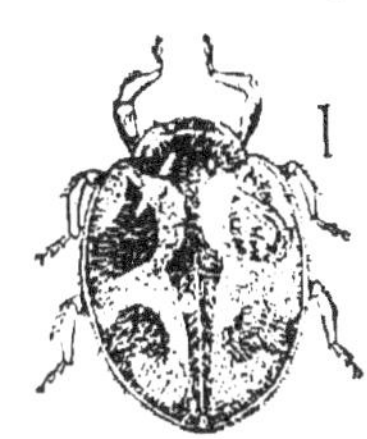

FIG. 10.—*Vedolia cardinalis*, adult; enlarged. (After Riley).

The adult beetles as well as the larvæ also feed upon the Iceryas, but with this difference, that the attack is usually made from above instead of from below.

I have never seen these Lady-birds in any of their stages feeding upon any other insect than the Icerya. On one occasion I confined six Lady-bird larvæ in a breeding-cage containing black scales (*Lecanium oleæ* Bernard), some of which were quite soft, but after the lapse of seven days none of these scales had been attacked, whereas three of the Lady-bird larvæ had been devoured by their comrades. At the same date I placed an equal number of these larvæ in another cage containing specimens of an undetermined species of *Lecanium* found on a peach-tree, several of the scales being still soft, but at the end of seven days none of them had been attacked, while four of the Lady-bird larvæ had fallen a prey to their rapacious brothers. I also tested these larvæ with a species of plant-louse found on orange-trees, but they did not attack them. It seems very evident, therefore, that the Iceryas are the natural food of these Lady-birds, and they feed upon these in all their stages, even attacking the winged males.

I have never seen any of our native insects attacking these Lady-birds, although Col. J. R. Dobbins informs me that on one occasion he saw a lace-winged fly larva (*Chrysopa* sp.?) in such a position that he thought it might have been engaged in feeding upon a Lady-bird larva. The ants do not molest them.

IMPORTATION AND SPREAD.

The first consignment of these Lady-birds reached me on the 30th of November, and numbered twenty-eight specimens; the second consignment of forty-four specimens arrived December 29; and the third consignment of fifty-seven specimens reached me January 24, making one hundred and twenty-nine specimens in all. These, as received, were placed under a tent on an Icerya-infested orange-tree, kindly placed at my disposal by Mr. J. W. Wolfskill, of this city. Here they were allowed to breed unmolested, and early in April it was found that nearly all of the Iceryas on the inclosed tree had been destroyed by these voracious Lady-birds. Accordingly, on the 12th of April, one side of the tent was removed, and the Lady-birds were permitted to spread to the adjoining trees. At this date I began sending out colonies to various parts of the State, and in this work have been greatly aided by Mr. Wolfskill and his foreman, Mr. Alexander Craw, both of whom were well acquainted with the condition of the orchards in this part of the State. By the 12th of June we had thus sent out 10,555 of these Lady-birds, distributing them to two hundred and eight different orchardists; and in nearly every instance the colonizing of these Lady-birds on Icerya-infested trees in the open air proved successful. The orange and other trees—about seventy-five in number—and also the shrubs and plants growing in Mr. Wolfskill's yard, have been practically cleared of Iceryas by these Lady-birds, and the latter have of their own accord spread to the adjoining trees to a distance of fully three fourths of a mile from the original tree.

Besides the three consignments of these Lady-birds referred to above I also received two later consignments. The first of these reached me February 21, and numbered thirty-five specimens; these I colonized on an Icerya-infested orange-tree in the large orange grove belonging to Colonel J. R. Dobbins, of San Gabriel. The last consignment of three hundred and fifty specimens arrived March 20; one-third of these I left with Colonel Dobbins, while the remainder I colonized on orange-trees in the extensive grove owned by Messrs. A. B. and A. Scott Chapman, in the San Gabriel Valley. All of these colonies have thrived exceedingly well. During a recent visit to each of these groves I found the Lady-birds on trees fully one-eighth of a mile from those on which the original colonies were placed, having thus distributed themselves of their own accord. The trees I colonized them on in the grove of Colonel Dobbins were quite large and were very thickly infested with the Iceryas, but at the time of my recent visit scarcely a living Icerya could be found on these and on several of the adjacent trees, while the dead and dry bodies of the Iceryas still clinging to the trees by their beaks, indicated how thickly the trees had been infested with these pests, and how thoroughly the industrious Lady-birds had done their work.

EXTRACTS FROM CORRESPONDENCE.

Enemies of Diabrotica.

With this I mail you a spider which I found with a *Diabrotica soror* in his jaws. Will you please send me the name of this spider, as also of the family to which it belongs? If new, would it not be well to describe it, or to have Dr. Marx do so if he will?

It may interest you to know that I have bred a Tachina fly from *D. soror*, but its wings never expanded, so it is not fit for study. I have just captured a large number of these beetles, and will try to breed perfect specimens of this fly.—[D. W. Coquillett, Los Angeles, Cal., June 19, 1889.

REPLY.—The spider which you found eating Diabrotica is *Xysticus gulosus* Keyserling. It belongs to the family Thomisidæ. Your note concerning the breeding of the Tachinid from the Diabrotica is very interesting.—[July 3, 1889.]

The New Zealand Latrodectus.

I take great interest in reading the periodical bulletin on "INSECT LIFE." In Vol. I, No. 7, January, 1889, I read an account of the spider called *L. mactans*, the description of which tallies with a spider I used frequently to see in New Zealand, North Island. I see that in your bulletin it is described as black with vermilion spot on abdomen. During my stay in the above country I saw many of these spiders, some black with a red *triangular* spot on back and some black with a yellow spot on back of same shape. Whether these are of the same species I am unable to say, but they frequented the same places, mostly banks of rivers, and were especially numerous on the banks of the Wanganui River. The Maoris told me that their bite was not fatal, but very painful. I knew of a case where a Maori was carrying wood from the river to his "whare", situated on the banks. He got bitten by one of these spotted spiders that was concealed in the wood—was bitten in the hand—and during the night the arm was paralyzed to the shoulder. Whisky applied externally and internally effected a cure. It is said that the pain is felt for weeks after, with perhaps a month or so of no pain between. I have mentioned these facts because I did not see in your bulletin any account of a *yellow* spotted *L. mactans.*—[C. Herberte Riley, Gabriella, Fla., April 27, 1889.

Chinch Bug Remedies.

In complying with your request, I will state that as early in the spring as the warmth of the season will permit the Chinch Bugs come out of their winter quarters and resort to their natural place of ravages, the wheat fields. They first gather in groups and burrow into the soil among the roots of the wheat, clearing the soil from around them and leaving a top crust. There they cohabit, and from the 1st to the 15th of May deposit their eggs by thousands on the roots of the wheat. They have previously cleaned the soil from the roots for that purpose. As soon as the sun shines hot enough to warm the ground sufficiently, the eggs hatch and the young bugs begin to suck at the roots of the wheat. As they grow and become older they crawl up to the top of the soil and up the stalks of wheat, and still suck as long as there is any life or sap in the stalks, when they begin to travel to other parts of the field for a new supply. All go in the same direction. The old bugs injure the wheat only by clearing the soil from the roots. As soon as they get through depositing their eggs they die. To evade their ravages I leave my wheat ground with as smooth a surface as possible in the fall. Then in the spring, as soon as winter is over, I put a heavy roller on my wheat ground, pack the soil firmly to the roots of the wheat, and thereby prevent the old bugs from burrowing about them. Besides, this gives the wheat an early, vigorous growth, and thus the bug is overcome, so far as its ravages in the wheat field are concerned. I sometimes roll my ground the second time, say, about the 15th of April or the 1st of May, or later. It depends upon how the bugs are progressing, which can be told by drawing the wheat and examining the roots for eggs. I never fail to make good wheat. Besides, the crop of bugs is so diminished that they never injure my other grain crops or grasses to any serious extent. My neighbors failed to make wheat in 1888, while I harvested a heavy crop. I knew one man who gave his crop of wheat (15 acres) to one hundred sheep in the month of April and turned them off the 1st of May. His wheat came out and made good grain, while all around him failed on account of the Chinch Bugs. The sheep packed the soil to the roots, and thus overcame the bugs. I hope that others will try the experiment, as I have done, and be convinced that the ravages of the bugs can be overcome.—[J. R. Adams, Goodland Mo. June 29, 1889.

Cut-worms.

The Cut-worms are very bad in this section. I am putting out quite a patch of sweet potato plants. I sprout largely for sale and for my own use, and also raise cabbage, tomato, pepper, and other plants. I find that the Cut-worms are working on most all of them. I have been making green clover traps and collecting them under the bunches of clover, then burning them under these and in the ground beneath. I have found as many as 64 worms under one bunch that we had placed between the ridges of sweet potatoes. I first soaked the green clover in Paris green, but I think I did not get it strong enough, as I found only a few dead ones under the traps. Pie-plant or rhubarb leaves are also good. They may be put under the half of a large drain tile split in two, and the south end stopped up with dirt so as to keep the leaves from drying out. The worms may then be hunted every day under the leaves and killed. Cabbage and turnip leaves are also good. Can you give me any other information on this subject? I have concluded now to try soaking the plants in a solution or tea made of red pepper just at setting them in the ground. I make it by boiling the pepper in soft water and then letting it get cold, when the plants may be dipped into it. In my next experiment I will try common kerosene (coal-oil) with soft water well agitated, and set the plants out immediately after dipping them in it.—[A. L. Thompson, Homer, Ill., May 20, 1889.

REPLY.— * * * The remedy which you have applied, viz, the poisoned clover, was first suggested in 1882 by Professor Riley, and was first experimented with so far as we know by Dr. A. Œmler, of Wilmington Island, near Savannah, Ga., who found it exceedingly successful, and who was enabled to almost entirely rid his land of Cut-worms. In our opinion you will find it a much better means of fighting the worms than either of the other remedies which you mention, and we would advise the greatest care in the trial of a kerosene remedy, lest the plants should be killed. It will be well in fact to emulsify the kerosene with soap and water and then dilute it considerably before dipping the young plants in it. Even then success can not be relied upon. * * * —[May 23, 1889.]

An Army-Worm Note from Indiana.

I recently had my first experience with the Army-Worms, which were discovered June 17 in a piece of rye growing on reclaimed swamp land commonly called "muck." We confined them to the rye, which they soon cleaned up. The piece contained about nine acres, and at one time, or when the worms were nearly developed, and about one and one-half inches long, as much as five acres were literally covered with them. An open ditch on one side, filled with swift-running water and ditching and pitting on another, turned them into a wood pasture of blue-grass, where I called in about 150 hogs, old and young, that quickly devoured those already in and all that came after. But I also began plowing under the rye stalks and stubble by encircling the whole piece, and they could not well travel across plowed land, so those confined within the circle became lank and lean soon after, but on about June 25 they disappeared somewhat suddenly, but how I do not know. Now, I would like to have some information as to the nature of the Army-Worm and the facts as to their origin in such immense quantities. If they had been propagating year by year along the fences and by-places adjacent to this field, which had been previously for four successive seasons cultivated in potatoes, no injury came from them and none were noticed about the field. The previous seasons were very dry and last winter was mild, with little freezing and scarcely any rain, and just suited the insect fraternity. Or might the fly have come in vast numbers during the very warm days of April from a southern region and deposited their eggs in the rank growth of rye growing in a loose porous soil that was laden with vapor like matter that may have been attractive to the fly? Scores of acres in rye in this vicinity, growing on the same kind of soils, were destroyed in the same manner, although the farms and fields were not contiguous to each other. No other crop was attacked or injured. A good deal of theory and speculation has been discussed by the Granger brethren hereabouts as to their origin, nature, disappearance, and reappearance. I have no complete works on entomology and can get no clear idea of their history. Chambers's Encyclopedia, which I have, says but little and nothing definite, though it says "they (the worms) go into the molting stage and re-appear again the same season, or produce two crops in the same season similar to the cotton-seed boll-worm of the South, which produces three crops in one season." A definite answer as to their origin and history will be thankfully received by myself and others. —[I. M. Miller, Upland, Ind., July 16, 1889.

STEPS TOWARDS A REVISION OF CHAMBERS' INDEX, WITH NOTES AND DESCRIPTIONS OF NEW SPECIES.

By Lord Walsingham.

[*Continued from page 54 of Vol. II.*]

Lithocolletis alnifoliella Hb.

Chamb. Bull. U. S. G. G. Surv., IV, 121 (1878).
Pack. U. S. Dept. Intr., Ent. Com. Bull. VII, 140 (1881).

This species was not included in the Index by Chambers, but in the List of Food-plants of Tineina (which was quoted by Packard in his List of Insects Injurious to Forest and Shade Trees) it is referred to as making a tentiform mine on the under side of the leaves of alder. For the present, at least, there is no evidence to justify its inclusion in the North American fauna.

Lithocolletis quercipulchella.

Chamb. Bull. U. S. G. G. Surv., IV, 120 (1878).
Pack. U. S. Dept. Intr., Ent. Com. Bull. VII, 53 (1881).

Packard is again only quoting from Chambers' List of Food-plants, where this species is referred to as feeding on the under side of oak-leaves. Chambers was acquainted with the larva of *quercibella*, which has a similar habit, but he omits this insect from his list of larvæ. In the Index he makes no reference to *quercipulchella*. There is, I think, no doubt that *quercipulchella* is a manuscript name for the species which he described as *quercibella*.

Lithocolletis lysimachiæella.

Chamb. Cin. Qr. Jr. Sc. II, 100 (1875): Bull. U. S. G. G. Surv. IV, 116, 154. (1878).

This name was given to a larva mining *Lysimachia lanuclata*, but until the perfect insect has been reared I can not feel justified in including it as a species in the revised Index.

The practice of publishing names for insects which are known *only* in the larval state is much to be deprecated. The following facts speak for themselves and will explain why these references will not be given in the revised Index.

Coleoptera.

Gn. ? *sp.* ? mining *Quercus alba*. Chamb. Can. Ent. IV, 123–4 (1872).
=*Lithocolletis tubiferella*. Chamb. Can. Ent. III, 165–6 (1871): IV, 123–4 (1872).
Gn. ? sp. ? mining *Quercus ilicifolia?* Chamb. Can. Ent. IV, 124 (1872).
=*Lithocolletis sp.?* Chamb. Can. Ent. III, 166 (1871): IV, 124 (1872).
Gn. ? sp. ? mining leaves of "Willow Oaks." Chamb. Can. Ent. (1872): IV, 124 (1872).
=*Lithocolletis* sp. ? Chamb. Can. Ent. III, 166 (1871): IV, 124 (1872).
Brachys æruginosa Say. Mining *Fagus ferruginea*. Chamb. Can. Ent. IV, 124 (1872).
=*Lithocolletis* sp. ? Chamb. Can. Ent. III, 166 (1871): IV, 124 (1872).

Gn.? sp.? mining *Acer saccharinum*. Chamb. Can. Ent. IV, 124 (1872).
    = *Lithocolletis* sp.? Chamb. Can. Ent. III, 166 (1871): IV, 124 (1872).
Metonius lævigatus Say. Mining *Desmodium*. Chamb. Can. Ent. IV, 124 (1872).
    = *Leucanthiza*? sp. Chamb. Can. Ent. III, 166 (1871): IV, 124 (1872).

Lithocolletis chambersella.

= *quinquenotella* Chamb.

Chambers describes a species of this genus as *quinquenotella*, this name being preoccupied by a European species. I would suggest the name of *chambersella* to replace it.

Lithocolletis umbellulariæ sp. n.

Antennæ, white, evenly dotted with brown along their upper sides, the five brown spots towards the apex being larger and more widely separated than the others.

Palpi, shining white.

Head, face shining white, frontal tuft yellowish in the middle, saffron-brown at the sides.

Thorax, golden saffron, whitish behind.

Fore wings, golden saffron, somewhat shining, a short white patch at the base of the dorsal margin reaches to the fold and is exteriorly dark margined; the dark margin, of a somewhat similar white spot on the costal portion of the wing, also reaches to the opposite side of the fold a little beyond it; at one-fourth of the wing-length is a waved white fascia running nearly straight from the dorsal margin to the fold and bulging outwards beneath the costa; this is distinctly dark-margined externally throughout and briefly so internally; immediately adjoining the costal margin at half the wing-length is a broad, very oblique white costal streak dark-margined on both sides and freely dusted with blackish scales around the apex; the black dusting is continued along the outer side of an opposite less oblique dorsal streak, the apex of which reaches as far as the edge of the costal streak; above it, at three-fourths the wing-length, is a white costal spot slightly margined with blackish atoms, and opposite to this is another white dorsal streak, very oblique, externally margined at the apex with dusky atoms, which are continued so as to form a large patch of black dusting at the apex of the wing, on the upper side of which patch lies a sickle-shaped white costal streak, concave towards the costal margin; cilia pale saffron, with a brown line running through the middle and reaching around the apex nearly to the anal angle, where they become paler, inclining to grayish.

Hind wings and cilia, pale grayish.

Abdomen, dark gray above, grayish-white beneath; anal tuft, yellowish.

Hind tibiæ, white, with two broadish black bars across their upper sides and a small black terminal spot.

Exp. al., 9mm.

Type, ♂ ♀ *Mus. Wlsm.*

Mendocino County, Cal., found and bred in the month of June, 1871. Three specimens, from large diffused blister-like mines on the upper side of leaves of *Umbellularia californica* Nuttal; the pupa being inclosed in a semi-transparent flat oval silken web within the mine, like that of *cincinnatiella* Chamb., to which species it is somewhat allied. Its nearest ally in America is probably *macrocarpella* F. and B., but it differs in the possession of a dark-margined costa-basal spot and in the comparatively straight first fascia.

These characters also serve to separate it from *cincinnatiella* Chamb. I think it is open to question whether *cincinnatiella* may not be a form of *macrocarpella*. The only differences I can detect in comparing authenticated specimens of each species are the somewhat larger size of *macrocarpella* and the less shining appearance of the ground-color of the wing; moreover, the white streaks appear to be duller and perhaps somewhat more oblique, their dark bordering being more marked towards the costal margin than in *cincinnatiella*. At the same time the differences are very slight, and those who have an opportunity of collecting larvæ of the oak-feeding species would do well to study the subject.

Lithocolletis gaultheriella sp. n.

Antennæ, closely annulated with white and brown, the brown annulations somewhat wider apart towards the apex.

Palpi, silvery white, with a small spot on the outer side.

Head, face silvery white; frontal tuft saffron mixed with white.

Thorax, golden-saffron, posteriorly whitish.

Fore-wings, golden-saffron, shading to golden-brown, no basal streak, three costal and three dorsal snow-white spots, the first two pairs internally dark-margined; the first costal spot is situated at about one-third the length of the wing, its internal dark margin passing around its apex; the corresponding dorsal spot commences nearer the base of the wing and sometimes reaches obliquely to, or near, the point of the costal spot; the second costal spot at half the wing-length is somewhat oblique, square ended, and as in the case of the first is placed somewhat beyond its smaller corresponding dorsal spot, which is pointed and has some dark fuscous scales running outwardly from its apex and merging in the darkened lower margin of the costal spot above it; the third costal spot at one-fourth from the apex is somewhat triangular and lies also farther from the base than the corresponding smaller spot on the dorsal margin; before the anal angle, between these spots, lies a cloud of fuscous scaling serving to throw up and make more conspicuous these white markings on the golden-brown ground-color of the wing; inclosing the apex of the wing is a narrow, outwardly concave white streak, not reaching through the cilia on the apical but only on the costal margin; beyond it are a few darkened scales and sometimes one or two whitish ones with them; cilia pale, golden-saffron, tending to golden; gray about the anal angle. The only conspicuous markings on the under side are two pale spots in the costal fringes, corresponding with the last two markings on the upper side.

Hind-wings, grayish, with golden-gray cilia.

Abdomen, gray, anal tuft slightly paler.

Hind-tarsi, grayish-white, with one or two darker bands above.

Exp. al., 10-11^{mm}.

Type ♂ ♀ *Mus. Wlsm.*

A single ♂ bred from mines on the upper side of *Gaultheria shallon*, found at Rouge River, Oregon, in May, and bred June, 1872. Three others taken in Mendocino County in May and June, and a single specimen also taken near Crescent City, Del Norte County, Cal. It is one of the largest species of *Lithocolletis*.

Lithocolletis ledella sp. n.

Antennæ, whitish, faintly barred above with brown.

Palpi, white.

Head, face white, frontal tuft saffron, mixed with whitish.

Thorax, golden-saffron with a few white scales.

Fore-wings, golden-saffron with a white medio-basal streak, somewhat expanding outwards on the fold and reaching to one-third the length of the wing above it, this is dark-margined on its upper edge; beyond it are four costal and four dorsal silvery-white streaks; the first dorsal commences beneath the point of the basal streak and extends obliquely outwards to the middle of the wing, it is dark-margined internally and around its apex; the costal streak above it is short, rather square, and also internally dark-margined; the second costal streak, scarcely longer than the first, is a little oblique and also inwardly dark-margined; beyond this are two more narrow costal streaks, the first curved outwards and dark-margined internally, the second pointing inwards from above the apex, with a few black scales at the extremity; the second dorsal streak is triangular, dark-margined internally and around the apex, commencing somewhat further from the base than the second costal streak, its point lies between the second and third; the two last of the four dorsal streaks are very slender and pointing inwards, with a few black scales at their ends; where they reach the points of the costal streaks above them a black elongate spot lies at the apex, separated from the dark apical line which lies at the base of the golden-gray apical cilia.

Hind-wings and cilia, gray, with a faint golden sheen.

Abdomen, gray, anal tuft paler.

Hind-tarsi, whitish-gray, unspotted.

Exp. al. 9–10^{mm}.

Type ♂ ♀ *Mus. Wlsm.*

Six specimens, bred from somewhat folded mines, occupying the whole upper side of leaves of *Ledum glandulosum*, found in June in Mendocino County, Cal., and bred the same month. I met with this species also on the wing at the same time and place. It appears to be nearly allied to *salicicolella* Sircom, among the European species.

Lithocolletis alnicolella, sp. n.

Antennæ, whitish, very faintly spotted above.

Palpi, white.

Head, face white, frontal tuft grayish saffron.

Thorax, pale grayish saffron touched with white at the sides.

Fore-wings, pale grayish saffron with three dorsal and four costal silvery-white streaks, all dark-margined on their inner sides and at their points; a somewhat broad but very indistinct white medio-basal streak extends above the fold to one-third the wing-length, and a shorter streak of the same color follows the dorsal margin from the base to half the length of the one above it; the first dorsal streak is broad, outwardly oblique, and reaching nearly to the smaller triangular costal streak above it; in some specimens it actually attains to it, forming an angulated fascia; the point of the second dorsal, also somewhat triangular, is directed a little beyond the point of the second costal streak above it; these are both nearly perpendicular; the third dorsal very small; arising opposite the space between the third and fourth costal streaks, it reaches to the apex of the former; the end of the wing is inclosed by a dark semicircular line at the base of the cilia, within which is an elongate blackish spot; the cilia are grayish, with a faint saffron tinge.

Hind-wings and cilia, also pale grayish.

Abdomen, gray above, anal tuft scarcely paler.

Posterior tibiæ, whitish, unspotted.

Exp. al. 6^{mm}.

Type, ♂ ♀ *Mus. Wlsm.*

Two specimens were bred from larvæ found mining the upper sides of leaves of *Alnus incana* on Mount Shasta, Siskiyou County, Cal., in August, 1871, in which month the perfect insects emerged. Three other specimens were met with on the wing, also in the neighborhood of Mount Shasta. Judging from Chambers' description, his *alnivorella* must be exceedingly close to this species. There are certain distinct differences in the position and extent of the dorsal streaks, but my chief reason for regarding it as distinct is that Chambers describes the larva of *alnivorella* as feeding on the under side of the leaf, whereas my species feeds on the upper side of another species of alder. I am not aware of any instance of an alder-feeding *Lithocolletis* feeding on both sides of the leaves.

Lithocolletis incanella sp. n.

Antennæ, whitish, faintly spotted above.
Palpi, shining white.
Head, face shining white, frontal tuft white, with a few saffron scales at the sides.
Thorax, bright reddish-saffron, with a thin whitish line running around its anterior margin and communicating with the basal streak on the fore-wing.
Fore-wings, bright brownish-saffron with a long slender medio-basal white streak without dark margins, four costal and three dorsal streaks of the same color, sometimes with a slight metallic sheen; the first costal streak is a little before the middle of the wing, oblique and pointed, with a scarcely perceptible dark dusting along its inner margin; the first dorsal streak commences a little nearer to the base; it is dark-margined internally, and is somewhat wider than and reaches a little beyond the costal streak beyond it; the second costal streak is small and points slightly outwards; the third is nearly perpendicular; the fourth points slightly inwards from a little before the apex; these three are all dark-margined on their inner edge; opposite to these are the second and third dorsal streaks; the second is triangular, wider at the base and dark margined internally, its black dusting communicating with a patch of similar blackish scales at its apex extending to the second costal streak above it; the third dorsal streak is short, pointing inwards and dark margined on *both* sides, its outer margin being continuous with a dark line at the base of the cilia which encircles the tip of the wing reaching to the exterior costal streak; within this line, but separate from it, is an elongate apical spot of somewhat disconnected blackish scales, the cilia pale greyish.
Hind-wings and cilia, pale grayish.
Abdomen, dark gray above, anal tuft somewhat paler.
Hind tarsi, white, tipped with grayish, and two grayish-saffron spots above.
Exp. al., 9mm.
Type, ♂ ♀ *Mus. Wlsm.*

The larva feeds in mines on the under side of *Alnus incana* towards the end of June in Colusia County, Cal., the perfect insects emerged in July, 1871. Seven specimens were bred, and the species was also met with on the wing at Burney Creek (near Pit River), Shasta County, Cal.

GENERAL NOTES.

THE AMENDED CALIFORNIA HORTICULTURAL LAW.

We take from the *Los Angeles Evening Express*, of July 12, the following amendments to the old act to protect and promote the horticultural interests of the State. The act embodying these amendments was approved March 20. Entomological legislation is so unusual in this country that these rulings will be read with interest:

SEC. 1. Section 1 of said act is hereby amended so as to read as follows:

"SEC. 1. Whenever a petition is presented to the board of supervisors of any county, and signed by twenty-five or more persons who are resident freeholders and possessors of an orchard, or both, stating that certain or all orchards or nurseries, or trees of any variety, are infested with scale insects of any kind, injurious to fruit, fruit trees, and vines, codlin moth, or other insects that are destructive to trees, and praying that a commission be appointed by them whose duty it shall be to supervise their destruction as herein provided, the board of supervisors shall, within twenty days thereafter, select three commissioners for the county, to be known as a county board of horticultural commissioners. The board of supervisors may fill any vacancy that may occur in said commission by death, resignation, or otherwise, and appoint one commissioner each year, one month or thereabouts previous to the expiration of the term of office of any member of said commission. The said commission shall serve for a term of three years from the date of their appointment, except the commissioners first appointed, one of whom shall serve for one year, and one of whom shall serve for two years, and one of whom shall serve for three years from the date of appointment. The commissioners first appointed shall themselves decide, by lot or otherwise, who shall serve for one year, who shall serve for two years, and who shall serve for three years, and shall notify the board of supervisors of the result of their choice."

SEC. 2. Section two of said act is hereby amended so as to read as follows:

"SEC. 2. It shall be the duty of the county board of horticultural commissioners in each county, whenever it shall deem it necessary, to cause an inspection to be made of any orchard or nursery or trees or any fruit-packing house, store-room, sales-room, or any other place in their jurisdiction, and if found infested with scale-bug, codlin moth, or other insect pests injurious to fruit, trees and vines, they shall notify the owner or owners, or person or persons in charge or possession of the said trees or place, as aforesaid, that the same are infested with said insects, or any of them, or their eggs or larva, and they shall require such person or persons to disinfect or destroy the same within a certain time to be specified. If within such specified time such disinfection or destruction has not been accomplished, the said person or persons shall be required to make application of such treatment for the purpose of destroying them as said commissioners may prescribe. Said notices may be served upon the person or persons owning or having charge or possession of such infested trees, or places, or articles, as aforesaid, by any commissioner, or by any person deputed by the said commissioners for that purpose, or they may be served in the same manner as a summons in a civil action. If the owner or owners, or the person or persons, in charge or possession of any orchard, or nursery, or trees, or places, or articles infested with said insects, or any of them, or their larva or eggs, after having been notified as above to destroy the same, or make application of treatment, as directed, shall fail, neglect, or refuse so to do, he or they shall be deemed guilty of maintaining a public nuisance, and any such orchards, nurseries, trees, or places, or articles thus infested, shall be adjudged and the same is hereby declared a public nuisance, and may be proceeded against as such. If found guilty, the court shall direct the aforesaid county board of horticultural commissioners to abate the nuisance. The expenses thus incurred may be a lien upon the real property of the defendant."

SEC. 3. Section three of said act is amended so as to read as follows:

"Sec. 3. Said county boards of horticultural commissioners shall have power to divide the county into districts, and to appoint a local inspector for each of said districts. The state board of horticulture, or the quarantine officer of said board, shall issue commissions as quarantine guardians to the members of said county boards of horticultural commissioners and to the local inspectors thereof. The said quarantine guardians, local inspectors, or members of said county boards of horticultural commissioners shall have full authority to enter into any orchard, nursery, or place or places where trees or plants are kept and offered for sale or otherwise, or any house, store-room, sales-room, depot, or any other such place in their jurisdiction, to inspect the same or any part thereof."

SEC. 4. Section four of said act is hereby amended so as to read as follows:

"SEC. 4. It shall be the duty of said county board of horticultural commissioners to keep a record of their official doings, and to make a report to the State board of horticulture, on or before the first day of October of each year, of the condition of the fruit interests in their several districts, what is being done to eradicate the insect pests and diseases, also as to carrying out of all laws relative to the greatest good of the fruit interest. Said board shall publish said reports in bulletin form, or may incorporate so much of the same in their annual reports as may be of general interest."

SEC. 5. Section five of said act is hereby amended so as to read as follows:

"SEC. 5. Each member of the county board of horticultural commissioners, and each local inspector, shall be paid for each day actually engaged in the performance of his duties under this act, payable out of the county treasury of his county, such compensation as shall be determined by resolution of the board of supervisors of the county, before entering into the discharge of his or their duties."

SEC. 6. Section six of said act is hereby amended so as to read as follows:

"SEC. 6. Said county boards of horticultural commissioners shall have power to remove any local inspector who shall fail to perform the duties of his office."

SEC. 7. Section seven of said act is hereby repealed.

SEC. 8. Section eight of said act is hereby amended so as to read as follows, and to be known as section seven of said act, viz:

"SEC. 7. If any member of the county board of horticultural commissioners shall fail to perform their duties of his office, as required by this act, he may be removed from office by the board of supervisors, and the vacancy thus formed shall be filled by appointment by the board of supervisors."

SEC. 9. Section 9 of said act is hereby amended so as to read as follows, and to be known as section 8 of said act, viz:

"SEC. 8. It shall be the duty of the county board of horticultural commissioners to keep a record of their official doings and to make a monthly report to the board of supervisors, and the board of supervisors may withhold warrant for salary of said members and inspectors thereof until such time as said report is made."

SEC. 10. A new section is hereby added to said act, to be known as section 9, and to read as follows, viz:

"SEC. 9. All acts and parts of acts in conflict with the provisions of this act are hereby repealed."

SEC. 11. This act shall take effect and to be in force from and after its passage.

NEW CODLING MOTH AND PEACH-BORER ENEMIES.

Prof. E. A. Popenoe, in the *Industrialist* for June 6, mentions an interesting new parasite of the Codling Moth, which he determines as a new species of the genus Bethylus. Of this parasite 5 larvæ were found in a group feeding externally upon the dorsum of one of the abdominal segments of an Apple-worm taken from the interior of an apple. The larvæ spun yellow cocoons after arriving at full growth, and in fact their habit seems to be quite similar to that of Chalcids of the genus Euplectrus.

He also describes the manner in which the larvæ of *Trogosita obscura* devour the pupæ of the Codling Moth under tree bands.

He also states that he bred a large number of specimens of a honey-yellow Braconid from larvæ and pupæ of the Peach-tree Borer. We are surprised to notice that he states that he has not been able to find in any of the entomological reports reference to any parasites of this insect, and we may call attention to the statement upon page 255 of the Annual Report of this Department for 1879, that Professor Comstock during that year bred 4 parasites—2 Chalcids, 1 Microgaster, and 1 Braconid. Professor Riley has reared from the Ægeria *Phæogenes ater* Cress. and *Bracon nigrifectus* Riley MS.

Professor Popenoe's article is illustrated by figures of the Bethylus and Trogosita, which have been admirably drawn by Mr. C. L. Marlatt.

SOME PACIFIC COAST HABITS OF THE CODLING MOTH.

We have recently learned of certain interesting observations which seem to indicate that the Codling Moth differs slightly in habit in California from its customs in the East. Mr. Koebele, writing us under date of July 24, states that, at the end of May of the present season, when the apples were about 1 inch in diameter, he noticed the moths appearing in numbers. Soon after sunset they began to swarm around the trees, chiefly near the top, and kept it up until dark. He noticed small bats feeding upon them abundantly. This, he thinks, is the time of oviposition. He noticed, however, that the eggs were chiefly laid on the upper side of the fruit, and with pears often upon the stem. Few of the larvæ, according to his observations, entered the fruit at the spot where the egg was deposited, but beginning a slight hole at this point, they left it after becoming slightly larger and entered the fruit at the lower end. These observations were made in the Santa Cruz Mountains.

Similarly, Mr. B. D. Wier, in his Codling-Moth Notes, published in the *Pacific Rural Press* of June 8, from which we have previously quoted, states that, according to his observations, the egg is, as a rule, laid elsewhere than in the calyx.

THE EFFECT OF ARSENICAL INSECTICIDES UPON THE HONEY BEE.

The prevailing opinion seems to favor the theory, that if arsenical mixtures are sprayed or dusted upon fruit trees while the latter are in bloom the bees which frequent them will be destroyed. With this idea in view fruit-growers have very properly been cautioned not to use these mixtures during the blooming season, and in fact this has been urged as an argument against the use of these substances as insecticides.

The writer, while in Louisiana, was told by planters that dusting Paris Green upon the cotton plants killed the bees which frequented the blossoms thereon for the purpose of securing the nectar which was contained in them.

There appears, however, to be some good negative evidence bearing upon the problem, which it will be well to consider before forming a decided opinion in this really important matter.

Mr. Edwin Yenowine, a fruit-grower near New Albany, Ind., is a very strong advocate of the use of arsenical mixtures, as against both Codling Moth and Plum Curculio, and is also, to a limited extent, engaged in apiculture.

Some time ago, while spending a day with Mr. Yenowine, he reminded me that several years ago he had written me as to the probable effects on bees of the use, during the blooming season, of these arsenical mixtures, and had received a very cautionary reply. It appears that instead of following my advice he sprayed all sorts of fruits freely, both in and out of the blooming season, and instead of destroying his bees they have increased from 8 to 17 strong, healthy colonies, and have furnished honey of which he and his family have partaken freely. This conversation with Mr. Yenowine took place on the 23d of June, so that the increase shown was practically that of an unfavorable season.—F. M. WEBSTER.

NEMATODE INJURY TO CANE-FIELDS IN JAVA.

In connection with the forthcoming Bulletin 20 of this Division, on Nematode Worms injurious to the roots of plants in Florida, may be given a short notice of an article by Dr. F. Soltwedei on Nematodes working in the roots of sugar-cane in Java, taken from the *Agricultural and Horticultural Review* of August 1, 1887, which was inclosed by Vice and Deputy Consul Horatio G. Wood, of the United States consulate at Batavia, with his report to the Department of State, and reprinted with the same in the reports of consuls of this Government for May of the present year. In the remarks on the sugar-cane disease in Java, which form the subject of the report referred to and bear date of March 13, 1889, Consul Wood states that a congress composed of planters, exporters, and persons interested in the sugar production of Java, has just closed its session at Samarang. The object of this congress was mainly to discuss the cause and cure of the Nematode attacks on the cane-roots, there called the "sereh" disease, which is now spreading most rapidly and disastrously through the cane-fields of western and central Java, having been first discovered on the island only three years ago in plantations near Cheribon, a sea-port town on the north coast 125 miles to the eastward from Batavia. The report further states that the congress has subscribed a fund of $90,000 for the purpose of engaging a bacteriologist from Europe to visit the island, investigate the disease, and propose its remedy. The Nematodes reduce not only the quantity of the sugar crop but its quality as well, and the subject is therefore of the utmost importance in cane-growing regions.

Dr. Soltwedei, in his article, mentions having discovered in the cane roots the following genera, which all belong to the family Anguillulidæ: *Dorylaimus*, only once; several species of *Tylenchus*, of which the one found almost always attendant upon the "sereh" disease, seems to be new, and is named by him *T. sacchari*; and one species of *Heterodera*, *H. javanica*, which also seems when it occurs to cause the "sereh" disease, but has so far been discovered in only a few plants. *Tylenchus sacchari* has been found there also in the roots of sorghum, while several forms of *Tylenchus* have besides been discovered in the roots of rice and maize, though it can not at present be said with certainty that *T. sacchari* is among these. Some few observations are made on the latter, and, as nearly as can be ascertained, it feeds only in the young and juicy rootlets which sprout directly from the stalk, these becoming its breeding places. A description is given of the male and female, with the size of the same and of the egg, and remarks on the various organs, including those of generation and the spermatic fluid. The parasite can not be introduced except in earth from infested regions, and it appears that a great deal of moisture is required to complete its development. Dr. Soltwedel's article is merely preliminary and does not suggest any remedies.—T. T.

THE IMPORTATION OF OCNERIA DISPAR.

We are greatly interested to learn from the *New England Farmer* of July 13 that the larvæ of this well-known European insect, which is a rather large bombycid moth, have made their appearance in the town of Medford, Mass., feeding upon "everything from garden vegetables to oak leaves." The identification seems to have been made by Mrs. Fernald, and consequently can not be questioned.

In the latter part of July we received from Mrs. N. W. C. Holt, of Winchester, Mass., some young caterpillars on Mulberry and Apple which we take to be the larvæ of this insect. The importation at this late date of such a conspicuous species is of great interest.

ANOTHER LEAF-HOPPER REMEDY.

Mr. George West, of Stockton, Cal., according to the *Vineyardist* of July 15, has given the plan of feeding off his grape leaves by sheep, as a remedy for the grape-leaf hopper, a full test. Last fall, after the crop had all been gathered, he turned 3,000 head of sheep into his 600-acre vineyard, and in a short time they had eaten every vestige of foliage off the vines, leaving them completely bare. This year there has been no sign of the hopper.

A CABBAGE-MAGGOT EXPERIMENT.

An experiment with lime and liquid manure for the Cabbage Maggot, made upon a large scale by Mr. D. M. Dunning, of Cayuga County, N. Y., has resulted in the perfect success of the liquid manure and a partial success of the gas-lime.

HOW OFTEN HAS THIS BEEN NOTICED?

In a half-grown Cecropia larva, found August 7 upon Birch in the grounds of the Department of Agriculture, the left-hand tubercle on the back of the first abdominal segment is entirely wanting. There is not the slightest trace of it. The right-hand tubercle is as large as usual, and in every other respect the specimen is normal.

OBITUARY.

We are pained to learn of the death of Dr. Anton Ausserer, which occurred July 20 at Graz, Germany. Dr. Ausserer was a prominent worker in arachnology, and, in addition to a number of shorter papers, produced the only authoritative monograph of the exceptionally difficult family Territelariæ.

DOES THE WHEAT-STEM MAGGOT, MEROMYZA AMERICANA, DISCRIMINATE BETWEEN DIFFERENT VARIETIES OF WHEAT?

In the literature of this species nothing seems to have been recorded relative to its preference for certain varieties of wheats or indicating that any such discrimination has been witnessed.

During the five years that I have been located at the Purdue Experiment Station the small experiment plats, comprising from 40 to 50 different varieties of wheats, have shown but little difference in the extent of injury, which has in all cases been rather slight.

In larger fields there seems to be a difference in the severity of the attack of the spring brood of larvæ, which has this year been extremely well marked, especially between Velvet Chaff and Michigan Amber. Two fields sown the same day in September, 1888, on the same kind of soil, and in fact every perceivable element being equal except variety of seed, one of which was Velvet Chaff and the other Michigan Amber, suffered very differently; the former, on the 14th of June, having fully four infested straws to one in the latter. In a long, narrow plat, extending some distance between the two fields and being composed of both of these varieties mixed in about equal proportions, the ratio of injury to each was about the same as in the larger fields. The difference between the attack in the two varieties was sufficiently marked to attract the attention of Prof. W. C. Latta, agriculturist of the station, who is neither an entomologist nor familiar with the insect itself.

It is with a view of learning if this partiality is general, and also if it has been observed to extend to other varieties, that the question is here propounded and the observations given.—F. M. WEBSTER.

THE ASSOCIATION OF OFFICIAL ECONOMIC ENTOMOLOGISTS.

In pursuance of the call issued by Mr. James Fletcher, president of the Entomological Club of the American Association for the Advancement of Science, a meeting of those interested in the formation of such an association as that described in the title to this note was held August 29 and 30, at Toronto, Canada.

The following constitution was first adopted:

This association shall be known as the Association of Official Economic Entomologists.

Its objects shall be: (1) To discuss new discoveries, to exchange experiences, and to carefully consider the best methods of work; (2) to give opportunity to individual workers of announcing proposed investigations, so as to bring out suggestions and prevent unnecessary duplication of work; (3) to suggest, when possible, certain lines of investigation upon subjects of general interest; (4) to promote the study and advance the science of entomology.

The membership shall be confined to workers in economic entomology. All economic entomologists employed by the general or State Governments or by the State

Experimental Stations or by any agricultural or horticultural association, and all teachers of economic entomology in educational institutions may become members of the association by transmitting proper credentials to the secretary, and by authorizing him to sign their names to this constitution. Other persons engaged in practical work in economic entomology may be elected by a two-thirds vote of the members present at a regular meeting and shall be termed associate members. Members residing outside of the United States or Canada shall be designated foreign members. Associate and foreign members shall not be entitled to hold office or to vote.

The officers shall consist of a president, two vice-presidents and a secretary, to be elected annually, who shall perform the duties customarily incumbent upon their respective offices. The president shall not hold office for two consecutive terms.

The annual meeting shall be held at such place and time as may be decided upon by the association at the previous annual meeting. Special meetings may be called by a majority of the officers, or shall be called on the written request of not less than five members. Eight members shall constitute a quorum for the transaction of business.

The mode of publication of the proceedings of the association shall be decided upon by open vote at each annual meeting.

All proposed alterations or amendments to this constitution shall be referred to a select committee of three at any regular meeting, and, after a report from such committee, may be adopted by a two-thirds vote of the members present: *Provided*, That a written notice of the proposed amendment has been sent to every voting member of the association at least one month prior to date of action.

The adoption of the constitution was followed by an election of officers, which resulted as follows: President, Dr. C. V. Riley, U. S. Entomologist; first vice-president, Prof. S. A. Forbes, State Entomologist of Illinois; second vice-president, Prof. A. J. Cook, Professor of Entomology in the Michigan Agricultural College; secretary, Prof. J. B. Smith, Entomologist to the New Jersey Agricultural Experiment Station.

The charter members are as follows: C. V. Riley, of Washington; S. A. Forbes, of Illinois; A. J. Cook, of Michigan; J. B. Smith, of New Jersey; J. A. Lintner, of New York; J. H. Comstock, of New York; F. L. Harvey, of Maine; M. L. Beckwith, of Delaware; C. M. Weed, of Ohio; F. M. Webster, of Indiana; J. P. Campbell, of Georgia; James Fletcher, of Canada; C. J. S. Bethune, of Canada; E. Baynes Reed, of Canada; William Saunders, of Canada; E. J. Wickson, of California; C. W. Woodworth, of Arkansas; H. Garman, of Kentucky; O. Lugger, of Minnesota; C. P. Gillette, of Iowa; H. Osborn, of Iowa; L. Bruner, of Nebraska; L. O. Howard, of Washington, and one or two others, whose names we are not able to announce at the present time.

The association adjourned August 30 to meet the coming winter at the time and place of meeting of the Association of Experiment Stations, presumably at Washington, the coming November.

THE ENTOMOLOGICAL CLUB OF THE AMERICAN ASSOCIATION FOR THE ADVANCEMENT OF SCIENCE.

This organization met at Toronto, Canada, August 28 to September 3. Among the members in attendance were Mr. James Fletcher, Prof. A. J. Cook, Prof. J. B. Smith, Prof. H. Garman, Mr. E. Baynes Reed,

Rev. C. J. S. Bethune, Mr. William Saunders, Dr. P. R. Hoy, Mr. C. M. Weed, Mr. L. O. Howard, Mr. J. Alston Moffat, Mr. H. H. Lyman, Rev. W. A. Burman, Prof. C. W. Hargitt, Mr. E. P. Thompson.

The address of the president, Mr. James Fletcher, dealt principally with the injurious insects of the year, and was of extreme interest. It was also warmly discussed. Other papers were read by Professors Cook and Smith and by Messrs. Weed, Lyman, Fletcher, and Howard. Papers were also read which had been received from Prof. C. H. Fernald, Mr. W. H. Edwards, and Dr. F. W. Goding.

The officers elected for the next meeting are: Prof. A. J. Cook, president; Rev. C. J. S. Bethune, vice-president; Prof. F. M. Webster, secretary.

DYNASTES TITYUS IN INDIANA.

Although a southern species, this insect is known to occur in the southern portions of some of the Northern States. Say recorded its occurrence, in an old cherry tree, near Philadelphia, lat. 39° 57′ N., and this is looked upon as its probable northern limit, from whence it can be traced westward through Indiana, Illinois, and Missouri, but only in localities considerably further southward.

In December, 1886, Prof. A. H. Graham, superintendent of the public schools of Columbus, Ind., lat. 39° 13′ N., showed me a specimen which had been found on the top of one of the school buildings, by workmen engaged in repairing the roof. Pennsylvania excepted, this seems to be the northernmost locality where the species has been found. Fruit-growers accuse the larvæ of destroying the roots of the grape.—F. M. WEBSTER.

THE FIELD CRICKET DESTROYING STRAWBERRIES.

Although this insect has not, so far as I am aware, been recorded as destructive to the fruit of the Strawberry, nevertheless it has long been accused of such depredations by strawberry growers. Several years ago a gentleman of Mississippi, whose name I have mislaid, complained of serious injury to his berries by these insects, stating that they first ate the seeds and then the pulp. More recently similar accusations have come from the fruit-growers of southern Indiana; but in none of these cases have the crickets been actually observed feeding on the berry.

While this might indicate that other insects were, perhaps, equally implicated, it is also true that this cricket is a shy fellow, and in order to observe him in the act of feeding one must use the utmost caution. Only once have I been able to detect them in the act of destroying the fruit as accused. This was on June 3, 1886, when I captured an individual which had made such progress in devouring a ripening berry as to leave no doubt regarding the capabilities of his race in that direction.—F. M. WEBSTER.

THE PLUM CURCULIO SCARE IN CALIFORNIA.

Recent advices from one of our California agents, Mr. D. W. Coquillett, show that the published statements in the California newspapers of late date to the effect that the Plum Curculio has made its appearance in Los Angeles County, are entirely unfounded. Fuller's Rose Beetle (*Aramigus fulleri*) has been mistaken for *Conotrachelus nenuphar*. The Rose Beetle has been found to be very destructive in that vicinity to the leaves of Evergreen Oaks, Camelias, Palms (*Washingtonia filli-fera*), *Canna indica*, and several other plants.

LACHNUS LONGISTIGMA ON THE LINDEN IN WASHINGTON.

The Linden Tree-louse, *Lachnus longistigma* Monell, described in Thomas' Third Report on the Insects of Illinois, pp. 119 and 120, and which bears a close resemblance to *L. platanicola* Riley, has so far been recorded on the Linden in only one locality, Monell having observed it a few miles west of St. Louis, as he states in his description of the insect. This instance is noted by Packard in the Seventh Bulletin of the U. S. Entomological Commission (p. 127), where it constitutes the sole mention of the occurrence of this Lachnus.

As a record for the Eastern United States, it may be mentioned that the species is abundant this year (1889) in Washington on trees of the European Linden, a number of which have been found infested in the northwest part of the city. The first tree was examined on August 18, when the insects were in abundance on the underside of the lower limbs, and some winged specimens were found amongst them, while the pavement beneath was stained with their exudations and held the honey-dew in little puddles; the same being observed under infested trees noticed later.

This species differs from *L. platanicola* in being larger, with the wings more dusky and the stigma black. It is also interesting to note that some experiments carried on by Mr. Pergande, of this Division, in transferring specimens of *L. platanicola* to Linden and *L. longistigma* to Sycamore, resulted in both cases in the failure of the colonies.—T. T.

U. S. DEPARTMENT OF AGRICULTURE.

DIVISION OF ENTOMOLOGY.

PERIODICAL BULLETIN. OCTOBER, 1889.

Vol. II. No. 4.

INSECT LIFE.

DEVOTED TO THE ECONOMY AND LIFE-HABITS OF INSECTS, ESPECIALLY IN THEIR RELATIONS TO AGRICULTURE, AND EDITED BY THE ENTOMOLOGIST AND HIS ASSISTANTS.

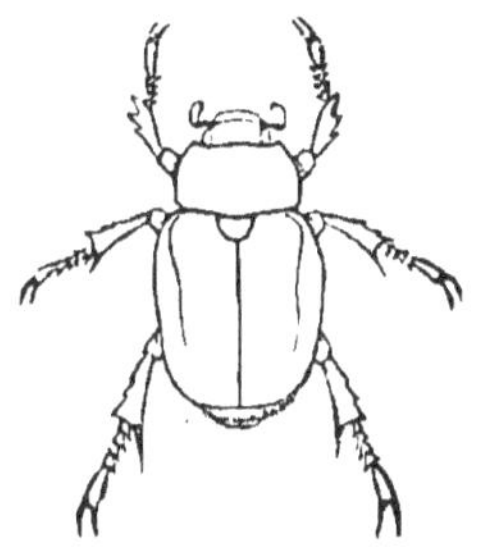

[PUBLISHED BY AUTHORITY OF THE SECRETARY OF AGRICULTURE.]

WASHINGTON:
GOVERNMENT PRINTING OFFICE.
1889.

CONTENTS.

II

SPECIAL NOTES.

Greeting.—The writer is pleased to greet more directly again the readers of INSECT LIFE, after an absence of five months, the most enjoyable portions of which have been the ocean voyages going and coming. Paris is proverbially beautiful, and we met many delightful people there, among them not a few entomologists; but America never looked more fair nor Washington more attractive to our eyes than upon our return, and, while it would be incorrect to say that we are more ready for work (which has not been intermitted, but was simply transferred to other scenes) we cheerfully relieve Mr. Howard from the Divisional harness and give him an opportunity for well-merited rest and vacation. In doing so we desire to publicly thank him, as also the rest of the Divisional force, for the manner in which his and their several duties have been discharged.—C. V. R.

Lestophonus or Cryptochætum—Professor Mik's Opinion.—In the August number of the *Wiener Entomologische Zeitung* Prof. Josef Mik, in commenting upon Dr. Williston's "Note on the Genus Lestophonus" in the May number of INSECT LIFE (Vol. I, p. 328), confirms Dr. Williston's placing of this form in the *Ochthiphilinæ*, and states that, in his opinion, there can hardly be any doubt regarding the identity of *Lestophonus* with Rondani's *Cryptochætum*. The figures of the wing, he states, agree perfectly, and so do the descriptions. He says that Rondani in his expression "Areola basali antica incompleta" does not refer to the anterior but to the posterior basal cell, as can be seen from the third part of the Prodromus (Fig. VII^{k} of the plate) of this author.

Entomology in Ohio.—We received September 13th the Annual Report of the Ohio Agricultural Experiment Station for 1888, which contains upon pages 122 to 176 the Report of the Entomologist, Mr. C. M. Weed.

The principal articles are upon experiments in preventing the injuries of the Plum Curculio; a practical preventive of Rose Bug injuries to grapes and peaches; on some insects affecting Currants and Gooseberries; notes on some Raspberry Insects; on the autumn life-history of certain little-known Plant-lice; notes on various insects affecting garden crops; heat as a remedy for Bean and Pea Weevils; the Chinch Bug in Ohio; on two Potato Insects (*Epicærus imbricatus* and *Doryphora 10-lineata*); on injuries of the Striped Grape-vine Beetle, and a list of the articles published by the entomologist during the year. The report is carefully prepared and well printed, the most valuable contribution to the knowledge being the account of the experiments with arsenicals against the Curculio, showing, as they do, the utility of the arsenicals for this purpose, and confirming the conclusions which we expressed in our last annual report. Many of the articles have been published elsewhere in advance.

Rosin Wash for Red Scale.—In accordance with instructions from the vision, Mr. Coquillett has been making experiments with this wash against the red scale (*Aspidiotus aurantii*), and after twenty different tests made with various preparations, from the 17th of July to the 8th of August, the one which gave the best results was found to be composed of rosin, 20 pounds, caustic soda (70 per cent. strength), 6 pounds, fish oil, 3 pounds, and water to make 100 gallons. In preparing this wash the necessary materials were placed in a boiler and covered with water and then boiled until dissolved and stirred occasionally during the boiling. After dissolving, the preparation was boiled briskly for about an hour, a small quantity of cold water being added whenever there was danger of boiling over. The boiler was then filled up with cold water, which mixed perfectly when added slowly and frequently stirred. It was then transferred to a strong tank and diluted with water to 100 gallons. Neither the leaves nor the fruit were injured, while a large proportion of the scales were destroyed. Those which escaped were either on the fruit or the underside of the leaves. The cost of the wash is 80 cents for 100 gallons or four-fifths of a cent per gallon. An orange tree 16 feet tall by 14 feet in diameter was given 14 gallons. This, however, seems to us to be an unnecessarily large amount, but upon this basis the cost of spraying per tree is 11.2 cents.

Meeting of Association of Economic Entomologists.—A notice from the secretary is published on page 123.

THE HORN FLY.

(*Hæmatobia serrata* Robineau-Desvoidy.)

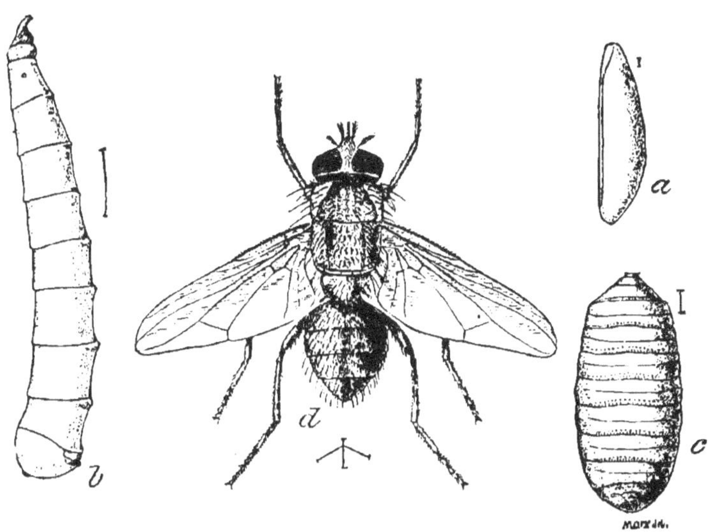

FIG. 11.—HÆMATOBIA SERRATA: *a*, egg; *b*, larva; *c*, puparium; *d*, adult in biting position—all enlarged. (Original.)

Our knowledge of this pest is now sufficiently far advanced to enable us to present a preliminary article giving the main facts ascertained. A more complete article will, however, be published in our annual report for the year.

FIRST APPEARANCE—SPREAD—INVESTIGATION.

Our attention was first called to this pest in September, 1887, when Mr. I. W. Nicholson, of Camden, N. J., wrote us under date of September 22, as follows:

> Herewith I send some specimens of flies which appear to have made their first appearance about the middle of August. They are very annoying to cattle, but rarely settle upon the horses or mules. They gather in patches or clusters particularly upon the legs, and are very active. I should like to know if they are common in other parts of the United States. They appear to be very numerous in all the counties near Philadelphia, yet I have seen no person who has observed them before this season.

Later letters the same season from Mr. Nicholson mentioned the common habit of clustering upon the horns, and the fact that after a severe frost in the middle of October the fly disappeared.

May 15, 1888, the same gentleman wrote us that the flies had promptly made their appearance May 10, or a little before, in great numbers. A few days later we heard of the same insect in Harford County, Md., through Mr. George R. Stephenson, who reported its occurrence in that locality the previous summer.

By the summer of 1889 the pest had extended in numbers much farther to the southward, and the Department was early informed of its occurrence in Harford and Howard Counties, Md., and Prince William, Fauquier, Stafford, Culpeper, Louisa, Augusta, Buckingham, and Bedford Counties, Va. The alarm became so great that we were anxious to learn all that was possible about the species, and arranged to have it investigated. Considerable time has therefore been devoted to the study of the habits and life history of the insect. This was done mainly by Mr. Howard, who made a number of short trips to The Plains, Warrenton, and Calverton during June and July. Later in the season Mr. Marlatt assisted in the work, which had been greatly facilitated by Mr. G. M. Bastable, Mr. David Whittaker, Mr. M. M. Green, and Mr. William Johnson, and particularly by Col. Robert Beverly. To the courtesies of these gentlemen we would acknowledge our indebtedness. August 20 Mr. Howard found the flies practically in Washington—in Georgetown—and the next day Mr. Marlatt found them in Rosslyn, at the Virginia end of the Aqueduct Bridge, so that further trips for material were not necessary.

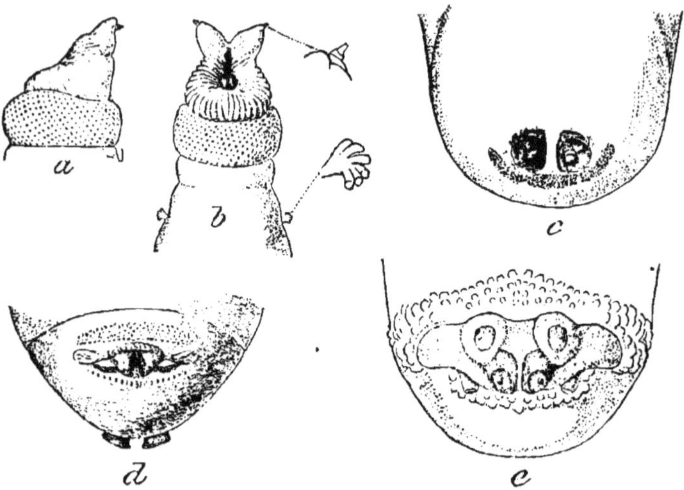

FIG. 12.—HÆMATOBIA SERRATA: *a*, side view of head of larva; *b*, ventral view of head of larva, showing antennæ and thoracic stigmata; *c*, dorsal view of anal end of larva, showing anal stigmata; *d*, anal plate of puparium; *e*, ventral view of anal end of larva, showing anal plate—still more enlarged. (Original.)

The result of the summer's observations by these two gentlemen is that the life history of the insect has been accurately made out from the egg to the fly through several consecutive generations, and that substances can be recommended which, from their experience, will keep the flies away for from five to six days, while from the life history a suggestion as to preventives is made, which, under certain circumstances, will prove undoubtedly of great benefit.

IS IT A NATIVE OR AN IMPORTED PEST?

Since this insect was first brought to our notice we have felt that it was an imported pest. Its first appearance in the neighborhood of Philadelphia and its gradual spread southward have favored this idea. Dr. Williston, to whom we sent specimens for name, wrote us that he thought it an introduced species, and very close to *Hæmatobia serrata* of Robineau-Desvoidy, differing only in color of legs and antennæ. He has since, however, described it as a new species (see *Entomologica Americana*, Vol. V, No. 9, September, 1889, pp. 180–181), under the name *Hæmatobia cornicola*, giving *H. serrata* as a questionable synonym. His published remarks on this point are:

> I can not resist the belief that the species is an introduced one, and suspect that it may be identical with *H. serrata* R.-Desv., occurring in France. Aside, however, from the discrepancies that his description shows in the color of the legs, an identification of this author's species is usually, at least, only a guess. Macquart's very brief description is better; but the palpi are distinctly enlarged, and he says they are not. Nothing but a comparison of the specimens will settle the question.

Meantime Dr. Lintner had sent specimens to Baron Osten-Sacken at Berlin, who determined them, as Dr. Lintner informs us in a letter dated September 16, as the European *serrata*, placing it in the closely allied genus or subgenus *Lyperosia* of Rondani. We are quite inclined to accept Baron Osten-Sacken's dictum in this matter and so also we feel assured will Dr. Williston, and we hence conclude that our species is the European *serrata*, whether it be ultimately placed in *Hæmatobia* or *Lyperosia* both of which genera were split off from *Stomoxys* and are considered by Schiner as subgenera of this last. At present we shall follow Dr. Williston in placing it in *Hæmatobia*.

We know little of the European geographical distribution of *H. serrata*. Robineau-Desvoidy described it from France and Schiner gives its location as south France, while Macquart gives it as inhabiting the south of France, and records it specifically from Bordeaux. The fact that in this country it has spread with much greater rapidity towards the south than towards the north would seem to indicate that it is a south European species.

The habits of Hæmatobia in Europe are given by Railliet* as follows:

> The Hæmatobias are very small flies which live in the fields and seldom penetrate into the stables. As their name indicates, they are at least as blood-thirsty as Stomoxys. They attack the animals in the pastures, particularly cattle, and they often collect in great numbers upon a single individual, with their wings expanded, working in through the hairs to pierce the skin. *H. stimulans* Meig, and *H. ferox* R.-D. are the principal species of our region.—[France.]

The exact time and place of the introduction, it is impossible to ascertain. Upon its first importation in small numbers it was probably for

* Éléments de Zoologie Médicale et Agricole.

some time unnoticed, and its first noticeable appearance may not have been at the point of importation.

All imported cattle from Europe pass through the quarantine stations of this Department at either Littleton, Mass., Garfield, N. J., or Patapsco, Md., and an examination of the records developes one or two points of interest. Since 1884 only ten head of cattle have been imported into the country direct from France. All of these have passed through the New Jersey station, but their ultimate destinations have in no cases been within the regions now infested with the fly. The other importations have been from points like Antwerp, London, Amsterdam, Hamburg, Glasgow, Liverpool, Southampton, Hull, Rotterdam, and Bristol. The year 1886, immediately preceding the appearance of the fly, was marked by quite an extensive importation of Holsteins from Amsterdam and Rotterdam and London, through the Garfield station, mainly for parties in New York City. Over three hundred were imported, and an interesting point to investigate will, therefore, be the occurrence or non-occurrence of this fly in Holland.

POPULAR NAMES AND POPULAR ERRORS.

The popular name which is here adopted—the "Horn Fly"—has the sanction of popular use. It is sufficiently distinctive and we therefore recommend its adoption. The name of "Texas Fly" and "Buffalo Fly" and "Buffalo Gnat" are also in use in some sections and indicate an impression that the insect came from the West. Dr. Lintner uses the term "Cow-horn Fly." Objections may be urged to all of these.

The most prominent of the popular errors is the belief that the fly damages the horn, eats into its substance, causes it to rot, and even lays eggs in it which hatch into maggots and may penetrate to the brain. There is no foundation for these beliefs. As we shall show later, the flies congregate on the bases of the horns only to rest where they are not liable to be disturbed. While they are there they are always found in the characteristic resting position, as shown in Fig. 14, and described later. Where they have been clustering thickly on the horns, the latter become "fly-specked" and appear at a little distance as though they might be damaged, and it is doubtless this fact which has given rise to the erroneous opinions cited.

LIFE HISTORY.

THE EGG.—*Place, Method, and Time of Oviposition.*—Mr. Howard's first impression upon entering the field, that the eggs would be found to be laid in freshly dropped dung, proved to be correct. He brought to Washington with him from Calverton dung dropped on the night of July 28 and exposed in the field during the 29th, and from this dung the first adult flies, five in number, issued August 7, only ten days from the laying of the eggs. This settled the point of place of oviposition and breeding. It seemed probable that this was the only substance in which the species breeds, as indeed it is the only likely substance which exists in sufficient quantity through the pastures to harbor the multitudes of flies which are constantly issuing through the summer. However, many living females were captured and placed in breeding cages with horse-dung and decaying animal and vegetable material of different kinds, each isolated, and it resulted that a few oviposited in the horse-dung and four flies were reared from this substance. There is no evidence, however, that in a state of nature the flies will lay their eggs in anything but cow-dung.

The time and manner of oviposition were puzzling at first. After hours of close watching of fresh dung in pastures close to grazing cattle not a single Hæmatobia was seen to visit the dung, much less to lay an egg. This close observation was made at all times of the day from dawn till dusk without result, while breeding-cage experiments were all the time proving that nearly all fresh droppings contained many eggs. With some hesitation, therefore, the inference was made that the eggs were presumably laid at night, as stated in the note upon p. 60 of the August number of INSECT LIFE.

The question was, however, considered by no means settled, and on the discovery of the fly at Rosslyn Mr. Marlatt was directed to make especial observations upon this point. The first result was that careful examination of dung dropped in the early morning (prior to 7 a. m.) showed very few eggs, not more than eight or ten to a single dropping, while that dropped between 4 p. m. and later in the night contained still fewer. On a dung dropped between 10 and 11.30 a. m. in the hot sunshine, however, examination, a few minutes after, showed a large number of eggs—estimated at three hundred and fifty. Other very fresh droppings were examined and the eggs were found to range from none at all to over three hundred. One animal was then fortunately observed, from close quarters, in the act of passing her dung. As the operation commenced, forty or fifty of the flies moved from the flank to the back of the thigh near the "milk mirror," and at the close of the operation they were seen to dart instantly to the dung and to move quickly over its surface, stopping but an instant to deposit an egg. The abdomen and ovipositor were fully extended and the wings were held in a resting position. Most of them had left the dung at the expiration of thirty seconds, while a few still remained at the expiration of a minute. Every individual had returned to the cow, however, in little more than a minute. This explains the previous non-success in observing the act of oviposition, for the Virginia cattle on the large stock-farms are comparatively wild, and although the dung was examined as speedily as possible after dropping, the flies had already left.

The results, therefore, indicate that the eggs are deposited during daylight, chiefly during the warmer time of the day, between 9 and 4, and mainly between 9 in the morning and noon. They are laid singly, and never in clusters, and usually on their sides on the surface of the wet dung; seldom inserted in cracks.

Description.—Length, 1.25mm to 1.37mm; width, 0.34mm to 0.41mm. Shape, irregular oval, nearly straight along one side, convex along the other. General color, light reddish brown, lighter after hatching. General surface covered with a hexagonal, epithelial-like sculpture, each cell from .027mm to .033mm in length by about half the width. In the unhatched egg, even in those just deposited, a long, rather narrow, ribbon-like strip is noticed along the entire length of the flattened side, rather spatuloid in shape. In hatching this strip splits off, remaining attached at one end, and the larva emerges from the resulting slit.

LARVA.—After the eggs hatch the larvæ descend into the dung, remaining, however, rather near the surface.

Newly-hatched Larva.—Length, 2.45mm, and greatest width, 0.48mm. Color, pure white. Joints of segments rather plainly marked, venter with slightly elevated ridges at ends of abdominal segments, the ridges with delicate sparse rugosities. Resembles in main full-grown larva.

Full-grown Larva.—Length, 7mm; greatest width, 2 to 2.5mm. Color, dirty white. Antennæ 3-jointed, last joint pointed. Head with a lamellar or ridged structure shown in figure: divided by cleft at tip; skin behind lamellar structure coarsely granulated, while that of thoracic and abdominal joints is nearly smooth. Thoracic stigmata pedunculate with six pedunculate orifices. Ridges on venter of abdominal joints not strong, fainter than in young larva. Anal stigmata large, slightly protruding, very dark brown, nearly round, flattened on proximal borders, slightly longer than broad, 0.14mm in length, with one central round opening, and a series of very delicate marginal tufts of cilia, four tufts for each spiracle, each issuing from a cleft, but none on the proximal edge. Anal segment below with a dark yellow chitinous plate showing six irregular paired tubercles; the surface of the skin surrounding the plate rather coarsely granulated.

PUPARIUM.—When ready to transform the larvæ evidently descend from the dung into the ground below from a half to three-quarters of an inch. Actual observations were made on larvæ in dung in breeding-cages where the soil was fine sand, affording ready entrance to the larvæ. Where the dung has been dropped upon hard ground the probabilities are that they will not enter so deeply, and may indeed transform upon the surface of the ground at the bottom of the dung.

Description.—The puparium is from 4mm to 4.5mm in length, by 2mm to 2.5mm in width, regularly ellipsoidal, the head rather more pointed; dark brown in color. The segments are plainly separated. The anal stigmata are darker in color than the rest of the skin: are slightly protruded and preserve the same shape as in the larva. The central opening is still visible, as are the slight indentations of the border. The ventral plate, noticed at the base of the anal segment of the larva is still noticeable as a series of tubercular elevations.

DURATION OF THE PREPARATORY STAGES AND CONSEQUENT NUMBER OF ANNUAL GENERATIONS.—The first flies reared at the Department issued August 7 from eggs deposited July 28. These were five or six in number. August 8 four more issued from the same lot. August 12 six flies issued, reared from eggs laid July 31; August 13 two more, and August 14 two more from the same lot. Delayed specimens issued from this lot August 20 and 23. August 26 seven flies were reared from two or three days' old dung, collected August 17. These observations show the bulk of the flies during late July and August to issue from ten to fifteen days from the laying of the eggs. In all cases the eggs hatched in less than twenty-four hours. Experiments a little later gave the following periods:

Aug. 21. Eggs deposited in confinement placed at 7 p. m. on cow dung free from eggs of other flies.
23. Larvæ one-fourth grown.
25. Larvæ one-half inch long.
27. Larvæ leaving manure and entering sand to pupate.
Sept. 5. Three flies issued.

Aug. 23. Eggs placed with isolated dung at 1.30 p. m.
24. (9 a. m.) Eggs have hatched.
25. Larvæ one-fourth inch long.
29. Apparently full grown.
30. Puparia found.
Sept. 5. Two flies issued.
6. Four flies issued.

1. Eggs deposited 10.25 a m.
2. Eggs were hatched when examined at 9 a. m.
5. Larvæ half grown.
7. Larvæ entering sand.
8. Five puparia taken from sand.
9. All in puparia.
15. Three adults. } All found at 9 a. m.
16. Twenty adults. } All found at 9 a. m.
17. Twenty-six adults. } All found at 9 a. m.
17. Twenty adults, issued between 12 and 4 p. m.

From these records it will be seen that from ten to seventeen days, say two weeks, is about the average time from the laying of the egg to the appearance of the flies, and with four active breeding months, from May 15 to September 15, there will be eight generations. The flies will undoubtedly breed later than September 15, but we may allow this time to make up for the time occupied in the development of the eggs in the abdomen of the female. With seven or eight annual generations the numbers of the flies are not surprising.

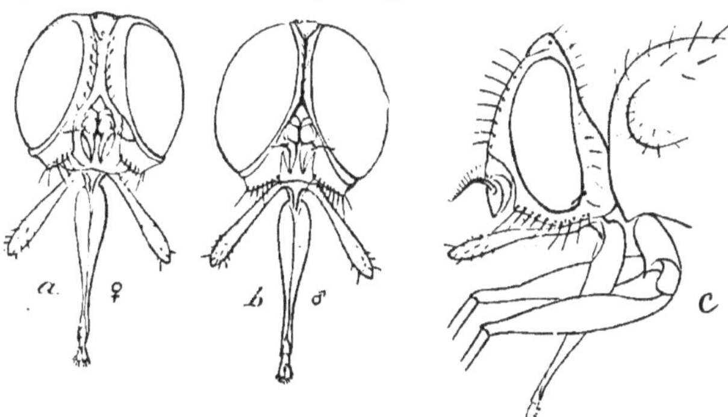

FIG. 13. HÆMATOBIA SERRATA: a, head of female, front view; b, head of male, front view; c, head from side—all enlarged. (Original.)

THE ADULT—*Its Habits.*—The flies were observed in the greatest abundance during July. They make their first noticeable appearance in Virginia early in May, and, from hearsay evidence, remain until "late in the fall" or until "right cold weather." At the date of the present writing, September 28, they are still as abundant as ever around Washington. The characteristic habit of clustering about the base of the horn seems to exist only when the flies are quite abundant. When they average only a hundred or so to a single animal, comparatively few will be found on the horns. Moreover, as a general thing the horn-clustering habit seems to be more predominant earlier in the season than later, although the flies may seem to be nearly as numerous. The clustering upon the horns, although it has excited considerable alarm, is not productive of the slightest harm to the animal. Careful study of the insects in the field show that they assume two characteristic positions, one while feeding and the other while resting. It is the resting position in which they are always found when upon the horns. In this position the wings are held nearly flat down the back, overlapping at base and diverging only moderately at tip (see Fig. 14). The beak is held in a nearly horizontal position and the legs are not widely spread. In the active sucking position, however, the wings are slightly elevated and are held out from the body, not at right angles, but approaching it, approximately an angle of 60 degrees from the abdomen. The legs are spread out widely, and the beak, inserted beneath the skin of the animal, is held in nearly a perpendicular position, approaching that in Fig. 13*c*. The fly, before inserting its beak, has worked its way through the hairs close to the skin. While feeding, however, the hairs which can be seen over its body do not seem to interfere with its speedy flight when alarmed, for at a fling of the tail or an impatient turn of the head the flies rise instantly in a cloud for a foot or two, returning again as quickly and resuming their former positions.

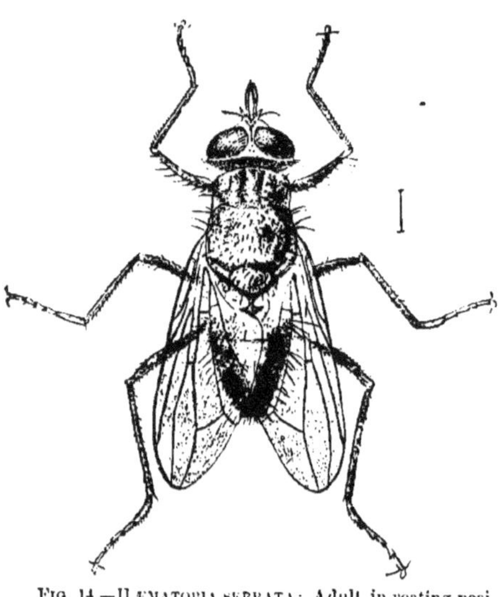

FIG. 14.—HÆMATOBIA SERRATA: Adult in resting position—enlarged. (Original.)

The horns are not the only resting places, for with the horns black for 2 inches above their base we have seen the flies towards nightfall settle in vast numbers upon the back between the head and fore shoulders, where they can be reached by neither tail nor head. When feeding they are found over the back and flanks and on the legs. During a rain-storm they flock beneath the belly. When the animal is lying down a favorite place of attack seems to be under the thigh and back belly, around the bag. With certain animals the dewlap seems to be badly attacked while with others this portion of the body is about exempt. Certain cattle again will be covered with flies and will lose condition rapidly, while others are but slightly troubled.

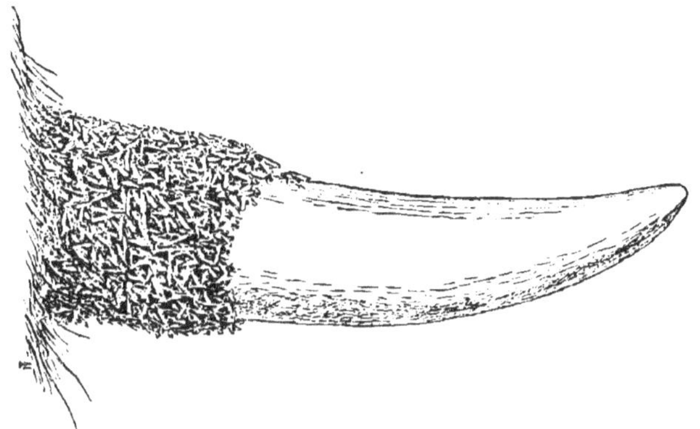

FIG. 15.—Cow-horn showing band of resting flies—reduced. (Original.)

On the horns the flies settle thickly near the base, often forming a complete band for a distance of 2 inches or more. (See Fig. 15.) They seem to prefer the concave side to the convex side of the curve of the horn, probably for the reason that the cow can not scrape them off so readily, and one cow was noticed in which they reached nearly to the tip of the horn on the concave side of the curve only.

Description.—For a description of the adult we may adopt that sent us by Dr. Williston, which was drawn up from Virginia specimens which we had sent to him, and which is substantially identical with that published by him recently in *Entomologica Americana* (*loc. cit.*).

Male.—Length 3.5 to 4^{mm}. Sides of the front gently concave, its least width about equal to one-fourth of the distance from the foremost ocellus to the base of the antennæ; in the middle a narrow, dark brown stripe; a single row of slender bristles on each side. Antennæ brownish red; second joint slightly tumid; third joint a little longer than broad, with its inferior angle rectangular; arista swollen at the base (which is black), the pectination long. The narrow sides of the front, and the still narrower facial and genial orbits silvery gray, with a slightly yellowish cast; facial foviæ and cheeks blackish, the latter clothed with yellowish hair. Palpi black, the inner surface and immediate base more yellowish; gently spatulate in shape, nearly as long as the proboscis, and extending two-thirds of their length beyond the oral margin. Mesonotum sub-shining black in ground-color, but mostly concealed beneath a brownish dust, which, on the pleuræ, is more grayish. Abdomen with similar dust; in the middle with a more brownish sub-interrupted stripe, and narrow darker posterior margins to the segment. Femora black, or very deep brown, first two pairs of tibiæ and tarsi brownish yellow or luteous, the hind tibiæ and tarsi blackish brown; hind tibiæ on the posterior surface with a noticeable, erect, subapical bristle; hind tarsi about as long as their tibiæ, the first three joints widened from their base to tip, so as to form a distinct serration on their inner, acute angles, each of which terminates in a long hair. Wings with a light blackish tinge (due to microscopic pubescence), the immediate base yellowish, the first posterior cell rather symmetrically narrowed to terminate broadly at the extreme tip of the wing.

Female.—Front straight on the sides, its width about equal to one-half of the distance from the foremost ocellus to the base of the antennæ; the median deep brown stripe about as wide as the pruinose sides. Palpi yellow, with the margins and tip blackish. Legs more yellowish; hind tarsi regular; pulvilli and claws small.

AMOUNT OF DAMAGE.

The amount of damage done by the fly has been exaggerated by some and underestimated by others. We have heard many rumors of the death of animals from its attacks, but have been unable to substantiate a single case. We believe that the flies alone will never cause the death of an animal. They reduce the condition of stock to a considerable extent, and in the case of milch cows the yield of milk is reduced from one-fourth to one-half. It is our opinion that their bites seldom even produce sores by themselves, although we have seen a number of cases where large sores had been made by the cattle rubbing themselves against trees and fences in an endeavor to allay the irritation caused by the bites; or, in spots where they could not rub, by licking constantly with the tongue, as about the bag and on the inside of the hind thighs. A sore once started in this way will increase with the continued irritation by the flies and will be difficult to heal. Those who underestimate the damage believe that the fles do not suck blood, but such persons have doubtless watched the flies only upon the horns or elsewhere in their resting position when the beak is not inserted, or have caught them and crushed them when their bodies contained little blood. In reality the flies suck a considerable amount of blood, however, and it is their only nourishment; if captured and crushed at the right time the most skeptical individual will be convinced.

REMEDIES.

Preventive Applications.—Almost any greasy substance will keep the flies away for several days. A number of experiments were tried in the field, with the result that train-oil alone, and train-oil with a little sulphur or carbolic acid added, will keep the flies away for from five to six days, while with a small proportion of carbolic acid it will have a healing effect upon sores which may have formed. Train-oil should not cost more than from 50 to 75 cents per gallon, and a gallon will anoint a number of animals. Common axle grease, costing 10 cents per box, will answer nearly as well, and this substance has been extensively and successfully used by Mr. William Johnson, a large stock dealer at Warrenton, Va. Tallow has also been used to good advantage. The practice of smearing the horns with pine or coal-tar simply repels them from these parts. Train-oil or fish-oil seems to be more lasting in its effects than any other of the substances used.

Applications to destroy the Fly.—A great deal has been said during the summer concerning the merits of a proprietary substance, consisting mainly of tobacco dust and creosote, known as "X. O. Dust," and manufactured by a Baltimore firm, as an application to cattle, and it has received an indorsement from Prof. J. B. Smith, Entomologist to the New Jersey Experiment Station. We are convinced that this substance has considerable merit as an insecticide, and know from experience that it will kill many of the flies when it touches them, although they die slowly, and a few may recover. The substance costs 25 cents per pound, and is not lasting in its effects. Where it is dusted through the hair the flies on alighting will not remain long enough to bite, but two days later, according to our experience, they are again present in as great numbers as before. A spray of kerosene emulsion directed upon a cow would kill the flies quite as surely, and would be cheaper, but we do not advise an attempt to reduce the numbers of the pest by actually killing the flies.

How to destroy the early Stages.—Throwing a spadeful of lime upon a cow dung will destroy the larvæ which are living in it, and as in almost every pasture there are some one or two spots where the cattle preferably congregate during the heat of the day, the dung which contains most of the larvæ will consequently be more or less together and easy to treat at once. If the evil should increase, therefore, it will well pay a stock raiser to start a load of lime through his field occasionally, particularly in May or June, as every larva killed then represents the death of very many flies during August. We feel certain that this course will be found in many cases practical and of great avail and will often be an advantage to the pasture besides.

OTHER FLIES REARED FROM COW DUNG.

Our observations on the life-history of the Horn-fly have been greatly hindered and rendered difficult by the fact that fresh cow dung is the nidus for a number of species of Diptera, some of about the same size and general appearance. We have in fact, chiefly this summer, reared no less than twenty distinct species of flies from horse and cow dung, mainly from the latter, and six species of parasitic insects. We shall give these some consideration in our final article in the annual report, but can not elaborate here. The plan finally adopted to secure the isolation of the Hæmatobias was to remove the eggs from the surface of the dung and place them with dung which was absolutely fresh and collected practically as it fell from the cow. Even in this way very great care was necessary to prevent the occurrence of other species.

SOME INSECT PESTS OF THE HOUSEHOLD.*

BED-BUGS AND RED ANTS.†

By C. V. RILEY.

There is a peculiar propriety in considering these two household pests in the same article, for it is a fact not generally known, and not, I believe, previously published, that the character of the red ant is not wholly bad. It has one redeeming trait, and that is that it will (although perhaps under exceptional conditions) destroy bed-bugs. Has any one ever known a house overrun with red ants in which bed-bugs were common at the same time? I think not. One of my assistants, Mr. Pergande, had an opportunity at Meridian, Miss., during the war, of seeing an old building used as a barracks and filled with bed-bugs, invaded by countless numbers of red ants. Several ants would attack a single full-grown bed-bug, pull off its legs and carry away the helpless body. They penetrated the closest cracks of the rough beds and dragged out old and young bugs and eggs. There is, then, some slight consolation in having the ants about one's house, but with care and cleanliness, especially at the North, there is no excuse for the occurrence of either pest.

THE BED BUG.

(*Acanthia lectularia* L.)

I have occasionally met with a favored individual who had never seen a bed-bug; in fact a well-informed entomologist recently sent me a specimen for name, indicating his non-familiarity with the species! But such fortunate people are rare, and there are very few housekeepers who have not, by accident perhaps, or through slovenly servants, made the intimate acquaintance of the ubiquitous pest delineated herewith.

The bed-bug (*Acanthia lectularia*) has found its way wherever man has pushed, and is too well known to need description. Its odor and the effects of its bites are as universally known, and the word "bed-buggy" has entered our literature as descriptive of a particular class of odors. The original home of the pest is probably Southeastern Europe and the Asiatic and African countries around the eastern end of the Mediterranean. It was introduced into England at least as early as 1503, and doubtless reached America soon after extensive settlement. Certain English writers have endeavored to father the pest

* On account of the inquiries that are continually made of the Entomologist for remedies for our commoner household pests, we have decided to reprint, with slight change or addition, certain articles recently contributed to *Good Housekeeping* (Springfield, Mass.).

† From *Good Housekeeping*, May 25, 1889.

on America, but there is strong evidence that it was known to Aristophanes, Dioscorides, Pliny, and Aristotle.

The adult bug (Fig. 16*b*) is well adapted, from its flattened shape, to entering narrow crevices in the joints of bedsteads or cracks in walls, or in other convenient places of concealment, and in such places the females lay their eggs. These eggs are white, of an oval form, slightly narrowed at one end, and are terminated by a cap which breaks off when the young escape. The young bugs are whitish, and at firts nearly transparent. The head is comparatively broader than in the old bug, and the antennæ are stouter. They molt several times before attaining full growth, and among the specimens in my possession I can distinguish about four distinct stages. The bug figured at 16 *a* has probably molted once, and the differences in the head, thorax, and antennæ, from the full-grown bug, will be readily seen. The disagreeable smell, characteristic of these insects, arises from certain minute odoriferous glands which in the young bug open on the back of the thorax, and in the adults on the lower side of the body.

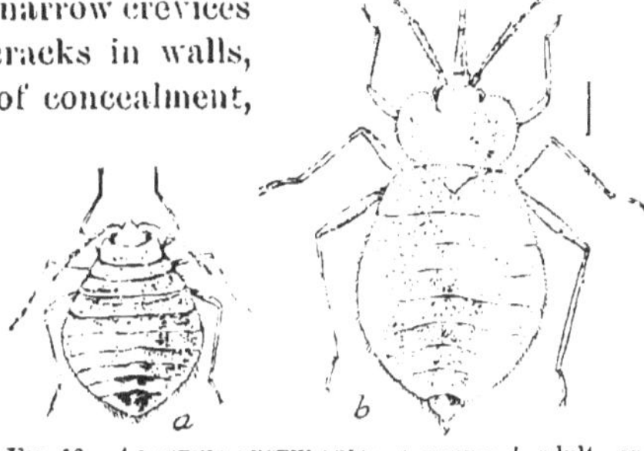

FIG. 16.—ACANTHIA LECTULARIA: *a*, young; *b*, adult—enlarged. (After Riley.)

The number of annual generations depends on conditions of food and warmth. With plenty of food and an even temperature they will multiply with great rapidity, while under contrary conditions reproduction may be greatly retarded. Adult bugs have been known to remain alive for more than a year without a single meal. It is this fasting capacity, together with its form so well adapted for hiding, which renders it so difficult to thoroughly disinfect an infested house.

Here again benzine must be our strongest weapon. Finely sprayed with a hand atomizer it will penetrate the minutest cracks, and is sure death to the insect in all its stages, including the egg. It is a certain remedy, and used thoroughly will destroy every bug in a house. Kerosene is almost as good and is a little more lasting in its effects. Many preventives have been advised, but none are permanent. One of the best formulas for a substance with which to paint the cracks in a bedstead or the wall is one ounce corrosive sublimate, half pint alcohol, and one-fourth pint spirits of turpentine.

It will be a work of supererogation to advise the experienced housekeeper to pay particular attention to the belongings of new servants, and even to the baggage of refined and cleanly guests who come from the South or West and have stopped on the way at hotels. Indeed, I feel that little of a practical nature can be written of this insect that experienced housekeepers will not know already. It may not be out of place, however, before passing to the red ant to say that the bed-bug has been found in the woods under the bark of trees, and that therefore in country houses in certain localities the occasional presence of the bugs is not necessarily a mark of uncleanliness.

It may be well also to state that there exist other allied bugs which possess much the same odor and whose bite is even more severe than that of the true Bed-bug. The Blood-sucking Cone-nose (*Conorhinus sanguisuga*, Fig. 17) is one of these. It is found occasionally in beds as far north as New Jersey and Illinois, but does not habitually breed in such locations. Its bite is very painful and it will absorb a considerable amount of blood. We show the adult bug and the nearly full-grown larva at 17. The colors are black and red.

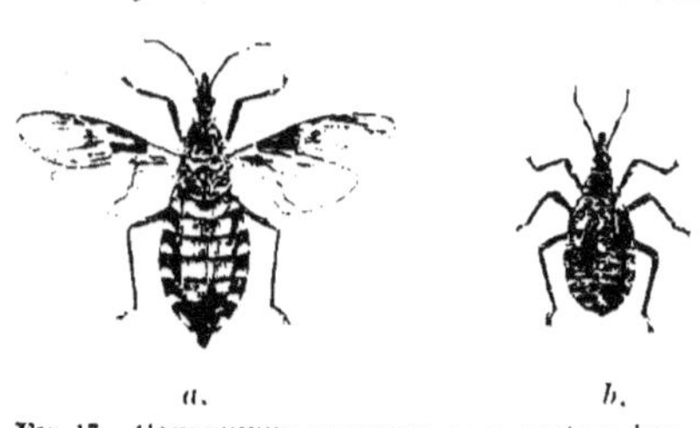

FIG. 17.—CONORHINUS SANGUISUGA: *a*, mature bug; *b*, pupa. (After Riley.)

THE LITTLE RED ANT.

(*Monomorium pharaonis* L.)

The "red ant," as this insect is almost universally called, is another of the household pests which we seem to owe to the older civilization of Europe, and, like other domestic pests, it has become almost cosmopolitan. It has been generally considered of North American origin, and as one of the few American species which has become wide-spread in Europe. It is often confounded in the literature of the subject with *Myrmica molesta* Say, which is, however, a synonym. In the larger cities of Europe it is as much of a pest to-day as it is in this country. It probably received the scientific name of "Pharaoh's ant" on account of a defective knowledge of Scripture on the part of its describer, who doubtless imagined that ants formed one of the plagues of Egypt in the time of Pharaoh, whereas the only entomological plagues mentioned were lice, flies, and locusts.

Ordinarily in households this insect is not a nuisance from the actual loss which it causes by consuming food products, but from its inordinate faculty of *getting into things*. It is attracted by almost everything in the house, from sugar to shoe polish, and from bath sponges to dead cockroaches. It seems to breed with enormous fecundity, and the incidental killing off of a thousand or so has little effect upon the apparent number. A house badly infested with these creatures is almost uninhabitable. They form their nests in almost any secluded spot, between the walls or under the floors or behind the base-boards, or among the trash in some old box or trunk, or in the lawn or garden walk just outside the door. In each of these nests several females will be found, each laying her hundreds of eggs and attended by a retinue of workers caring for the larvæ and starting out from dawn till dawn on foraging expeditions in long single files like Indians on the war-path.

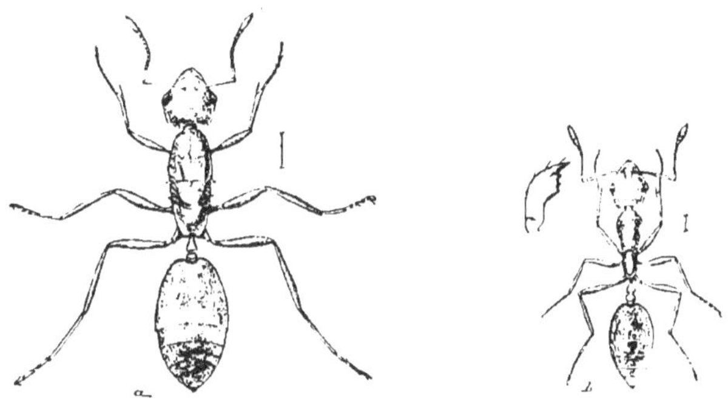

FIG. 18.—MONOMORIUM PHARAONIS: *a*, female; *b*, worker enlarged. (After Riley.)

I have shown at figure 18 the female and worker greatly enlarged, and there is nothing in their structure to which I need call especial attention. Nor need I speak further of the habits of the species, and the matter of remedies is soon disposed of. Our first recommendation is to find the point from which they all come. They may have built the nest in some accessible spot, in which case a little kerosene will end a large part if not all of the trouble. If the nest is in the wall or under the floor and taking up a board will not bring it within reach, find the nearest accessible point and devote your energies to killing the ants off as they appear. Where the nests are outside nothing is easier than to find them and to destry the inhabitants with kerosene or bisulphide of carbon. The nests are almost always in the immediate vicinity of the house. The ants are peculiarly susceptible to the action of pyrethrum in any form, be it Persian or Dalmatian powder or buhach, and a free and persistent use of this powder will accomplish much.

A great number of remedies have been proposed in the household columns of various journals, but nearly all depend upon the use of a mixture of some sort for trapping the ants, and at the best are slow and tedious means of warfare. The best of these with which I have had any experience consists in placing small bits of sponge moistened with sweetened water in the spots where the ants most do congregate, collecting the sponges once a day or so, soaking them in hot water and then replacing them. Small bits of bread and poisoned molasses or small vessels of lard in which a few drops of oxalic acid have been put have also been recommended, as well as the free use of borax, so often advised for roaches. The people of the Southern States suffer more from these pests than we do at the North, and a Floridian of experience (Mr. C. G. Cone, of Crescent City) recommends a mixture of borax and sugar, well mixed with boiling water, and left here and there on bits of broken crockery. If any one tries this I should be glad to learn the result.

A much larger black or brownish ant (*Camponotus herculeanus* var. *pennsylvanicus*) often builds its nests in door-yards so close to the houses that it becomes a great nuisance, overrunning the rooms, and even getting into the clothes, so as to be a personal discomfort. A case was brought to my notice two years ago in Washington, where a fine old homestead was on the point of being sold on account of the annoyance caused by these ants. An investigation showed one enormous nest several feet in diameter in the back yard, and several colonies here and there in other parts of the premises. The large colony was completely destroyed by the use of bisulphide of carbon. A teaspoonful was poured down each of a number of openings, and a damp blanket was thrown over them for a few minutes. Then, the blanket being removed, the bisulphide was exploded at the mouth of each hole by means of a light at the end of a pole. The slight explosions drove the poisonous fumes down through the underground tunnels, killing off the ants in enormous numbers. The mainsource of the trouble being thus destroyed, the nuisance was greatly lessened, and all talk of selling the old place has ceased.

IDENTITY OF SCHIZONEURA PANICOLA Thos. AND S. CORNI Fab.

By Herbert Osborn.

Hitherto the species of *Schizoneura* infesting grass roots and dogwood leaves, respectively, have been considered strictly distinct species, and, so far as I can learn, no suspicion has been expressed that they bore any relationship to each other.

My observations the present season establish, I think, beyond question the identity of the species, and that the insects migrate by a winged viviparous brood during the first frosty weather of autumn from the roots of grasses to the leaves of Dogwood, where they establish colonies in great numbers.

Mr. Clarence M. Weed has described the autumn viviparous form and the sexual generation and eggs produced on *Cornus* leaves by what is evidently the same species, though he refers it to *cornicola* Walsh. It is reasonably certain, therefore, that eggs deposited on *Cornus* twigs by the sexual autumn form hatch in spring, producing broods which in early summer give rise to a winged brood making the return migration to roots of grasses.

The full record of observations and evidence establishing this connection can best be presented with observations on the farther habits of the species and when certain other points are determined, but the connection of the two species seems a point of sufficient interest to merit the immediate attention of entomologists.

It may be stated here, however, that winged individuals of *S. pani-*

cola bred from grass agree very perfectly with individuals of *S. corni* found on Dogwood establishing colonies directly after the time of migration. Previous to the migration, Dogwood has been free from aphides, as evidenced by condition of leaves and absence of moulted skins or other indication, and finally winged *panicola* reared from grass roots and transferred to *Cornus* leaves, establish colonies agreeing entirely with those of *corni* on the same plant. My specimens agree perfectly with the description of Fabricius (Ent. Syst., IV, 214), but this description is so brief and general that it might not be sufficient for determination. Mr. Oestlund, however (Aphididæ of Minn., p. 28), states that specimens collected in Minnesota agree in all respects with the description and figure by Buckton, and, as my specimens agree perfectly with Mr. Oestlund's description, I adopt his reference to the European *corni*: Walsh's *fungicola* (Proc. Ent. Soc. Phil., I, 304) is apparently a fresh *corni* that he found resting on fungus; and as he describes *cornicola* as like *fungicola*, except abdomen black, I am inclined to think he had simply older or contracted specimens for the description of the latter. Passerini's *Schizoneura venusta* (Gli Afidi, p. 38), infesting roots of grasses in Europe, is evidently the equivalent of *panicola* Thos. in the United States, and I find by turning to Passerini's original description that he mentions its similarity to *corni* Fab., without, however, suggesting any relation between them. He says:

Valde similis, *Schizoneuræ corni*, quæ autem diversa dorso omnino nigro in apteris, et abdominis basi et apice tantum albido in alatis.

All discrepancies in the descriptions (which are very slight) seem to me to be accountable on the ground of difference in appearance of the recently issued and more mature individuals, along with a considerable variation in extent of the black patch on the disk of the abdomen and the number of sensoria on the third joint of the antennæ.

NOTES ON THE BREEDING AND OTHER HABITS OF SOME SPECIES OF CURCULIONIDÆ, ESPECIALLY OF THE GENUS TYLODERMA.

By F. M. Webster.

Speaking from an economic point of view, public interest in the species of the genus *Tyloderma* is at present centered in *fragariæ* Riley,* from the fact that its larvæ burrow into and destroy the thick bulbous root of the strawberry.

The life-history of the insect, from the time the immature larva is found in the plant, has been quite fully studied; but its history up to

*I received this species from Mr. C. N. Ainslie, of Rochester, Minn., in 1880, who stated that it was of rare occurrence.

this period, including the time, place, and method of oviposition, remains a complete blank, so far as published observations are concerned.

About the 20th of November, 1888, I captured a number of adult beetles in an old strawberry field in southern Indiana. Taking them home with me and placing them on plants transplanted to a warm room where the temperature was from 65° to 70° Fahr., they immediately began pairing. A few days later one of the females was observed to eat a hole through one of the bud scales, which at this time enveloped the crown of the plant (all leaves and leaf-stalks having been previously removed), and afterwards reverse her position and push the tip of her abdomen into the hole, dropping, as I supposed, her egg down among the young unfolding leaves. Leaving home on the 6th of December, I did not return again until the last of the following March, during which time both plants and beetles perished.

On the 4th of April I received from Mr. J. C. Beard, of New Albany, Ind., a fresh supply of beetles, composed of both sexes, and from the same field from which I had taken my previous supply. These were not placed on growing plants, but in a glass jar and fed each day with fresh leaves. The sexes were pairing when received, but I secured no eggs until the 7th, when a single egg was found on the bottom of the jar. No more eggs were found until the 17th, when two additional ones were found, also on the bottom of the jar.

The Egg.—The egg is 0.9^{mm} in length, and 0.6^{mm} in breadth, with the ends, each, equally obtusely rounded; color very light yellow, often covered with a whitish, glutinous substance.

I now had over a dozen females in the jar, and, notwithstanding they were pairing with the males constantly, there appeared to be no inclination towards ovipositing. Wishing to learn (1st) whether or not this was due to a lack of favorable conditions, and (2d) if there was any particular part of the plant more favorable than another, I planted three strawberry plants in as many flower-pots, placing the first so deep in the soil that only the leaf stalks were exposed, the second in such a way that it was exposed to the base of the leaf stalks, and the third so as to leave nearly all of the bulbous root exposed above ground. A single female was taken from the jar and placed on the first plant, and covered with a glass. All leaves had been cut away, leaving two leaf stalks each about one and one-half inches in height. After being placed in the jar the female remained perfectly quiet for a few minutes and then began an inspection of the stems. An excavation was made in one of them about an inch above the soil, but was abandoned without being used as a nidus. The second stalk was then taken under consideration and critically examined, after which, with head downward, she began excavating a cavity about one-fourth of an inch above ground, and, after finishing it, she reversed her position and deposited her egg at the mouth of the cavity. The labor of oviposition over, she again turned about, and, after carefully pushing the egg in place, began collecting the down from the stem, pulling it off with her jaws and tucking in, over and about the egg, effectually filling the cavity, the whole operation occupying about one hour and a half.

The second female conducted herself in much the same manner, except that she constructed her egg cell just at the surface of the soil, which was also exactly at the base of the leaf stalk, and, to my astonishment, after placing her egg, filled the cell with mud, and besides drawing the damp earth up about the plant in such a manner as to conceal the spot entirely. The time occupied was about as with the first.

The third began her labor as promptly as either of the others, but punctured the bulbous root about half an inch from the surface of the soil and about the same distance below the base of the leaves, and filled the cavity, after ovipositing, with the loose material on the outside of the root. Time nearly the same as in the others.

The foregoing seemed to indicate to me that the females were withholding their eggs on account of their environment, and as a rule they continued to do this until they died, after the 1st of May. There also appeared to me to be a partiality for ovipositing in the plants somewhere near the surface of the earth, which would ordinarily, and in the fields, be near or just below the juncture of the leaf stems with the root. Dissection of females revealed but few eggs in the ovaries, and these about as large as previously indicated. I, of course, know nothing of the movements of these beetles before they came into my hands, but, judging from my own observations, should not expect them to deposit above a dozen eggs each, and that, under favorable conditions, these eggs might be deposited during March and April or withheld until May, if necessary.

All of the eggs which were deposited in the plants, under my observation, were sacrificed in the attempt to determine the egg period. Two eggs were, however, deposited by other females, about the 3d of May, on the inside of glass tubes, in which they were confined. I watched the development of the larvæ in these eggs, it being a very easy matter to do so through the glass, and that portion of the shell which adhered to the walls of the tube. The larvæ did not reach maturity until nearly the middle of June, and ate through the shell, where the latter was attached to the tube, od the 18th of same month. It must be borne in mind that these eggs were in an unnatural environment, and the results are to be taken for just what they are worth.

The species *foveolatus* Say oviposits in the stems of the Evening Primrose, (*Enothera biennis* L., in June. The method of oviposition is very much the same as in the preceding, the mother beetle covering the cavity, after depositing the egg therein, by raking the epidermis of the stem together, and fixing it in and over the hole, where it dries and forms a sort of scab, remaining until after the wound has wholly or in part healed. The eggs are rather larger than those of *fragariæ*, but shaped and colored much the same. The insect, in all its stages except the egg, may be found in the stems of this plant during the month of August, the more advanced stages nearest to the ground. The main stem in the one selected and the work of the beetle may be readily detected by their scarred and pitted appearance. Except from being larger, the larvæ of this species do not differ materially in general appearance from the preceding. The punctures which are so apparent in the adult beetle are also to be observed in the pupæ.

Of the breeding habits of *variegatus* Horn, I know nothing, and only refer to the species here in order to record its occurrence in an ant-hill in the month of December.

I have observed *æreus* Say about plants of Evening Primrose, but have not observed them ovipositing. Moreover, have observed them of various sizes and in great numbers in localities where there were no plants of the Primrose.

Rhyssematus lineaticollis Say breeds in the seed pods of *Asclepias incarnata*, the larva feeding upon the seeds and transforming to the adult in the late autumn. The larva is white, robust, and much wrinkled, with sparsely-placed, short bristles distributed over the body; the head is much smaller than first segment, yellow, with mouth parts darker. Length when extended 6^{mm}. In the vicinity of La Fayette, Ind., the larvæ are preyed upon by a species of *Bracon*, the larvæ of which leave the body of their host and spin small brown cocoons within the seed pod, several parasites inhabiting a single larva of *Rhyssematus*.

June 18, 1889.

EXTRACTS FROM CORRESPONDENCE.

The Spread of the Australian Lady-bird.

The Vedolia has multiplied in numbers and spread so rapidly that every one of my thirty-two hundred orchard trees is literally swarming with them. All of my ornamental trees, shrubs, and vines which were infested with white scale, are practically cleansed by this wonderful parasite. About one month since I made a public statement that my orchard would be free from "Icerya by November 1," but the work has gone on with such amazing speed and thoroughness, that I am to-day confident that the pest will have been exterminated from my trees by the middle of August. People are coming here daily, and by placing infested branches upon the ground beneath my trees for two hours, can secure colonies of thousands of the Vedolia, which are there in countless numbers sucking food. Over fifty thousand have been taken away to other orchards during the present week, and there are millions still remaining, and I have distributed a total of sixty-three thousand since June 1. I have a list of one hundred and thirty names of persons who have taken the colonies, and as they have been placed in orchards extending from South Pasadena to Azusa, over a belt of country ten miles long and six or seven in width, I feel positive from my own experience, that the entire valley will be practically free from Icerya before the advent of the new year. You will be as much pleased to read this as I am to write it.—[J. R. Dobbins, San Gabriel, Cal., July 2, 1889.

Wasps in India.

A tin trunk belonging to Mrs. Sidney Preston, wife of a gentleman in Her Majesty's civil service, was packed with wearing apparel, etc., in Hoti Murdan, and brought to Jhelum, Punjab, India, in March, 1889. It was left in a veranda for two months and opened in May. It contained, to the surprise of the owner, four large nests of wasps, the ordinary *Vespa* of the district. A small hole was at last discovered near the hinge, affording a possible clue to the entrance of the parent or parents. One of the nests was so large as entirely to fill up a baby's hood. After getting rid of the paper-like nests and the living wasps, which were numerous, the remainder of the clothing in the box was found to be covered with dead wasps in quantities; in fact, with several hundred of them. The contents of the box had been carefully camphored and peppered when packed.—[A. O'D. Taylor, Newport, R. I.

Injurious Insects in New Mexico.

I have forwarded to you by same mail this day a square tin box containing inclosed two small boxes. The larger square box contains a number of specimens of the bean or frijole bug, also two small pupæ of the same insect, and further, a single specimen of a bug said by the sender to prey on his grape-vines. Having no means of killing the insects I forward them as I received them, most of them alive. In the small round box you will find a few specimens of another bug resembling the first somewhat in its markings and general shape, but larger and evidently a different insect. These are all dead, and were collected by myself personally on a plant of the Convolvulus or Ipomæa family, near Bernalillo, in the Rio Grande Valley. Not having a Gray's Manual I am unable to give the plant its name in botany. It is named by the Mexicans, calabaza (gourd) on account of its enormous root, which is supposed to resemble a large, warty species of native gourd. Its flowers, of a pale purple color, resemble very large morning-glories. The plant, which is found in all New Mexico, but especially in the sandy wastes which border the valley proper of the Rio Grande River, is an upright bush with long, narrow leaves. The stems and leaves die out every year, but the root is perennial, and must live many years, for it becomes very hard and woody. The seeds resemble those of the morning-glory, but are much larger. I have described this plant so particularly because the larger of the two species of bugs, which is of a paler color and with fewer and less marked black dots (the one in the small round box), is found in large quantities on the plant; and the Mexicans have an idea, whether correct or not (of this I am no judge because I am not an entomologist), that the frijole chinch (the smaller bug in the square box), which is the destructive bug that preys on the beans, originates from the other.

The convolvulus bug appears early in the spring; I gathered it on the plants myself in May. The Bean bug appears in July. Although I felt satisfied that the two insects are different, and that a bug that preys on the Convolvulus family could not equally prey on beans, I thought this matter of sufficient interest, and brought a handful of convolvulus bugs, which I put in the midst of a small patch of beans growing in the garden, but within ten minutes they had all left, and for two weeks I looked carefully through the beans, but never saw a bug of any kind on them. The Bean bug commits great depredations on bean fields, often destroying them entirely. The only means the Mexicans have found to somewhat prevent its ravages is to plant their beans late, about the middle of July, the bug appearing to swarm in smaller numbers later in the season. The chief season of the Mexican bean bug seems to be from the middle of July to the first of September. The Phaseolus grown by the Mexicans belongs to the same family as our string beans; the pod can be eaten as a string bean, and the bean is of a yellowish brownish color, of ordinary size, somewhat flatish. When cooked and prepared in the Mexican way it is the best bean I have ever eaten, far superior and better flavored than our so-called navy bean, and it would be a real acquisition to the American bill of fare. The Mexicans eat their beans three times a day—at every meal the year round, if they have them. In a few days I will endeavor to go myself to the place from which these bean bugs (I think you ought to call them Mexican bean bugs if not already named) were sent to me, some 20 miles from Las Vegas, to examine them myself on the vines, and will then send you another lot and describe what I see.—[J. F. Wielandy, Springer, N. Mex., July 23, 1889.

REPLY.—I have your letters of the 22d, 23d, and 24th of July, and also all of the specimens which you mention. I am very much obliged to you for your full information and for the specimens which you send. The insect which you call the New Mexico bean bug is *Epilachna corrupta*, one of the few plant-feeding lady-birds. A congeneric species feeds upon the leaves of squash in the more northern States, and is mentioned by Professor Riley in his fourth Missouri report. The larger beetle found upon Convolvulus is one of the leaf beetles known as *Chelimorpha cribraria*. Your long account in your letter of the 23d is very interesting, and unless you send me something to supersede it after your visit in person to examine the insects in the field, I shall publish it in INSECT LIFE. Among your specimens we also found the common rose bug of the Northern States (*Macrodactylus subspinosus*). The application of an arsenical poison early in the season should be an effective remedy against the bean bug. Your locality is a very interesting one, and I trust you will keep your eyes open for injurious insects for us.—[July 31, 1889.]

SECOND LETTER.—In order to investigate the Mexican bean bug more fully (there being no beans in this immediate neighborhood) I went last Sunday to Watrous, some 50 miles south of this place, on the Atchison, Topeka & Santa Fé Railroad, where I examined them on the farm of Mr. William Kroenig, who is, with me, one of the *very* few persons who take an interest in such matters in New Mexico. The result is that I am enabled to send you to-day the insect in the egg stage, the larva stage, and the imago stage. The pupae I am not able to procure, for reasons apparent enough. In conversing with Mr. Kroenig I find the following facts: That he has known the insect since he has been in this region, which is about forty years; that it was then just as bad as now; that it is found chiefly on beans cultivated in old fields, and on land newly cultivated is comparatively scarce, or even unknown, for the first few years; that frequently it destroys the entire crop; that the only way to keep down its ravages to some extent is to plant the beans during the interval between the first appearance of the bugs and their second appearance in the fall. The question with me is now to find out if they have more than one brood, and if so, how many. During my visit I examined a new field of beans in which there were no insects. From that we went to a corn field in which there were beans planted among the corn. We there found chiefly larvæ, and only 4 bugs. The bugs had apparently laid their eggs and died. The larvæ were nearly all of the same size. I also found 3 bunches of eggs, which, together with the larvæ, I put in the little vial with a mixture of alcohol and water. The parent bug appeared about the 15th of July for the first time in this locality, possibly a few days sooner. On the 28th, they, as well as the eggs, were nearly all gone, I finding, as stated, only 4 bugs and 3 bunches of eggs. I found among them two varieties of lady-bugs, which seemed engaged in preying upon the eggs and small larvæ, and of which I inclose a couple of specimens. I do not know whether the larger, paler colored of the two insects which I take to belong to the lady-bug family is really one; I never saw it before. You will know. The 4 Mexican bean bugs and the lady-bugs are together in one box, and the larvæ in the bottle together with the eggs. I am positive that another appearance of the full-grown bug occurs in September and October, because I saw some of them at that time last year myself. You have no doubt received some of the bugs I have sent you last week inclosed in letters; one being a bug found on a species of Ipomæa or Convolvulus; the other being the notorious Mexican bean bug, which is the brown bug of the Coleoptera order—sixteen spotted. I will continue my observations on this insect. I send you a few bean leaves to show you the manner in which its depredations are committed. You will notice that it does not eat the leaf, but only the parenchyma on both sides. It also eats the flowers and the very small young pods.

I also send you another box with a bug of the Hemiptera order, which I found in a garden at Las Vegas, preying upon young cabbage plants, which it sucks, causing the leaves to dry and the young plants to wilt and die entirely, in the same manner as the squash bug preys upon Cucurbitæ. This very pretty harlequin-colored Hemipteron appears frequently in immense numbers, living on various plants of the genus Brassica, such as cabbage, mustard, turnip, etc., and sometimes appears in immense numbers, destroying everything and causing very great havoc. It is also said to have existed in this region from time "immemorial." I am told that it has originated on a native plant of the Brassica family, which has purple or bluish flowers, but I have never seen the plant and do not know how the insect propagates itself. I also send you a third, grayish insect, which abounds in immense quantities on the farm of Mr. Kroenig. It is omnivorous, at least apparently. It does especially great damage on young apple trees. I inclose two apple leaves to show how it works, eating the parenchyma, some young trees being entirely denuded in appearance, although none of them die from the effect. They are not entirely killed, only greatly retarded in growth. I have seen this bug on apple trees, pear trees, plum trees, apricots, grape vines, on a native wild species of willow, even on beans, but it does not appear to touch the peach. It abounds in millions, very much like the May bug (hanneton) of Europe. I know nothing about its mode of multiplication. * * * [J. F. Wielandy, Springer, N. Mex., July 30, 1889.

SECOND REPLY.—Thank you very much for your long and interesting letter of the 30th ultimo, concerning the New Mexican Bean Bug. I shall be glad to publish this letter nearly in full. The two Lady-birds which you found feeding upon the eggs are *Hippodamia convergens* and *Coccinella transversoguttata*. The bug which you found upon cabbage is the common Harlequin Cabbage-bug (*Murgantia histrionica*). The beetle which you found upon young apple trees is congeneric with our Rose Bug of the North. It is *Macrodactylus uniformis*. The beans which you inclose have been handed to the head of the Seed Division with the request that they be planted.—[August 5, 1889.]

The Corn-feeding Syrphus-fly.

A few days ago, while passing through a corn-field, I noticed that most of the lower leaves of the plants were brown, yellow, and dried up. My first idea was that this was due to the Chinch Bug. Of course I set to work at once to investigate, and found only a solitary bug here and there, not sufficiently numerous to do any damage. On carefully stripping down the leaves that were partially discolored I found, snugly feeding between the base of the leaf and the stem, many lively but delicate-looking larvæ, sometimes five or six at the base of one leaf. The larvæ seem to be all of one species, but of various sizes, or ages, and here and there in the same places where the larvæ were feeding I found pupæ of different ages, some black and some only recently changed. The stems under the enfolding base of the leaf, where the larvæ feed, are bathed in or covered with the juice of the plant, and the effect produced is exactly the same as that produced by the Chinch Bug. To-day I mailed you a canister, in which I hope you will find plenty of larvæ and pupæ of different ages, if they are not dried up before they reach you. You will also, perhaps, find a few small insects that I found in the same places with the larvæ. No corn can successfully contend with this pest. At this time, although there has been an unusual amount of rain this summer, the leaves of the corn are "sere and brown" half way up the stalk.—[J. G. Barlow, Cadet, Mo., August 9, 1889.

REPLY.—Your letter of the 9th instant with specimens has been received. The insect in corn is a very interesting thing, and you will find it figured and described under the caption of the Corn-feeding Syrphus-fly (*Mesograpta polita*) in No. 1, Vol. I, of INSECT LIFE. Your letter is therefore of considerable interest, and will go on record among our notes.—[August 14, 1889.]

Larvæ of Cephenomyia in a Man's Head.

I was called to see a case to-day, who had just come from Swartbout Cañon, 30 miles from here, the messenger stating that his father had Screw Worms in his nose and wanted me to get them out. I found the patient at the home of his son, in bed. His name is E. P. Fowler; age, 61; occupation, a carpenter; native of New York; raised in Ohio. I found him breathing hard, accelerated pulse and temperature, a bloody mucus issuing from the nose, the passages nearly closed from dried blood and mucus, nose swollen and pain between the eyes, as well as reddened looking in the mouth, with the back parts of a leaden color and covered with mucus. I procured warm water, carbolized it, and took forceps and small plugs of cotton and removed the dried secretions as far as I could. I then came on to the maggots and removed 40 of them with the forceps from the nose. I used a powder-blower and blew into each nostril in different directions an impalpable powder of calomel, after which several maggots came away of themselves. I send you a sample of five of them in this mail. Mr. Wright, my neighbor, being an entomologist, I gave him a number of the maggots. He reports them feeding on a bony piece of raw beef, they having refused cooked beef. I hope to gain some information of the fly, whether it is identical with the Sheep Grub, Green Bottle fly, or is it an individual species. The patient has had nasal catarrh for many years, and it is probable the secretions formed a suitable field for the deposit and development of the maggot.—[Wesley Thompson, M. D., San Bernardino, Cal., August 7, 1889.

REPLY.—Your very interesting letter of August 7 has just come to hand, and the specimens also arrived in good condition. The larvæ which you send do not belong to the species which is ordinarily known as the Screw Worm, but to a different group. Instead of being Muscids they are Œstrids, and although it is impossible to determine the precise species from the larvæ, the genus is *Cephenomyia*. The larvæ of those species of this genus of which we know the larvæ, are found in the nasal passages of deer, and within the last two months we have received from Mrs. Bush, of San José, larvæ taken from the deer which may be the same species as the one which you send. The occurrence of this larvæ in the head of your patient was of course more or less accidental, although not without precedent. I hope that Mr. Wright will succeed in rearing the fly, although the larvæ are evidently not more than half grown, and success seems doubtful.—[August 15, 1889.]

STEPS TOWARDS A REVISION OF CHAMBERS' INDEX, WITH NOTES AND DESCRIPTIONS OF NEW SPECIES.

By LORD WALSINGHAM.

[*Continued from page 81 of Vol. II.*]

Lithocolletis nemoris sp. n.

Antennæ, white, spotted above with fawn brown.

Palpi, white.

Head, face white, frontal tuft whitish, much mixed with saffron-brown, especially at the sides.

Thorax, saffron.

Fore-wings, rather shining saffron with snow-white markings consisting of two transverse fascia, slightly oblique, and angulated beneath the costal margin, beyond which are one dorsal and two costal streaks; there is no basal streak; the first fascia at one-fourth the wing-length is but slightly angulated, margined with scattered blackish scales, widely on its outer and very indistinctly on its inner side; the second fascia at the middle of the wing is rather more strongly angulated than the first; this is also slenderly dark-margined internally and more widely so externally, the black dusting on its outer side being produced backwards at the angle in the direction of the first costal streak; this is at the commencement of the costal cilia, rather further from the base than the first dorsal streak, which is oblique, its point terminating below the point of the first costal streak; from the points of these two streaks a cloud of black scales proceeds outwards along the middle of the wing, forming a dark patch below and beyond the second costal streak which is situated just before the apex; the cilia are saffron, shading to pale grayish-saffron beyond their faintly darker median line.

Hind wings and cilia, pale grayish, with a very faint saffron tinge.

Abdomen, pale gray, anal tuft saffron-yellow.

Hind tarsi, white with two grayish-fuscous bars above.

Exp. al. 8mm.

Type ♂ ♀ *Mus. Wlsm.*

The puckered mines of this species were found in some abundance in June, 1871, in Mendocino County, California, on the upper sides of leaves of *Vaccinium ovata*, the mine occupying the whole surface of each leaf and causing the margins to approach each other. I took the species also on the wing at the same time and place. This species belong to the same group as *cincinnatiella* Chamb.

Lithocolletis oregonensis sp. n.

Antennæ, closely annulate with white and brown.

Palpi, whitish, dusted with gray externally.

Haustellum, yellow.

Head, face grayish, frontal tuft grayish-fuscous.

Thorax, golden-saffron.

Fore wings, golden-saffron, with four rather shining white fasciæ and a semi-circular white apical streak inclosing a black apical spot and reaching through the cilia on the costal and dorsal margins; the first fascia is situated within one-fourth the wing-length, the dorsal portion of it commencing nearer to the base than the costal portion and proceeding obliquely outward to a little above the fold, the shorter costal portion only being conspicuously dark margined internally; the second fascia, just before the middle, is distinctly curved, almost angulated outwards, and has a conspicuous margin of black scales on its inner side; the third fascia, commencing before the costal cilia, is less curved than the second, but its black inner margin interrupts it in the middle by a short line of black scales; the fourth fascia, at the apical fifth of the wing, is also internally black-margined, but the black scaling is almost interrupted, becoming very slender at the middle of the wings; the apical spot is black, encircled by white as already described; the cilia are grayish, tinged with fuscous about the anal angle, and with a short golden-saffron dash from the black apical spot; there is no line along their base.

Hind wings and cilia, pale grayish.

Abdomen, gray.

Hind tarsi, whitish, thickly spotted with fuscous above.

Exp. al. 7mm.

Type ♀ *Mus. Wlsm.*

Two specimens taken on the wing near Fort The Dalles, on the Columbia River, in northern Oregon, in April, 1872.

A beautiful and distinct species, somewhat allied to the European *scabiosella*. I have unfortunately no knowledge of its food-plant.

Lithocolletis insignis sp. n.

Antennæ, yellowish, unspotted.

Palpi, white.

Head, face white, frontal tuft white with a few saffron scales.

Thorax, white.

Fore-wings, pale saffron, with a rather golden tinge; a broad white basal streak on the upper half of the wing, running parallel to the costal margin for one-third the wing-length, thence deflexed and confluent with the middle of the upper edge of the first very broad white dorsal streak. The basal streak is sometimes extended at the base across the fold reaching to the dorsal margin, thus leaving between itself and the first dorsal streak a small curved, oblique saffron streak; sometimes it is not thus projected across the fold, but upon the dorsal margin beneath it is found a separate short dorso-basal white dash. Above and slightly beyond the point at which the broad basal streak is deflexed there is a very oblique costal streak, somewhat triangular, with its apex reaching nearly to the apex of the much larger first dorsal streak below it; beyond this the second streak, situated just beyond the middle of the costal margin, is of about the same size, also triangular, a little less oblique, and corresponding with a wider and more conspicuous white dorsal patch opposite to it. The third and fourth costal streaks, of which the former points slightly outwards. The latter is perpendicular, reaching nearly (or in some specimens quite) to a white patch on the dorsal margin before the apex, which seems to consist of two confluent white dorsal streaks. At the extreme apex is a minute black apical spot, surrounded by a semi-circular dark line at the base of the apical cilia, which are tinged with golden saffron at the extreme apex. Beneath the apex the cilia are white, blending into saffron-gray about and before the anal angle; all the white markings are distinctly dark-margined on all sides. The white streaks on the fore wings of this species are so large and conspicuous as in some cases to almost obliterate the pale saffron ground-color, and different specimens vary much in the proportionate space occupied by one and the other.

Hind wings and cilia, pale gray.

Abdomen and anal tuft, grayish-white.

Hind tarsi, whitish, spotted above with gray.

Exp. al. 9mm.

Type ♂ ♀ *Mus. Wlsm.*

I met with this very beautiful and distinct species in June, 1871, in Lake and Mendocino Counties, California, and again on Mount Shasta, Siskiyou County, in August of the same year. It is evidently a scarce species, as I met with a single specimen only on each of the four different occasions. I am unable to give any information as to its larval habits. It seems to belong to the same group as *fitchella* and the European species *roboris*, but differs very greatly in the form of its markings.

In addition to the known American species of this genus, I have received two more, which are undescribed, from Dr. Riley, one feeding on *Grindelia robusta*, the other on *Betula*. I prefer to leave their description to my distinguished friend, who has probably a better series of specimens to refer to than I have.

I am indebted to the late Professor Bolander, of San Francisco, and to Mr. W. Carruthers, of the British Museum, for the identification of some of the plants mentioned in this paper.

The following is a list of plants, with the species of *Lithocolletis*, which feed upon them, so far as they are known to me. I have published this in the hope that it may facilitate the collection of further information concerning the life-histories of the very numerous species belonging to this interesting genus.

North American species of Lithocolletis.

Food plants.	Larvæ. Superior.	Larvæ. Inferior.
Tiliaceæ:		
Tilia americana	Tiliella *Chamb*	Lucetrella *Clem.*
Anacardiaceæ:		
Rhus toxicodendrum	Guttifinitella *Clem*	
	Toxicodendri *F. & B.*	
Sapindaceæ:		
Æsculus glabra	Guttifinitella	
	Var. Æsculella *Chamb*	
Aceraceæ:		
Acer saccharinum	Aceriella *Clem*	Clemensella *Chamb.*
		Lucidicostella *Clem.*
Leguminosæ:		
Desmodium viridiflorum		Desmodiella *Clem.*
Phaseolus pauciflorus		Desmodiella *Clem.*
Amorpha fruticosa		Uhlerella *Fitch.*
Robinia pseudacacia	Ostensackenella *Fitch*	Ostensackenella *Fitch.*
	Robiniella *Clem*	Robiniella *Clem.*
Robinia viscosa	Robiniella *Clem*	Robiniella *Clem.*
Robinia hispida	Ostensackenella *Fitch*	Ostensackenella *Fitch.*
	Robiniella *Clem*	Robiniella *Clem.*
Robina *sp.?*	Gemmea *F. & B*	(? Superior and inferior.)
Amphicarpaæ monoica		Morrisella *Fitch.*
Rosaceæ:		
Cerasus serotina		Pomifoliella *Z.*
Prunus americana		Pomifoliella *Z.*
Crataegus tomentosa		Pomifoliella *Z.*
Pyrus coronaria		Pomifoliella *Z.*
Pyrus malus		Pomifoliella *Z.*
Cydonia vulgaris		Pomifoliella *Z.*
Cydonia japonica		Pomifoliella *Z.*
Hamamelideæ:		
Hamamelis virginica	Aceriella *Clem*	
Caprifoliaceæ:		
Lonicera albida		Affinis *F. & B.*
		Fragilella *F. & B.*
Lonicera sempervirens		Fragilella *F. & B.*
Symphoricarpus vulgaris		Fragilella *F. & B.*
		Mariella *Chamb.*
		Symphoricarpella *Chamb.*
Symphoricarpus *sp.?*		Affinis *F. & B.*
Compositæ:		
Solidago patula		Solidaginis *F. & B.*
Grindelia robusta	*Sp.?*	
Ambrosia trifida		Ambrosiella *Chamb.*
Helianthus giganteus		Ambrosiella *Chamb.*
		Ignota *F. & B.*
Elephantopus carolinianus		Elephantopodella *F. & B.*
Actinomeris squarrosa		Elephantopodella *F. & B.*
		Amoena *F. & B.*
	Actinomeridis *F. & B*	(? Superior and inferior.)
Verbesina virginica		Elephantopodella *F. & B.*
Ericaceæ:		
Gaultheria shallon	Gaultheriella *Wlsm*	
Ledum glandulosum	Ledella *Wlsm*	
Vacciniaceæ:		
Vaccinium ovatum	Nemoris *Wlsm*	
[*Primulaceæ:*		
Lysimachia lanceolata		Lysimachiella* *Chamb.*]
Laurineæ:		
Umbellularia californica	Umbellulariæ *Wlsm*	
Ulmaceæ:		
Ulmus americana	Ulmella *Chamb*	Argentinotella *Clem.*
Ulmus fulva	Ulmella *Chamb*	Argentinotella *Clem.*
		Occitanica *F. & B.*
Celtis occidentalis	Celtifoliella *Chamb*	Celtifoliella *Chamb.*
Juglandaceæ:		
Juglans nigra	Caryæfoliella *Clem*	
Juglans cinerea	Caryæfoliella *Clem*	
Carya alba	Caryæfoliella *Clem*	Caryalvella *Chamb.*
Carya olivæformis	Caryæfoliella *Clem*	
Carya *sp.?*	Eppelsheimii *F. & B*	(? Superior and inferior.)
Cupuliferæ:		
Quercus alba	Bifasciella *Chamb*	Æriferella *Clem.*
	Cincinnatiella *Chamb*	Albanotella *Chamb.*
	Hamadryadella *Clem*	Argentifimbriella *Clem.*
	Tubiferella *Clem*	Basistrigella *Clem.*

* This species has not yet been bred.

North American species of Lithocolletis—Continued.

Food plants.	Larvæ. Superior.	Larvæ. Inferior.
Cupuliferæ—Continued.		
Quercus bicolor	Conglomeratella *Z*	Argentifimbriella *Clem.*
		Basistrigella *Clem.*
Quercus castanea		Argentifimbriella *Clem.*
		Basistrigella *Clem.*
		Fitchella *Clem*
		Hagenii *F. & B.*
Quercus macrocarpa	Hamadryadella *Clem*	Quercibella *Chamb.*
	Macrocarpella *F. & B.*	
Quercus nigra		Æriferella *Clem.*
Quercus obtusiloba	Cincinnatiella *Chamb*	Rileyella *Chamb.*
	Conglomeratella *Z.*	
	Hamadryadella *Clem*	
	Lebertella *F. & B.*	
	Quercivorella *Chamb*	
Quercus prinoides		Basistrigella *Clem.*
Quercus prinus		Fitchella *Clem.*
		Hagenii *F. & B.*
Quercus rubra		Minutella *F. & B.*
		Rileyella *Chamb.*
Quercus tinctoria	Bethuniella *Chamb*	Æriferella *Clem.*
	Unifasciella *Chamb*	Basistrigella *Clem.*
		Obstrictella *Clem.*
Quercus *sp.?*	Castanella *Chamb*	Diaphanella *F. & B.*
Castanea americana	Castanella *Chamb*	
	Caryliella *Chamb*	
Fagus sylvatica		Faginella *Z.*
Corylus americana	Coryliella *Chamb*	
Ostrya virginica	Coryliella, *var.* ostryella *Chamb.*	Obscuricostella *Clem.*
	Tritæniella *Chamb*	Ostryæfoliella *Clem.*
Carpinus americana	Coryliella *Chamb*	
Betulaceæ:		
Alnus incana	Alnicolella *Wlsm*	Incanella *Wlsm.*
Alnus serratula		Auronitens *F. & B.*
Alnus *sp.?*	Alnivorella *Chamb*	
Betula *sp.?*	*Sp.?* (superior and inferior)	
Salicaceæ:		
Salix alba		Salicifoliella *Chamb.*
Salix babylonica		Salicifoliella *Chamb.*
Salix longifolia		Salicifoliella *Chamb.*
Salix *sp.?*		Atomariella *Z.*
		Scudderella *F. & B.*
Populus grandidentata		Atomariella *Z.*
Populus tremuloides		Atomeriella *Z.*
Populus *sp.?*		Populiella *Chamb.*
		Salicifoliella *Chamb.*
Food plants unknown	Alniella (*Z.*) *F. & B*	(?Alnus.)
	Australisella *Chamb*	
	Bostonica *F. & B*	
	Chambersella *Wlsm*	
	Insignis *Wlsm*	
	Obsoleta *F. & B*	
	Oregonensis *Wlsm*	
	Sexnotella *Chamb*	

(*To be continued.*)

GENERAL NOTES.

THE CABBAGE PLUTELLA IN NEW ZEALAND.

In the last number of INSECT LIFE we mentioned the occurrence of this cabbage pest in South Africa and referred to our previous statement (Annual Report for 1883) concerning its occurrence in Australia. We have now to record the fact that it seems to be well known as a cabbage pest in New Zealand. The *New Zealand Farmer* for August, 1889, states that information is recorded by more than one of its readers concerning this insect and quotes at length from the *New Zealand Country Journal* for May, 1887, an article concerning its habits and damage. The article is illustrated by a reproduction of Curtis's well known figure, and treats of the pest under the English name of "The Diamond Back Turnip Moth." The *Country Journal* we have not had the pleasure of seeing before, and we may mention the fact that the turnip crops of 1886–'87, in the vicinity of Canterbury, suffered to a very serious extent from the ravages of the larvæ of this insect, while the moths might be seen in countless thousands during March and April. So great were the ravages during 1887 that in some instances the turnip crop was reduced to 25 per cent. of its normal condition. This is a serious thing, because in New Zealand of late years the culture of the turnip is increasing enormously, and the author of the article states that without it it would be difficult to profitably carry on the work of bringing into cultivation large areas of new land, and the fertility of areas already under cultivation could not be so well maintained. Without the turnip, moreover, the trade in frozen mutton could not be carried on to such an extent as it promises by the aid of this crop. Many cruciferous plants would also suffer. According to Mr. Fereday, the insect has been known in New Zealand for years past.

CANNIBALISM WITH COCCINELLA.

Apropos of the note from *Science Gossip* in the August issue of INSECT LIFE, concerning the cannibalism of *Coccinella dispar*, I desire to record some observations made in southern Illinois four or five years ago, showing an even more reprehensible habit of some members of this group than the eating of the pupæ. I was studying apple insects for Professor Forbes at the time, in early spring, and some species of Coccinellidæ were very abundant in the orchards of Mr. Parker Earle, at Anna, Ill. Many of them were ovipositing, and the clusters of bright yellow eggs were not uncommon upon the trunk and larger limbs. One species in particular, *Coccinella 9-notata*, I believe, though as I have not my notes with me, I am not certain, was laying eggs abundantly and *was also eating them with avidity.* I caught adult beetles in the act a number of times, and afterwards proved by observations on specimens in confinement that they are not at all averse to eating eggs presumably of their own species.—[Clarence M. Weed.]

RHODE ISLAND POPULAR NAMES FOR CORYDALUS CORNUTUS.

We are indebted to Prof. W. W. Bailey, of Brown University, Providence, R. I., for the following list of names used in Rhode Island for *Corydalus cornutus* or Hellgramite Fly: Dobsons, Crawlers, Amly, Conniption Bugs, Clipper, Water Grampus, Goggle Goy, Bogart, Crock, Hell Devils, Flip Flaps, Alligators, Ho Jack (locally in Scituate, R. I.), Snake Doctor, Dragon, and Hell Diver.

SOUTHERN SPREAD OF THE COLORADO POTATO-BEETLE.

Apropos to the note on page 22, current volume of INSECT LIFE, allow me to state that there are good reasons for the belief that *Doryphora 10-lineata* occurred at Jackson, Miss., in April, 1888. While at Vicksburg, late in April, last year, I was told of their appearance on potatoes, in the vicinity of Jackson, and took pains to question my informer as to their looks, and his replies left no doubt as to the identity of the species.—[F. M. Webster, La Fayette, Ind., July 25, 1889.]

THE GAS PROCESS FOR SCALE INSECTS.

While at Orange I learned of four persons who had used the gas process for ridding their trees of the red scale, and they much preferred it to spraying. Dr. W. B. Wall, the county treasurer of Orange County, told me that it cost him about one and a half times to fumigate what it would to spray the trees with a wash costing one cent a gallon, and that one fumigation accomplished as much good as *three* sprayings, besides leaving the tree in a better condition. There is still considerable injury to the leaves of trees fumigated in very hot weather, but I hope to overcome this by using a tent constructed from a different material than those heretofore used, as there is reason for believing that it is the rays of *light* rather than of heat that decompose the gas.—[D. W. Coquillett, Los Angeles, Cal., July 22, 1889.]

A SAD BLUNDER IN NO. 2.

Unfortunately I allowed a very careless error to appear in print in No. 2 in the item entitled "A Peculiarity of Certain Caddis Flies." The title should read instead of "Caddis Flies," "Trichopterygid Beetles." In reading the German article in the *Entomologische Zeitung* the word "Trichopterygier" impressed me as referring to the Trichoptera and I allowed the item to go to press before discovering the blunder. Professor Riley was absent and about to leave France, so that copy of the item was not sent him, as the mistake would otherwise never have occurred.—[L. O. H.]

ARSENICALS AND THE HONEY BEE.

In the last number of INSECT LIFE, pp. 84–85, in his note on the effect of arsenical insecticides upon the honey bee, Mr. Webster desires to state that it was *during a period of two years* that Mr. Yenowine sprayed all his fruits freely, so that the increase in his bee colonies was practically that of *one* unfavorable season, the season of 1888.

FIRST ANNUAL MEETING OF THE ASSOCIATION OF OFFICIAL ECONOMIC ENTOMOLOGISTS.

The Association of Official Economic Entomologists will hold its first annual meeting in the city of Washington, D. C., on November 12, 1889, at 11 o'clock a. m., in the Entomological rooms of the U. S. National Museum.

According to the resolution of the Association at the Toronto meeting, the annual meeting was to be held on the date and at the place where the Association of Agricultural Colleges and Experiment Stations should next meet. The date and place for the latter meeting having been fixed, the above notice is hereby given to all members of the Association of Economic Entomologists. All titles of communications to be read should be sent to the secretary as soon as possible, and those desiring enrollment as members will also please communicate with the secretary.

JOHN B. SMITH,
Rutgers College, New Brunswick, N. J.

ENTOMOLOGICAL SOCIETY OF WASHINGTON.

September 5, 1889.—The society opened with an informal discussion, in the course of which Mr. Schwarz's list of Myrmecophilous insects, read before the last meeting, was increased by the addition of two spiders belonging to the genera *Synemosyna* and *Synageles* by Dr. Marx, and a beetle (*Microrhopala melsheimeri*) by Mr. Ulke.

Mr. Schwarz read a note on the spread of *Sitones hispidulus*, a European clover insect, which has probably been recently imported. Its sudden appearance in great numbers in Washington and the likelihood of its becoming a dangerous enemy to clover in this country were discussed. Additional observations on this insect were made by Messrs. Ulke and Linell.

In a note on a new food plant of *Pieris rapæ*, Mr. Schwarz stated that he had found the eggs, larvæ, and pupæ on *Cakile americana* in July at Cape May, N. J., and Virginia Beach, Va. He questioned whether this plant, which grows abundantly all along the Atlantic coast, has not been instrumental in the spread of the Cabbage butterfly from north to south.

Mr. Schwarz exhibited an exceptionally large specimen of *Lymexylon sericorne*, calling attention to a remarkable secondary sexual character, viz, the flabellate maxillary palpi. These beetles have been found near Washington in and about decaying wood of the red oak.

C. L. MARLATT,
Acting Recording Secretary.

o

U. S. DEPARTMENT OF AGRICULTURE.
DIVISION OF ENTOMOLOGY.
PERIODICAL BULLETIN. NOVEMBER, 1889.
Vol. II. No. 5.

INSECT LIFE.

DEVOTED TO THE ECONOMY AND LIFE-HABITS OF INSECTS,
ESPECIALLY IN THEIR RELATIONS TO AGRICULTURE,
AND EDITED BY THE ENTOMOLOGIST
AND HIS ASSISTANTS.

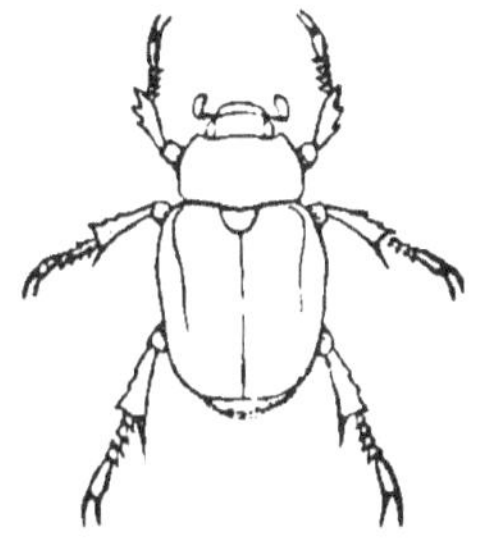

[PUBLISHED BY AUTHORITY OF THE SECRETARY OF AGRICULTURE.]

WASHINGTON:
GOVERNMENT PRINTING OFFICE.
1889.

CONTENTS

SPECIAL NOTES.

Work of the Division on the Pacific Coast.—During the past summer Prof. W. A. Henry, director of the Wisconsin Agricultural Experiment Station, was sent by the Secretary of Agriculture to the Pacific coast to report upon certain matters connected with agricultural research in that part of the country, and, incidentally, to look into the work of the agents of the Department and to ascertain the popular feeling regarding the character and importance of their work. Professor Henry has just submitted his report to Secretary Rusk, and that portion relating to the work of the Entomological Division has been referred to us. The several paragraphs reproduced therefrom in another part of this issue will have interest as the testimony of a man of established reputation as an original investigator in practical agriculture.

Food Habits of Snowy Tree-crickets.—We publish in this number an article by Miss Mary E. Murtfeldt, in which she gives the results of some detailed observations which we desired her to make, showing that these insects, particularly *Œcanthus latipennis*, Riley, are insectivorous through all of their stages, and that when deprived of animal food they invariably perish rather than partake of vegetable food. These experiments will undoubtedly interest all entomologists. We have already stated (see Fifth Report on Insects of Missouri, p. 120), that during their early life the young crickets subsist principally upon plant-lice, eggs of other insects, and even upon each other; but that as they grow larger they are often content with a vegetable diet. This statement, however, was made in reference to the common *niveus*. We may, perhaps, infer from Miss Murtfeldt's observations that *Œ. latipennis* is more strictly carnivorous than *niveus*, or possibly that the strictly carnivorous habits were exceptional for this season. Full as her observations are, they require verification by others, and in different seasons, to enable us to lay down the law that the broad-winged species is always an animal feeder.

The Chinch Bug Entomopthora.—In a number of the agricultural journals during the past summer, items have appeared referring to the experiments being conducted by Prof. F. H. Snow, of the Kansas State University, in the intentional dissemination of this disease. We notice in the October 2d issue of the Lawrence (Kan.) *Daily Journal* a long account of the success of the experiments, in which letters to Professor Snow are quoted at length and which thus bear the impress of his sanction. It is stated in this article that Professor Snow obtained some bugs killed by the Entomopthora, and mixed them with live bugs which were soon attacked and died. Repeating this experiment until he had a sufficient number of dead bugs on hand he distributed them in small batches to various farmers, agricultural experiment stations, naturalists, and others—in all, to about fifty persons. Each lot was accompanied with directions to collect ten to twenty times the number of healthy bugs and mix them with the diseased bugs for thirty-six or forty-eight hours, and then turn them loose in the field and watch closely for the result. The letters published are mainly from agriculturists and are favorable. In other words, all the published answers state that the disease seemed to have been communicated.

Ever since Prof. O. Lugger published his apparently favorable results in the same direction, something more than a year ago, we have watched the accounts of subsequent attempts, and endeavored to ascertain whether any thoroughly scientific evidence of the spread of the disease has been established. The matter is of sufficient importance to require the most careful weighing of the evidence, as the apparent evidence is so easily misconstrued, and the danger of unjustified statement and assertion is so great. In this particular article we notice that no dates are given to the letters, and that the correspondents in no way show that the supposed healthy bugs were examined critically, the evidence of life being assumed to mean healthfulness. The chief difficulty is that at the time when the disease is prevalent in one locality the same climatic and zymotic conditions are liable to—and in fact usually do—prevail through a wide extent of country, and that the disease, if it has not already appeared, may be about to appear over the whole area. This at once establishes the necessity of the most careful observations by means of check experiments. If the diseased bugs are simply placed among apparently healthy bugs and the latter subsequently become diseased, the proof of direct transmittal by contagion is but negative. If, however, healthy bugs are isolated from the imported diseased bugs and remain healthy, then a probability is established in favor of the contagion by contamination. The disease is always most prevalent in cool, wet weather, from midsummer on, when large numbers of the older bugs are naturally dying from other causes, and are probably more liable to fall victims to any scourge of this kind.

The subject is of extreme interest, and while there are reasons which would make us doubtful of any tangible and practical results following the attempted artificial spread and propagation of the disease, and which make us accept with caution the more sanguine views of men like Professors Lugger and Snow, yet there is sufficient promise of such results to justify the fullest and most careful experimentation. This will doubtless be had in the next year or so by the co-operation of the entomologists connected with the different experiment stations. The full life history of the particular Entomopthora is of extreme importance in this connection.

SOME INSECT PESTS OF THE HOUSEHOLD.

By C. V. Riley.

[*Continued from page* 108.]

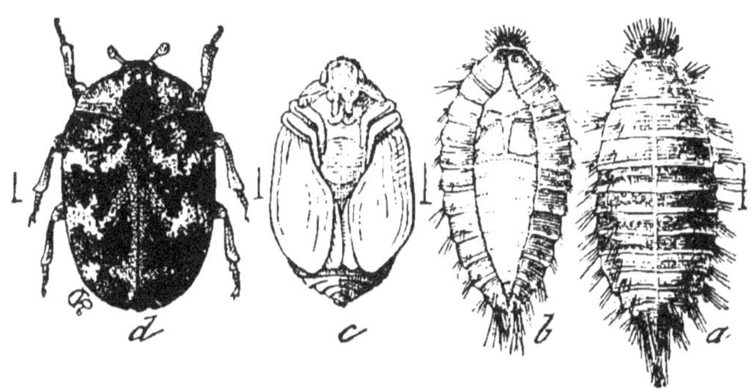

FIG. 10.—ANTHRENUS SCROPHULARIÆ: *a* larva, dorsal view; *b*, *do.*, ventral view; *c*, pupa; *d*, adult—all enlarged (after Riley).

THE CARPET BEETLE, OR SO-CALLED "BUFFALO MOTH."*

(*Anthrenus scrophulariæ* L.)

This destructive insect, the despair of the good housekeeper, has been known in the eastern United States since 1874, when newspaper articles began to appear complaining of its ravages. In 1876, it was first brought to the attention of entomologists by Prof. J. A. Lintner, of Albany, who found it at Schenectady, N. Y. Between 1874 and 1877, it had been found at various points in New Jersey, at Schenectady, Albany, Syracuse, and Buffalo, N. Y., and Boston and Cambridge, Mass. Within this range of cities it has since flourished and done great damage, but has not greatly extended. It is found, however, in all the New England States, and as far west as Illinois, and as far south as Washington, though not a troublesome pest at this last named point.

Like a number of other important insect pests it is a European species, but, although occurring commonly abroad, it is not known as a car-

---

* Reprinted substantially from *Good Housekeeping*, April 13, 1889.

pet pest, for the obvious reason that carpets are rare in most European countries. Rugs, which are frequently taken up and shaken, do not offer a comfortable dwelling-place for this insect, which is of a secreting and retiring disposition. It seems probable that the pest was imported almost simultaneously by carpet-dealers in New York and Boston, and thence shipped in goods to inland cities. Dr. H. A. Hagen, in 1875, for instance, was able to trace three-fourths of the infested carpets brought to his notice to a particular line of goods sold at a single establishment in Boston. At the present day this insect is the greatest household pest in our northeastern States. It ruins carpets and all stored woolen goods, while furs do not escape its attacks. Let us then briefly consider its life history and summarize the best remedies to be used against it.

The accompanying figures (Fig. 19 *a* to *d*), which I prepared some twelve years since, illustrate three of the stages of the insect (all except the egg), and the natural sizes are indicated by the hair lines at the side.

The larva, which is the stage in which the insect is most familiar to the housekeeper, is shown at *a* from above, and *b* from below. This is the active feeding state in which it does the damage. The full-grown larva is rather longer than the beetle and is brown in color, clothed with stiff brown hairs, which are longer around the sides than on the back, and still longer at the extremities. Both at sides and extremities they form tufts, the hinder end being furnished with three tufts of long hair, and the head with a dense bunch of shorter hair.

The quiescent state between the larva and the beetle is called the *pupa*, and is shown at *c*. It needs no furthur description, but it should be stated that the *pupa* is seldom seen, being formed within the last partly split skin of the larva.

The perfect beetle, *d*, is three-sixteenths of an inch long, nearly as broad, and broadly elliptical in outline. It draws in its legs and feigns death when disturbed. The figure will enable the housekeeper to recognize it when we explain that its colors are white, black, and scarlet. The black and white are indicated in the figure, while the red is confined to a stripe down the middle of the back, widening into projections at three intervals, and meeting the irregular white bands.

The beetles begin to appear in the Fall and continue to issue through the winter and spring. They soon pair and the females deposit their eggs, probably upon the carpet itself and not in floor-cracks, as is sometimes supposed. The eggs, with favorable temperature, soon hatch, and the larvæ grow apace, molting some six or more times. Under ordinary circumstances there is probably but one annual generation, although there may be more; but, as I have shown by experiment with related species, the larvæ are able to remain for a long time without food, in which case the growth is very slow and the number of molts great. When full grown the larva seeks to hide itself in a crack in the floor or some other convenient shelter and transforms to pupa within the larval skin. After a time the larval skin cracks along the back, showing the pupa, which later splits open and the beetle emerges.

The beetles fly to the windows during the day-time and may often be caught upon the panes. They are also to be captured outdoors upon the flowers of composite and scrophulariaceous plants, but probably do not voluntarily leave the house until their eggs have been deposited.

As already indicated in the mention of the fact that this insect is not noted as a pest in Europe, the use of rugs instead of carpets is highly to be recommended in localities where it abounds. Rugs are more often shaken out and the pest is thus discouraged.

Where carpets are used, however, and only taken up once a year at "house-cleaning," the conditions are very favorable for the insect's increase, particularly where the house-cleaning is hurriedly and carelessly done. When a house has once become infested nothing but the most energetic measures will completely rid it of the pest, and in complete riddance is the only hope, as in a year a very few individuals will so increase as to do great damage. At house-cleaning time, then, as many rooms should be bared at once as possible, and the housekeeper should go carefully over the rooms, removing all dust, and with a hand-atomizer charged with benzine should puff the liquid into all the floor-cracks and under the base-boards until every crevice has been reached. The carpets themselves, after thorough beating, should be lightly sprayed with the same substance, which will quickly evaporate, leaving no odor after a short time. The inflammability of benzine should be remembered, however, and no light should be brought near it. This done, before relaying the carpets, it will be well to pour into the cracks a moderately thick mixture of plaster of Paris and water, which soon sets and fills them with a solid substance into which the insects will not enter. Then lay around the borders of the room a width of tarred roofing-paper and afterward relay the carpets. This thorough treatment should answer in the very worst cases, and in a house so cleaned the insect will probably not regain a foot-hold during the ensuing year. Cloth-covered furniture which may have also become infested should be steamed or also treated with benzine, and chests or drawers in which infested clothing has been stored should be thoroughly sprayed.

Another method of treatment, and one which I have frequently recommended, was indicated by me in a former communication to *Good Housekeeping* in rendering my decision in the competition for best remedies for household pests. It can be used to advantage whenever the work of the larva is noticed or suspected. It consists in laying a damp cloth (an old towel or a folded sheet will do) smoothly over the suspected part of the carpet, and ironing it with a hot iron. The steam thus generated will pass through the carpet and kill all the insects immediately beneath. If not too laborious, an entire room could be treated to advantage in this way.

Camphor, pepper, tobacco, turpentine, carbolic acid, tallow, pyrethrum powder, and many other substances have been recommended from time to time, but all must be considered as inferior to the plans here just outlined.

It has been said that the best housekeepers are the most uncomfortable people in the world, always on the lookout for dirt or indications of insect pests; but if the somewhat elaborate treatment I have given is gone through with once a year, the good housekeeper may then sit down and placidly fold her hands for all the trouble *Anthrenus scrophulariæ* will give her.

THE CARNIVOROUS HABITS OF TREE CRICKETS.

By Mary E. Murtfeldt.

From observations and experiments on the Snowy Tree Crickets (*Œcanthus niveus* De Geer and *Œ. latipennis* Riley) during the past two summers I incline strongly to the opinion that they should be classed with the beneficial rather than with the injurious species. They are accused of cutting into and sipping the juices of various fruits, of severing the berries from grape clusters, and even of cutting the latter from the vines. In the process of oviposition also they are charged with the destruction of grape and raspberry canes and the twigs of various fruit trees by their punctures and by crowding the pith with their eggs. The latter charge is irrefutable; but when we consider the amount of wood that it is necessary to remove from vines and trees annually, the few twigs punctured by these insects should not be allowed to count against them. As to their injuries to growing fruit, I have never been able to verify any observations of the kind. During the present season I colonized a considerable number—mostly *Œ. latipennis*—on a portion of a grape vine and watched them at all hours of the day, without ever detecting them in the nefarious work of snipping off either berries or bunches. Nor was there any circumstantial evidence of their having done anything of the kind at night. Furthermore all my observations upon them in the rearing cage prove that at no stage of their existence can they subsist on vegetable food, either fruit or foliage. When deprived of other insects for their sustenance, they invariably perished.

Early in June of last year I had a colony of *Œ. niveus* hatch from apple twigs that had also been badly punctured by *Ceresa bubalus*. At hatching each tiny cricket left at the aperture of the bark through which it emerged the filmy pellicle in which it had been inclosed in the egg. There were about a dozen in all, and I kept them under constant observation on my writing-desk. During the day they remained almost motionless in one position, if possible concealed from light and sight on the under side or in the folds of a leaf. They were, from the first, supplied with various berries and tender leaves, but evidently never touched them for food. On the morning of the fourth day two or three were dead, and showed signs of having been nibbled by their hungry brothers. Some leaves of plum infested with a delicate species of yellow aphis were then put into the jar, but attracted no immediate attention. As twilight deepened, however, the crickets awakened to greater activity. By holding the jar against the light of the window or bringing it suddenly into the lamp-light, the little nocturnal hunters might be seen hurrying, with a furtive, darting movement over the leaves and stems, the head bent down, the antennæ stretched forward, and every sense apparently on the alert. Then the aphides provided for their food would be caught up one after another with eagerness and devoured with violent action of the mouthparts, the antennæ meanwhile playing up and down in evident expression of satisfaction. Unless I had provided very liberally not an aphis would be found in the jar the next morning, and the sluggish crickets would have every appearance of plethora. Later on in their lives, by reducing them to the point of starvation, I repeatedly made them feed in the daytime, so that I might the more distinctly observe the process, which is certainly very interesting.

The growth of the insects is rather slow. Three larval moults take place at intervals of about two weeks. In the case of those reared in the jar the habit of devouring the exuviæ was not very strictly adhered to, although in some instances it was partially eaten. Probably owing to the abundance of legitimate food there was no cannibalism, after the first few days, among my pets, and while they did not seem to seek each other's society they hunted over the same leaves and twigs without injuring each other, though it was amusing to observe the alacrity with which both would retreat if two chanced to come in contact.

Wings were not acquired until late in August, and at this time I again attempted to change their diet to fruit, grapes, plums, etc., an experiment that resulted in the death of all but three of my specimens. Those which remained fed for about two weeks longer upon oak *Tingis*, *Aphis populi*, and on a brownish aphis which infested the new shoots of grape, but neither of the two males essayed any musical performances, nor would the single female that reached its perfect state puncture any of the twigs that were furnished her, and all three died long before those out of doors had ceased to sing.

During the present summer my attention was again attracted to these insects by finding them so constantly and numerously on oaks infested with *Phylloxera rileyi*. Every leaf dotted by the aphis would have its tree cricket in addition to various smaller foes. The species most commonly seen was *Œ. latipennis*, distinguished to casual observation by its somewhat larger size and by the brilliant orange red or red and yellow dorsal stripe of the pupæ. The size and the broader wings sufficiently characterize the mature insect. A close examination reveals many less obvious distinctions between the two.

I found that one specimen of *Œcanthus* would clear the *Phylloxera* from a large oak leaf in the course of a single night when confined to one leaf. On one occasion one of the crickets ate two saw-flies which had emerged in the jar; I am not positive that it killed them, but it certainly devoured all the softer parts of the body. I have also had them feed upon various kinds of small leaf-hoppers and tingids, and am convinced that they are thoroughly and constantly carnivorous and therefore a valuable ally in reducing the numbers of our smaller insects.

LIFE HISTORY OF ONE OF THE CORN BILL-BUGS.

(*Sphenophorus ochreus* Lec.).

By F. M. Webster.

Although its method of attack is somewhat unlike, this insect is closely allied to the species figured in Vol. I, p. 186, of Insect Life, and there described as destroying sugar-cane in the Sandwich Islands.

While by no means rare, and diffused over the country from Canada to Arizona, the species under discussion has but recently come to the front as a destructive insect, the first published notice of its depredations appearing in the monthly report of the Illinois State Board of Agriculture for June, 1888. It was there accused of puncturing the stems of young corn, and feeding on the tender folded leaves in the center of the plant, near the surface of the ground, its depredations being confined to fields planted on newly-drained swamp lands, which had previously been grown up with rushes (*Scirpus*) and reeds (*Phragmites*), its supposed food plants.

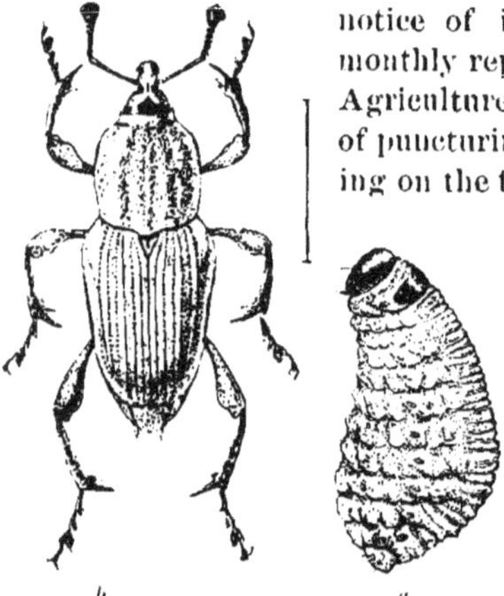

Fig. 20.—*Sphenophorus ochreus*: *a*, larva; *b*, adult—enlarged (original).

There is the best of evidence that this pest has for several years been working serious injury to the corn crop planted on recently-drained swamp lands in Indiana, hundreds of acres being thus destroyed. Until quite recently, however, I have not been able to work up the matter thoroughly enough to get an insight into the life history of the depredator, and though there are yet a few minor points lacking, still I am able to give its probable habits during the entire year.

The insect passes the winter in the adult stage, coming forth from its hiding places in spring, and feeding upon the tender portion of the stems of reeds and rushes, and later on the same parts of the young corn plants, if the field has been planted to that grain. Late in May and early in June the female burrows down into the earth and deposits her eggs in or about the bulbous roots of *Scirpus*, the roots of this plant consisting of bulbs connected by smaller slender roots. The larvæ burrow in these bulbs, which are many of them the size of an ordinary hen's egg and very hard, and transform to the adult beetle therein, appearing on the rushes, reeds, or corn in August and September, and feeding after the manner of their ancestors. The large size of the larvæ and the diminutive size of the corn at the period of oviposition, renders it very unlikely that this species will ever breed in the roots of corn, and, indeed, no trouble has been experienced after the natural flora of the land has been eradicated.

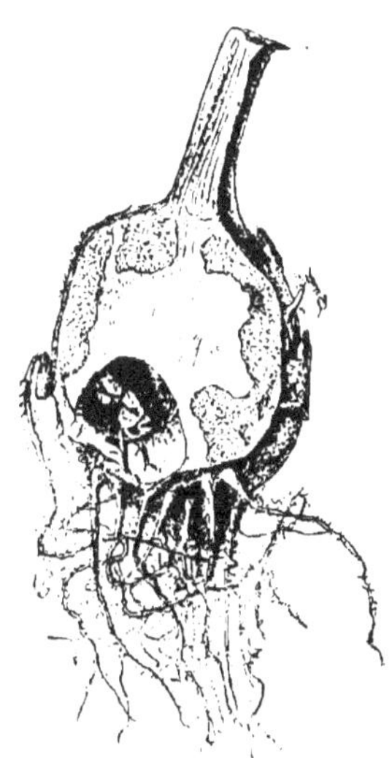

FIG. 21.—Work of *Sphenophorus ochreus* in roots of *Scirpus*—natural size (original).

At the commencement of my investigation, and after learning the habits of the larvæ, it looked as though breaking the ground in June or July and throwing roots and larvæ up to the scorching rays of a midsummer sun might destroy the pest. But having reared adults from the egg in bulbs kept in dry earth from the middle of June until the 25th of August, it would seem that little can be accomplished in that direction, and the only plan which now promises success, is to destroy all trace of their native food plants long enough before planting to corn to starve the adults, or compel them to seek other uncultivated localities. A field of 75 acres, in the vicinity of La Fayette, which was nearly a total loss this season, is being fall-plowed, and the result will be seen another year.

The egg I have not been able to identify with certainty, except as dissected from the ovaries of the female, but it is in all probability quite large, elongate, and white.

The larva is white with brown head, the latter small, the body becoming very robust posteriorly, so much so that it appears to be fully two-thirds as broad as long, and very much wrinkled. Feet wanting.

The adult is black beneath but varying in color above from pale ochreous to plumbeous and cinereous. The size varies from less than one-half to nearly three-fourths of an inch in length.

In some instances I find that the work of these snout beetles has been confused by unentomological farmers with that of a cut-worm which eats into the young corn a short distance above the roots and then works upwards in the stem, after the manner of *Gortyna nitela*, above ground. This last depredates on corn in newly-broken lands, both of native and timothy sod; but I have failed to find them in blue-grass sod. The worm is the larva of *Hadena stipata* Morr, a species not previously known to injure corn. Their method of work is such that an attacked plant never recovers, and one worm may destroy a whole hill of corn, going from one plant to another without coming to the surface. Larvæ continued to work up to the 1st of July, and the moths appeared about the 25th of that month. Serious damage has been reported in various parts of the State, specimens accompanying the complaints. I found them the most abundant in low, recently-drained, and newly-broken lands.

THE NEW ZEALAND KATIPO.

By R. Allan Wight, Auckland, New Zealand.

The Maori name of this spider is "*Katipo*," the proper name, *Latrodectus scelio* and it belongs to the family Theridiidæ. All old colonists, natives, and scientific men in New Zealand are agreed that it is dangerously poisonous. The poison is of an extraordinarily virulent nature, and fatal cases are not wanting. The habitat of this spider is strictly confined to the sea-shore. There are no other poisonous spiders known in New Zealand. Mr. A. T. Urquhart, who is a very old colonist, and our best arachnologist, says that there are species of Agalenidæ and Tegenaria, which inhabit gardens and old houses, but they have no resemblance to the Katipo. The only way to account for Mr. Taylor's statement that there are two species of Katipo is by supposing he must have taken the male and female for distinct species, and that by the term "red spider" he must have meant "spider with a red spot."

As for the mistake Dr. Wright makes in saying that there is an inland species that inhabits gardens and spins a "slight web," it is easily accounted for. Before Dr. Wright came to New Zealand the natives were more industrious (*i. e.*, they had more slaves), and they used to convey many canoe loads of sea-shells and sand far inland to form beds for the *Kumera*, or sweet potato. When I first saw these beds in deserted gardens, I was told the sea had left them there, but geological reasons did not bear the idea out, and I soon found the natives had transported them for the *Kumera* beds. My further doubts, as to whether the mollusk had been brought in them, for manure, were settled by the presence of the Katipo, which was proof of the shells having been dry and brought from above high-water mark. In these days before the Pheasant and some other birds were imported, the coast was full of the spiders, the natives used to burn the grass before sleeping on it, and when they removed the shells, large numbers of spiders were transported with them. This accounts for the majority of cases of persons bitten by Katipoes being native women and old women, because the work of the *Kumera* beds generally falls to them. And moreover the most fatal cases are in summer, because at that season the old women are constantly engaged picking off the larvæ of the Bind-weed Hawk-moth (*Sphinx convolvuli*). Removed from the shore the Katipo seems even more venomous than in its native habitat, and the Maories will burn down a house and all that is in it where a person has been bitten, if they do not find the spider, sooner than let it escape, because they think that upon this depends the recovery of the sufferer. As for the "thin web," the spiders on the beach weave the same web, and even those packed by me for Washington had done so before the box as fastened down and they were captured on the sea-shore.

The poison is generally treated as a narcotic, with stimu.ants, but it seems peculiar that no one ever seems to press a ring over the fresh-made wound to keep the poison from spreading. To give some idea of the effects and nature of the poison, I will condense a few cases out of a great many kindly sent to me, for some of the best of which I have to thank Mr. Urquhart, and, to save repetition, I may as well say that I select only those upon reliable evidence, and where the sufferer was in good health and condition at the time. also where the Katipo was recognized.

Mr. King, of Waimate: Bitten in the leg; violent pain; considerable swelling and inflammation; treatment, hot vinegar; lasted three hours; imputes cure to having been driven into great and sudden excitement from other causes.

The Rev. Mr. Mathews: Bitten on the shoulder; great pain; punctured wound; slight swelling; inflamed 3 inches around: had to walk sharply for 20 miles; dull, heavy pain for three days.

Archdeacon Clarke and party: Bitten by a brood of very young Katipoes; great irritation for some hours.

Captain Burleigh: Twice bitten, arm and shoulder· great irritation and rash *on neck and head* for some hours.

Dr. Shortland, one of our oldest and most esteemed settlers, gives cases as far back as 1842, from which I select.

Particularly powerful, healthy young man, bitten on the leg, brought in dying condition; wound like that of a large flea; intense pains all over the body for twelve hours, then violent pains in the soles of the feet; in violent perspiration all the time; all the body covered with a rash like the measles; skin all came off; ammonia injected into the wound; large dose of brandy; duration of illness not given.

Another case: Wound "like the bite of a sand-fly," intense cold and shivering for three days; great difficulty in keeping up the pulse; violent "pins and needles" all over the body; profuse perspiration; swelling not great; violent pains lasted a week; weak and depressed for "a long time after."

Case of a Maori woman bitten on the thigh whilst tending *Kumera* beds: No better means being at hand sweet oil used and recovered in three days; at first seemed to be dying.

Dr. Shortland adds that he has often placed Katipoes on his hand, of both sexes and all ages, and never was bitten, from which he infers that they do not bite unless hurt. There is a case given by Dr. Trimnell, on the authority of the resident magistrate of Nelson (Mr. Bishop), of the death of a child. The fact is beyond doubt, but the particulars are not given.

The Rev. Mr. Meek gives a very circumstantial accou of his son's case, and, as it is a curious one, I may here state that the reverend gentleman's word is beyond doubt. Dr. Mohbeer was also in attendance. It must be severely condensed. Bite on shoulder, "excruciating" pain; pain found its way down to the groins, then up the spine and into arms and chest; moaned with pain day and night; patient very strong and healthy young man. "I never saw any one in such agony in all my life"; veins very much swelled; wound punctured, ammonia injected, turnip poultice applied; "when removed, quantity of black matter exuded; when legs rubbed, quantity of inky-black fluid emitted;" severe pains lasted three or four days; depression not over after a month; treated with frequent doses of brandy.

Mr. Meek adds that fatal cases are frequent amongst the natives in his district. Besides these I have many other similar cases, and amongst those that have not been published otherwise are one of a girl and one of an old man suffering severely, much in the same way; ammonia and spirits were used and recovery took place in about a week. One case of a boy is recorded who did not recover for *many months*, and never perfectly. Several there are of women bitten in the legs and abdomen, seized with cold and shivering and suffering great pains generally for three or four days, and then taking a month to recover, and there is one of a woman which proved fatal, and another of another woman who, brought in apparently dying, was taken with the usual symptoms of narcotic poisoning, but who recovered, although treated with nothing but doses of *laudanum*. I must say, however, that from my knowledge of natives, some of the primary symptoms are not improbably caused by intense fear, as they have a terrible dread of the Katipo; but this observation would not apply to the white man.

A CATERPILLAR DAMAGING THE CORK-TREE.

We learn in a roundabout way (through the Consular Report of the Province of Victoria) that the cork-tree in the Province of Cataluna, District of Gerona, Spain, has recently been suffering from the attacks of an undetermined larva, which in a few days strips a tree of its leaves, giving it the appearance of having been burnt. The caterpillar first made its appearance in the woods of Llagostera in 1886, and has rapidly increased in numbers. It is described as being of the size of the silk worm, of a dark gray color, and covered with down, and to produce "small white butterflies."

ANOTHER STRAWBERRY SAW-FLY.*

Monostegia ignota (Nor.).†

By F. W. Malley, Champaign, Ills.

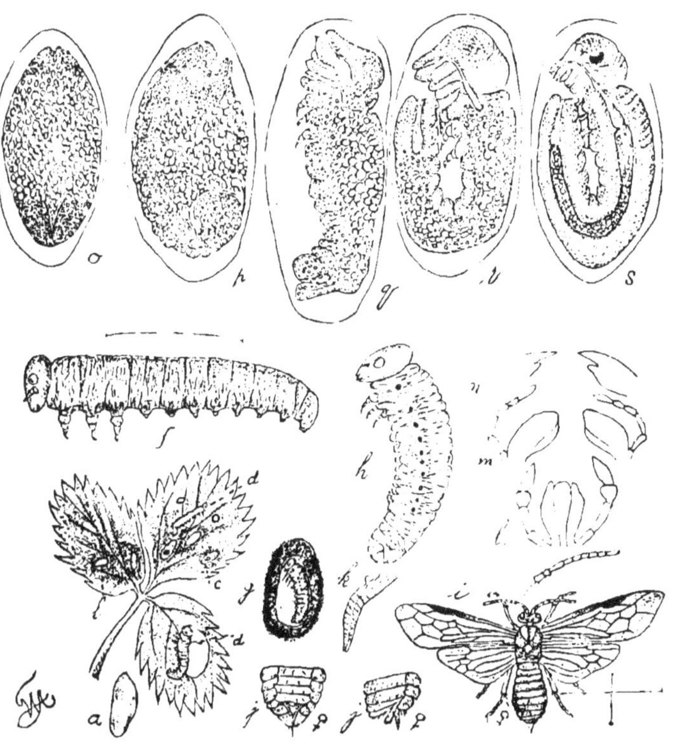

Fig. 22.—Monostegia ignota: *a*, egg; *b*, blisters containing eggs; *c*, blisters from which larvæ have issued; *d*, *d*, young larvæ; *f*, full grown larva; *g*, cocoon containing larva (natural size); *h*, shows *g* enlarged; *i*, adult female, *jj*, ventral and lateral view of abdomen of female; *k*, saw; *l*, labium and labial palpi; *m*, maxillæ and maxillary palpi; *n*, mandibles; *o*, ventral view of embryo after segmentation; *p*, embryo, lateral view, ventral surface outermost; *q*, embryo, lateral view, ventral surface curving inward; *r*, embryo, lateral view, ventral surface doubled upon itself, and showing beginnings of alimentary canal; *s*, embryo, showing alimentary canal completed, eyespots, muscles of mouth-parts, &c. (Drawn by the author.)

The adults of this species are black four-winged saw flies (*Tenthredinidæ*), about .28 inch long. By displacing the wings, characteristic dull whitish spots are seen on the back of the abdomen. However, the

*This article is a brief extract, giving the more important results of the study of the above-named species, and included in a Thesis prepared for the degree of Master of Science at the Iowa Agricultural College, Ames, Iowa.

† *Selandria ignota* (Nor.). Trans Am. Ent. Soc., I, page 257.

Monostegia ignota (Nor.). Cresson's Synopsis N. Am. Hymen., page 162.

casual observer who depends on this character alone is liable to be misled, as there is another species of saw-fly *Harpiphorus maculatus* (Nor.)‡ closely resembling it, and having similar markings on the back of the abdomen. The most certain method of distinguishing the two species is to note the number of submarginal cells in the fore wings, *M. ignota* having four, and *H. maculatus* only three.

The larvæ of *M. ignota* have infested the strawberry beds on the college grounds for several years, feeding on the leaves, and would, if numerous enough, threaten the crop. This has not been the case here, but reports from other parts of the State say that "the worms are simply ruining our plants." Drawings of this species in all its stages are given in Fig. 22.

Adult saw-flies of this new strawberry pest were found depositing eggs from the 1st to 25th of April. the period of greatest deposition being from the 10th to 20th. Adult females were captured, confined, and eggs obtained that have furnished larvæ which have been carried through all the larval stages and their habits studied in connection with observations in the field. The eggs are deposited singly on the under side of the leaf, just beneath the epidermis. In no case were the eggs found deposited in the petiole of the leaf as is said to be the habit in *H. maculatus*, but frequently alongside or in the angle between two veins; seldom more than three or four eggs are found deposited in a single leaflet.

When first deposited the eggs (Fig. 22*a*) are pure white, tapering towards both ends, one side slightly concave, the other quite convex; are .475^{mm} wide by .875^{mm} long. The point of deposition can hardly be seen at first, but the swelling of the eggs, due to the developing embryo, causes light-colored blisters of 0.5–.75 by .75–1^{mm} in size. During embryonic development the transverse diameter of the egg is doubled or trebled, lengthens about one diameter but does not thicken much. In Fig. 22 are shown a few of the more important changes taking place during the embryonic growth of the larvæ. Its embryology has been traced in detail, but only a suggestive outline can here be given.

First. Segmentation of the yolk and partial differentiation of the anterior and posterior embryo lobes. Fig. 22*o*.

Second. Division of the anterior lobes and the differentiation of the ventral surface which at this stage occupies the outer circumference.

Third. The folding of the embryo upon its ventral surface and the differentiation of the two lower anterior lobes. Fig. 22*r*.

Fourth. Beginnings of the alimentary canal; anteriorly, the œsophagus; posteriorly, the rectum and colon. Fig. 22*r*.

Fifth. Continued development, forming the remainder of the alimentary canal; appearance of the eye-spots and muscles of the head and mouth parts. Fig. 22*s*.

---

‡ *Emphytus maculatus* (Nor.). Bost. Proc., VIII, 1861, pages 157, 158. Trans. Am. Ent. Soc., I, page 232.

Harpiphorus maculatus (Nor.). Cresson's Synopsis N. Am. Hymen., page 160.

Sixth. Division of the outer wall into distinct segments and hatching of the embryo.

When ready to issue the young larvæ eat a small hole through the inclosing epidermis and emerge. At first they are slender 22-footed slugs; bodies white, translucent, much wrinkled; granular; 2–2.3^{mm} long; upper part of the head cream colored; claws of the pectoral legs eyes, labrum, mandibles, brown; remaining mouth parts, whitish brown; ring around the eyes black. The young worms begin their ravages at once, eating small holes through the leaves. After feeding six or seven days they pass through the first molt, are about one-half larger, the dorsal and lateral surfaces yellowish green, ventral surface pale. At each of the three succeeding molts, all of which occur within the next eight or ten days, the color is of a deeper green. The larvæ when full grown are between .55 and .65 inch long. Head and mouth parts, claws, and first joints of the pectoral legs are of a more distinct brown; body a beautiful deep green, much wrinkled, with one dorsal and two lateral obscure blackish stripes. Anterior segments but slightly larger than the posterior ones.

By the 1st of May the worms begin maturing and entering the earth, and by about the 1st of June all have entered the ground. Entering the earth to the depth of an inch or so, a frail earthen cocoon is formed, on the inside of which there is a thin silken lining. Larvæ in cocoons formed May 1 have shrunk to one-half of their original length, but up to date (August 22) have not pupated. The shrunken larvæ still retain their green color, but the stripes are more distinct, due no doubt to the fact that they have been crowded into about one-half their original length.

As yet no second brood has been obtained. However, if the larvæ should pupate and issue any time in August or forepart of September there would yet be time enough for oviposition, hatching of eggs, and maturing of larvæ before frost would interfere. This que tion will soon be determ'ned, and, indeed, will prove to be an interesting one, since there has been much confusion and controversy as to the number of broods of the old pest, *H. maculatus.* It seems barely possible that the two species have infested the same beds and have been confused with each other in some of the observations made. It is hoped that, with our present knowledge of the species, a further study of them in their respective localities will determine questionable points. In this locality there is slight evidence that both species are present. The evidence is very slight, however, in that no adults of *H. maculatus* were captured, and but one immature larva in one hundred alcoholic spec mens bears the unmistakable markings on the head which characterize the larvæ of that species. (See Fig. 23 for comparison of the heads of the larvæ of *M. ignota* and *H. maculatus.*)

Numerous specimens of the adults of *M. ignota* were examined, neuration of the wings especially noted, and no variation found. Some slight variation in the size of adults and depth of coloring of the legs was discovered. It was also found that the description of *Monostegia obscurata* Cress. applied very closely, and accordingly specimens of adults were sent to Mr. E. T. Cresson, Philadelphia, Pa., for comparison and determination. His reply was that "your specimens seem to agree with *Selandria ignota* Nor. As to *S. obscurata*, I think on an examination of more abundant material it will prove to be the same as *ignota*." Also, "I would not like to say that your *Selandria* is a new species without an examination of a larger series of *ignota* than we have in our collection." Hence the best that can be done at present is to say that the species is *Monostegia ignota* (Nor.).

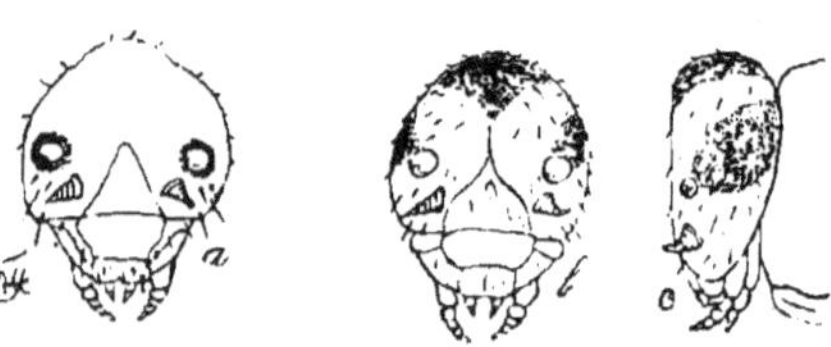

FIG. 23.—*a*, head of larva of *Monostegia ignota* (Nor.); *b* and *c*, front and side view of head of *Harpiphorus maculatus*. (Drawn by the author.)

As to the geographical distribution of this species, little can be said just now. Among the specimens from which Norton described *M. ignota* was one from Illinois, and *M. obscurata* was described by Cresson from material collected in Colorado.

The period of greatest abundance of the worms is from about the 25th of April to 5th of May, though they begin appearing about the middle of April. Hence most of the worms have hatched before the vines are well in bloom, feed, mature, and again disappear by the last of May, before much fruit has ripened. It will therefore be perfectly safe to apply any of the arsenical poisons, with great efficiency, as early as April 20 to 25, and with comparative safety about the 1st of May.

Of the insect enemies preying upon the worms, *Coriscus ferus* was found to be very beneficial indeed. No parasites have as yet been reared.

In conclusion, I could not honorably fail to give due credit to and acknowledge the needed guidance and instruction of my kind and worthy instructor, Prof. Herbert Osborn, without whose suggestions and friendly criticisms of the work while in progress it must have been less accurate and complete. To Prof. F. M. Webster for "genuine *H. maculatus* larvæ," and to Mr. E. T. Cresson for determination of specimens sent him, I wish to tender my sincere thanks.

PACIFIC COAST WORK OF THE DIVISION OF ENTOMOLOGY.*

By Prof. W. A. Henry, Madison, Wis.

Several days were spent in company with Mr Coquillett, of Los Angeles, in visiting fruit farms at various points in that vicinity and noting the destructive effects of the white scale and red scale, and the efforts in progress to check their ravages. At Orange, in Orange County, the destruction of citrus trees by the red scale has been great, and only a few more years would suffice to leave that section without any such trees if remedies to check the destruction had not been put in operation the present season. The Santa Anna vine disease has destroyed most of the grape-vines, and every orange orchard shows in a greater or less degree the attacks from the red scale. Every stage from thriftness to death itself was noted. In some orchards only the yellow-spotted character of the leaves showed the presence of the scale just beginning its fatal work; in others the ends of the branches were leafless and dead, the interior portions of the top yet carrying leaves, though little or no fruit. Still other orchards had but the stumps of the orange trees left, all of the limbs to the size of one's arm having been killed by the scale and removed with the saw. From these stumps green shoots showed signs of life, and if care was given promised to renew the value of the orchard. The careless treatment of the land showed as plainly as the trees themselves the discouragement of the people.

Usually an orange orchard in southern California receives the best of care, and the carefully-tilled soil lying loose without a weed in sight and as level as a floor delights the lover of thrift and good tillage. In many orchards weeds cover the ground and form thickets 5 or 6 feet high, so dense that a man can hardly get through them. The dead and dying orange trees among these weeds stand like monuments marking the deadly march of the insidious, insignificant, but wonderfully fatal scale. In company with Mr. Hamilton we visited the orchard in which Mr. Coquillett was conducting spraying experiments with resin-soap solutions. I will refer to these experiments again later on. We also visited many other groves in all stages of thrift and decay, from those bearing heavy crops to those with nothing but the stumps standing. It was very apparent that those who had fought this scale the most vigorously, even though very imperfectly heretofore, are coming out the best in the end, and that those who early gave up and neglected their orchards will suffer far the most heavily. One orchard near the California Central Railroad station, at Orange, of 850 seedling trees, showed the ends of the branches already dead, and there were scales enough on the leaves to so reduce the vitality of the trees the present season that

*Extracted from a report submitted to the Secretary of Agriculture (see the special notes in this number, p. 125).

by next spring most of the trees would have to be cut back to mere stumps. A few weeks before our visit the owner plucked up courage and sprayed the trees with the resin-soap compound in a very thorough and systematic manner, the whole operation costing for the 850 trees $200. We spent an hour in observing the effects of the wash, and estimated that more than 95 per cent. of the scale had been destroyed, while not one leaf in ten thousand had been injured in the least by the wash. Mr. Hamilton informed us that resin was now being brought to Orange by the car-load for the purpose of making the resin soap. For the first time people are really taking heart, and were going at their orchards in dead earnest to make them profitable once more. The plow had been set to work to reduce the weeds and bring back the old-time thrift in many cases, though some orchards were yet as desolate as ever. Before speaking further in regard to remedies for the red scale, the destruction of the cottony-cushion scale should be noted.

In studying this insect we first visited the place of Mr. William Niles, in Los Angeles, where the "lady-bug" (*Vedalia cardinalis*) was being propagated by the county insect commission for dissemination among the orange groves infested with the cottony cushion or white scale. We found five orange trees standing about 18 feet high inclosed by walls of cheap muslin supported by a light frame-work of wood. The orange trees inside this canvas covering had originally been covered with the white scale, but the Vedalia which had been placed on these trees were rapidly consuming the last of the pests. Entering one of these canvas houses we found the Vedalia, both larvæ and adults, busy consuming the scale; here and there on the canvas were the beetles endeavoring to escape to other trees. These insectaries were in charge of Mr. Kircheval, one of the county insect commissioners, who kept a record of the distribution of the beetle. It was indeed a most interesting sight to see the people come, singly and in groupes, with pill-boxes, spool-cotton boxes, or some sort of receptacle in which to place the Vedalias. On application they were allowed within the insectaries and each was permitted to help himself to the beetles, which were placed in the boxes and carried away to be placed on trees and vines infested by the white scale at their homes. Mr. Kircheval kept a record of the parties and the number of beetles carried off. The number coming for the Vedalia was surprisingly large—scores in a day—and each secured at least a few of the helpful beetles. That the supply should hold out under such a drain was a great surprise, and speaks better than words the rapidity with which the Vedalia multiplies when there are scale insects enough to nurture the young.

We visited other points: Lamanda Park, Santa Anita, Sierra Madre Villa, Pasadena, etc. At the time of our visit to Sierra Madre Villa, August 23, the white scale had already disappeared before the Vedalia. At Santa Anita, the ranch of Mr. E. J. Baldwin, we examined a 350-acre orange orchard, in which the white scale had started a most destructive course. Mr. Baldwin began an equally vigorous defense, going personally into the orchard and superintending the work of fighting the white scale. There was every sign, however, that the scale was going to be the victor. Some of the trees were almost ruined by the severity of the application made. Happily, before the pest had gone far in its work, the Vedalia was heard from, and Mr. Baldwin secured a number, which were placed in the hands of one man specially detailed to look after its welfare. This individual spent six weeks in colonizing the Vedalia in various parts of the orchard. After that time a careful examination showed the superintendent that the work of colonizing was so complete that further effort in that line was unprofitable. It was predicted at the time of our visit that a few weeks more would leave the orchard entirely free from the white scale. At Chapman's we found the citrus orchard, formerly so famous, entering the death stages from the white scale, which was now fortunately being so effectually checked. At Pasadena, on the grounds of Prof. Ezra Carr, we found that some of the shrubbery had been seriously injured by the white scale, but thanks to the Vedalia, not a single pest was alive at the time of our visit. Mrs. Jennie Carr pronounced the Vedalia "a miracle in entomology."

A word in relation to the grand work of the Department in the introduction of this one predaceous insect. Without doubt it is the best stroke ever made by the Agricultural Department at Washington. Doubtless other efforts have been productive of greater good, but they were of such character that the people could not clearly see and appreciate the benefits, so that the Department did not receive the credit it deserved. Here is the finest illustration possible of the value of the Department to give people aid in time of distress. And the distress was very great indeed; of all scale pests the white scale seems the most difficult to cope with, and had no remedy been found it would probably have destroyed the citrus industry of the State, for its spreading to every grove would probably be only a matter of time. It was the Department of Agriculture at Washington which introduced the Washington navel orange into south California, and the Department has now given an effective remedy for the worst scale insect. The people will not soon forget these beneficial acts.

At Sierra Madre Villa, in the orchard of W. D. Cogswell, a chalcid fly was found to be parasitic on what is there called the red scale. In company with the county insect commissioners and Mr. Coquillett we visited this orchard. It is quite evident that the so-called red scale of this orchard has been greatly checked and may yet be entirely destroyed by the chalcid. At E. J. Baldwin's the commission also found the same scale being destroyed by the same parasite. In this case each parasite destroys but a single insect, and the commissioners were very solicitous and also skeptical as to its ability to rapidly destroy the red scale. Furthermore, they questioned whether the chalcid would destroy the true red scale, as they did not believe that the scale on the orchards mentioned was identical with that about Orange. The Vedalia has brought the people a simple, rapid, and effective remedy for the white scale, and the commission was very solicitous lest the people should give up the use of washes for the red scale and wait for the spread of the chalcid parasite. If the parasite should multiply but slowly, which seems probable, the red scale would be enabled to spread and do great harm before overtaken. It is of the highest importance, at this time, that a constant fight against this scale should be made, and there should be no halting, even if imperfect means of holding the pest in check are only at hand.

I carefully examined the experiments conducted by Mr. Coquillett with resin washes, and consider that he has used excellent judgment in the manner in which he has conducted them. I think he plans his spraying experiments carefully and with good judgment, and carries them through with thoroughness to the end.

It seems to me of the highest importance that experiments with washes be prosecuted, and that the great advance of the last year be followed up vigorously. With the resin washes for the red scale, and the Vedalia for the white scale, the citrus industry will again move forward and people have the confidence in it of former days.

CICINDELA LIMBATA Say.

By Lawrence Bruner.

Recently, while walking over the sand-hills lying to the south of the Dismal River in Thomas County, Nebr., I found a few specimens of Say's *Cicindela limbata*. This very interesting beetle is, so far as I am at present aware, confined to the sand-hill region of central and northern Nebraska. In this region it is also restricted in its distribution to certain peculiar localities.

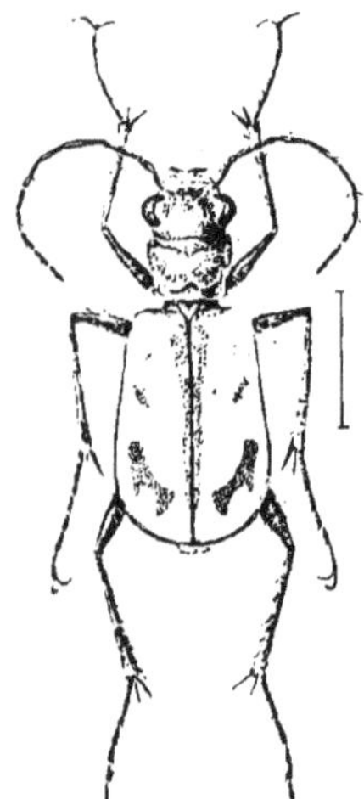

Fig. 24. — Cicindela limbata—enlarged. (Original.)

The species was first observed by me on the afternoon of the 11th of July, at about 6 o'clock p. m., while walking through a large "blow-out," two sides of which were almost perpendicular, while the others were sloping and composed entirely of loose white sand. Three of the beetles were taken, two of them *in coitu*.

The next day until 2 p. m. was spent in looking for more of them. In all two dozen specimens were taken—every one of them in "blow-outs" of a similar type to that in which the first were seen, *i. e.*, with one or more nearly perpendicular sides and in which little or no vegetation occurred.

In habits this tiger beetle resembles *Cicindela lecontei* so ar as the run and flight are concerned. It is not so active an insect as some others of the sandy-soil frequenters, nor does it run or fly as quickly as they, no doubt depending more or less upon its color for protection. But little variation is noticeable in the markings of the different individuals; in fact, the few specimens taken tend to show a much more pertinent adherence to a typical pattern in this respect than is usually the case with the species of the genus. *C. lepida*, *C. formosa*, *C. venusta*, and *C. punctulata* were also taken in similar places. Of these latter the *C. lepida* alone was restricted to the bare white sands of blow-outs of considerable size, while the other three were also to be encountered away from these locations indiscriminately among the sand-hills.

The larval burrows of *limbata* are evidently placed in the somewhat solid upright banks upon the sides of the larger "blow-outs." Of these burrows none were seen that could be definitely said to belong to this beetle, although some search was made for them. Evidently the season was too early for them. My reason for thinking that the larvæ are to be found here is that the parent beetles are most frequently seen about these banks when *in coitu*; and also because the material composing the walls of these "blow-outs" is too fragile and loose at every other point save here for sustaining the burrows.

EXTRACTS FROM CORRESPONDENCE.

Injury by Xyleborus dispar in England.

The beetle which is considered one of the rarest of the British Coleoptera, the *Xyleborus dispar* Fabr. (formerly known as *Bostrichus* or *Apate*) has appeared in such great numbers in plum wood in the fruit grounds at Toddington, near Cheltenham, as to be doing very serious injury. I found on anatomizing the injured small branches that one of the galleries which the horde of beetles (packed as closely as they can be) forms or enlarges passes about two-thirds round in the wood more or less deeply beneath the bark, whilst another of the tunnels, likewise occupied with its closely-packed procession of beetles, was in possession of about 2 inches of pith, so that the rapid destructoin of the tree was fully accounted for. The attack appears, as far as I see, to disappear usually very rapidly; but I am advising burning to make sure. This disappearance I conjecture may arise from excessive rarity of the male of this species; amongst about 60 females which I extracted from the tunnels I found only one male. * * * —[Eleanor A. Ormerod, Torrington House, St. Albans, England, August 22, 1889.

Insect Pests in Colorado in 1889.

Here is a short summary of the insect pests in Colorado for 1889, so far as they have come under my notice:

There has been considerable immunity from the attacks of insects in Colorado this year, so far as I can learn. Neither *Eurycreon* nor *Caloptenus* have molested in this section, at any rate. Warble-flies (*Hypoderma bovis* DeG.) have been complained of in some parts, and *Chrysops* and *Culex* have been troublesome as usual in Wet Mountain Valley. Mr. H. G. Smith, jr., has sent *Anthomyia brassicæ* from Denver, with a note that it injured turnips; and earlier in the year the same crop at Denver was reported to suffer from the attacks of *Phyllotreta pusilla* Horn, specimens of which were sent. In Wet Mountain Valley *P. pusilla* is common, but seems to confine itself to wild plants. *Aphis brassicæ* has been complained of in some parts as injuring cabbages. *Carpocapsa pomonella* is apparently well established and destructive in southern Colorado, to judge from apples in the market, though it is not so injurious here as it has been in other parts of America.

Of Orthoptera, Mr. W. P. Lowe has sent *Diapheromera* from Pueblo County, but it seems to be rare. The sparrow-hawk does excellent service in keeping down Orthoptera. One shot on the Sangre de Cristo Range had its stomach full of what appeared to be *Camnula pellucida* var. *obiona*, and one from Pueblo County had remains of *Anabrus* in its stomach.

Musca domestica ranks as a first-class nuisance in Wet Mountain Valley, swarming in houses and getting into everything. A blow-fly (*Lucilia*) is a great pest in the earlier part of the year.

Heliothis armigera is abundant in Custer County, but apparently harmless.

Agrotis saucia is also common.

A box of crackers from Denver was found badly infested with small larvæ, almost certainly of a species of *Ephestia*.—[Theo. D. A. Cockerell, Westcliffe, Custer County, Colo., October 2, 1889.

Spraying for Black Scale in California.

* * * Since I wrote you last I have taken up the study of scale insects—not very scientifically, but in an extremely practical and disagreeable way—that is, experimenting with a solution for their destruction and the disinfecting of orchards on contract. I have already sprayed and contracted to spray over half the orchards in the country, and people have actually begun to consider me an authority on "bugs." * * * I only took up the subject with the view of clearing our own orchard, *and I did it.* There is a little satisfaction in clearing black scales off an orchard so black and covered with scales that you can scarcely see wood or leaves, and the fruit so smutty it has to be scrubbed before you can market it—and that is the condition of some of the orchards close to the sea. I have a good machine, one I built myself (I could not buy one large enough), but am not satisfied with it. I believe a small petroleum engine might be made to do the pumping cheaper and better than a man. It is a powerful force-pump, mounted on a sheet-iron tank, on a wagon, and has four sprays on the ends of 8-foot rods; so it takes six men to keep it going. * * *—[Harvey C. Stiles, Helix, San Diego County, Cal., September 26, 1889.

The Australian Ladybird in New Zealand.

I have been searching closely in places where Icerya were and where they were devoured by swarms of these beetles, and I can not find a trace of them in any stage. The specimens I brought here with me refused every scale insect I could find and every aphide, and they all died of starvation except those I turned loose, and these I can not find now. At first I thought they would eat *C. cacti*, but they merely tasted it. It is very likely these beetles came over from Australia in the ovisacs, as we import a good many trees from thence; and, if so, it would account for two things: First (as but few would come in that way), for their being so long in gaining head against their prey, and, second, for their existence in districts only, many districts having imported independently of the others. But what is occupying me just now is not being able to find what else they feed upon, and Mr. Koebele now repeats what he told me in Auckland, that he found them feeding upon Icerya and nothing else. I think you will find that Icerya will not be easily eradicated altogether, and will occasionally break out again in places, and if these beetles, whose extraordinary rapacity can not long be supported by Icerya, can eat nothing else, they must die out, and then the pest will again gain head. I would, therefore, take great care of your Lestophonus, which, although slow, is sure, and has done untold good in Australia, besides having the advantage of living upon other hosts. These, together with your own native parasites, may yet be of great service to you, and quite able to keep Icerya in check after the beetle has reduced it to a minimum.—[R. Allan Wight, Te Komata, Paeroa, Auckland, New Zealand.

A Museum Pest attacking Horn Spoons.

I mail you to-day an insect which is destroying our horn scoops, spoons, combs, etc., in the drug store. I also inclose a piece of horn scoop upon which the insect has been feeding. Will you be so kind as to inform me what the insect is, by what means to get rid of it, etc.? * * * —[J. P. Brashears, Fort Worth, Tex., October 3, 1889.

REPLY.— * * * The insect in question is *Anthrenus varius*, one of the common museum pests. This insect feeds upon almost any dry animal substance, museums being especially subject to its attacks. It has also been reported as feeding on whalebone. Bisulphide of carbon will destroy it in all stages, and if your goods are in a comparatively tight show-case or box, this substance can be used easily and with good results. The odor of camphor or naphthaline will probably prevent their attacking non-infested material, and these substances are, especially the latter, constantly being employed in museums for this purpose.—[October 9, 1889.]

Some Notes from England.

C. destructor has certainly spread over a more extended area this year, so far as can be judged by reports, and I only note those (except from qualified observers) that are accompanied by corroborative specimens. But, withal, the injury does not seem (excepting in the case of one field) to be of importance.

Specimens of what I think may prove to be attack of *Diplosis equestris* Wagner, have been sent me, but the very peculiar "saddle-like" growths consequent on the larval injuries were on barley stems *not* wheat, so until we make some advance or rear the imago I can not feel sure that we have the true "Sattlemarke."

The *Pulvinaria ribesiæ* Signoret, is a newly observed trouble to *Ribes* in this country so far as identification goes, but appears to have been here in one, possibly two localities for a few years. * * *—[Eleanor A. Ormerod, St. Albans, England, September, 1889.

A Note on the Lady-bird Parasite.

To-day while re-reading some of the articles which appeared in "*Insect Life*" (Vol. I), I was interested in the one on page 101 *et seq.*, entitled "A Lady-bird Parasite," as it called to mind a similar observation made by me in 1885. During a part of that year I assisted Professor Forbes, and transmitted to him in my report in substance the following: "May 29: Attached to the underside of a clover leaf was observed a small cocoon, possibly one-fourth of an inch long. Upon this cocoon a Lady-bird (*Megilla maculata* De G.) was found apparently watching the cocoon. The beetle remained in the same position until death came to her relief a day or two after the imago appeared, which occurred June 5." As this was the same beetle observed by you, and as your illustration represents the appearance exactly as observed by me I presume that the observations were parallel, though I did not carefully study the parasite, having only a very poor microscope.—[F. W. Goding, Rutland, Ill., October 10, 1889.

Nezara puncturing Bean Buds.

Yours received relating to the insect described as *Nezara hilaris.* I have closely observed the habits of it since, and think I can not be mistaken when I say that this particular insect has abandoned its predatory habits and taken to a vegetable diet.

I send herewith another batch, thinking you will be able to tell by dissecting that he is filled with the juices of the bean; you can distinctly smell the bean odor. In addition you will discover a sucker, which he keeps closed against his under body. In his operation of feeding he lowers it with an apparent joint like the elbow; this is straightened as it is inserted into either the base of the bean flower or into the tender pods. While working on the young buds or flowers he goes from one to another, not satisfying himself until he has exhausted a good many; he seems very greedy. * * *—[George G. Curtiss, Brooks, Stafford County, Va., September 30, 1889.

REPLY.—* * * The insect in question is a common plant bug, probably *Nezara hilaris*. The species can not be certainly determined in the absence of adult specimens. This insect is ordinarily predaceous and feeds on other insects, but it is also known to feed on the juices of plants. It has been found puncturing the pods of the Trumpet Creeper in a manner very similar to your description of its work on bean pods. It may, therefore, be a question whether the damage it thus causes to plants is not greater than the benefit derived from its feeding on and destroying the larvæ of other insects. An application of kerosene emulsion will probably be effective against it.—[October 1, 1889.]

Beetles in a Pin-cushion.

I send you by mail a sample of the bugs found in the pin-cushion at Phenix. The facts were as stated in the paper which you read. The bug is one of the smallest, but the only one which I could get.—[D. O. King, M. D., Pontiac, R. I., July 8, 1889, to H. R. Storer, M. D., Newport, R. I.

"In the Phenix House a guest was entertained the other night who in the morning averred that the room he occupied was haunted. This he told the host, who made a cursory answer. But the guest went on to explain how the haunts and bogies plagued him. He said they were scratching their hands over everything around the dressing-case, and kept him awake the greater part of the night. The host and hostess went to investigate. Sure enough, there was the scratching, sharp noise, without ceasing. It seemed to come from a large toilet cushion on the dressing-case, but there was not a break or crack in its satin covering. So certainly did the noise proceed from the interior of the cushion that it was ripped open, and from its inner covering of cotton cloth the filling was shaken. It was filled with coarse shorts, such as used in stables for feed, and from this tumbled and rolled dozens of black bugs, known as 'snapping bugs' of an inch long. These were what had made the scratching noises as they crawled about against the lining of the cushion. The cushion had been made about four years ago, and as it had never been opened the insects must have germinated in the grain."—[*Providence Journal*, July 3, 1889.

The inclosed history, with specimens (living), may interest you. I was sufficiently amused by the newspaper jotting to request my friend, Dr. King, of Warwick, who lives in the locality indicated, to look the matter up. He seemed to think, with the people in question, that the case was one of prolonged gestation and artificial delivery, while I am inclined to think that there must have been some minute opening in the cushion which escaped notice.—[H. R. Storer, president Newport Natural History Society, Newport, R. I., July 12, 1889.

REPLY.—The specimen which you send is the adult beetle of the common mealworm (*Tenebrio molitor*). The story as given in the newspaper clipping is not unreasonable and the shorts used as filling for the pin-cushion may have contained the eggs of the beetle when the cushion was originally made. The larvæ developed in the shorts and transformed to beetles, and there is no reason why several generations might not have lived in the cushion, providing there was sufficient food.—[July 19, 1889.]

Texan Digger Wasp.

I send you to-day box with specimens by mail. One of them is a large insect of, I presume, the Hornet species which I received in a damaged condition.—[J. F. Wielandy, Springer, N. Mex., September 26, 1889.

REPLY.— * * * The specimens last sent are the large Texan Digger Wasp or Hornet (*Pepsis formosa*), one of the largest and most showy of the fossorial or sand wasps. It is commonly known as the Tarantula-killer and is reported to attack that enormous spider, *Mygale hentzii*, stinging it and inserting an egg in its body, after which the spider is introduced into a hole or nest in the sand some 5 inches deep. The wasps emerge in June and are common until Fall. It is a southwestern species but occurs as far north and east as central Kansas at least. There is a full illustrated account of it in Vol. I of the old *American Entomologist*. * * *.—[October 2, 1889.]

Abundance of Datana angusii.

I wish to call the attention of the Department to a new and very destructive species of caterpillar—at least new to us. As nearly as I can ascertain, this caterpillar made its appearance here about three years ago, but perhaps longer. It prefers for its abode hickory and walnut shade trees in pasture fields, meadows, and grain fields; and I believe also apple trees. When they have once taken possession of a tree they never quit it so long as the semblance of a green leaf remains upon that tree. They leave not a skeleton leaf, as does the well-known orchard caterpillar. The petiole and a portion of the axis or midvein is all that remains to show that a leaf once existed there, whether simple or compound. I have been observing this pest with a view to ascertain some of its characteristics and habits, and experimenting as to the most effective means for its destruction. It is distinct from the web caterpillar, in that it is large and more voracious. It does not spin a web, nor does it draw the leaves together, but devours them bodily, net, veins, and all, except as above stated, the petiole and the heavier portion of the axis.

General Appearance.—In color it is dark purple, with four well marked white lines on each side; the lowest being the heaviest, and the second from below being lightest, while the two uppermost lines are of about uniform size, and about half as wide as the lowest. Its head is black, and armed with powerful mandibles. It is partially covered with thin rows of white hair.

Size when full grown.—When full grown it is probably 6 centimeters in length and 6 millimeters in diameter. Its body is now a little darker and its hair a little longer and whiter than in the young of 2 centimeters length.

Habits observed in feeding and Manner of Repose.—These caterpillars travel up the tree from the ground, single file, each one leaving a thread behind it, which every other carefully follows doing likewise until all camp upon the same leaf until it is literally covered, and which they do not leave until there remains only a melancholy ruin, not having the semblance of a leaf, when they turn and follow back the thread to a point a foot or often several feet above the ground, where they pile upon each other like bees for repose, to the number of many thousands, and the bulk of a pint or more. They hold fast by the middle, turning the two extremities out. Several such bunches are often seen upon the body of the same tree. Just beneath the limb as it leaves the trunk of the tree is a favorite resting place of these very peculiar organisms. When the leaves of one branch are devoured (and they usually select the lowest branches first), one of them strikes out in a new direction, laying his thread, which all the rest follow till they arrive in pastures new upon another branch; and so they go from branch to branch till not the semblance of a green leaf remains upon the tree. They have now completed their work—verified the teaching of Malthus. They retire to their camps for repose, where they perish for lack of more leaves to devour. Here their remains are bound together by an almost imperceptible fiber or thread, and are not dislodged by the peltings of hail or by winter storms. The crops of several years past are distinctly seen upon the trunks of the trees they have stripped of their foliage and of their glory.

These caterpillars are rapidly increasing in numbers. In an adjoining county an entire orchard is reported as destitute of leaves as in midwinter. I have seen no account of this new pest; probably it has not been reported. I have never seen this caterpillar elsewhere, and not here till this year. It travels from one tree to another; some trees in the same field may escape for several years, but they will reach every tree in time.

Means applied for its Destruction.—Coal oil is promptly fatal to this pest. A few drops poured onto some of these colonies is speedily fatal, especially if ignited. But this is a very slow means of destruction and dangerous to the life of the tree. I will try carbolic acid as less injurious to the tree.—[A. D. Binkard, Peru, Miami County, Ind., July 23, 1889.]

REPLY.— * * * The insect is one of the rarer of the forest caterpillars, and it consequently has been given no common name. Its scientific designation is *Datana angusii*. The caterpillar has long been known to us, and has been reared to the imago. It is a rather large, brown moth inconspicuously marked. The facts which you give concerning its extraordinary abundance with you are very interesting, and unless you have objections we shall be glad to publish a note on the subject. From your account these caterpillars will be very easy to kill by spraying with an arsenical mixture.—[August 12, 1889.]

STEPS TOWARDS A REVISION OF CHAMBERS' INDEX, WITH NOTES AND DESCRIPTIONS OF NEW SPECIES.

By LORD WALSINGHAM.

[*Continued from page 120 of Vol. II.*]

CRYPTOLECHIA Z. AND ITS ALLIES.

The following tabulation may enable students more easily to assort and recognize the species belonging to the genera noticed in this paper. It must be taken to apply especially to the North American forms as it is obvious that in dealing with a more extended geographical series many other divisions and subdivisions would be required.

A. Veins 7 and 8 of the fore-wings from a common stem; 6 and 7 of the hind-wings separate and parallel.
   1. Veins 2 and 3 of the fore-wings adjacent at origin, =*Cryptolechia* Z.
   2. Veins 2 and 3 of the fore-wings remote at origin.=*Machimia* Clem.

B. Veins 7 and 8 of the fore-wings separate; 6 and 7 of the hind-wings from a common stem.
   1. Veins 2 and 3 of the fore-wings separate. =*Stenoma* Z., and *Menesta* Clem.
   2. Veins 2 and 3 of the fore-wings from a point or from a common stem; 4 very close, =*Ide* Chamb.

CRYPTOLECHIA Z.

=*Psilocorsis*, Clem.
=*Hagno*, Chamb.

Chambers (Bull. U. S. G. G. Surv., IV, 84) rightly places his genus *Hagno* (equivalent to *Psilocorsis*, Clem.) in a section of the genus *Cryptolechia*. It is indeed similar in neuration, palpi, and antennæ to *Cryptolechia straminella*, a South African species described by Zeller (Handl. Kong. Svensk. Ak., 1852, 107), as the type of the genus then created. Zeller subsequently (Hor. Soc. Ent. Ross., XIII, 259) removed *straminella* to *Machimia*, adopting Clemens' genus for a large section of the then extended genus *Cryptolechia*, but *straminella* differs from *Machimia tentoriferella* Clem. in the proximity of veins 2 and 3 of the fore-wings, as in the case of *Psilocorsis*, which was distinctly pointed out by Clemens (Proc. Ac. Nat. Sc., Phil., XII, 212). Thus if we retain the name *Cryptolechia* for the original type *straminella*, and those species which correspond to it, *Psilocorsis* must be dropped as a synonym and *Machimia* be retained for *eutoriferella* Clem., and others in which vein 2 of the fore-wings is remote from vein 3.

Cryptolechia quercicella Clem.

Psilocorsis quercicella Clem.
=*Depressaria cryptolechiella* Chamb.
=*Cryptolechia cressonella* Chamb.
=*Hagno faginella* Chamb.
=*Psilocorsis dubitatella* Z.

Chambers himself (Bull. U. S. G. G. Surv., IV, 86) recognized the probability that the first four of these forms would turn out to be varieties of one species, although a specimen of his *C. cressonella* was sent for comparison with Clemens' type, at Philadelphia, with the following result: "Mr. Cresson informs me that it is not *Psilocorsis quercicella* Clem., which differs by having a rather broad, distinct, dusky border on the apical margin of the anterior wings, otherwise they look very similar." A good, fresh specimen has the dusky border plain and visible, a worn specimen scarcely shows it, but so far as I have seen, variation alone is sufficient to account for Mr. Cresson's opinion.

Specimens received from Miss Murtfeldt (presumably the same as those referred to by Chambers, (l. c., p. 84), as having been bred by Miss Murtfeldt and Professor Riley, in Missouri from *Ambrosia*, and compared with the Texan specimen sent to Mr. Cresson), are now before me and are undoubtedly Clemens' species *quercicella*, corresponding with my specimen compared with his type in the collection of the American Entomological Society at Philadelphia. Chambers (l. c., 85-86) thinks a specimen identified by Zeller as *quercicella* Clem. must be his *cressonella*. Zeller's specimen labelled "quercicella" is in my cabinet, but it is not rightly identified; it is a dark form, not separable from *reflexella* Clem.

I have seen the type of *Psilocorsis dubitatella* Z. (Hor. Soc. Ent. Ross., XIII, 262-3, 1887) in Dr. Staudinger's collection. It is a pale variety of the true *quercicella* Clem., with a slight transverse shade beyond the middle and the double dark line on the apical margin and cilia.

Cryptolechia obsoletella Z. of which I have the type, is very like a small *reflexella*, but shows no indication of the transverse darker striæ on the fore wings. I should regard it as distinct for the present. It is darker than *ferruginosa* Z. (of which I have also the type), having none of the ochreous tint of that species, but the discal and marginal dots are very similar, although somewhat more pronounced. Further investigation is required to clean up the life-history of these species. If one of them feeds on *Ambrosia* it seems improbable that this can be the species bred by Clemens from oak. Possibly the species I have from Miss Murtfeldt may not be the one referred to by Chambers.

Cryptolechia reflexella Clem.

Psilocorsis reflexella Clem. = *Cryptolechia quercicella* Z.

Zeller's collection contains a female of this species labeled *Psilocorsis quercicella* Clem., and it is evident that this is the specimen referred to by him (Ver. Z.-b. Ges. Wien., XXIII, 242) when in describing *obsoletella* he remarks, "Viel kleiner als *quercicella.*" The species varies a good deal in size and in the distinctness of the distal and marginal spots. Apart from the color of the fore wings, which is distinctly darker and therefore less contrasted with the superficial speckled markings, the longer palpi, the darker color of the hind wings, and its lacking the distinct double blackish line in the cilia of the fore wings appear to be the chief distinguishing characters by which to separate it from *quercicella* Clem.

A specimen from Dr. Riley bred from Birch (*Betula* sp. ?) is only to be distinguished from *reflexella* by its smaller size and shorter palpi, wherein it approaches dark varieties of *quercicella*. I shall not venture to describe it as distinct.

Should an extended series of bred specimens of any of these darker forms establish a reliable means of distinguishing them, it is yet possible that one of Chambers' names may be hereafter revived for this more probable variety.

Cryptolechia concolorella Bent.

Mr. Beutenmüller has lately published (Ent. Am., IV, 30), a description under this name. He is probably right in referring it to this genus, but as he gives no description of the neuration, or of the form of the wings, nor any details of structure it is impossible to place it correctly.

STENOMA SCHLAEGERI Z., AND ITS ALLIES FROM THE UNITED STATES.

Cryptolechia schlaegeri was first described by Zeller in 1854, in the ninth volume of the Linnea Entomologica, pages 372–3, and figured on Plate 3, Fig. 18 of the same volume. The description was taken from specimens of both sexes from New York in his own collection; the hind wings are described as gray with whitish cilia. In the tenth volume of the same publication, pages 158–9, he supplements his description of the species, and compares it with the Mexican *Cryptolechia frontalis* there described. Here he remarks that the hind wings of the female are usually whitish. He goes on to describe a variety of the same species from Georgia, "var. *b.* ♂ parva, alis ant. breviusculis," in the King's Museum, at Berlin, of which he writes that the hind wings are lighter gray than in the male of var. *a*, but darker than in the female of that variety. A careful examination of about twenty specimens (including Zeller's type) from various localities in the United States proves that at least two distinct forms exist. These two forms are easily separable by the shape of the uncus in the males, and usually by the color of the hind wings; the commonest form having pale hind wings, especially in the ♀—*schlaegeri* of Zeller—has the uncus simple, scarcely enlarged towards its apex and ending in an obtuse point (Fig. 25*a*). The other having dark cinereous hind wings in the ♂, has the uncus dilated and distinctly notched or furcate at the apex (Fig. 25*b*). The form of the lateral claspers is approximately the same in both.

FIG. 25.—CRYPTOLECHIA SCHLAEGERI; *a*. Uncus of the common form. *b*. Uncus of the less common form. Enlarged (original).

The small variety (var. *b*. of Zeller's supplementary notice) from Georgia and Texas has the hind wings and simple uncus of the true *schlaegeri* and is apparently undistinguishable from it, except in size, since the markings on the anterior wings are subject to some variation in position and intensity of coloring in specimens of all sizes. The shorter ciliation of the antennæ, noticed by Zeller, is scarcely more than proportionate to the reduced size of each individual.

C. frontalis is described as having the hind wings gray, but broader than in *schlaegeri*. So far, so good. It is noticeable that in the supposed form of *schlaegeri*, described in Zeller's supplementary paper (Lin. Ent. X, pages 158–9) with paler hind wings (and simple uncus) there is considerable variation in the shape and position of vein 2 of the fore wings, both in the large variety (Zeller's var. *a*) and in the small form (var. *b* of Zeller). In some specimens vein 2 arises from the same point as vein 3 and proceeds with a slight bend to the margin above the anal angle. In others it arises either from the same point as vein 3, or extremely close to it, and is abruptly bent backwards in the first instance before taking its ordinary direction. In others again this vein arises quite separate from vein 3, being more or less bent in its outward course; in one specimen before me, which is undistinguishable from Zeller's var. *b*, these veins are separated at their origin by even a greater distance than that which separates veins 3 and 4, but this appears to be exceptional.

Stenoma leucillana Z.

Specimens in my collection, taken by Belfrage in Texas, agree very closely with Walker's type of the Nova Scotian *algidella*, but comparing it with a series of what I take to be *leucillana* Z., it can only be regarded as a dark variety of the female of that species. I think it extremely doubtful whether *leucillana* Z. is really distinct from the well-known *schlaegeri*, which varies sufficiently in size and color to connect it with this somewhat smaller and paler form. Indeed if I have rightly determined Zeller's *leucillana* it would be imposible to draw the line between them in a lengthening series. Until an opportunity may occur for examining the type specimen in the Berlin Museum, I prefer to err on the side of caution rather than to treat the name as a synonym. All my specimens of this smaller form have the uncus simple as in the true *schlaegeri*.

Stenoma algidella Wlk.

For the present I shall adopt the same course with regard to *Cryptolechia algidella* Wlk., which is probably also only a small form of *schlaegeri*, although occurring so far northward as Nova Scotia. Should the acquisition of further material enable me to express a more decided opinion the alteration can be made in the final revision of the index.

Stenoma furcata sp. n.

Antennæ in the ♂ brownish, finely ciliated on both sides; in the ♀ the color is much paler.

Head and palpi white.

Thorax slightly tinged with brownish-gray on the upper and central parts, without a patch of dark scales behind it.

Fore wings elongate, narrow, produced, but somewhat depressed and rounded at the apex; the costa very slightly arched at the base, scarcely convex beyond it; apical margin oblique; dorsal margin straight, almost parallel with the costal, but slightly diverging to the anal angle, which is ill-defined; white, with a slight tinge of brownish-gray, commencing near the base of the dorsal margin and extending to the anal angle below the discal cell, and very faintly in a narrow line along the base of the cilia in the apical margin; cilia white, tinged with grayish towards the anal angle and along their tips. In the ♀ there is a faint indication of pale, grayish clouds and spots at the end of the cell, and of a pale grayish transverse line between this and the apical margin on the lower half of the wing, and in the abdominal angle are some raised scales, as in *schlaegeri* (these would probably be found also in better specimens of the ♂); there are also a few divided black scales in the middle of the cilia; *under side* strongly clouded with brownish-gray; the costal and apical margins narrowly paler.

Hind-wings very broad, evenly rounded, but somewhat produced at the apex; dark cinereous in the ♂; pale grayish-ochreous in the ♀; cilia whitish; *under side* cinereous.

Abdomen cinereous; uncus abruptly bent over from the base, distinctly divided into two short forks at the apex; lateral claspers produced into two angular points, of which the lower one is smaller and sharper than the upper.

Legs whitish, unspotted.

Exp. al.: ♂ 27, ♀ 30 mm.

Habitat, Arizona. (Two males and two females collected by the late H. K. Morrison.)

Type, ♂ ♀, *Mus. Wlsm.*

This species differs from *Stenoma schlaegeri* Z. in its narrower and more elongate fore wings, which in the specimens before me have little or no indication of the gray clouds and blotches prevalent in that species, and very noticeably in the form of the uncus; also in the absence of the dark patch of scales at the back of the thorax. I have a single specimen, collected by myself in California in 1871, which might be regarded as an intermediate link between this species and *Stenoma schlaegeri*. It has the uncus distinctly dilated and notched at the apex, a faint thoracic spot, and a few raised scales at the abdominal angle of the fore wings; there are no spots in the cilia, but a narrow gray line runs along the middle; the hind wings are nearly as dark as those of *furcata*, and the fore wings are somewhat more clouded with gray.

It will probably be found, when more material comes to hand, that the form of the uncus is a more reliable character for separating this species from *schlaegeri* than any distinction in the intensity of markings, which will probably be found to vary as in that species.

Stenoma crambitella sp. n.

Antennæ ciliated in the ♂; shining ochreous beyond the basal joints, which are white

Palpi white, slightly shaded with pale brownish-ochreous externally on the second joint, except at the apex.

Head white; face smooth, shining yellowish-gray.

Thorax white, with a faint ochreous tinge.

Fore wings elongate, narrow at the base, very slightly convex at about the basal third of the costa, straight beyond; apex rather pointed; apical margin straight, oblique, rounded at the anal angle; dorsal margin straight, white, rather shining, with a suffusion of faint ochreous scales (only visible under a lens) along the veins and nervules; on the extreme costal margin at the base are a few grayish-fuscous scales, and a single dot of the same color lies at the end of the discal cell in the middle of the wing; cilia white.

Hind wings grayish-white, with a faint ochreous tinge; cilia white.

Abdomen agreeing in color with the hind wings; uncus simple, blunt, bent over, not notched at the apex (being much shorter than in *schlaegeri* or any of its allies with which I am acquainted): lateral claspers upturned, rounded at the apex, with a triangular excrescence on the lower edge near the base.

Legs whitish; posterior tarsi tinged with grayish-ochreous.

Exp. al.: 22mm.

Habitat, Arizona (received from the late H. K. Morrison).

Type, ♂ ♀, *Mus. Wlsm.*

Stenoma humilis Z.

=*Cryptolechia humilis* Z.

=*Cryptolechia nubeculosa* Z.

=*Harpalyce canusella* Chamb.

Zeller, in describing *Cryptolechia nubeculosa* (Ver. Z-b. Ges. Wien., XXIII, 245–6, Pl. III, 12), does not refer to his previous description of *humilis* (Lin. Ent. X, 156–8, Pl. I, 6). A comparison of the figures would perhaps not lead to the conclusion that they were identical, but with five or six specimens undoubtedly *nubeculosa* before me, I am strongly inclined to the opinion that his older description of the species in the Berlin Museum had escaped his memory. The range of variation in the species is not great, but quite sufficient to account for the slight differences of markings detailed in the descriptions and figures.

MENESTA Clem.

=*Hyale*, Chamb.

The genus *Menesta* of Clemens is undoubtedly allied to *Cryptolechia* Z.; its neural and structural characters are the same as those of *Stenoma*, and notwithstanding its diminutive size and more abruptly rounded fore wings, it is doubtful whether in any tabulation of these genera it can be rightly separated from it. For the present it may be well to retain the genus as represented by a single species.

Menesta tortriciformella Clem.

=*Gelechia liturella* Wlk.

=*Hyale coryliella* Chamb.

This species has been redescribed by Walker (Cat. Sp. Ins. B. M., XXIX, 591) under the name of *Galechia liturella*, as already pointed out by me (P. Z. S. 1881, 319). *Hyale coryliella* Chamb. (Cin. Qr. Jr. Sc., II, 242), which Chambers, in the Index (Bull. U. S. G. G. Surv., IV, 150), refers with a "?" to *Menesta tortriciformella*, is without doubt another name for this species, and consequently the genus *Hyale* sinks as a synonym of *Menesta*.

IDE.

The genus *Ide* is distinguished by having veins 7 and 8 of the fore wings separate, 2 and 3 from a point, or from a short common stem, and 4 very close to the base of 2 and 3; in the hind wings 6 and 7 arise from a common stem.

Ide lithosina Z.

Cryptolechia lithosina Z.
=*Harpalyce tortricella* Chamb.

I have several specimens of *lithosina* Z.; some from Texas (Belfrage), others from Florida (Morrison), and one from Boll's collection. They vary in the ground-color of the fore-wings from bone white, as described by Zeller, to yellowish or straw-color, as described by Chambers, and in the presence or absence of one, or sometimes two, brownish dots at the end of the discal cell. In one specimen these are quite conspicuous. A careful examination of the genital appendages shows that these forms are not specifically distinct; the uncus is single, with a long narrow stem beyond the dilated base; overarched and spatulate at the apex, the end of the spatulate being notched; the lateral claspers are scarcely more than half the length of the uncus; also somewhat narrowed at the base. their ends dilated and notched posteriorly, the upper lobe being rounded at the apex, the lower slightly longer than the upper, and acutely triangular.

Ide osseella sp.n.

Antennæ, pale bone-color.
Head and palpi, pale bone-color, the latter somewhat darker on the second joint.
Thorax, bone-gray, slightly darker than the head.
Fore-wings, shining, unicolorous bone-color, with scarcely paler cilia, along the base of which is a very slender almost undistinguishable grayish line; at the end of the disk is a reduplicated bone-gray spot, the larger portion of it being above the smaller, with which it is sometimes confluent. *Under side*, very pale bone-gray. Neuration: The veins are all separate, except 2 and 3 which in one specimen are from a common point, in the other from a short stem.
Hind-wings, pale, shining bone-gray, with scarcely lighter cilia. *Under side*, very pale bone-gray.
Abdomen, pale shining bone-gray.
Legs, pale bone-gray, the posterior tarsal joints with the slightest tinge of ochreous.
Exp. al., 24^{mm}.
Habitat, California. (Two females from the Zeller collection.)
Type, ♀, *Mus. Wlsm.*

This species is apparently allied to *lithosina* Z., but it is of larger size.

Ide vestalis Z.

Cryptolechia vestalis Z.
=*Harpalyce albella* Chamb.

Zeller in describing *vestalis* (Ver. Z.-b. Ges. Wien, XXIII, 247), says that it is closely allied to *albella*, but as Chambers' *Harpalyce albella* was not then published, it is obvious that his reference must have been to a species described by himself, under this name, received from Surinam.

(To be continued.)

GENERAL NOTES.

THE BOT-FLY OF THE OX, OR OX WARBLE.

In INSECT LIFE, Vol. I, p. 383, we noticed the investigation recently undertaken by the *Farmers' Review*, of Chicago, of the damage to the cattle interests of this country resulting from the attacks of the Bot-fly of the Ox.

We have had considerable correspondence with the editor, as also with Miss Ormerod, on this subject; and as preliminary to a statement of our own views in the next number, we give here a summary of the articles mentioned and of the results reached in the several lines of investigation followed out.

The objects which the Farmers' Review hoped to attain are given in the issue of that journal of July 17, 1889, as follows:

(1) To impress upon the farmers of the country the seriousness of the loss they are annually suffering as a result of the work of the "grubs" in the backs of their (*a*) beef stock and (*b*) dairy cows.

(2) To arouse them to a recognition of the good policy and actual necessity of fighting (*a*) the Ox Warble-fly and (*b*) the grubs produced in cattle from eggs deposited by the fly.

(3) To show them plainly that the fly and its noxious product may be successfully fought and eventually reduced to perhaps uninjurious numbers.

(4) To interest all concerned and secure their help in (*a*) disseminating throughout the country facts going to show how serious is the damage done by these grubs in cattle, and (*b*) finding a demonstrating medium for the prevention and cure of the trouble.

(5) To instigate a national investigation of the matter by the Department of Agriculture.

In the introductory articles the life-history of the fly has been outlined, quoting for this purpose the short account in Packard's *Guide to the Study of Insects*, and the more important articles on the subject from the various reports of Miss Ormerod, of England, where the attacks of this fly have attracted greater attention than elsewhere, and where much attention has been paid to the means against it.

A host of letters from farmers and stockmen were published, which, so far as they related to the habits and natural history of the fly, were, as a rule, pretty badly mixed, and added little if anything to that already known. Reports were also received from professors of agriculture, entomologists, and veterinarians, which give, as did also those of farmers and stock-raisers, valuable data concerning its abundance in various States, the loss in value to hides, effect on quantity and quality of beef and milk, and also the effect of the attacks on the animals themselves.

From the reports received the approximate percentage of grubby cattle and the average loss on grubby hides for the principal stock-raising States of the Mississippi Valley have been estimated as follows (August 7, 1889):

Illinois.—Seventy-three per cent. of the cattle marketed in the grubby season are infested with grubs. The average loss on a grubby hide is one-third.

Iowa.—Seventy-one per cent. of the cattle in the majority of counties are grubby in the season specified. Loss on grubby hides one-third.

Indiana.—Forty-eight per cent. of the cattle grubby. Loss on hides one-third.

Wisconsin.—Thirty-three per cent. of cattle grubby. Loss on hides one-third.

Ohio.—Fifty-six per cent. of cattle grubby. Loss on hides one-third.

Missouri.—Fifty-seven per cent. of cattle grubby. Loss on hides one-third.

Kansas.—Sixty per cent. of cattle grubby. Loss on hides one third.

Kentucky.—Fifty-seven per cent. of cattle grubby. Loss on hides one-third.

In *Minnesota* and *Dakota* grubs are practically unknown among cattle.

In *Nebraska* they are not very bad where found; twelve counties report an average of 40 per cent. The rest heard from are free of the pest. Grubby hides are "docked" one-third of their value.

In *Michigan* 61 per cent. of the cattle are infested with grubs in the southern and middle counties. In the northern counties they are unknown or very scarce. Grubby hides sell for one-third less than sound ones.

The amount of this loss can be better appreciated perhaps by reproducing in condensed form the approximate estimate of the loss on the hides of cattle received at the Union Stock-Yards of Chicago during the grubby season, which includes the months from January to June. Using the reports by States above given as a basis it is estimated that 50 per cent. of the cattle received are grubby. The average value of a hide is put at $3.90; and while from the report referred to one-third value is the usual deduction for grubby hides in this estimate, but $1 is deducted, or less than one-third. The number of cattle received in 1889 for the six months indicated was 1,335,026, giving a loss on the 50 per cent. of grubby animals of $667,513. When to this is added the loss from depreciated value and lessened quantity of the beef, the amount for each infested animal is put at $5, indicating a total loss on these animals from the attack of the fly of $3,337,565.

Without considering the lessened quantity, the inferiority of the beef of animals infested by the grub is strikingly shown in an article on the subject in which the testimony of retail butchers and buyers of meat in Chicago and other cities is given. It is shown that the buyers of the highest class of meat, who supply hotels and restaurants, will not on any account purchase carcasses showing traces of Warble attack. Such beef has to be sold, therefore, at a price below that obtainable for good beef, free from grub damage, and the lessened value per animal was put at from $2 to $5.

The appearance known as Licked-Beef, which, resulting from the presence of the grub, may be described as a moist or running surface of a greenish-yellow color, is certainly unwholesome in look, if not in fact. The description of such meat as given in the *Farmers' Review*, quoting again largely from Miss Ormerod, is almost sufficient to turn one against beef altogether.

"The Effect of the Warbles in the Dairy" is the title of an interesting article by T. D. Curtis, in which the loss in the quantity of the flow of milk, as well as its deterioration in quality, resulting from the annoyance of the animals by the flies while the latter are depositing eggs and later by the grubs, is very conclusively shown, and he estimates the shrinkage at 10 per cent. and the loss in quality at the same rate, making a total of 20 per cent.

There is finally a discussion of remedies, including those employed in England and in this country, and the expression of a wish that the Division of Entomology of the U. S. Department of Agriculture should take up the investigation, with a view of clearing up such points as may yet be obscure both as to the life history of the insect and the means against it. We shall take up these points more fully in our future remarks.

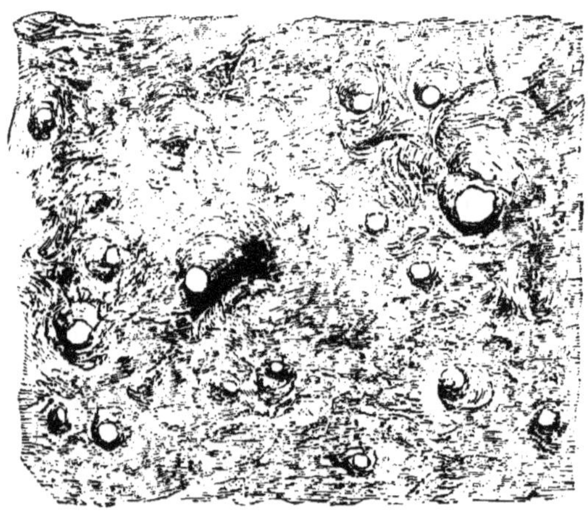

FIG. 26.—Portion of inside of tanned warbled hide (after Ormerod.)

We have recently received from Miss Ormerod a leaflet of eight pages, dated September, 1889, entitled "Notes on 'Licked-Beef' and 'Jelly' and Injury to Hides from Attacks of Ox Warble-fly, or Bot-fly," *Hypoderma bovis*, DeGeer, supplemented by correspondence, in which is described very fully the condition of the beef resulting from grub attack, commonly known as "licked-beef" or "jelly" from the supposition generally held that the loosening of the hide and the discoloration and inflammation of the subcutaneous flesh about the grubby places, and also the frothy or jelly-like appearance of the flesh, results in part from the licking by the animal of such places. Letters from butchers are quoted, giving further details of the exact nature of the injury and the amount of depreciation in value of the beef.

The loss is shown to fall largely on the cattle-owners by waste of food not formed into beef or milk, and also, but to a less extent, on butchers in the deficiency of receipt per pound on the carcass and on the hide. We reproduce a single instance given by Miss Ormerod to indicate the extent of the loss so resulting. A heifer which turned out a much lighter weight than was expected proved to be badly warbled. "The loss on the hide at 1*d.* per pound would be about 5*s.* ($1.25); the loss on the beef, the animal being sold by the stone, fell on the owner. This was estimated at least six stone less than it should have been, and deficiency in weight on hide and beef was put at 50*s.* to 60*s.* ($12 to $15).

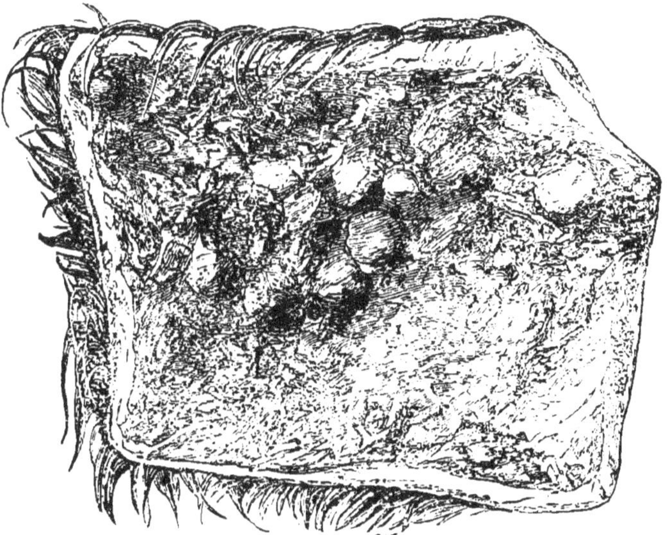

FIG. 27.—Piece of warbled hide, warbles about half size (after Ormerod).

The mischief done to the hides in the decreased value of the tanned product is also discussed by Miss Ormerod, and figures are given, which we reproduce, showing a portion of the under side of a warbled hide, warbles about half size, and a portion of inside of tanned warbled hide.

The aggregate loss in England from warble attacks as estimated by different practical men is put at from £2,000,000 to £7,000,000 sterling, at least, per annum, or perhaps as much as £1 per head of horned cattle.

THE MINNESOTA LOCUST OUTBREAK.

The report of Prof. O. Lugger, Entomologist of the Minnesota Agricultural Experiment Station, on the Rocky Mountain locusts, in Otter Tail County, Minnesota, in 1889,* is of especial interest.

As we have long ago shown in our Reports on the Insects of Missouri, and in the Reports of the U. S. Entomological Commission, plow-

* Bulletin No. 8, University of Minnesota, Agricultural Experiment Station, pp. 17–36.

ing in winter-time or early spring is the most effectual means of preventing grasshopper injury the coming summer; but this recommendation has rarely been carried out on a co-operative scale. In the grasshopper-infested section of Minnesota, however, Professor Lugger has shown the present year what can be accomplished by timely and energetic co operation.

In the fall of 1888 it was ascertained that in the infested region of Otter Tail County enough eggs had been deposited by late swarms of locusts to seriously endanger the crop of 1889. It was found that the eggs were preferably laid in stubble-fields abandoned by their owners, and also in certain spots in the timothy fields and pasture lands. In winter-time the governor and State legislature were appealed to, money was promptly appropriated and rendered available immediately, and competent persons appointed to superintend operations, which were executed in early spring with energy and circumspection. At first the larger of the abandoned stubble-fields in the immediate vicinity of cultivated fields were plowed, and then the worst infested places in the timothy fields and pastures. The whole area thus plowed in this single county, at the expense of the State, embraced no less than 6,361 acres. The farmers in the mean time plowed the fields intended for the use of corn, and largely assisted the State authorities in plowing at their own expense the smaller fields which were ascertained to contain a dangerous number of eggs.

The success of this operation was complete. Not a single grasshopper egg hatched on the plowed fields wherever the plowing was done carefully and to a sufficient depth. When, in the month of May, the grasshoppers hatched on the timothy fields, the farmers, knowing that there was now no danger of an invasion of grasshoppers from the neglected or abandoned fields in their vicinity, willingly set to work to assist the authorities in the warfare against the young locusts. A large number of "hopper dozers" (coal-oil pans*), previously prepared, were at hand, and were operated on a large scale. Burning stubble wherever practicable, and, in one instance, a judicious use of London purple, was also resorted to.

"About the middle of June," says Professor Lugger, "it became quite plain that the crops were saved, and that most of the locusts had been killed."

This gratifying result was obtained at a comparatively trifling expense. and we congratulate Professor Lugger on the success of his efforts in this direction.

THE WEEPING-TREE MYSTERY.

Prof. Herbert Osborn has called our attention to an article in the Dallas (Texas) *Morning News* of October 9, in which a very well written

*First described and recommended in Riley's "The Locust Plague in the United States."

and humorous account is given of the solution by the reporter of that paper of the mystery of the so-called "weeping trees," reports of which from Grayson County and other parts of Texas are said to have "set the State agog with various explanations of the phenomenon, ranging from the superstitious credence of the supernaturally inclined to the positive denial and derisive laugh of the constitutionally skeptical." The brave reporter, however, upon the discovery of one of these remarkable trees in Dallas, laying aside all superstition, climbed courageously up the trunk and discovered that the tears were shed by a multitude of small insects "of dark green color with gold under the wings, which adhered to the bark and scampered about when disturbed, and flew away when pressed too closely." Prof. G. W. Curtis, of the Texas Agricultural and Mechanical College, secured specimens and sent them to Professor Osborn, who recognized them as the common little leaf-hopper, *Proconia* (*Oncometopia*) *undata*, which we have referred to in previous writings and on pages 53 and 54 of vol. 1 of INSECT LIFE as occurring upon the Orange in Florida and upon cotton-plants in other Southern States, and which we have there stated is remarkable for the distance to which it ejects drops of honey-dew.

We frequently met with this species in the cotton-fields in the summer of 1879, and noted the extraordinary abundance of the secretion. Professor Curtis in his letter to Professor Osborn stated that in Dallas they made the tree present a decided appearance of weeping quite profusely, the drops being small but coming quite thick and fast. Each insect would eject a drop at intervals of two seconds during a period of several minutes, and would then stop for a little while.

AN EARLY OCCURRENCE OF THE PERIODICAL CICADA.

Dr. J. C. Ridpath, the historian, has very kindly sent me the following extract from one of the many valuable works contained in his private library. The writer had the State of Virginia under consideration when the excerpt was written, and therefore it is quite probable that the third prodigy was an occurrence of what is now known as Brood VIII of *Cicada septendecim*.—F. M. WEBSTER.

[Stedman's Library of American Literature, Volume I, pages 462, 463. Excerpt from the writings of T. M., supposed to have been Thomas Matthews, son of Samuel Matthews, governor of Virginia. Written in 1705.]

About the year 1675 appeared three prodigies in that country, which, from the attending disasters, were looked upon as ominous presages.

The one was a large comet every evening for a week or more at southwest, thirty-five degrees high, streaming like a horse-tail westwards until it reached almost the horizon, and setting towards the northwest.

Another was flights of pigeons, in breadth nigh a quarter of the mid-hemisphere, and of their length was no visible end; whose weights break down the limbs of large trees whereon these rested at nights, of which the fowlers shot abundance and eat them; this sight put the old planters under the more portentous apprehensions, because the like was seen, as they said, in the year 1640, when the Indians committed the last massacre, but not after until that present year, 1675.

The third strange appearance was swarms of flies about an inch long and big as the top of a man's little finger, rising out of spigot holes in the earth, which eat the new-sprouted leaves from the tops of the trees without other harm, and in a month left us.

LAPHRIA CANIS Will.: A CORRECTION.

On page 43 of the present volume of INSECT LIFE the statement is made that *Laphria canis* Will. was very abundant in Michigan in May, 1886. The writer has since felt that this statement admitted of doubt, as the habits of the fly there described are unquestionably those of *Bibio albipennis*, which was, in all probability, the species under observation. The specimen of *Laphria canis* which I sent to Dr. Williston for determination was taken some months afterward from among alcoholic specimens of flies, and believed at the time to be one of the individuals that had been so numerous in the spring, but in this I fear that I was deceived. *Laphria canis* should, of course, be recorded for Michigan, on the authority of one specimen of uncertain date of capture, determined by Dr. Williston.—T. TOWNSEND.

ENTOMOLOGICAL SOCIETY OF WASHINGTON.

October 3, 1889.—Fifty-fifth regular meeting. Prof. James Fletcher, Entomologist to the Dominion of Canada, was elected a corresponding member of the society.

Dr. Fox made some remarks on "Malformations in Spiders," exhibiting two specimens (*Epeira sclopetaria* ♂, and *Dictyna* sp. ♀), in which one or more of the eyes were absent. He also exhibited a table showing the relative position of the eyes as normally found in different families of spiders. The subject was further discussed by Dr. Marx.

Mr. Schwarz then read a communication from Dr. G. H. Horn on the food-habits of a rare Cerambycid beetle (*Cœnopœus palmeri*), which lives in its early stages in the stems of *Opuntia bernardina*. These food-habits are the more remarkable from the fact that all the other known species of this group (*Acanthocinini*) live beneath the bark of dying or dead trees. Mr. Schwarz also read a note on the peculiar flight of a specimen of the flying locust, *Dissosteira carolina*, while observed to be pursued by an English sparrow, its flight, in escaping the bird's attacks, veering directly up or down, but never to one side; and presented for record an observation on *Chalybion cæruleum*, a blue wasp, which in catching the spiders that form its prey, pretends to be caught in their webs and easily captures them when they appear. These papers were discussed by Dr. Marx and Mr. Ashmead.

Mr. Townsend read a paper on some interesting flies from Virginia, noticing and exhibiting specimens of: *Holcocephala abdominalis*, to Say's description of which he made some additions; four species of *Trichopoda* (*T. radiata* Loew, *T.? hirtipes* F., *T.? ciliata* F., and *T.* sp.), two of which have not been recorded for this locality; and *Palloptera superba* Loew, with some notes on its habits.

Dr. Marx read by title a revision of Hentz's Spiders of North America. The meeting then adjourned.

WM. H. FOX, M. D.,
Recording Secretary.

U. S. DEPARTMENT OF AGRICULTURE.

DIVISION OF ENTOMOLOGY.

PERIODICAL BULLETIN. DECEMBER, 1889.

Vol. II. No. 6.

INSECT LIFE.

DEVOTED TO THE ECONOMY AND LIFE-HABITS OF INSECTS, ESPECIALLY IN THEIR RELATIONS TO AGRICULTURE, AND EDITED BY THE ENTOMOLOGIST AND HIS ASSISTANTS.

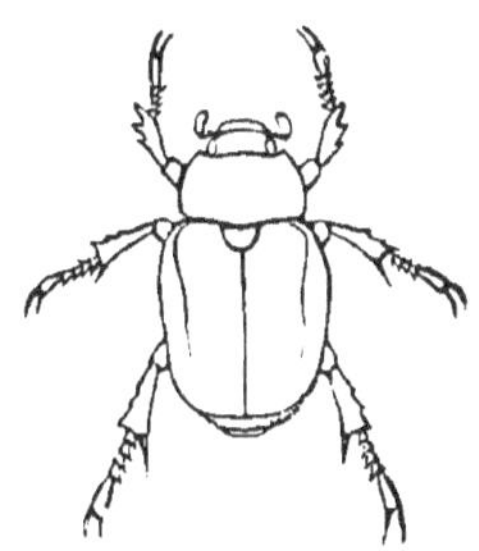

[PUBLISHED BY AUTHORITY OF THE SECRETARY OF AGRICULTURE.]

WASHINGTON:
GOVERNMENT PRINTING OFFICE.
1889.

CONTENTS.

II

SPECIAL NOTES.

The Official Association of Economic Entomologists.—We devote considerable space in this number to the official minutes of the first annual meeting of this association, which was held November 12, 13, and 14 in this city, as announced in Vol. II, No. 4. The meeting was very successful, both in point of attendance and in the character of the discussions and the papers read, and it was especially gratifying to have so many who were not in attendance apply for active and associate membership. The general sentiment, as expressed in discussing such questions as that of co operation, seemed to be that the association should retain the broader character originally designed, which would include in its membership others engaged or interested in economic entomology who are not necessarily connected officially with agricultural colleges or experiment stations, though it is very evident that the more active members will consist of those officially employed in one way or another. In view of the simultaneous meeting in the same museum building of the Association of Agricultural Colleges and Experiment Stations, the question of how the entomological organization could best co-operate with the other was one of the most interesting, and one which it seemed at first difficult to solve. The fact that the Association of Colleges and Experiment Stations decided to organize committees in different specialties—one being in entomology—to a certain extent limited the co-operation, and at the same time facilitated it, since said committee, working with a similar committee from the entomological association, will be able to perfect plans of co-operation and help to carry them out.

Studies in Embryology.—We have just received from the author a valuable contribution to our knowledge of the embryology of insects in a paper by William M. Wheeler, curator of the public museum, Milwaukee, Wis., entitled "The Embryology of *Blatta germanica* and *Doryphora decemlineata*," reprinted from the *Journal of Morphology*, Vol. III, No. 2, September, 1889.

After describing his method of work and the manner in which the eggs and other material were prepared for study, the author discusses the formation of the eggs in the ovaries; oviposition and the subsequent development of the embryo, including a discussion of the formation of germ layers and embryonic envelopes, together with a description of the external change in the embryo of *Blatta* and the subsequent stages in the evolution of the embryo in *Doryphora*.

The author concludes with a list of the authorities referred to in the course of his work. The article covers 92 pages and is illustrated with 16 text figures and 7 large lithographic plates.

Willow and Poplar Insects.—In Bulletin No. 9 of the Agricultural Experiment Station of the University of Minnesota, the first portion of which contains a consideration of Russian willows and poplars, we find some sixteen pages by Prof. O. Lugger, treating of insects affecting poplars and willows. He describes at some little length, with figures, *Cimbex americana*, *Nematus ventralis*, *Lina scripta*, *L. tremulae*, *L. lapponica*, *Saperda calcarata*, *S. concolor*, *Vanessa antiopa*, *Hyphantria cunea*, *Acronycta lepusculina*, *Platysamia cecropia*, and *Telea polyphemus*. The articles are brief and popular, and contain for the most part restatements of well-known facts; but the author mentions that *Cimbex americana* is attacked by a Tachinid fly in Minnesota. The work of the Poplar Girdler (*Saperda concolor*) is for the first time illustrated, and a number of different species of parasitic Ichneumonidæ are reported to have been bred from it. *Acronycta populi*, Riley, is made a synonym of *A. lepusculina*, Guenée, following Grote; but this is an error, the latter species, known to us, being different in both larva and imago, and occurring on the Pacific coast.

Another Importation from Europe.—Prof. J. H. Comstock, in Bulletin No. 11 of the Agricultural Experiment Station of Cornell University, has given in detail an account of the life-history of the well-known European Corn Saw-fly (*Cephus pygmaeus*), which, curiously enough, he finds very abundantly in wheat on the university farm. This insect has not previously been recorded in this country. Professor Comstock finds that the adults emerge early in May, oviposit about the middle of the month, and that in a very short time the larvæ work through nearly the entire length of the straw, descending early in July to the root. Here, after cutting the straw nearly through an inch above the ground, they spin silken cocoons and remain dormant until early the following spring, when they complete their transformations.

He finds that their presence in the stalk reduces the abundance of the grain little, if any, and that the principal damage is the lodging of the grain. He has found the species in wheat alone. He has seen parasites in two cases, but has not been able to secure good specimens. He thinks that the insect is not confined to the vicinity of Ithaca, but that it will be found elsewhere. Experiments made to ascertain the amount of damage by weighing the grain from the infested and the non-infested heads showed in every case a decided superior weight in favor of the heads of the infested stalks. The explanation offered—undoubtedly the correct one—is that oviposition takes place early and that only the largest stalks are chosen.

Professor Smith's Bulletin on the Horn Fly.—In bulletin No. 62 of the New Jersey Agricultural Experiment Station, Prof. J. B. Smith summarizes his observations on the Horn Fly (*Hæmatobia serrata*). We notice from the date that the bulletin was submitted just about the time our article on this insect in No. 4 of INSECT LIFE appeared, and, as a result, neither our observations nor our conclusions are referred to. Professor Smith has also succeeded in tracing the life history. He secured eggs in confinement August 6, from which the imagos issued August 20 and 22. The bulk of the bulletin is taken up with extracts from extensive correspondence, and some fifteen pages more with descriptions of the different states and with anatomical details accompanied by figures. He suggests the use of plaster instead of lime for the manure heap on chemical grounds and for the preservation of the fertilizing qualities of the manure. He further suggests that, by sending a boy through the pasture with a shovel and with instructions to thoroughly spread all cow droppings so that they may rapidly dry out, the larvæ and eggs will be destroyed—a suggestion of value only in dry and sunny weather. He erroneously supposes that the eggs are largely laid at night, while our latest observations prove plainly that this is not the case, and this vitiates the discussion of remedies as applied to the manure pit or the interior of the stable wherever cattle are pastured during the day.

Entomology at the Paris Exposition.—The record of the fact that two grand prizes for the United States were awarded at the Paris Exposition (one to the Department of Agriculture and one to the Entomologist) in class 76, which comprises useful and injurious insects, will not be out of place in these pages. Only one other grand prize was awarded in this class, and that was to Japan. This exceptional recognition of our exhibit at Paris is, *çela va sans dire*, gratifying, but not more so than the fact that the agricultural exhibit, included in fifteen classes, received seven grand prizes, forty gold, sixty-eight silver, and fifty-four bronze medals, and thirty-nine honorable mentions. This is a relatively larger percentage of medals, than was awarded to the United States in the other seventy-one classes, and a very much larger percentage of awards in the agricultural groups, as compared with those obtained by the United States, either at the Paris Exposition of 1867 or of 1878.

THE SO-CALLED MEDITERRANEAN FLOUR MOTH.

(*Ephestia kühniella* Zeller.)

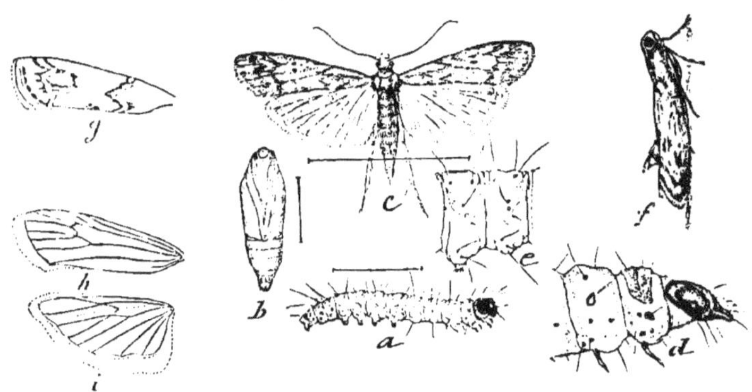

FIG. 28.—*Ephestia kühniella*: *a*, larva; *b*, pupa; *c*, adult—enlarged; *d*, head and thoracic joints of larva; *e*, abdominal joints of same—still more enlarged; *f*, moth from side, resting; *g*, front wing, showing more important markings; *h*, venation of fore-wing; *i*, venation of hind-wing—somewhat enlarged (*a*, *b*, *c*, and *e*, original; *d*, *f*, *g*, *h*, and *i*, after Snellen).

This insect, which during the last few years has been doing so much damage in mills in England, Belgium, and Germany, has during the past summer appeared in destructive numbers on this continent. During August the attention of Mr. James Fletcher, Dominion entomologist of Canada, was called to a serious outbreak of this pest in a Canadian city,* which has recently been written up by Dr. P. H. Bryce, secretary of the provincial board of health in Ontario, and issued in pamphlet form in Bulletin No. 1 of this organization. We publish in this number, under the head of "Extracts from Correspondence" a letter from Mr. Fletcher referring to this outbreak, which has suggested the desirability of bringing together in condensed form a summary of the known facts concerning this pest, and a few points suggested by our notes and collections.

It will be remembered that in INSECT LIFE for March (Vol. I, p. 315) we published a long letter from Miss Ormerod, in which she described the damage done by this pest in England, and that in our reply (*loc. cit.*) we stated that the species does not occur in the United States. In the hurry of getting ready to leave for Paris we allowed this statement to be made, notwithstanding the fact that we had had in the National Museum collection for some time specimens of a moth indistinguishable from this species from A. W. Latimer, of Eufaula, Ala. On referring to our notes we find also that we had seen specimens from North Carolina in the collection of M. Ragonot in Paris. These facts undoubtedly prove the occurrence of the insect in North America for at least some years back. Up to the present time the species seems to have been

* We omit the precise locality by request.

rare here, for every case of serious damage to grain by Lepidopterous larvæ which has been carefully investigated has shown that the author of the damage was either the Angoumois Moth (*Gelechia cerealella*), the Grain Moth (*Tinea granella*) or *Ephestea interpunctella* (=*zeæ* Fitch), a congeneric insect which was treated by Dr. Fitch under the common name of the "Indian-meal Moth."

As will be seen by the following digest of recent European writings on the subject, the insect is supposed to be of American origin, but admitting that it has been known for a few years in America, and that during the summer of 1889 it made a destructive appearance in Canada, the point as to its origin still remains obscure. It has, in fact, really been known longer in Europe than in America, and the first specimens from which Professor Zeller described the species were reared in Germany. It seems to be simply another instance of the extreme readiness with which Europeans attribute all new pests to this country.

That the insect is with us now, however, in destructive numbers, and that it is a pest of no small magnitude, cannot be doubted. The condition of affairs in Canada, as stated by Mr. Fletcher in his letter, is by no means exaggerated. Mr. Howard was in Canada the latter part of August, and accompanied Mr. Fletcher on a tour of inspection to the worst infested establishment, and the entire building was completely overrun by these creatures. Hardly a crack or a nail hole was to be found

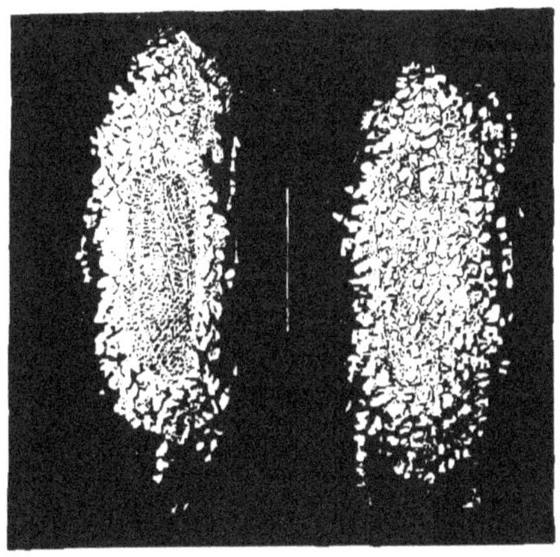

FIG. 29.—*Ephestia kühniella*: *a*, cocoon from below, showing pupa through the thin silk attaching the cocoon to a beam; *b*, same, from above—enlarged (original).

without the cocoons (Fig. 29), and every bit of flour or grain remaining was spun together by their webs. The moths were still flying about in numbers, although great efforts had already been made to destroy them. The government of Ontario made strenuous efforts to stamp out the pest, as can be seen from the bulletin already mentioned. The machinery was taken down and steamed, the walls were scraped down, and the elevator spouts and loose wooden work, together with pipes, bags, and quantities of stock were burned up; belts, caps, and cloth bags were boiled and the whole place was subjected to sulphur fumes. Every inch of space about the machinery was subjected to the flame of a kerosene torch. For a long time before this energetic treatment was commenced (for the pest was noticed as early as March) the moths were flying freely about the building and hundreds must have escaped through the open windows to enter other mills and feed-stores, and by depositing their eggs commenced the ruin of other millers and dealers.

The insect in question appears to have been first brought to the attention of an entomologist in 1877, when the moths and larvæ were sent to Prof. P. C. Zeller from a flour mill in Halle a. S. Professor Zeller described the species in the *Stettiner Entomologische Zeitung* for 1879, pp. 466–471, naming it after the gentleman who sent him the first specimen, Kühn, and stated that in the mill in which they appeared American wheat is much used. The flour is spun up by the larvæ into a kind of felt, and in this felt they dwell in silken tubes. The moth appears in the greatest numbers in May and June, and a second generation appears in August. Professor Zeller had never seen it before in any collection of European or exotic insects, but did not hesitate to state that it came, in all probability, from North America; why, nobody knows. P. C. T. Snellen, in the *Tijdschrift voor Entomologie* for 1881, pp. XX to XXII of the proceedings, has mentioned Zeller's paper.

In 1883, Professor Zeller wrote to us, under date of February 20, as follows:

> I send herewith *Ephestia kühniella* in order to ascertain positively whether it is really of North American origin. This predaceous domestic insect, the natural history of which is described in the *Stettiner Zeitung*, appears to have died out here at Grünhof. * * *

Preudhomme de Borre, in the *Comptes Rendus de la Société Entomologique de Belgique*, July 5, 1884, gave an account of the injury done by this insect in a noodle factory in Belgium, where the insect was supposed to have been introduced with American corn. Various plans for disinfecting the mill proved useless, the only effective remedy being a thorough cleaning.

Dr. F. Karsch in the *Entomologische Nachrichten* for May, 1884, under the caption "*Ephestia kühniella*, Zeller, Eine Nord Amerikanische Phycide am Rhein," records the appearance of this moth at several places along the Lower Rhine. The specimens bred by him have fore-wings of a glossy lead gray, whereas in the typical specimens raised by Zeller the ground color is pure yellow or nearly brownish. He refers them unquestionably, however, to one species. He had looked in vain through American literature for an account of this moth. Fitch's *Tinea zeæ* is the only one that approaches it, but his description does not agree with *kühniella*. Dr. Karsch, nevertheless, thought *zeæ* might prove to be a variety of *kühniella*. In the same month (May, 1884) M. Maurice Girard, (*Bulletin des Séances de la Société Entomologique de France*, pp. LXXIII,

LXXIV) read a note on the ravages of this moth which had appeared in enormous numbers in a flour mill at Lodelinsarte, Belgium. He added a short description of the moth and larva. M. E. Ragonot stated in the discussion of this note of M. Girard that the insect had been first noticed in Europe in 1879 by Zeller, and was supposed to have been imported with American flour. Ragonot himself had specimens coming from North Carolina, Mexico, and Chili.

In an editorial note in the *Entomologische Nachrichten* for 1885, pp. 46, 47, mention is made of reports of the appearance of this insect in mills near Bremworde. The insect multiplies with incredible rapidity. The application of bisulphide of carbon and the burning of sulphur were useless. All that could be done was to stop the mill and thoroughly clean out the pipes and screens. It is positively asserted in this note that in this locality it had been ascertained that the insect was introduced with American wheat. In another editorial note in the same periodical for the same year (pp. 239, 240) a review is given of a communication by Prof. H. Landois to the *Braunschweiger Tageblatt*, in which it is stated that this pest is by far the most annoying and dangerous of all the insects affecting wheat or flour. Moving and airing the wheat is said to have no effect against this species, which is fond of a draft. Countless numbers of webs were found in a pipe through which the flour was lifted by air pressure. For many days they were forced to shut down in order to clean the pipes and screens. The larvæ preferably gnawed the fine miller's gauze. An anatomical examination showed the number of eggs in a single female to be 678.

Prof. P. C. T. Snellen in the *Tijdschrift voor Entomologie*, Vol. 28, 1885, pp. 237–251, gives quite an extended article on this insect, which is illustrated with Plate 8, in all the different stages and in colors. The figures were drawn by Prof. Dr. J. Van Leeuwen, jr. The author states that the main object of his article is to introduce the illustration, as it is made up chiefly of a summary of Zeller's article already referred to. He makes some remarks on the color of the larvæ in correction of Zeller, gives a short account of the mode of pupation, and a careful description of the pupa. The bulk of the article, however, is taken up with a comparison of *kühniella* with other European species of *Ephestia*.

There follow now five articles published in English periodicals, two by W. Thompson, one by J. W. Tutt, one by Charles G. Barrett, and one by Sidney Klein. Mr. Thompson, on pages 66 and 139 of *The Entomologist*, Vol. 20, 1887, records the breeding, during November and December, of specimens of this insect found feeding on rice-cones. Mr. Tutt, on page 212 (*loc. cit.*), records the breeding of larvæ found feeding on flour in a cargo at the London docks, giving a short account of the feeding habits. Mr. Barrett, on pp. 255–256 of *The Entomologist's Monthly Magazine*, Vol. 23, April, 1887, summarizes Zeller's observations, and refers to Mr. Thompson's experience. Mr. Klein's article is published in the *Transactions of the Entomological Society of London*, 1887, monthly proceedings, pp. LII to LIV. His observations were made from May to September, 1887, on an immense colony of larvæ which had over-run some large warehouses in the east end of London. Fumigating with sulphur and hot-liming the floors, ceilings, and walls for several days did not prevent their spread. The flour was mingled with silk threads so as to be useless. The eggs appeared to be laid on top of the sacks, and hatched within a few days. The larvæ burrowed through the sacking, spinning long galleries through the flour, generally not penetrating to a greater depth than three inches. When full grown they leave the flour, crawl to the floor and up the wall, and spin their compact cocoons at the angle of the wall with the roof. They are difficult to keep in breeding cages on account of this migratory habit when full grown, and because they escape through the smallest orifices. Chickens were introduced into the warehouse and gorged themselves with the larvæ. A small ichneumon fly destroyed the pest by September.

The principal English article, however, is by Miss Ormerod. In her twelfth report, for 1889, she reviews the previous accounts of the pest in England and refers to a new case in the north of England, where they made their appearance in 1888. The larvæ entered the spouts and machinery, destroying the silks, and stopped the flow of flour through the spouts by their webs. Remedies were tried as follows: The mill was stopped for a week, the machinery was thoroughly cleaned, hot steam was run into the machines and all through the mill. The walls and floors were whitewashed with freshly slacked lime and paraffine (the English term for what we call kerosene in this country), and all moths that were seen were captured and killed. This heroic treatment failed to destroy the pest. It was supposed that this north of England case was due to the importation of eggs and young larvæ in returned empty sacks from London. Miss Ormerod thinks that the insect came to England from Europe or the East rather than from America, although the sole reason which she gives for this supposition is the fact that the name of the moth does not occur in Grote's check list of the moths of North America in 1882.

Dr. Bryce's bulletin, elsewhere referred to, and quoted by Mr. Fletcher, we will not mention in detail. It is prepared with care, but the figures could not well be poorer or more characterless.

Our own studies of *kühniella* have been made upon material brought us by Professor Panton, of the Guelph Agricultural College, last summer; others in the National Museum collection, which contains the rubbed specimen from Eufala, Ala., five from Europe from M. Ragonot, and others received from Zeller in 1883.

Ephestia interpunctella we have bred upon a number of occasions. We first raised it upon wheat at St. Louis, in October, 1870. Larvæ have been sent to us from a meal-sack at Boylston, Mass.; we have reared it from corn from Guatemala; larvæ and moths were received from a firm of manufacturing chemists from Detroit, Mich., who had found them crawling about over sacks containing roots of dandelion—moths, in fact, being found in the bags; we found numerous larvæ infesting wheat in the Atlanta Exposition building in 1884; large numbers of larvæ were also found in a jar containing Chickasaw plums at the same exposition; larvæ were received from Ripley, Miss., on two occasions in 1885, some of which were said to have been found feeding on sugar in barrels; one specimen was bred from dry Opuntia from Texas; larvæ were received from Detroit, found among old books; larvæ of all sizes were found infesting Pecan nuts in St. Louis, in September, 1872; moths were bred by Dr. A. W. Hofmeister in Iowa from Cinnamon bark; moths were bred from English walnuts in St. Louis in 1876, and the species in all states was found abundantly in a wheat warehouse in Alexandria, Va., in 1883. Moreover, in 1873, at St. Louis, one of these moths was bred from old woolen stuff in company with *Tineola biselliella*, but there is some doubt connected with this case.

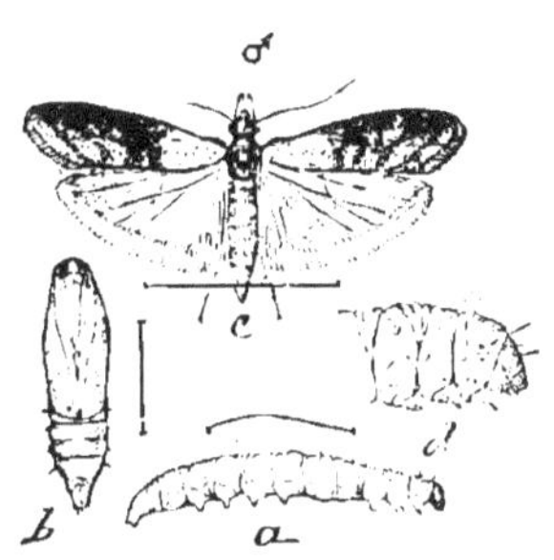

FIG. 30.—*Ephestia interpunctella*: *a*, larva; *b*, pupa; *c*, adult—enlarged; *d*, head and thoracic joints of larva—still more enlarged (original).

We have figured the states of *interpunctella* (Fig. 30) in comparison with those of *kühniella* (Fig. 28), in order that both may be readily recognized. The early states are quite similar in appearance, but the larvæ may be distinguished by the following characters:

The larvæ of *kühniella* are more slender and of a more uniform diameter than those of the other species. The abdominal legs are longer, cylindrical, with a circular fringe of hooklets at the crown. In *interpunctella* the legs are short, conical, with the fringe of hooklets at the crown oval. All piliferous warts in *kühniella*, most of which are rather minute, are still rather prominent, readily observed, and of a black or brown color. Those most conspicuous are the lateral ones, in front of the first spiracle; the subdorsal one, each side of the meso-thorax, almost completely encircled by a narrow black ring interrupted only at its upper margin (Fig. 28*d*). In *interpunctella* all the warts, while present, are concolorous with the rest of the body, and can be distinguished only with great difficulty. The surface of the body of *kühniella* is almost perfectly smooth, while that of *interpunctella* is somewhat granulate.

THE OX WARBLE.

(*Hypoderma bovis* De Geer.)

With each of the recent, and withal valuable, articles in the *Farmers' Review** relating to the above-named insect, appeared the running headline. "The First Investigation of the Subject in this Country," and this rather boastful announcement was coupled with certain reflections on the study of this insect by entomologists of this country, which were scarcely justified and added nothing to the otherwise excellent results obtained. While it is true that no careful estimate of the amount of damage occasioned by the fly in this country had been previously made, and the data relating to this phase of the subject is the most valuable outcome of the work of the journal referred to, it is also equally true that the life-history and habits of the fly, and the means against it which the *Farmers' Review* recommends to its readers as of most value, have been frequently given in various agricultural and scientific journals of this country.

Indeed, the chief characteristics and habits of this common cattle pest, which occurs all over the civilized world, have been known, together with some of the means now recognized as of the most avail against it, from the earliest times. One of the best accounts appeared nearly one hundred years ago in the *Transactions of the Linnæan Society of London*, 1796, Vol. III, page, 289 in a paper read by Mr. Bracy Clark, entitled "Observations on the Genus Œstrus," in which the habits and means against the Ox Bot were detailed practically as they are known to-day. Vallisnieri, Réaumur, Geoffroy, De Geer prior to Clark, and Fallen, Joly, Brauer and Schiner subsequently, have each published careful observations.

This insect has not attracted so much attention in the United States as in England, especially since Miss Ormerod began to investigate and publish upon the subject. Nor is its work so important with us as it is in England, on account of the relatively higher price of cattle and hides there. Yet in our scrap books we have a considerable number of articles clipped from American journals during the past twenty years, and in January, 1877, we published in the *Scientific American* an article on Bots which was quite widely quoted, and which, while dealing with bots in general, gave briefly the habits, ravages, and means against *H. bovis.*

We may here reproduce that article as far as it refers to the insect under discussion, and add such further details as may be necessary to a full understanding of the subject:

* * * Almost all cloven-footed animals, and many other herbivorous species, are infested with bots. These are legless grubs which fall into three categories: (1) Gastric, or those which are swallowed by the animal infested, and which live in the stomach in a "bath of chyle." (2) Cephalic, or those which crawl up the nostrils and inhabit the frontal sinuses. (3) Cutaneous, or those which dwell in tumors just beneath

* See INSECT LIFE, Vol. II, No. 5 (Nov., 1889) pp. 156-158.

the skin. They are all the larvæ or early states of two-winged flies (Diptera) belonging to the family Œstridæ, characterized by having the mouth parts entirely obsolete, and popularly called gad-flies or bot-flies. * * * In the third kind, the parent lays the egg on those parts of the body which can not well be reached by the mouth of the animal attacked, and the young grub, which soon hatches, burrows in the flesh, and subsists upon pus and the diseased matter which results from the wound inflicted, and the irritation is constantly kept up. The well-known wormal or ox bot (*Hypoderma bovis*), so common along the backs of our cattle, and especially of yearlings and two-year-olds, and dreaded as much by the tanner as by the animal infested, is typical of this kind. Residing in a fixed spot, we no longer find in this species the strong hooks at the head, and the spines around the body are sparse and very minute, while the parts of the mouth are soft and fleshy.

All bot-larvæ breathe principally through two spiracles placed at the blunt and squarely-docked end of the body, and in the ox bot these are very large, and completely fill up the hole to the tumor in which the animal dwells. When ready to transform, it backs out of its residence, drops, and burrows into the ground, and there, like the other species, contracts and undergoes its final change to the fly. The eggs of this ox bot are elliptic-ovoid, slightly compressed, and have at the base a five-ribbed cap on a stout stalk with which to strongly attach them to the skin of the animal. (See Fig. 33*a*.)

The perfect insect (see Fig. 31) is something over one-half inch in length, black, banded with yellow, as indicated in the figure, and is not unlike a bee in appearance. The flies issue during the entire summer, but are particularly abundant during the months of July and August. The individual life of each fly is, however, comparatively brief, not exceeding a month. The time between the deposition of the egg and its hatching has not been definitely observed, but, from what is known of other species of the family, will be found to last but a few days. During the fall and winter months the young larvæ develop very slowly; but in spring and early summer growth is much more rapid and the characteristic hard swellings with central opening, now large and prominent, exuding a yellowish matter, may easily be discovered. Fig. 33*b* represents the full-grown larva, together with the figures of the anal breathing pores, all enlarged. (The lines at the side of the larva, puparium, and egg indicate their natural size.)

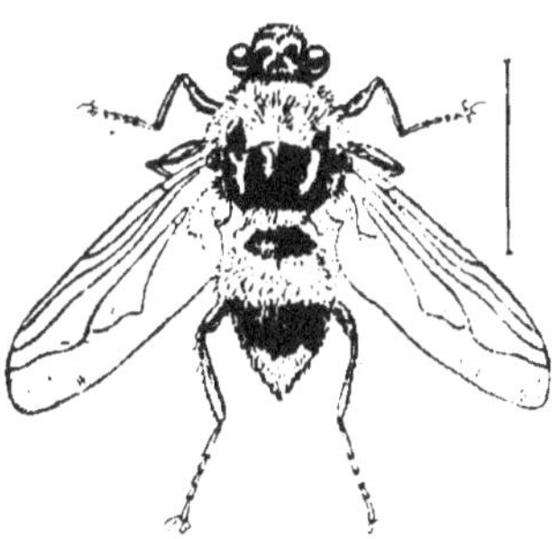

FIG. 31.—*Hypoderma bovis*—enlarged (after Brauer).

FIG. 32.—*Hypoderma bovis*; head of female fly from the front—enlarged (after Brauer).

On escaping from the back of the animal the larva, which in the earlier stages is yellowish white, is of a gray color, which rapidly darkens until in the contracted puparium the color becomes very dark brown, almost black. The pupa state lasts about thirty days, the time depending somewhat upon the weather, and the perfect insect escapes by forcing open a peculiar subtriangular lid at the anterior extremity of the puparium, a figure of which showing the lid detached we reproduce from Clark's earliest paper (see Fig. 34).

The facts in the life-history above given are for the most part well understood, and there has been little difference of opinion among authorities except as relating to the exact manner of the deposition of the egg. Those who believe that the eggs are thrust into or beneath the skin express a belief admittedly not based upon observation, and contrary to all analogy. That there should be differences of opinion upon a question where observation is so difficult is, perhaps, not to be wondered at. It is extremely difficult to follow the movements of the parent fly on an animal rendered restless or frantic by her presence or her attacks, and it is further quite difficult to discover a single egg concealed by the hair of the animal's back. The manner of placing the egg given by us in the article quoted above is based on experience with warbled cattle in Illinois from 1860–'63, when we were interested directly in stock raising, had charge of some three hundred head of cattle, and had frequent opportunity to examine and study the grubs *in situ* and the habits of the perfect insect.

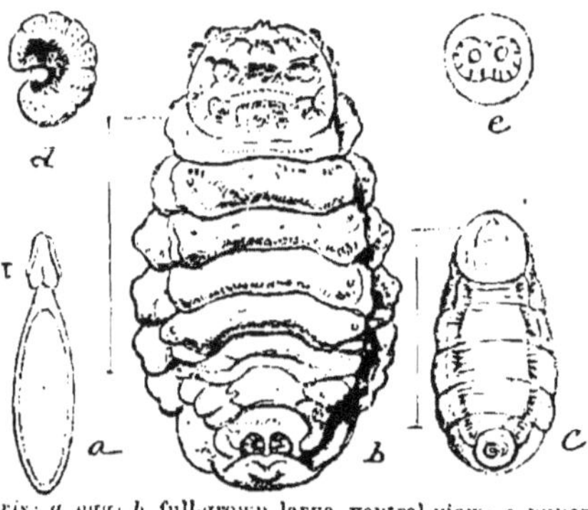

FIG 33.—*Hypoderma bovis*: *a*, egg; *b*, full-grown larva, ventral view; *c*, puparium, ventral view; *d*, newly hatched larva, side view; *e*, anal stigmata of larva—all enlarged (after Brauer).

It is a long time ago and we made no definite notes at the time, but we believe that we can trust our recollection. Analogy, unity of habit in the family, and structure all confirm it and are against the belief in insertion.

A careful study of the structure of the egg (Fig. 33 *a*), which we have seen in this and in a very closely allied species, the so-called Heel-fly (*Hypoderma lineata*), as well as the descriptions and figures by other authors, show that the grooved and slightly pediceled enlargement of the end which is attached is admirably adapted for being strongly fastened to the skin and to the base of the hairs, and all observations that have been recorded point to the fact that the young larva works its way directly from the egg under the skin, as is the case with other parasitic

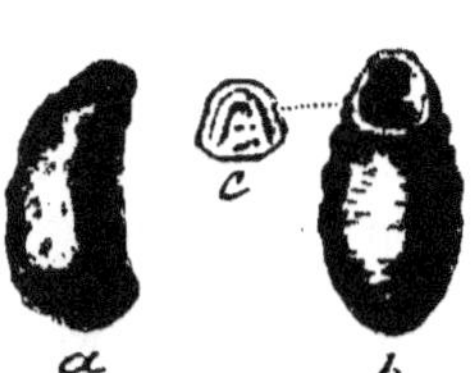

FIG. 34.—*Hypoderma bovis*: *a*, puparium, from side; *b*, same, from below, showing exit hole of adult; *c*, cap which splits off to allow the adult to issue—natural size (after Clark).

Diptera. The structure of the ovipositor clearly excludes the possibility of puncture, for, though horny, it has a blunt trifid tip, and is beset at the end with certain minute hairs, and structure of this character is a very safe guide to habit. Figure 35 is drawn so as to show the telescopic and extensile nature of this organ.

The excitement, amounting often to frenzy, which has been noticed in cattle when the bot-flies are ovipositing, and which has probably led to the idea of stinging, results from the instinctive dread of the fly rather than from any real pain, though no doubt the secretion which so firmly fastens the eggs is accompanied by an irritating sensation. This will account for most of the supposed cases of stinging, including the case of the man-infesting bots. (*Vide* INSECT LIFE, Vol. 1, pp. 76 and 226.) In the case of the horse Bot-fly or the sheep Bot-fly, where it is well known that the eggs are not inserted, the animals exhibit a similar dread and nervousness. The fact that the egg has been observed partly extruded from the fly about to oviposit also bears on this point.

Mr. Bracy Clark, in "An Appendix or Supplement to a Treatise on the Œstri and Cuterebræ of Various Animals"* (*Transactions Linnæan Society, London*, 1843, Vol. XIX, pp. 81-94), which treatise was but an elaboration of the paper already mentioned by us, after describing the peculiar noise of the parent fly which is apparently so frightful to cattle, says:

> We may also further observe that there can not be any very painful affliction, as the fly has really no instrument fitted for such a purpose, the feminine ovipositor being a mere tube, made of flexible materials, piece inserted in piece, exactly as in the common telescope. However, it is possible on reaching the skin or cuticle of the beast, which is always highly sensitive in these hairy animals, that it might produce a degree of uneasy tickling, which, added to the noise, and perhaps an instinctive fear, always impressed upon them, is altogether sufficient for the extraordinary alarm we see.

F. Brauer, in his *Monographie der Œstriden* (1863), while stating that the manner of placing the egg is still obscure, does not think that the egg is inserted into the hide. He has found also what he supposed to be the newly hatched larva in the first layers of the skin near the exterior surface.

Miss Ormerod was at first strongly inclined to believe that the eggs are deposited below the skin, but in her latest pamphlet on the subject she says that the egg is probably deposited on the surface, and that the newly hatched maggot makes its way through the skin by means of the sharp, cutting hooks clothing its body surface. In support of this she says:

FIG. 35.—*Hypoderma bovis*. ovipositor of female: *a*, from side; *b*, tip, from below—enlarged (original).

*An essay on the Bots of horses and other animals, London, 1815.

That the jagged-sided channel (not smooth-sided as it would be if pierced by an egg-laying tube) leads in a slanting or straight or curved direction from a little opening at the bottom; * * * also I have found the tunnel partly cut down from the outside, and I have found a small, soft body in it.

The injury occasioned by the presence of these grubs to hides and the diminished quantity and inferior quality of the beef and dairy products were perhaps sufficiently indicated in our notes on this subject in the last number of INSECT LIFE.

The value of the application of various oils both to prevent the oviposition of the fly, and especially to destroy the larvæ, has been long known; and, aside from the discovery that certain substances are more effective than others, little has been added to our knowledge of remedies of late years. Clark, in the articles already cited, fully indicates the good of such applications, and states that Pliny, who was acquainted with these flies, "has recommended for protecting animals from their attacks to annoint them with fats and oils."

In our article, which we have already quoted at length, the use of kerosene was particularly recommended to destroy the larvæ, as well as to deter the fly from ovipositing. In the discussion of remedies by Miss Ormerod, in her various reports, a number of strong-smelling oils are recommended, with which to smear the animals' backs to protect them from the fly. Of these, train-oil or fish-oil—the same that has proved of advantage against the Buffalo Gnat in the South and recently against the Horn Fly in Virginia and Maryland—has been especially recommended. A similar application is the simplest and easiest method of destroying the warbles, which it does by closing the breathing pores on the posterior end of the body. The destruction of the larvæ in this way may be effected by one or two applications in autumn, and is the most satisfactory method of controlling this pest. The appearance of the flies during the entire summer renders operations against these difficult and expensive.

Additional means of protection against the flies are: the use of kerosene emulsion, rancid butter or tar-oil mixed with sulphur, or dry sulphur alone; against the grubs, any of the oily preparations mentioned above, and in addition, the placing of a small quantity of mercurial ointment on the hole in the skin, or of spirits of tar, or carbolic acid; finally, piercing the grub with some sharp instrument or removing it by pressure.

This condensed account of what is known of the habits of this Bot Fly is given partly in compliance with an earnest request from Mr. Alexander, of the *Farmers' Review*, that we take up the question, and partly with a view of showing how little there is to be done by the Department of Agriculture except by extending the inquiry in statistical lines somewhat similar to those followed by him. Even admitting that some more careful observations might be made bearing on the actual mode of oviposition and duration of the egg state, these are points of biologic interest rather than of economic importance.

The point, therefore, to be considered is whether the question of fuller statistical information as to damage done is sufficient to justify national investigation. This can best be answered by stock-raisers and breeders themselves, and where they themselves have not sought or urged such an investigation we should hardly feel justified in spending time and means therefor, considering the large amount of work on hand for which there is pressing demand.

Being thoroughly familiar with the stock interests of the country, we know how difficult it is to get farmers to care for their stock so far as this warble is concerned, and we are satisfied that where self-interest does not dictate better attention, we can do little more than point out the means of avoiding injury and the desirability of so doing.

ASSOCIATION OF ECONOMIC ENTOMOLOGISTS.

FIRST ANNUAL MEETING.

NOVEMBER 12, 1889.

The second meeting of the Association of Official Economic Entomologists was opened by a session held at 11 o'clock in the rooms of the Department of Insects at the U. S. National Museum, the president, C. V. Riley, occupying the chair. The following members were present: C. V. Riley, Washington; S. A. Forbes, Illinois; A. J. Cook, Michigan; J. A. Lintner, New York; Lawrence Bruner, Nebraska; William Saunders, Ottawa; J. P. Campbell, Georgia; C. P. Gillette, Iowa; R. Thaxter, Connecticut; H. Garman, Kentucky; W. B. Alwood, Virginia; Otto Lugger, Minnesota; M. H. Beckwith, Delaware; W. H. Ashmead, E. A. Schwarz, Th. Pergande, M. L. Linell, C. L. Marlatt, Tyler Townsend, and L. O. Howard, Washington.

In the absence of the secretary, Mr. J. B. Smith, Mr. L. O. Howard was nominated and elected secretary *pro tem.*

The minutes of the previous meeting were read, and, with a single exception, approved. Mr. Howard, as a member of the committee on by-laws, read the report of his committee. The report was accepted, and the by-laws read by paragraphs, amended, and adopted, as follows, with the exception of section 2, of Article III, which was laid upon the table for future action:

BY-LAWS.

ARTICLE I.—*Of Members.*

SECTION 1. The classes of members are defined in the constitution, as are their rights to vote or hold office. Members of all kinds have equal privileges as to presentation of papers and in the scientific discussions at the regular meetings, and may, by permission of the presiding officer, speak on business questions before the association.

SECTION 2. All members have equal rights to the published proceedings of the association and to any publications controlled by or distributed by the association, save that should any publications of economic interest be distributed by the association, the distribution lists furnished by the active members are first to be regarded.

ARTICLE II.—*Of Officers and their Duties.*

SECTION 1. It shall be the duty of the president, in addition to the ordinary duties of a presiding officer, to prepare and deliver an annual address, to be delivered at the annual meeting over which he presides.

SEC. 2. It shall be the duty of the secretary to provide the necessary stationery and such books as he may be directed to provide, the expenses for which shall be met by an assessment of not less than 25 cents on the members in attendance at the meetings. The sum so collected shall be used by the secretary to re-imburse himself for advances made and to meet the ordinary expenses of the association. An account shall be rendered at each annual meeting, and if needed, an additional assessment shall be imposed.

SEC. 3. All officers shall be elected by ballot after open nomination, and this by-law shall not be suspended except by unanimous consent of the voting members present.

ARTICLE III.—*Of Meetings.*

SECTION 1. Notice of the time and place of meetings shall be published in all the American entomological periodicals and in INSECT LIFE.

SEC. 2. Special meetings shall be called as provided for in the constitution, and notice of such meetings shall be given by the secretary by mailing to each voting member a formal specification of the time and place of meeting at least two weeks before the date fixed in the notice. The notice shall state the reason for such meeting, and shall specify the business to be transacted, and no other business shall be transacted.

SEC. 3. The order of business at regular meetings shall be, at the first session:

(1) Calling the meeting to order by the president.
(2) The annual address by the president.
(3) Reports of officers.
(4) Reports of committees.
(5) Proposal and election of members.
(6) Written business communications.
(7) Verbal business communications.
(8) New business.
(9) Programme of papers and discussions.
(10) Adjournment.

On the following sessions:

(1) Reading and action on the minutes of previous meetings.
(2) Unfinished business.
(3) Proposal and election of members.
(4) New business.
(5) Programme of papers and discussions.
(6) Adjournment.

At the last session of the meeting the order of business shall be as at other sessions except that after order 5 will come:

(6) Election of officers for the next meeting.
(7) Fixing time and place of next meeting.
(8) Reading and action on rough minutes of the entire session.
(9) Final adjournment.

ARTICLE IV.—*Amendments to By-laws.*

SECTION 1. Changes in these by-laws may be made at any regular meeting in the same manner and on the same notice as prescribed in the constitution for amendments to that instrument.

The association then adjourned until 1.30 p. m.

AFTERNOON SESSION.

The meeting was called to order at 1.50. The following names were added to the list of active members: C. L. Marlatt and Tyler Townsend, of Washington, D. W. Coquillett, California; E. A. Popenoe, Kansas; J. M. Stedman, New York; C. H. Fernald, Massachusetts. The application of A. S. Packard of Rhode Island was referred to a committee consisting of the president and the secretary, with power to act after it shall have been ascertained whether Dr. Packard at present teaches economic entomology.

The following were elected associate members of the association: F. W. Goding, Illinois; T. D. A. Cockerell, Colorado; George D. Hulst, New York.

Arthur E. Shipley of Cambridge, England, was elected a foreign member.

The secretary was instructed in the case of the applications of F. H. Chittenden, of New York, C. L. Eakin, of West Virginia, and George F. Whittemore, of Massachusetts, to inform the applicants that according to the present information of the association they are not entitled to associate membership.

Upon the nomination of Mr. Cook, Mr. F. H. Hillman, of Nevada, and upon the nomination of Mr. Forbes, Mr. John Marten, of Illinois, were placed on the rolls as active members. Upon motion, a committee of three upon programme was provided for, the president appointing Messrs. Howard, Cook, and Lugger.

The secretary, on behalf of the Entomological Society of Washington, invited the visiting entomologists to attend a meeting Tuesday evening at the residence of Dr. William H. Fox, 1826 Jefferson Place.

Mr. H. Garman read two papers entitled: (1) "Notes on a Corn Root-worm in Kentucky;" (2) "The Bordeaux Mixture as an Insecticide." The writer had discovered that the Corn Root-worm of Kentucky is not *Diabrotica longicornis*, but *D. 12-punctata*. He has studied its life history at some length and has discovered that it is double brooded in Kentucky, and surmises that it hibernates as an adult. It affects moist lands much more severely than dry lands, and the previous crop seems to have little relation with the amount of damage, which is contrary to the state of affairs in Illinois with *D. longicornis*. Its work is like that of the allied species, and many fields were found to be severely injured. He described briefly the different stages of the two species and suggested remedies.

In his second communication he described the effect which treating potatoes with the Bordeaux mixture had upon the Flea Beetle and upon the Margined Blister-beetle. His experiments showed that potatoes treated with it were damaged much less by both species than were portions of the crop which were untreated.

In discussing these papers Mr. Riley stated that the transformations of *D. 12-punctata* and its corn-root feeding habits had been known to him for some years, the species being included among the divisional notes at the Department. Mr. Forbes had experienced the same thing in Illinois and stated that in small fields the yield had been reduced 20 per cent. He also had found only one brood in Illinois. He stated the curious fact that occasionally larvæ were found of a red color, in which microscopic examination revealed a Bacillus, which he had succeeded in cultivating and is now growing in culture tubes. The culture medium is stained red, and this is due to a diffusion of color and not to penetration of the Bacillus. Mr. Riley stated that the adults of *Diabrotica* unquestionably hibernate. Mr. Pergande stated that he had found *D. 12-punctata* in the neighborhood of Mount Vernon some years since, feeding very abundantly upon the roots of corn.

Mr. Lintner, in discussing Mr. Garman's second paper, called attention to the distinction between "insecticide" and "preventive measures," and hardly thought that Mr. Garman could call the Bordeaux mixture an insecticide in the case he had mentioned.

Mr. Garman stated that he considered Mr. Lintner's point well taken, and that he was really not certain that the insecticide effect of the mixture was as great as its preventive effect. He had proved, however, upon a small scale in confinement that it had an undoubted insecticide effect upon the Colorado Potato-beetle. Mr. Riley stated that recently in France he had seen the Bordeaux mixture used upon a large scale, and that its effect could be distinguished at a distance, as it gave the vines a bluish or glaucous appearance. He stated also that it had been found in France that many insects are destroyed by this mixture. He stated that the discovery of the use of the Bordeaux mixture was an excellent illustration of accidental discovery, since, long before the appearance in France of the *Peronospora viticola*, vine-growers in the neighborhood of Bordeaux had used the mixture on the outer rows to deter thieves, and when the disease appeared it was found that the vines thus treated were not affected. He stated that were he a vine-grower he should certainly mix some other insecticide with the mixture, in order to more surely accomplish two results at once.

The meeting then adjourned until 11 o'clock Wednesday morning.

NOVEMBER 13, 1889.

The association met at 11 a. m., President Riley in the chair. The minutes of the previous day were read and approved.

By a special motion it was resolved to omit all personal titles in the minutes.

Under the head of "new business" it was moved and carried, in obedience to a suggestion that such action was desired, that the association co-operate with the Association of Agricultural Colleges and Experiment Stations in so far as to meet with them at 4 o'clock and to report progress.

S. A. Forbes then read a paper entitled "Office and Laboratory Organization." Premising that every laboratory should start with a well-considered and elastic scheme, he described at some length the circumstances of his own work, and his own plan of organization.* In discussing this paper Mr. Riley dwelt upon the subject of simplicity in methods and described the system which he had introduced into the Division of Entomology and the Department of Insects of the National Museum. He believed that, in the case of rapidly accumulating material sent in from all parts of the country, to keep such a record of all accessions and a system of cross-references as described by Mr. Forbes would involve an amount of clerical work hardly justified by the results, and described his methods, particularly in regard to the recording of biologic material.

Parallel with Mr. Forbes' paper, he discussed methods of keeping and cataloguing letters, newspaper clippings, and books.

The topic was then announced for discussion, "Where shall we publish descriptions of new species and results of non-economic observations?" Mr. Lugger stated that at his station he found it impossible to publish anything in the bulletins which was not of self-evident practical importance, and that he was accumulating a great deal of valuable information which thus could not see light.

Mr. Bruner stated that a different condition existed at his station, and that he was obliged to publish such observations and even descriptions of new species, but to insert them as foot-notes.

Mr. Riley read a letter from Mr. C. H. Fernald, of Amherst, objecting strongly to the publication of descriptions of new species in bulletins. Mr. Gillette stated that he was obliged to publish descriptions of new species in the bulletins of his station, his director insisting that the results of all the original work done at the station should first see light in its bulletins. Mr. Cook expressed agreement with Mr. Fernald's letter and offered the following resolution:

"*Resolved*, That it is the opinion of this association that the bulletins of the Experiment Stations and Agricultural Colleges should not contain descriptions of new species."

The resolution was unanimously adopted.

Mr. Forbes moved that the secretary represent the association at the 2 o'clock meet-

---

* This paper appears in full in this number, p. 185.

ing of the Experiment Station section and convey the resolution of this association to meet with them at 4. Carried. The association then adjourned until 2.30.

AFTERNOON SESSION.

The meeting was called to order at 2.50.

Mr. Cook offered the following resolution:

"*Resolved*, That a committee be appointed to act in connection with the Association of Agricultural Colleges and Experiment Stations."

After some discussion the resolution was adopted.

Mr. Howard moved that a committee of three, with the president as chairman, should be appointed to report to the section of Experiment Stations at 4, and also to attend the discussion of amendments to the constitution of the general association in order to explain the status of the Association of Official Economic Entomologists.

The topic, "How far shall we recommend patent insecticides and machinery," was announced for discussion. Messrs. Cook, Lugger, Bruner, Forbes, Riley, and Lintner discussed this topic at some length, the joint opinion being embodied in the following resolution, which was offered by Mr. Cook and adopted by the association:

"*Resolved*, That in our opinion we, as officers of the Experiment Stations, should be slow to recommend even by mention any patent insecticide until by analysis and test we find it worthy of recommendation."

The general opinion seemed to be that in case a patent insecticide proved to be thoroughly efficacious and *sufficiently cheap* there should be no hesitation in recommending it.

Mr. Gillette read a paper entitled "Spraying Points," in which he gave certain conclusions which he seemed to have reached by recent experiment. He stated that white arsenic freshly mixed with cold water did less damage to foliage than Paris green, while London purple brought about greater damage than Paris green. Arsenic, however, prepared by boiling, produced a more injurious effect than either of the other substances, which would indicate that it is the arsenic in solution that is to be feared.

Mr. Marlatt called attention to the fact that the different sides of the same tree, according to sun exposure, and difference in the ages of trees, tend to produce different results from spraying. The matter was discussed at some length by Messrs. Riley, Lintner, Cook, and Gillette.

Mr. Gillette read a paper entitled "Codling Moth Experiments," in which he gave the results reached at the Iowa Experiment Station the past season by using a dry application of Paris green in plaster, in the proportion of 1 of the poison to 100 of the plaster; an application of carbolized plaster prepared by thoroughly mixing 1 pint of the crude acid in 100 pounds of plaster, and an application of London purple in water in the proportion of 1 pound of the poison to 128 gallons of water. By estimating the protection in the usual manner it was found that the Paris green and plaster application saved 94 per cent., the carbolized plaster 34 per cent., and the London purple 68 per cent. of the fruit that would have been wormy in the absence of any treatment.

Mr. Gillette then called attention to the fact that nothing like correct results could be expected by figuring out the protection in the ordinary manner except in northern latitudes where the insect is single brooded. The results obtained would be too small. In order to get accurate results the two broods must be kept separate, otherwise the results will be greatly vitiated by the great number of eggs that will be laid upon the sprayed trees by moths flying in from the checks and also by the smaller number of eggs that will be laid on the checks because of the great number of larvæ of the first brood destroyed upon the treated trees in their vicinity.*

*This paper will appear in Bulletin No. 7 of the Iowa Agricultural Experiment Station.

As the time for adjournment had arrived, the discussion of this paper was postponed until the next session.

The association adjourned to meet at 9 o'clock, November 14

NOVEMBER 14, 1889.

The meeting was called to order at 10 o'clock by Vice-President Cook. The minutes were read and approved.

The secretary read a letter from D. S. Kellicott, who stated that at present he considered himself not eligible to membership.

Under the head of "programme" the chair announced that the discussion of Mr. Gillette's paper of the previous day was in order.

Mr. Forbes expressed himself as of the opinion that, from our present knowledge of the use of the arsenites as insecticides, they can be recommended for use on the peach. In spraying for codlin moth he had not found that any special benefits resulted from spraying for the second brood.

Mr. Cook had found that injury resulted to the peach from the use of white arsenic stirred in cold water.

Mr. Riley, regarding the apparent revulsion of feeling concerning London purple, stated that in his opinion we must be very slow in reversing judgments, carefully formed, of years of experience, and that both London purple and Paris green varied in quality; that their effects varied on different plants, and even in different kinds of weather.

Mr. Bruner presented some notes on *Diabrotica longicornis*, which he had found very abundantly in the city of Lincoln, Nebr., as late as the middle of October. The species is to a certain extent nocturnal in habit, as he had collected 250 at one electric light. The adults feed upon the foliage of radishes and turnips, and have been found about the roots of the wild sun-flowér. He has not found it breeding at the roots of corn, but knows that it does so occur in his State. He thinks that it must have some other larval food plant.

In discussing this paper Mr. Forbes stated that he had failed to find this insect breeding upon anything else than corn, although extensive search had been made by himself and his assistants for other larval food plants. He admits that there is a strong possibility that it has other food plants, and Mr. D. S. Harris thinks that he has found it upon Purslane. Mr. Forbes considers the species as normally inhabiting the far West and spoke of its extraordinary increase in Illinois in late years. Twenty years ago Walsh mentioned the finding of three specimens in Illinois as worthy of remark.

Mr. Lugger had found three pupæ of *D. 12-punctata* at the roots of Rudbeckia in a field which had been grown in corn the previous year.

Mr. Garman stated that Mr. D. S. Harris thought that he had also found it upon the roots of Lambs-quarter.

Mr. Riley stated that years ago in Missouri it was very rare, and may be considered as belonging to the class of insects which have changed their habits of late years.

Mr. Bruner had never seen a specimen in Nebraska until within the last two or three years.

The topic of "Co-operation" was then taken up for discussion.

Mr. Lugger suggested that the distribution of beneficial insects was a subject which might enter into a co-operative scheme.

Mr. Forbes stated that he had formulated no distinct plan of co-operation, but that in his opinion there was no objection to duplication of work, but that there were, rather, arguments in favor of it.

The question resolved itself into two heads: How can State workers help each other, and how can the General Government help State workers?

On the latter point he stated that in his opinion the assistance will be comparatively of a technical character in the way of determination of specimens and references to literature. As this side of the work is more likely to be overlooked, he would be glad to see a resolution passed commending the technical side of the Government work in entomology to Congress.

Referring to Mr. Lugger's suggestion, he further suggested that the distribution of diseased insects afforded an opportunity for co-operation.

Mr. Riley stated that he felt strongly that an opportunity for co-operation existed in special lines. He thought that a standing committee on co-operation might be appointed to plan definite experiments on mooted questions and to send out authoritative suggestions to station entomologists and to members of the association. He suggested uniform standards and uniform and better correlated results. In regard to the gathering of statistical information, he instanced the case of *Hypoderma bovis*, stating that the work of the Department at Washington could be greatly facilitated by the assistance of different entomologists in their respective localities. The case of the spread of a new pest affords another field, as accurate information of the rate and extent of the spread could be more easily gained by co-operative work.

Mr. Forbes spoke of the concert of observations and report in regard to outbreaks over a wide area, but considered that all arrangements should be flexible and that the work of a formal committee might be cumbersome and slow. He thought that the work might be accomplished by mere suggestion, by letters either from individual workers to one another or from the Department at Washington to the members of the association.

Mr. Lintner thought that it would be desirable and that the members of the association had a right to ask that the Division of Entomology should formulate a plan of co-operation and that the Division itself should also have the right to call for aid on the members of the association. He instanced the Rose Bug as a case where co-operation would be advisable. He had learned from a correspondent in Virginia that the Rose Bugs of a given neighborhood came from a swamp, and he urged that all members of the association in localities where this insect is abundant should endeavor to find whether its breeding places were restricted to sandy or swampy localities.

Mr. Alwood rather dissented from the proposition that the co-operation should be left to correspondence. He thought that the particular charging of a committee with the planning of work would be more effective.

Mr. W. O. Atwater, by invitation, addressed the association and said that the plan adopted by the horticulturists seems to him a very good one, and thought that it would be advantageous to extend the scheme of co-operation beyond the experiment stations and to interest all practical workers in the subject. He dwelt at length upon the necessity of a high scientific ideal.

Mr. Lintner offered a resolution which, after amendment, was adopted in the following form:

"*Resolved*, That a committee of five be appointed by the president, of which he shall be chairman, and which shall consider and report to the next annual meeting upon a method or methods to secure co-operation among the members of the association. It is also authorized to represent the association in conference with any committee on entomology which may be appointed by the Association of Agricultural Colleges and Experiment Stations."

The topic of "amendments to the constitution" was then brought before the association for discussion.

Mr. Forbes moved that the paragraph relative to meetings be amended to read as follows:

"The annual meeting shall be held at such place and time as may be decided upon by the association at the previous annual meeting, and special meetings may be called by a majority of the officers. Eight members shall constitute a quorum for the transaction of business."

Mr. Lintner proposed that the opening paragraph of the constitution be amended so as to read as follows:

"This association shall be known as the Association of Economic Entomologists."

It was moved and carried that section 2 of Article III be taken from the table, and upon motion it was adopted in the following form:

"Special meetings shall be called as provided for in the constitution, and notice of such meetings shall be given by the secretary by mailing to each voting member a formal specification of the time and place of meeting at least two weeks before the day fixed in the notice. The notice shall state the reason for such meeting and shall specify the business to be transacted, and no other business shall be transacted at the special meeting."

The meeting then adjourned to 4 p. m.

AFTERNOON SESSION.

The association reconvened at 4 o'clock; President Riley in the chair.

The committee appointed to report to the Association of Agricultural Colleges and Experiment Stations reported that they had taken no action, as no opportunity had been allowed.

The following resolution was proposed by Mr. Cook and unanimously adopted:

"The Association of Official Economic Entomologists desire to express their hearty appreciation of the generous support afforded the Entomological Division of the Department of Agriculture, as is shown by the publication of bulletins, reports, and INSECT LIFE, no less than the aid which we receive individually through this Division of the Department. We also recognize the great opportunity of the Division to publish monographs, and especially to advance the technical part of entomology. Therefore we wish to express to the Secretary of Agriculture our great desire that all possible aid be given this Division, that such publications may be increased and such valuable work further extended."

The following resolution was offered by Mr. Alwood and adopted by the association:

"*Resolved*, That the committee on co-operation appointed by the Association of Economic Entomologists express a desire to co-operate with the committee on entomology of the Association of American Agricultural Colleges and Experiment Stations."

It was moved, seconded, and carried, that the association hold its next annual meeting at the same time and place at which the Association of Agricultural Colleges and Experiment Stations next meets.

The president appointed as his colleagues upon the committee to submit a plan of co-operation, S. A. Forbes, J. H. Comstock, A. J. Cook, and J. A. Lintner.

It was moved, seconded, and carried, that it is the sense of this meeting that the officers elected at the preliminary meeting should hold office until the second annual meeting.

It was moved and carried that the Department of Agriculture be requested to publish the proceedings of the present meeting in INSECT LIFE.

On motion of Mr. Lintner a vote of thanks was given to the acting secretary.

The association then adjourned.

L. O. HOWARD,
Secretary pro tempore.

OFFICE AND LABORATORY ORGANIZATION.*

By S. A. Forbes, *Champaign, Ill.*

With the sudden establishment of a large number of new offices and laboratories of investigation in a field hitherto very slightly occupied, the subject of special office organization and equipment becomes highly important and interesting, and will become more so as the work of each station increases in scope, difficulty, and complexity. Although I have never been a station worker, in an experience of fifteen years in the gradual development of a natural history institution, in which I began ignorant and alone and which now commonly employs six to eight assistants, I have learned, among other things, the very great importance of having from the first a well-considered and *elastic* scheme of organization, under which the work may grow freely from year to year without *outgrowing* any of the more or less costly equipment of its earlier periods. While an investigator works alone, or with mechanical aids at most, he needs little else, perhaps, but helps to memory; but as soon as he finds himself able and obliged to call in the aid of more or less skilled assistants, the results of whose labors he must be able to command and collate rapidly at will, he finds an elaborate system indispensable. A future of this description I hope we may all at least look forward to; and it is on this ground that I have thought it profitable to describe my own system—tested now by several years' use in a field somewhat more trying, probably, than the average station worker will need to occupy.

The institution to which I refer combines under one management a natural-history survey of Illinois, the work of the official entomologist of that State, and the instruction work of the department of zoölogy and entomology in the State University; and the object of its organization is such a co-ordination of the collections (both determined and undetermined, technical and economic), the collection records, the notes of observations and experiments (whether my own or those of my assistants), the correspondence of the office, and the literature accessible to us, that each and all of these may be readily drawn upon and made completely available for the treatment of any subject whatever which comes within our field.

The essentials are the collections (classified and unclassified), the records, the notes and correspondence, and the library; and the organization consists in an arrangement and orderly analysis of each of these, with a complete system of cross-references from one to another. The collections are, as usual, the reference collections (determined, labeled, and precisely arranged in the zoölogical order) and the miscellaneous, duplicate, and undetermined material, including the economic series; the records are the accessions catalogue and the species catalogue, with card index to each; the notes are on slips, in labeled boxes, classi-

* Read before the second meeting of the Association of Economic Entomologists, November 13, 1889.

fied in zoölogical order; the correspondence is alphabetically arranged by half years; and the library is arranged in order of subjects and catalogued on cards, article by article, under authors' names, this card catalogue having subordinate subject indexes.

The reference collection in entomology is in excellently made double boxes, usually four specimens representing each species, one bearing a species label, which shows, beneath, the date and locality of the specimen and the name of the person responsible for the determination. The other three specimens have date and locality only, with sometimes a species number, where it is possible that specimens of different but similar species may get mixed by inadvertence in returning specimens to the boxes.

The miscellaneous, duplicate, and undetermined specimens are also in labeled boxes (if dry), all classified, at least to families, each winter, and all bearing a number corresponding to an entry in the accessions catalogue. If the species has been determined, the specimen will also bear a species catalogue number. The alcoholic economic and miscellaneous material is in vials and bottles, closely stored in racks, each vial bearing at least an accessions catalogue number, this series being arranged in numerical order.

The accessions catalogue contains an entry for each time and place at which collections have been made, showing date, place, collector's name, and the general character of the collection, as nearly as it can be conveniently described without determination. This catalogue has also a broad column for cross references to the species catalogue. These accessions catalogue numbers must be placed on every package of specimens received, and, as packages are broken up and the contents mounted, on each specimen, except where these are put into the reference collection, when the data indicated are written out on a label, as above described. All note slips referring to these collections must also make a cross reference to this accessions catalogue; that is, must bear the proper accessions catalogue number. In brief, every specimen, every note, and every entry in the species catalogue must show a reference to the accessions catalogue, and every entry in the latter must finally refer to the species catalogue by as many numbers as there were species in the collection represented by it. These latter references enable one to learn in a moment what any given collection consisted of.

Material intended for the breeding cages is likewise entered and numbered on the accessions catalogue, and this number is placed at the head of the breeding-cage record, kept on slips like the other notes. Whatever specimens are bred are similarly entered, references being made by number to these entries in the body of the notes.

The species catalogue is simply a numbered list of specific names, with references against each entry to all the accessions catalogue numbers representing collections in which the species was found. These references enable one to determine for each species all the dates and localities of its collection. This catalogue is indexed on cards, alphabetically arranged, each name on a card being followed by numbers corresponding to the various entries of that number on the species catalogue. We also keep up an accessions catalogue index made on a similar plan, intended to give us access to the miscellaneous and unclassified material in our collections.

The result of this arrangement is that no matter at what point one takes up a topic, whether he has before him a specimen, a note slip, an accessions catalogue entry representing date and locality of collection, or a species catalogue name, he can rapidly bring together from the other sources all the material, information illustrating it.

Our notes are all made on single slips of uniform size, suitable for either ink or pencil entries, and each has at the head the accessions catalogue number of the collection to which it refers, followed commonly, for convenience, with a brief general remark sufficing to show the nature of the object mentioned. These notes, as already explained, are in paper boxes, labeled on the edge with the name of the family or other group to which the notes contained apply, and arranged in systematic order, the scheme being a perfectly elastic one, requiring only the insertion of now and then a few new boxes, as the notes under any head become so numerous as to make subdivision necessary. In these boxes are also placed slips bearing brief abstracts of letters which contain important scientific information, with references to the places of these letters in the file.

The library has as the basis of its organization the authors' card catalogue already mentioned, with subject indexes, also on cards, the degree of analysis varying according to the needs of our work. The entries under each author's name being numbered, the references in the subject index are to the author's name and the number of his article.

If I were now to begin a new work, I would at once begin an accessions catalogue of collections, and an authors' catalogue to my library, and would keep my notes on slips, with references to the accessions catalogue entries. The other features of the scheme of organization I have outlined above could then be added as they were needed and as they could be provided for.

EXTRACTS FROM CORRESPONDENCE.

The Mediterranean Flour-moth.

* * * I know of no better means of obtaining information upon economic entomology than through the pages of your most valuable publication. I shall be obliged if you will insert the following notice of the appearance in Canada of the Mediterranean Flour-moth, *Ephestia kühniella*, with the double purpose of putting those concerned upon their guard against this troublesome and extremely injurious insect, and at the same time eliciting from your correspondents as much information as possible as to its occurrence in America. For a year or two it has been giving trouble in some of the large mills and feed-stores in England, and Miss Ormerod has published

a valuable notice and warning to English millers in her last report. During the past summer it has been brought to my notice as a most serious pest in one of our Canadian cities. The outbreak was so serious that our provincial government of Ontario took the matter in hand, and through Dr. P H Bryce, the secretary of the provincial board of health, have just issued a bulletin upon its operations, appearance in the different stages, and the means which have been adopted to eradicate it before it spreads further. This bulletin, which is written in a manner which will be understood by every one, is most timely, and will, I believe, be attended with very beneficial results.

The milling interests of America are, however, so enormous that it becomes important to make known its appearance here as soon as possible, so that prompt action may be taken immediately a new occurrence takes place.

The following extracts from Dr. Bryce's bulletin will show the gravity of the case. The first is condensed from the account given by the firm in whose mill the insects were observed.

"The first appearance of the Flour-moth we remember seeing was during the month of March, 1889. The moth was seen flying about in the basement of the mill, but little attention was paid to it. In April there was an appearance of a few moths on the different floors of the mill, even at the top. In the month of May we were troubled with a few worms in some of our goods, and in June more of them appeared. In July they increased rapidly. About the middle of July we shut down for a day or so; took the clothing from our bolting reels and cleaned it and washed the inside thoroughly with soft lye soap and lime. We did the same with the elevators. When we started up again every corner and part of the mill had been thoroughly cleaned, as we supposed, and we commenced to work again; but after about four days we found our bolting reels, elevators, etc., worse than before. They were literally swarming with webs, moths, and worms, even inside the dark chambers of the reels. We shut down again and made a more thorough cleaning by washing, etc. While this was going on we found there was no use to try and clear ourselves of the pest, as the mill walls, ceilings, cracks, crevices, and every machine was completely infested with moths, cocoons, and caterpillars, and there was no use going on."

Eventually the firm had to vacate their premises and build a new mill.

Dr. Bryce continues upon page 11 of the Bulletin, after detailing its habits, as follows:

"From the foregoing it will be apparent that the moth may not only be transported from one place to another in any one of its various stages, but that search for its presence in any one or all of these must be made where its presence is suspected. It will at once be seen how great are not only the dangers of its transmission from one mill to another and one locality to another, but also how many are the difficulties attaching to its detection, while as yet only a few individuals may have been introduced into a warehouse or mill. With what rapidity the *Ephestia kühniella* develops under favorable conditions, nothing will better illustrate than the correspondence of a sufferer therefrom already published. When it is stated that a large warehouse, some 25 feet wide, 75 feet long, and four stories high, became literally alive with moths in the short course of six months, while thousands upon thousands of the cocoons were found adherent to the walls, joists, posts, ceilings, and in every nail-hole, cracks in floors, partitions, machinery, and furniture throughout the whole building; while in sample boxes of cardboard, in small and large bags, in flour stored anywhere throughout the building, it was abundantly present, it will be understood what millers have to expect to encounter if they neglect the most vigorous measures to destroy the first moths which at any future time may appear on their premises. To illustrate further the difficulty of overcoming the pest, once introduced, it may be stated that several men have been at work in the building from which our correspondent has removed his machinery, for over a fortnight in burning all woodwork, as flooring, fixtures, etc., sweeping down walls and destroying the rubbish, the walls thereafter having to be washed down and the floors scrubbed with disinfectants; while during the process many pounds of sulphur have been burned in order that the fumes may aid in the work of destruction."—[James Fletcher, Ottawa, Canada, October 31, 1889.

Spider Bites—Two Ceylonese Cases.

Since reading your several notices of spider bites in America, two cases have come under my own observation. In both cases the patients (Tamil coolies) were bitten on the hand by the large, hairy spider, *Mygale fasciata*, while working in the field. Both patients complained of recurring spasms followed by soreness and muscular pains extending through the leg, arm, and neck on the affected side. The local medical officer applied, in one case, fuming nitric acid to the puncture, and in the second case injected permanganate of potassium. This second treatment seems to have been the most successful, the painful symptoms abating in a much shorter period.—[E. Ernest Green, Eton, Punduloya, Ceylon, October 5, 1889.

Scent in Dung-beetles.

I have just returned from gathering a load of moss (*Sphagnum*) out of a swamp miles in extent, where I saw a most remarkable illustration of the power of smell in insects. The day was mild and still, and there in the midst of the swamp the excrement of my horse attracted a large number of the small dark scavenger beetle, about the size of a horse-fly, so common in cleared lands at this season of the year. They all came from the direction of the higher land. I have long been of the opinion that the power of scent was stronger in insects than in any other department of animated creation. * * * —[W. W. Meech, Vineland, N. J., October 18, 1889.

Beetles from Stomach of a "Chuck-wills-widow."

I send by mail some "bugs" taken from the stomach of a Chuck-wills-widow. Please state name, and whether injurious to agriculture.—[G. H. Ragsdale, Gainesville, Cook County, Tex., May 12, 1886.

REPLY.— * * * I beg to acknowledge the receipt of yours of recent date, accompanied by insects taken from the stomach of the Chuck-wills-widow (*Antrostomus carolinensis*). This bird has a curious habit of bolting these large beetles whole while on the wing. There are two species in your sending. One is *Ligyrus gibbosus*, a species the larva of which feeds upon the root of sunflower, and which has been recorded as doing considerable damage in Nebraska, where the sunflower is grown as a crop; the other is *Lachnosterna rugosa*, a southern representative of the common May beetle of the north. The larva of this insect is a white grub and doubtless feeds on the roots of grass and similar vegetation.—[May 18, 1886.]

A Harvest-mite Destroying the Eggs of the Potato-beetle.

I send you inclosed in small box a specimen of an insect found by me feeding upon the eggs of the Colorado Potato beetle. I have been troubled every year a great deal by the ravages of the slugs, but this summer there are none upon my vines, though the usual number of old beetles are seen depositing their eggs. This insect may be as common as the house-fly, but he is new to me, and has won my gratitude. Perhaps you may be interested in him, but if not no harm will be done in placing him before your notice.—[Charles C. Bryant, Silver Lake P. O., Kingston, Mass., June 18, 1886.

REPLY.— * * * The insect which you found feeding on the eggs of the Colorado Potato-beetle is a Harvest-mite of the genus *Trombidium*. It appears, so far as I can ascertain, to be a new species, and consequently we should be very glad to receive further specimens. Is it at all common with you? I think that no record has been published of the work of any Harvest-mite upon the eggs of the Potato-beetle, and in consequence your letter possesses considerable interest. * * * —[June 22, 1886.]

Supposed Injury to Grass from Gastrophysa polygoni.

Please to inform me as to the inclosed insects. A few days ago they made their appearance in great numbers in the court house yard, and are destroying the grass very rapidly.—[N. R. Smithson, Winchester, Ill., June 2, 1887.

REPLY.— * * * This beetle is known as *Gastrophysa polygoni.* It is a perfectly harmless species, injuring no crop and feeding solely upon the weeds of the genus *Polygonum* (knot-weed, joint-weed, goose-grass, door-weed, smart-weed, etc.) which grow among the grass in lawns. * * * I know of no recorded instance of such a habit, and your observation therefore becomes interesting if true. Will you therefore please advise me whether you are not mistaken, and whether the insect does not feed upon some one of the weeds mentioned among the grass, rather than upon the grass itself?—[June 7, 1887.]

Damage to dead Trunks of Pine by Rhagium lineatum.

I send you by mail to-day specimens of the Pine-tree Borer, as requested in your letter, which bids fair to exterminate our pine trees. If you have any remedy to advise, would be glad to hear from you.—[E. R. Memminger, Flat Rock, N. C., September 8, 1888.

REPLY.— * * * The insects sent are *Rhagium lineatum.* This species does not kill the pine trees, but simply bores beneath the bark and into the decaying wood of trees that have been killed by some other cause, or dead portions of live trees. It also attacks spruce and fir logs, stumps, and dead standing trees. In case it should become destructive to logs which have been cut for timber, it can be destroyed by stripping off the bark and portions of the sap-wood infested.—[September 13, 1888.]

Some Vedalia Letters.*

* * * The Vedalias that you brought to my place about the 20th of last March, and which we colonized on four large orange trees that were covered with Fluted Scale, have spread in all directions, although to begin with they followed the direction of the wind most readily. From those four trees they have multiplied so rapidly that in my orchard of 3,000 trees it is seldom that we can now find a Fluted Scale. I find a few of them on some weeds in spots, but I can also find the beetles there. The trees have put on a new growth and look altogether different; even the black fungus on the old leaves has loosened its hold and begins to fall to the ground. Besides having cleaned my orchard, they spread also to the orchard of my cousin and to my father's orchard; the latter was also re-enforced by colonies from Mr. J. W. Wolfskill and from Col. J. R. Dobbins. As my father has some 10,000, trees, and most all were more or less infested, the Vedalias had a grand feast ahead of them, and they have done their work most wonderfully. What I have said of my orchard applies to my father's also, and really to all our neighbors. When the Vedalias first began to multiply we took colonies of fifty or more in the pupa state and placed them in different portions of the orchard, and even had we not done so the Vedalia unaided would itself have reached there in almost the same time.

On the Chapman place the Vedalias have cleaned the Fluted Scales off of the 150 acres of land. They have taken more than an oppressive burden off of the orange grower's hands, and I for one very much thank the Division of Entomology for the *Vedalia cardinalis*, the insect that has worked a miracle.—[A. Scott Chapman, San Gabriel, Cal., October 18, 1889.

* * * The Vedalia had practically freed my orchard of Iceryas on the 31st of July. It was on that date that I was obliged to post a notice at the entrance to my place, saying that I had no more Vedalias for distribution. The scale and lady-bird

*These were addressed to Mr. Coquillett, at Los Angeles.

had fought out the battle, and while the carcasses of the vanquished were everywhere present to tell of the slaughter, the victors had disappeared almost entirely from the field. I have 35 acres in orchard—some 3,200 trees in all. I never colonized any Vedalias in my grove, excepting the two consignments which you brought to me yourself—one box on February 22 and two boxes March 20. I noticed the first increase from the lot No. 1 on the 15th of April, and from lot No. 2 on the 24th of the same month. On the 25th of April I found larvæ upon several adjacent trees. These facts are from memoranda made at the time. I have a list of the names of fruit growers, 226 in number, to whom I personally distributed over 120,000 Vedalias in colonies of various sizes between May 31 and July 31. * * * —[J. R. Dobbins, San Gabriel, Cal., October 22, 1889.

I am glad to report that the lady-birds you sent me are doing good work and increasing in this neighborhood, and as soon as all are supplied I will establish some on the mountain where the brush is full of them, also a small patch near the Ocean, and hope the Cottony Cushion-scale will soon be a scarce article in this section.—[Joseph Sexton, Goleta, Cal., August 12, 1889.

On Hæmatobia serrata.

I have just received INSECT LIFE, No. 4, Vol. II, for which please accept my most sincere thanks.

On page 95 I find a passage which calls, on my part, for the following statement:

On receiving the specimens of *Hæmatobia serrata* from Dr. Lintner in September, 1888, I at once suspected that they might be specifically identical with some European Stomoxid, and I communicated them for identification to my friend, Mr. Kowarz. He answered as follows:

"Ich habe mir alle Mühe gegeben, aber ich vermag in dieser Fliege nichts anderes als *Hæmatobia serrata* R. D. (*Lyperosia* Rnd.) zu erkennen. Sie unterscheidet sich von den europäern nicht im Geringsten."

Translation: "I have taken great pains with this fly and can not recognize in it anything but the *H. serrata* R. D. (*Lyperosia* Rnd.). It does not in the least differ from the European specimens."

It is important, in such a case, to have it distinctly stated that the identification is based upon an actual comparison of specimens by the best authority. Mr. Ferdinand Kowarz, in Franzensbad, Bohemia, I consider as the entomologist who, at present, possesses the most extensive knowledge and experience of European Diptera, especially so far as the discrimination of species is concerned. I take, therefore, his decision as trustworthy and final, and I regret that Dr. Lintner did not mention Mr. Kowarz's name in the first publication which he made upon receiving my answer (in the *Country Gentleman*, Albany, N. Y., November 29, 1888).

My own knowledge of European Diptera is very insufficient, and in all doubtful cases I apply either to Mr. Kowarz or to Professor Mik, or, for Cecidomyiæ, to Dr. Franz Löw (the two latter in Vienna).

You will do me a favor by the publication of this letter in one of your next numbers.—[C. R. Osten Sacken, Heidelberg, November 20, 1889.

GENERAL NOTES.

OVIPOSITION OF TRAGIDION FULVIPENNE.

A desirable addition to our knowledge of the life-history of *Tragidion fulvipenne* is made by Prof. E. A. Popenoe, in a paper entitled "Note on the oviposition of a Woodborer," read at the Wichita meeting of the Kansas Academy of Science, and published in the Manhattan *Industrialist* for November 2, 1889. The *Cerambycidæ*, as Professor Popenoe points out, ordinarily oviposit in cracks of bark or in fissures made by the parent insect, and hence the striking variation in this habit in the case of the above-named beetle is the more interesting.

Female beetles only were observed about a wood-pile on warm days about the end of September, and after considerable search they were seen ovipositing on sticks, probably on the chestnut oak. The habit of the insect in this particular is described as follows:

When detected in oviposition, the females were standing on the smooth bark, transversely to the stick, their bodies close to the surface, their antennæ bent under at the tips, which were touching the bark, and the broad tip of the abdomen closely appressed to the surface over which the insect stood. The close contact of the motionless tip of the abdomen to the bark prevented my noting the exact mode of placing the egg, and presently, becoming somewhat impatient, I lifted a beetle from position, and, to my surprise, instead of an opening in the bark as I had anticipated, I saw a tubercle simulating so closely in appearance and color the corky outgrowths common on the bark of the chestnut oak that I was at first inclined to believe it one of these, and to question the purpose of the female in maintaining so long the position described. On an examination of this tubercle, however, I found it to be hollow, and within it, lying on the bark, with no puncture or abrasion in the latter to be seen, was an oblong egg of a translucent, dull white surface, smooth and without markings, so far as I could see with a pocket triplet of good definition. This egg was sufficient in size nearly to fill the hollow tubercle, or egg-case, as I may now call it. The egg-case is rather regular, ellipticle, strongly convex, measuring about one-sixteenth of an inch in length. Under the microscope, the case appears on the surface to be made up of scales of the thin external layer of the oak bark, intermingled with glistening particles, as of dried mucus.

INSECTS INJURING THE TEA-PLANT IN CEYLON.

We have recently received from Mr. E. Ernest Green of Eton, Punduloya, Ceylon, a series of nine short articles on the "Insect Pests of the Tea-plant" published in the *Ceylon Independent*, July 3 to October 3. The papers are illustrated by engravings made by a native from drawings by Mr. Green and, while naturally not of a high state of art, are plain and characteristic. The pests treated are as follows:

THE FAGGOT WORM (*Eumeta carmerii*).—This insect is one of the Bag-worms, and its popular name is derived from the fact that its case resembles a bundle of minute faggots. The life history is very similar to that of our common Bag-worm (*Thyridopteryx ephemeraeformis*). Mr.

Green quotes a quaint native legend concerning these insects, to the effect that in a previous life they existed in the human form, when amongst other crimes they made a regular trade of stealing fire-wood; at their death their souls were sent into the bodies of insects and condemned to perpetually carry about with them a faggot of wood. This species is also found on the coffee plant.

THE BORER (*Zeuzera coffeæ*).—This insect which has been so frequently treated as a coffee enemy and so known to planters as the "Red Borer" is by no means uncommon as a borer of the tea-plant. It belongs to the *Cossinæ*.

THE TEA BARK-LOUSE (*Aspidiotus theæ*).—This is one of the most serious enemies of the plant and is very noticeable at the time of pruning.

THE YELLOW BARK-LOUSE (*Aspidiotus flavescens*).—This is a smaller species than *A. theæ*, but is much more readily recognized on account of its yellow color contrasting with the bark, while *A. theæ* is of the same color as the bark.

THE TRANSPARENT-SCALED BARK-LOUSE (*Aspidiotus transparens.*)—This species has been noticed only in small numbers and prefers the leaf to the bark. The scales are small, round, and colorless, and the insects can be plainly seen beneath them.

THE LOBSTER CATERPILLAR (*Stauropus alternus*).—This is a large leaf-feeding species, and when five occur upon a single plant the leaves become completely devoured. It is a close ally to the Lobster Caterpillar of Europe, *S fagi.*

THE RED TEA-MITE OR RED SPIDER (*Tetranychus biaculatus*).—This mite produces a copper sunburnt appearance of the leaves and it will be remembered as having previously been described by Mr. Wood-Mason as affecting the tea-plant in Assam. Mr. Green thinks it identical with the species described by Mr. Nietner as the "Red Spider of the coffee tree (*Acarus Coffeæ*)."

THE FIVE-LEGGED TEA-MITE (*Typhlodromus carinatus*).—This species, Mr. Green says, is closely related to the Rust-mite of the orange (*T. oleivorus* Ashm.) which feeds on both sides of the leaf, while the Red Spider is confined to the upper surface. He advised one part of kerosene emulsion to eighty parts of water, or one part of Phenile to two hundred and forty parts of water.

THE YELLOW TEA-MITE (*Acarus translucens*).—This mite produces the condition called "sulky" and feeds upon the buds. The living insects can be found only upon the bud and the underside of the two following leaves, and as each fresh bud opens the colony moves higher up, deserting the lower leaves, but these remain injured and always retain the marks of the insects. Excepting the Tea Bark-louse Mr. Green considers this to be the most serious pest to the plant. He thinks that the systematic destruction of all tea prunings while still green would prove an immense check to this pest and others.

A NEW WAY OF USING CARBON BISULPHIDE.

We have not yet seen any notice in this country of the point brought out by the president of the Lyons Viticultural Society in a recent address to the effect that vaseline is not only an excellent solvent of bisulphide of carbon, but that it also produces the power of penetrating the soils and of woody tissues in a most remarkable manner. Bisulphide after having been taken up by vaseline liberates itself progressively and then vaporizes. The action of the vapor is thus prolonged through many days. The strength of these vapors is far less than if the bisulphide be used alone, but the effect is of much greater duration. In warm climates, where if the bisulphide were used alone the vaporization would be exceedingly rapid, its use with vaseline will be of great benefit, although adding somewhat to the expense.

RANGE OF PYRALIS FARINALIS.

As is the case with other insects of similar habits, this common Mealworm Moth is very widespread. The British Museum Catalogue in 1858 records it from England, Germany, the whole of Europe, Madeira, United States, Nova Scotia, South Africa, Cape of Good Hope, and Australia. We mention it at this time for the reason that Mr. J. G. O. Tepper in his papers on "Common Native Insects," published in the *Garden and Field* of Adelaide, South Australia, states that this moth is very commonly met with in out-houses, kitchens, and even on trees in the field. He says:

> Whether it is native or introduced is hard to say, as the writer already met it about April, 1854, as commonly as now in the country.

It seems to us that it is without much question an European species imported into Australia and the rest of the globe, as it was noticed by the older Geoffroy and by Linnaeus.

KIND WORDS FROM ABROAD.

Mr. A. M. Pearson, chemist to the Department of Agriculture of Victoria, in a lecture on "Science and Farming," published in Bulletin No. 3, Department of Agriculture at Melbourne, makes use of the following expression:

> Science has also lent its aid in the direction of overcoming plant diseases and insect pests, and I think it must be acknowledged that the Americans, more especially the Department of Agriculture at Washington, have taken the lead in this direction.

ON SOME GALL-MAKING INSECTS IN NEW ZEALAND.

Under the above caption Mr. W. M. Maskell has published a short paper in the *Transactions of the New Zealand Institute* for 1888, in which he describes certain galls upon *Olearia furfuracea*, a native shrub, known by the settlers as "Ake-ake," and by the Maories as "Ake-piro." Mr. Maskell has reared from the galls a dipterous insect and a hymenopteran. In referring to the latter insect, which he calls *Eurytoma oleariæ*, he states that while it seems likely that this insect is a gall-producer, it may be only a messmate of a Cecidomyia, as its larvæ and pupæ are found mixed indiscriminately with those of the Cecidomyia, although in separate cells. He inclines to the belief that the Cecidomyia produces the galls and that the Eurytoma makes use of them as a residence. In considering this question he refers to the Joint Worm as *Eurytoma hordei*, and states that it is not certain that it is phytophagous, but that it may be only parasitic upon the larva of Cecidomyia. In this remark Mr. Maskell is behind the times, as there is no longer any question of the phytophagous nature of this species, but his greatest mistake occurs in the identification of the insect which he considers a Eurytomid. As his figures show, it is not an Eurytoma, and does not even belong to the family *Chalcididæ*. Specimens which he has kindly sent us show that it is a Proctotrupid of the subfamily *Platygasterinæ*, and belonging to the genus *Monocrita*. This identification of the insect renders it quite certain that it is a parasite.

VERTEBRATE ENEMIES OF THE WHITE GRUB.

Prof. C. W. Hargitt, of Miami University, in an article on the White Grub, contributed to the Oxford (Ohio) *News* of April 6 last, gives from his personal observations some interesting notes upon the subject of this note. He finds that the crow is among the most active and constant enemies of this insect.

> His presence in flocks, promenading pastures and meadows, is almost wholly due to his taste for this pest, as has been abundantly proved by an examination made upon the stomach and crop.

The robin and the blackbird he states to be hardly less active as devourers of the grub.

He also cites the sparrow-hawk, king-bird, jay, and the golden woodpecker as of less importance.

Among mammals he cites the mole and the skunk, while dissections of frogs showed several grubs and many adult beetles. In a single frog stomach six full-grown May-beetles were found.

NEW METHOD OF DESTROYING SCALE-INSECTS.

We understand that a patent has been issued to Mr. Edwin P. Fowler, of National City, Cal., for a process of dislodging and destroying scale-insects by means of a sand blast. We have been acquainted with the fact that this application was before the Patent Office for some time, but have been unable to publish anything concerning it pending its consideration. The plan is an ingenious one, but whether it will pay or not is a matter for future experiment. A fan-blower or other apparatus capable of creating an artificial current of air is employed; the current is directed against the tree, and in its transit from the fan is charged with sand. The force of the current is carefully gauged and the sand may be heated.

DR. FRANZ LÖW.

It is with profound regret that we have just received from his brother the sad news of the death of Dr. Franz Löw, which took place at Vienna, Austria, November 22, after a long and painful illness. With him entomological science loses a conscientious worker, whose labors have greatly added to the common stock of knowledge, and his premature death (he died in his sixty-first year) will everywhere be felt as a calamity.

His first entomological paper was published in 1857, and treats of the larvæ of the Coleopterous genus *Nebria*, but he soon became more interested in the life-history of gall-producing insects, especially *Diptera*, *Homoptera*, and *Acarinæ*. Of his numerous papers on this subject, published mostly in the Proceedings of the Zool. botan. Society of Vienna, every one marks an addition to our knowledge. This is especially true of the classification and life-history of the *Psyllidæ*, and he became the recognized leading authority on this intricate group of insects. Notwithstanding the works by Flor and Thomson, the classification of *Psyllidæ* had remained practically where Förster left it in 1848, and Löw's paper, "Zur Systematik der Psylloden," published in 1879, marks the first genuine progress since that time. Some years previously he had pointed out the great importance of the study of the earlier stages of *Psyllidæ* to a thorough understanding of this family, and his numerous contributions to this subject show how indefatigable he was in tracing and describing them.

Dr. Löw will also be remembered as the author of several valuable papers on Myiasis, and as one of the collaborators on the Zoologischer Jahresbericht from 1883 to 1885. Personally we shall greatly miss him as one of our most valued European correspondents, always ready to assist with suggestions and criticisms given in the most amiable and unpretentious way. He took a keen interest in American entomology; and it was a delightful (if often difficult) task to answer the many knotty questions he plied us with in his letters regarding all sorts of insects, especially those treated of or described by the older authors.

EUGÈNE MAILLOT.

We also deeply regret to learn of the death of another valued friend and correspondent, Maillot, director of the silk station at Montpellier. Maillot was a man of great scientific ability, and was, at the same time, an eminently practical man. He was studying the different races of silk-worms from all parts of the world at the time of his death, and had contributed in a large measure to the general adoption in France of the microscopic selection of silk-worm eggs as a preventive against pébrine. He was a student of Pasteur's, and a comparatively young man. His work entitled "*Leçons sur le ver à soie du Murier*," from a theoretical and practical point of view, is one of the best treatises upon sericulture which has been written up to the present time.

ENTOMOLOGICAL SOCIETY OF WASHINGTON.

The fifty-sixth regular meeting of the Entomological Society of Washington, D. C., held November 12, 1889.

Mr. F. M. Webster and Dr. John Hamilton were elected corresponding members of the society.

Mr. Howard exhibited a specimen of *Xylonomus rileyi* Ashm., taken on the Washington Monument.

Mr. Lugger read some notes on "The migration of the Archippus butterfly," and gave an interesting study of their spring and fall movements. He also noted a similar migration of *Vanessa cardui*. Dr. R. Thaxter stated in discussion that he had found Archippus wintering along the Gulf of Mexico in immense numbers.

Mr. Howard read a paper on "A few additions and corrections to Scudder's Nomenclator Zoologicus."

Mr. Marlatt gave some "Notes on the abundance of oak-feeding lepidopterous larvae this fall," and named twelve species of macrolepidopterous larvae taken in the course of about an hour.

Mr. Schwarz read a paper entitled "Caprification," and gave a thorough résumé under the following heads:

(1) The flower and fruits of the Capri fig and the wild species of Ficus.

(2) Enumeration of the fig insects and difficulties of study.

(3) Life history of true fig-insect (Blastophaga) and fertilization of wild species of Ficus and the Capri fig; and

(4) The true fig tree and the process of caprification.

Mr. Townsend read a paper on "The fall occurrence of Bibio and Dilophus," in the discussion of which it was conceded that the autumnal occurrence was simply due to an acceleration of development, as they hibernate in a nearly developed state.

WM. H. FOX, M. D.,
Recording Secretary.

U.S. DEPARTMENT OF AGRICULTURE.

DIVISION OF ENTOMOLOGY.

PERIODICAL BULLETIN. (Double number.) January and February, 1890.

Vol. II. Nos. 7 and 8.

INSECT LIFE.

DEVOTED TO THE ECONOMY AND LIFE-HABITS OF INSECTS, ESPECIALLY IN THEIR RELATIONS TO AGRICULTURE.

EDITED BY

C. V. RILEY, Entomologist,

AND

L. O. HOWARD, First Assistant,

WITH THE ASSISTANCE OF OTHER MEMBERS OF THE DIVISIONAL FORCE.

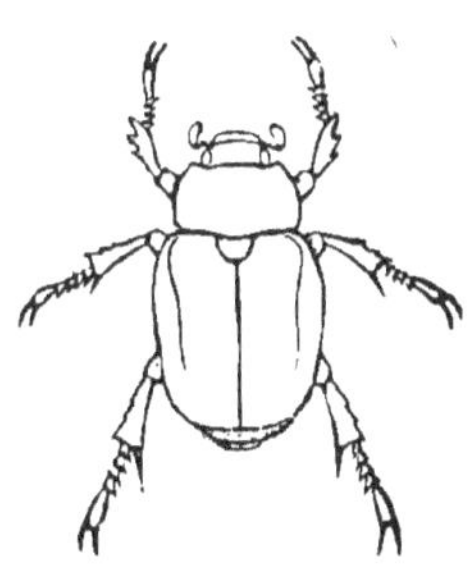

[PUBLISHED BY AUTHORITY OF THE SECRETARY OF AGRICULTURE.]

WASHINGTON:
GOVERNMENT PRINTING OFFICE.
1890.

CONTENTS.

SPECIAL NOTES.

A descriptive Catalogue of the Phalangiinæ in Illinois.—We have just received from Mr. C. M. Weed a paper with the above title published as a bulletin of the Illinois State Laboratory of Natural History (December, 1889), and also a partial bibliography of the same group as represented in North America. The descriptive catalogue includes the consideration of three genera and ten species, two of the species being new. *Liobonum dorsatum*, *L.* (?) *formosum* and *Oliogophus pictus* are figured. The bibliography includes five titles and fifty-eight references to descriptions of species. We are glad to see this neglected group worked up so satisfactorily.

Entomological News.—The Entomological Section of the Academy of Natural Sciences, of Philadelphia and the American Entomological Society announce the publication of a new journal to be devoted to notes and news, queries and answers, exchanges and doings of societies. It is edited by Mr. E. M. Aaron, assisted by an advisory committee consisting of Dr. G. H. Horn, Mr. E. T. Cresson, Dr. Henry Skinner, and Mr. Ph. P. Calvert. The subscription price is $1 a year, and ten numbers will be published, one for each month, with the exception of July and August. It began publication January 1, 1890. The main object of the journal, as stated in its circular of announcement, will be to keep entomologists acquainted with what is being published in serials at home and abroad, and it will also give news items concerning explorations and collectors. The journal will meet a present want and will be welcomed by American collectors. Backed by the American Entomological Society its success would seem to be assured.

Dr. Lintner's latest Report.—Dr. Lintner's fifth report on the injurious and other insects of the State of New York has been received. It is extracted from the forty-second report of the New York State Museum of

Natural History. It comprises nearly two hundred and fifty pages of very interesting matter and is illustrated by fifty text figures. The matter is prepared with Dr. Lintner's usual great care and contains valuable summaries of our information upon a large number of injurious insects. The consideration of each species is prefaced by a synonymical and bibliographical table which is of great value to the working entomologist. The principal articles are upon Remedies and Preventives, the Larch Saw-fly (*Nematus erichsonii*), the Cow Horn-fly, the Elm Leaf-beetle, and the Grain Plant-louse. Short accounts are given of other species, and under the head of "Insect Attacks" and "Miscellaneous Observations" many interesting notes are collocated. A small section of the report is devoted to Acarina and Myriapoda in which several injurious and beneficial mites are mentioned. In an appendix a list of the principal publications of the Entomologist during 1888 is given. We can commend Dr. Lintner's writings for the care with which quoted information is credited, and wish we could say the same regarding his illustrations, which are often used with no such regard for authority or source.

The Little Red Ant.—We publish in this number a free translation of an interesting article by M. A. Bellevoye on this insect. It will be interesting in connection with our article (Vol. II, No. 3) on the occurrence of this insect in America. Mr. Bellevoye's suggested inference that inasmuch as he was unable to observe that the ants carried any food to their nests this might be considered a result of domestication, as they always find something to feed upon in our houses, will hardly hold for this side of the water, as in our experience these ants are often seen carrying particles of food into cracks in walls and floors which probably lead to their nests.

Technical Entomology in Ohio.—The Ohio Agricultural Experiment Station has started an innovation in the line of a series of technical bulletins. The director explains in an obscure foot-note that the series is intended to embody the technical results of the work of the station, but that it is not expected that they will be of direct service to farmers in general. It is hoped, rather, that they may be found useful by workers in other stations, and thus indirectly serve the cause of agriculture. It comprises three articles by the entomologist, Mr. C. M. Weed, entitled (1) "Preparatory stages of the 20-spotted Lady-bird," (2) "Studies in Pond Life," and (3) "A Partial Bibliography of Insects affecting Clover." Of these articles, the one upon "Studies in Pond Life" is naturally of the greatest interest and value, and a number of new points are brought out. The "Larger Typha-borer" (*Arzama obliquata* G. and R.) is figured in larva, pupa, and imago, and he records a number of dates of transformation, and describes the larva and pupa.

"The Toothed-horned Fish-fly" (*Chauliodes rostricornis* Ramb.) is figured in the larva, pupa, and adult, and notes upon its life-history are given, adding, however, little to the observations recorded by Walsh in the second volume of the proceedings of the Entomological Society of Philadelphia. The Sagittaria Curculio (*Listronotus latiusculus* Boh.) is also figured in the larva, pupa, and adult, and its breeding habits, and the leaf and stalk and seed heads of the common arrow-leaf are described. The Lesser Water-bug (*Zaitha fluminea* Say) is stated to feed mainly upon the early stages of Dragon Flies. It also uses as food univalve snails and May-fly larvæ. *Notonecta undulata* is recorded as feeding upon May-fly larvæ and upon a species of Boatman (*Corisa alternata* Say). *Donacia subtilis* Kunze feeds upon a number of aquatic plants and pollinizes *Nuphar advena*. His observations indicate that the 13-spotted Lady-bird (*Hippodamia* 13-*punctata*) has aquatic tendencies, as he has commonly found it upon the leaves of aquatic plants. This accords with our own experience, and Mulsant mentions the same thing of this species in France. The stages of *Benacus griseus* and *Belostoma americanum* are described and those of the former species are figured. Altogether this is one of the best and most interesting (entomologically) of the experiment station bulletins so far issued.

Ultimate Larva of Platypsyllus.—We copy in the present issue from *Entomologica Americana* for February the description of an interesting larval form of this curious and anomalous beaver parasite, and would again call attention to the wonderful superficial resemblance to certain Mallophaga of the genera *Nirmus* and *Trichodectes*. In some species of the latter genus the mandibles are bidentate, as in this larva, while the caputal characters, the loss of the anal cerci, and the general form of body so depart from the earlier larva that the resemblance to the Mallophaga is still more striking. But none of the lice have the mouth-parts, otherwise, as in this larva, nor the single jointed tarsus.—C. V. R.

Oviposition of Hypoderma bovis.—The interesting facts narrated in this number by Dr. Cooper Curtice bring unexpected confirmation of what we stated in the last issue as to the eggs of this Ox Wormal being fastened externally, and would seem to indicate that, exceptionally at least, if not normally, the newly-hatched larva is taken in through the mouth and can live internally during the first stage. Whether these young larvæ in the œsophageal walls and under the pleura eventually perish or succeed in working beneath the skin is as yet to be ascertained, but we see nothing improbable in the latter course. These young larvæ are doubtless taken from one animal to another through the habit which cattle have of licking each other, and it is possible that in older cattle in which the hide is thick this mode of entrance of Hypoderma is more common than in younger animals. We have examined Dr.

Curtice's material and can corroborate the correctness of the determination. In this connection we also draw attention to the interesting communication of Dr. Elizabeth R. Kane (p. 238) relating to the traveling propensities of the young Hypoderma larva.

THE USE OF HYDROCYANIC ACID GAS FOR THE DESTRUCTION OF THE RED SCALE.

By D. W. COQUILLETT, *Los Angeles, Cal.*

In my reports to Professor Riley for the years 1887 and 1888, published in the annual reports of this Department for these years (pages 123 to 142, and 123 to 126 respectively), I gave an extended account of the use of hydrocyanic acid gas for the destruction of scale-insects (family *Coccidæ*); and I am not aware that anything has been published upon this subject since the appearance of the above-mentioned report for the year 1888. On page 126 of this report I gave an account of treating several orange and lemon trees with this gas, and the latest report given of the condition of these trees was under date of August 15, 1888; under date of February 17, 1889, the owner of the trees, Mr. I. L. Collins, wrote me as follows in regard to them:

DEAR SIR: I received yours of the 15th inst. asking about the condition of the lemon and orange trees treated with the gas. These trees are in a much better condition than those around them, as they have a full foliage while the others are nearly bare; what fruit they have on is comparatively clean, there being but few red scales on them. They already show that the coming season they will bear quite heavily, but now they have enough red scales on them to ruin them in a year. We expected that the scales would come on them again from the other trees, these not having been treated with the gas; I did not think the tops worth saving, so did not treat them with the gas. I will cut off the tops, as almost everybody else is doing, and will wash the stumps with a solution composed of 30 pounds of resin, 7 pounds of caustic soda or potash, and one gallon of fish oil to 100 gallons of water. The trees treated with the gas remained perfectly clean for over a month; then we found scales on the outside branches, having apparently been carried there by the horses in cultivating.

In accordance with a written request from several of the orange growers of Orange, I went down to that place in the latter part of September of the present year, and conducted a series of experiments with hydrocyanic acid gas for destroying the Red Scale, with the view of trying to discover some simpler and less expensive method for producing and manipulating this gas than the one heretofore in use. The lemon trees experimented upon and also the fumigating outfit used in making these tests were kindly placed at my disposal by their owner, Mr. A. D. Bishop; and the latter gentleman, in conjunction with Mr. A. H. Alward, also aided me in moving the outfit from tree to tree when making the tests. Among the different methods tried was one that gave very satisfactory results, and which, both in regard to expense and labor, is a great improvement upon any heretofore tried. It consists in using one part by weight of dry or undissolved potassium cyanide, with one part sulphuric acid and two parts of water. The generator is made of lead and is somewhat in the form of a common water-pail. After the tent is placed over the tree the necessary quantity of the dry cyanide is placed in the generator, the proper quantity of cold water added, and the generator placed under the tent near the trunk of the tree; the acid is then added to the materials in the generator, a barley sack thrown over the top of the latter, after which the operator withdraws and a quantity of earth is thrown upon the lower edge of the tent where it rests upon the ground to prevent the escape of the gas. After the expiration of fifteen minutes the tent is removed and placed upon another tree. I tested this method on several lemon trees and found that when the proper quantity of material had been used neither the foliage nor fruit on the trees were injured, while neither myself nor several other persons were able to find a living red scale upon the trees treated in this way.

The following table, based upon several of the tests referred to above, will aid in determining the proper quantity of each ingredient to use in treating orange and lemon trees:

Height of tree.	Diameter of tree.	Cyanide of potash.	Water.	Sulphuric acid.
Feet.	*Feet.*	*Ounces.*	*Fluid ozs.*	*Fluid ozs.*
10	8	2¼	4½	2¼
12	10	4½	9	4½
12	14	8¾	17½	8¾
14	10	5½	11	5½
14	12	7½	15	7½
16	14	12	24	12
18	14	15	30	15

It will be noticed that the proportions are 1 ounce by weight of the cyanide to 1 fluid ounce of the acid, and 2 fluid ounces of water; or in the proportion of cyanide one, acid one, water two. This being borne in mind, it will be very easy to ascertain how much acid and water to use when once the proper quantity of the cyanide required for treating any given tree has been ascertained.

In making the tests referred to above, I used commercial sulphuric acid and a medium grade of potassium cyanide, manufactured by Powers and Weightman, of Philadelphia, Pa. It is the same grade of cyanide as that which Mr. O. H. Leefeld purchased at the rate of 44 cents a pound, freightage included, as described in my report for 1888, page 125.

By comparing the table given above with the one given on page 125 of my report for the year 1888, it will be noticed that but little more than one-third the quantity of each ingredient is required for a tree of a given size by this new method, as compared with that required by the old one. In the third column of the table given in the previous report, each fluid ounce of the cyanide solution contains half an ounce by weight of the dry cyanide. At this rate, by the old process, a tree 14 feet high by 12 feet in diameter required 21½ ounces by weight of the dry cyanide, whereas by the new process it will require only 7½ ounces. At the present prices of the cyanide and acid, the cost of the materials necessary to treat an orange tree of the size given above, by this new method will amount to about 26 cents, as compared with 76 cents, the price when the old process is used.

Not only is the new process much cheaper than the old, but it is also attended with much less labor. By using the cyanide dry we are saved the trouble of first dissolving it; the dry cyanide is also easier to transport and safer to handle than the solution is, and if the vessel containing it should be accidentally overturned on the ground, the dry cyanide will not be lost, as it certainly would if dissolved. By thus using the cyanide dry it is not necessary to first pass the gas through sulphuric acid in order to render it harmless to the trees, thereby saving a great deal of labor, and admitting of the use of a much simpler and less expensive generator. By placing the latter beneath the tent there is less liability of the gas escaping while being generated and introduced into the tent from without, thereby also insuring the operator greater immunity from inhaling the gas. I also found that by thus placing the generator under the tent the blower heretofore used for distributing the gas inside of the tent could be done away with, thereby still further reducing the original cost of a fumigating outfit, besides doing away with the labor necessary in operating the blower. The time during which it is necessary to confine the tree in the gas has also been reduced one-half as compared with that heretofore allowed for destroying the Fluted Scale (*Icerya purchasi* Maskell), thereby rendering it possible to treat twice the number of trees in a given time that could be treated in the same time by the old process. I found by experiment that about five minutes were consumed each time in generating the gas.

The treatment with hydrocyanic acid gas is the only method known to me whereby the scale-insects located upon the fruit can be destroyed by a single operation. My own experience, and that of every other person with whom I have conversed upon this subject and who has had any considerable experience in the matter, indicates that no liquid preparation at present known will by a single application prove fatal to more than 90 per cent. of the number of red scales located upon the fruit, and when it is remembered that the supervisors of many counties in this State have passed laws making it a misdemeanor to sell or expose for sale fruit infested with scale-insects, the value of the gas treatment to our fruit-growers is made apparent.

The following is an account of the experiments I made with hydrocyanic acid gas as referred to above. The trees operated on were all of them lemon trees containing fruit, and were in a comparatively healthy condition, although very thickly infested with the Red Scale. Before making these tests, I had the experimental tent painted black, and am strongly of the opinion that when a tent of this color is used the foliage of the trees will be injured less when by inadvertence an overdose of the materials has been used than would be the case if a light-colored tent were to be used; the light rays, more than the rays of heat, serve to decompose the gas, and on this account any medium that will intercept the rays of light will, in a great measure, prevent the decomposing of the gas. In all cases where a blower was used for distributing the gas inside of the tent, the gas entered the blower direct from the generator and was forced into the lower part of the tent through a tin pipe, and the pipe which conducted the air and gas from the tent to the blower also entered the lower part of the tent and then turned upward, terminating near the top of the tent. By this means the gas and air in the upper part of the tent were drawn out and after passing through the blower again entered the lower part of the tent. This was for the purpose of more thoroughly circulating the gas inside of the tent; but, as will be seen by the later experiments this arrangement was found to be entirely unnecessary when the generator was placed under the tent. In nearly all of the later experiments too large a quantity of the materials was used, resulting in more or less injury to the tree or fruit, the injury being always the most severe on the topmost portion of the tree. The cyanide solution used in a few of these experiments consisted of 5 pounds of cyanide dissolved in 1 gallon of water, each fluid ounce of the solution containing an ounce by weight of the cyanide. The diluted sulphuric acid was composed of two fluid parts of the acid and three of water, and was allowed to become cold before being used.

(205) Took 10 fluid ounces of the cyanide solution and added in three minutes 12 fluid ounces of the diluted acid. 12.30 to 12.45 p. m., September 23, sun shining, light breeze. Scarcely turned the blower at all. Tree 12 feet high by 10 in diameter. When the tent was removed about half a dozen leaves on the new growth had perceptibly wilted. October 19, about three dozen leaves were dead; found eight live red scales, equally distributed on the leaves and fruit.

(206) Took 4 fluid ounces of the cyanide solution, and added in a minute and a half 2½ fluid ounces of pure sulphuric acid; turned the blower three minutes after adding the acid. 1.10 to 1.25 p. m., September 23, sun shining, light breeze. Tree 7 feet tall by 6 in diameter. When the tent was removed several of the leaves had wilted. October 19, about three dozen leaves and a large portion of the twigs on which they grew were dead; found only one live red scale, which was located upon a leaf.

(207) Took 6 ounces by weight of the dry cyanide and added in four minutes 12 fluid ounces pure sulphuric acid; turned the blower five minutes. 3.10 to 3.30 p. m., September 23, sun shining, light breeze. Tree 10 feet tall by 7 in diameter. October 19, leaves and fruit uninjured; found four live red scales, all of them located upon the leaves.

(208) Took 7 ounces dry cyanide and added in four minutes 16 fluid

ounces of the diluted acid; turned the blower five minutes. 3.55 to 4.15 p. m., September 23, sun shining, light breeze. Tree 9 feet tall by 8 in diameter. Two small pieces of cyanide remained in the generator unacted upon when the tent was removed from the tree. October 19, five dozen leaves and many of the young lemons were either dead or were more or less injured; found no live red scales.

(209) Took 7 ounces dry cyanide, set generator under the tent and added at once 14 fluid ounces pure sulphuric acid, placing a board over, but slightly above, the generator. 4.40 to 5 p. m., September 23, sun shining, light breeze. Tree 9 feet high by the same in diameter. October 19, no leaves or fruit were injured; found four live red scales, located mostly on the leaves.

(210) Took 2 ounces dry cyanide and 2¼ fluid ounces of water, added in a few seconds 2¼ ounces pure sulphuric acid. Turned the blower five minutes. 1 to 1.20 p. m., September 25, sun shining, light wind. Tree 8 feet high by 5 in diameter. October 19, about one-fourteenth of the leaves were killed; found no live red scales.

(211) Took 4 ounces dry cyanide and 4½ fluid ounces of water, added in a few seconds 4½ fluid ounces of pure sulphuric acid. 4.10 to 4.30 p. m., September 25, sun shining, light breeze. Turned the blower five minutes. Tree ten feet high by 9 in diameter. October 19, leaves and fruit uninjured; found no live red scales.

(212) Took 5 ounces dry cyanide and 10 ounces of water, added in a few seconds 5 ounces of pure sulphuric acid. Turned the blower five minutes. 5.10 to 5.30 p. m., September 25, sun shining, light breeze. Tree 11 feet high by 9 in diameter. October 19, leaves and fruit uninjured; found no live red scales.

(213) Took 7 ounces dry cyanide and 14 ounces water, added at once 7½ fluid ounces pure sulphuric acid. Turned the blower five minutes. 9.30 to 9.50 a. m., September 26, sun shining, light breeze. Tree 12 feet high by 10 in diameter. A piece of loose cotton batting a quarter of an inch in thickness was placed over the opening in the generator, through which the gas passed on its way from the generator to the tent. October 19, one-eighteenth of the leaves were killed and several of the green lemons were injured; found no live red scales.

(214) Took 5½ ounces dry cyanide and 22 fluid ounces of water, added at once 5¾ fluid ounces of sulphuric acid. Turned the blower five minutes. 10.30 to 10.50 a. m., September 26, sun shining, light breeze. Tree 10 feet high by 9 in diameter. Placed some cotton batting over the opening in the generator as described in the preceding experiment. October 19, one-eighth of the leaves were killed and several of the green lemons were injured; found no live red scales. (Two cats were confined in a barley-sack and placed on the ground beneath the tent before the latter was charged with the gas, and when the tent was removed from the tree both of them were dead.)

(215) Took 5 ounces dry cyanide and 10 ounces of water, added at

once 5¼ fluid ounces of sulphuric acid. Turned the blower five minutes. 11.25 to 11.40 a. m., September 26, sun shining, light breeze. Tree 10 feet high by 9 in diameter. Placed a piece of cotton batting over the opening in the generator as before. October 19, one-fifth of the leaves were killed; found no live red scales. Before being operated on this tree was in a very unhealthy condition.

(216) Took 3½ ounces dry cyanide and 8 ounces of water, added at once 4 ounces of pure sulphuric acid. Turned the blower five minutes. 1.50 to 2.05 p. m., September 26, sun shining, light breeze. Tree 11 feet high by 8 in diameter. Placed cotton batting over the opening in the generator as before. October 19, about eight dozen leaves were killed; found three live red scales.

(217) Took 5 ounces dry cyanide and 10 ounces of water, placed the generator under the tent and added at once 5½ ounces pure sulphuric acid and placed a barley sack over the generator. 2.35 to 2.50 p. m., September 26, sun shining, light breeze. Tree 12 feet high by 10 in diameter. October 19, leaves and fruit uninjured; found no live red scales.

(218) Took 6 ounces dry cyanide and 12 ounces water, placed the generator under the tent and added at once 6½ ounces of pure sulphuric acid, after which a barley sack was placed over the generator. 3.25 to 3.40 p. m., September 26, sun shining, light breeze. Tree 12 feet high by 10 in diameter. October 19, a few leaves at the top of the tree were killed; found no living red scales.

(219) Took 7 ounces dry cyanide and 14 ounces of water, placed the generator under the tent and added at once 7½ ounces of pure sulphuric acid, after which a barley sack was placed over the generator. 4.10 to 4.30 p. m., September 26, sun shining, light breeze. Tree 11 feet high by the same in diameter. October 19, a few leaves at the very top of the tree were killed and some of the green lemons were injured; found no live red scales.

THE LARVÆ OF HYPODERMA BOVIS, De Geer.

By Cooper Curtice, *Veterinarian.*

In the course of investigations of the Bureau of Animal Industry made during December, 1889, and January, 1890, I have been collecting the larvæ of *Hypoderma bovis* from cattle. I found larvæ of the first stage* (1) in the œsophageal walls, (2) one specimen under the pleura near the eleventh rib, (3) in the subcutaneous tissue of the back, and (4) in subcutaneous tumors which opened by an orifice upon the external skin. Larvæ of the second and third stages have been discovered

*By *first stage* I mean the earliest stage found. They were from 10-15^{mm} long and 1.5^{mm} thick, and were similar to the first stage of *Hypoderma diana*, as figured by Brauer (Mon. d. Œstriden).—C. C.

only in tumors. Molts of the first stage were found in the tumors with the second and were the means of connecting the three stages. Larvæ of the first stage were more abundant in the earlier part of the collection; in the latter part but few could be found, and later stages were more abundant. Hinrichsen, 1888 (Archiv. f. wiss. u. prak. Thierheilkunde, Bd. XIV, p. 219), found the first stages of a larva he hesitatingly referred to *H. bovis* in the spinal canals of ten out of twenty-five head of cattle examined. The presence of these larvæ of the first stages in the œsophagus, back, subcutaneous tissue and tumors, suggests that the life history of a certain portion of the larvæ, if not all, has been overlooked. It is possible that the eggs or young larvæ are licked by the cattle from the backs; that the larvæ make their way into the œsophageal walls, and from thence, during the proper season, through the back in the neighborhood of the eleventh rib, to the skin.

Further observations of this parasite will be made throughout the year in order to definitely establish the life history of the youngest stage, which hitherto seems to have been neglected. Illustrations of the various stages of the parasites and the injuries they produce will accompany the detailed report of the investigations which will appear in the publications of the Bureau of Animal Industry.

THE IMPORTED GIPSY MOTH.

(*Ocneria dispar L.*)

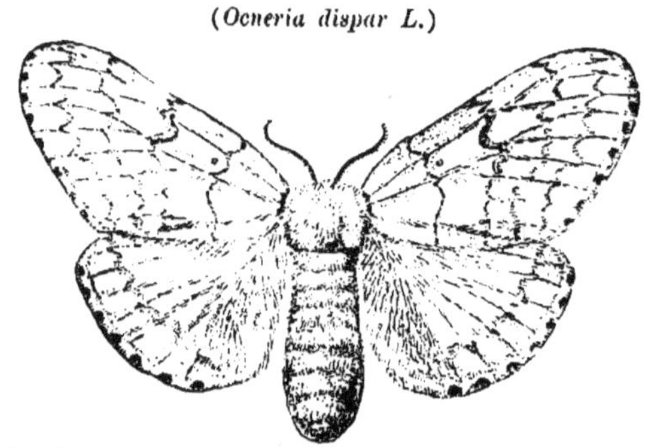

FIG. 36.—*Ocneria dispar*, female—natural size (after Ratzeburg).

This conspicuous insect, although not recorded in any of our check-lists of North American Lepidoptera, has undoubtedly been present in a restricted locality in Massachusetts for about twenty years. It was imported by Mr. L. Trouvelot in the course of his experiments with silk-worms recorded in the early volumes of the *American Naturalist*, and certain of the moths escaping, he announced the fact publicly, and we mentioned it in the second volume of the *American Entomologist*, p. 111 (1870), and in our second report on the insects of Missouri, p. 10. It is, indeed, a curious fact that during these twenty years the insect has not become a pest until last season, and still more curious that the moth does not seem to have found its way into the collections and is not mentioned in the check-lists. Last summer, however, it attracted considerable attention, and specimens were sent from Medford to the agricultural experiment station at Amherst, where Mrs. C. H. Fernald, in the absence of her husband, recognized the species. Several newspaper articles were published during the season, notably those in *The New England Farmer*, for July 13, and *The Boston Transcript* of October 31 and November 14.

Professor Fernald on his return from Europe undertook a thorough investigation of the matter, and in a special bulletin of the experiment station of the Massachusetts Agricultural College, published by the assistance of the secretary of the Board of Agriculture, and received by us November 29, has published an eight-page account of the species, with illustrations of the larva and pupa taken from Ratzeburg, and both sexes of the moth drawn from nature. Professor Fernald gives popular descriptions of the different states, and as a remedy recommends spraying all trees in the infested region with Paris green (1 pound to 150 gallons of water) soon after the hatching of the eggs in the spring, for two or three years under competent direction, and predicts the entire destruction of the pest if this course is followed. In Europe it is generally held in check by its natural enemies, but occasionally it becomes very destructive. In 1817 the cork-oaks of southern France suffered severely, and in 1878 the plane trees of the public promenades in Lyons were nearly ruined. Last summer Professor Fernald saw the moth in immense numbers on the trees of the Zoölogical Garden in Berlin, where the caterpillar had done great injury, and the opinion was expressed to him by prominent entomologists in Europe that if the species should get a foothold in this country it would become a far greater pest than the Colorado Potato-beetle on account of its prolificness, and the great number of its food-plants. The European food-plants are, among others, Apple, Pear, Plum, Cherry, Quince, Apricot, Lime, Pomegranate, Linden, Elm, Birch, Beech, Oak, Poplar, Willow, Hornbeam, Ash, Hazel, Larch, Fir, Azalia, Myrtle, Rose, and Cabbage. It is found in nearly all parts of Europe, and in southern and western Asia, extending as far as to Japan.

FIG. 37.—*Ocneria dispar*, male—natural size (after Kirby).

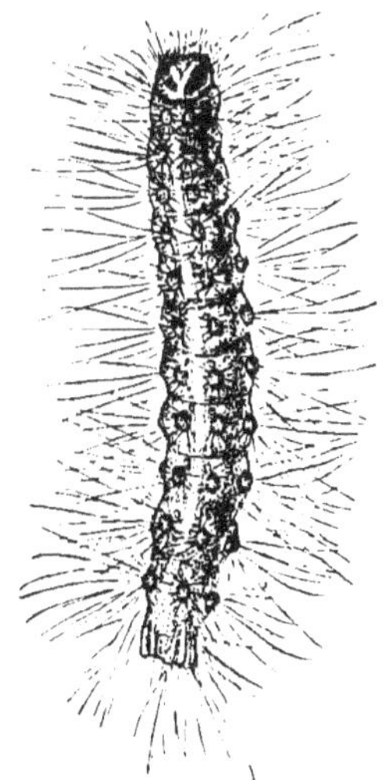

FIG. 38.—*Ocneria dispar*, larva—natural size (after Ratzeburg).

Prof. W. P. Brooks found it abundant at Sapporo in Japan in 1883, where it fed upon strawberry in addition to other plants. In Massachusetts it is reported as feeding upon the leaves of Apple, Cherry, Quince, Elm, Linden, Maple, Balm of Gilead, Birch, Oak, Willow, Wistaria, Norway Spruce, and Corn. Professor Fernald states that in this country it occurs only in Medford, Mass., where it occupies an area in the form of an ellipse about 1½ miles long by one-half mile wide. We have just learned, however, from Mr. Lewis E. Hood, of Somerville, that it was observed in that town last summer. The instance mentioned in INSECT LIFE, Vol. II, p. 86, of its occurrence at Winchester is still uncertain, as only partly grown larvæ were sent by our correspondent, Mrs. Holt.

FIG. 39.—*Ocneria dispar*, pupa—natural size (after Ratzeburg.)

Regarding its natural enemies, Professor Fernald states that none have been noticed in this country, but that eleven species of *Ichneumonidæ* and seven species of *Tachina* flies have been noticed in Europe. This statement is evidently taken from Ratzeburg, who mentions this precise number of eleven Hymenopterous parasites (not all Ichneumonidæ, by the way). By means, however, of a manuscript catalogue of the relations of parasitic Hymenoptera, which Mr. Howard has in preparation, we are able to more than double this list, and as a matter of general interest we publish the following:

1. *Pimpla flavicans* Rtz., Rtz. W. S.
2. *Pimpla instigator* Grav., G. et L. 409, Rtz. W. S.
3. *Pezomachus hortensis* Gr., (hyper) Brischke A. W. T. 128.
4. *Limneria difformis* Gr., Kirch., 94.
5. *Hemiteles fulvipes* Gr., Kirch. 66, Brdg. Ent. XVI, 106, Brischke, A. W. T., Rtz. W. S.
6. *Campoplex conicus* Rtz., Kirch 90, Rtz. W. S.
7. *Campoplex difformis* Gr., Rtz. W. S. = *Limneria*.
8. *Mesochorus pectoralis* Rtz., Rtz. W. S.
9. *Mesochorus gracilis*, Brischke A. W. T. 128.
10. *Mesochorus splendidulus* Gr., Brischke A. W. T. 128.
11. *Apanteles glomeratus* L., G. et L., 413.
12. *Apanteles fulvipes* Hal., Brischke A. W. T. 128.
13. *Apanteles melanoscelus* Rtz., Kirch. 121, Rtz. W. S. = *Apanteles difficilis* Nees.
14. *Apanteles solitarius* Rtz., Kirch. 122, Rtz. W. S., Brischke A. W. T. 128.
15. *Microgaster calceatus* Hal., Marsh. M. B. B. 246.
16. *Microgaster* (?) *tenebrosus* Wesm., Brischke A. W. T. 128.
17. *Microgaster tibialis* Nees., Brischke A. W. T. 128.
18. *Microgaster* (?) *liparidis* Ratz., Ratz. W. S., Kirch, 121.

19. *Microgaster pubescens* Rtz., Kirch. 121, Rtz. W. S. = *calceatus* Hal.
20. *Eurytoma abrotani* Panz., Rtz. W. S., Kirch. 155, Brischke, A. W. T. 128.
21. *Pteromalus halidayanus* Rtz., (hyper) Brischke, A. W. T. 130.
22. *Pteromalus pini* Hartig, (hyper) Brischke, A. W. T. 128.
23. *Pteromalus bouchéanus* Rtz., (hyper) Brischke, A. W. T. 128, G. et L. 428.
24. *Eupelmus bifasciatus* Giraud, G. et L. 420. On eggs.

Among the twenty-four species above mentioned there will undoubtedly be a few synonyms, and from the known generic habits there are unquestionably a number of secondary parasites. Brischke has called special attention to the fact that Nos. 17, 21, 22, and 23 are hyper-parasites, and to these we may unquestionably add 13, and in all probability, 14, 15, and 16, as *Mesochorus* has often been reared from *Microgaster* cocoons, and as we are not familiar with any cases of primary parasitism in this genus. There is also some little doubt about the species of *Campoplex*, so that only fourteen undoubted primary parasites are left. The majority of these insects are not confined to *Ocneria dispar*, and some of them are well-known and widely-spread beneficial insects. The *Apanteles glomeratus*, for instance, is a well-known European parasite of the common Cabbage Worm, and occurs quite abundantly in this country. It is almost incredible that the caterpillar should have no American parasite, and we imagine that careful study will show that some of our American species of the *Microgasterinæ*, at least, will be found to infest it, while predatory insects, of course, are not so strictly confined as to the character of their prey.

In conclusion we may state that if Professor Fernald's recommendations are carried out at all strictly we have little fear of the spread of this pest, and agree with him that it can be entirely killed out with the expenditure of a little time and money.

SOME INSECT PESTS OF THE HOUSEHOLD.

By C. V. Riley.

[Continued from page 130.]

THE TRUE CLOTHES-MOTHS.*

"And he, as a rotten thing, consumeth, as a garment, that is moth-eaten."—Job, xiii, 28.

The true clothes-moths are the housekeepers' dreads, in parts of the country where the Buffalo-bug is not known, and they flourish, though with diminished prominence, through comparison with the Buffalo-moth, in all sections. They are cosmopolitan insects, having been carried in clothes to all parts of the world, and no one of them is indigenous in the United States, so far as we know. The greatest confusion existed

*Reprinted substantially from *Good Housekeeping*, April 27, 1889.

until within recent years as to the proper nomenclature of the species noted for their damage in this country, and as a striking example I may state that Dr. Packard, in his well-known *Guide to the Study of Insects*, under the head of "The Common Clothes-moth," describes the larva, case, and pupa of one species, the moth of a second, and gives it the name of a third. Some years ago I sent a number of specimens to Lord Walsingham of Merton Hall, England, a world-famous authority upon these small insects, and cleared up, with his assistance, the confusion then existing. About the same time Prof. C. H. Fernald, then of Orono, Me., now of Amherst, Mass., also performed the same task with Lord Walsingham's assistance.

From these investigations we learn that there are three distinct species of clothes-moths common in this country, all of which are of European origin. They are somewhat similar in the larva and pupa states and all lay minute pale yellowish ovoid eggs or nits on the stuffs which they attack and injure; but they differ somewhat in the moth

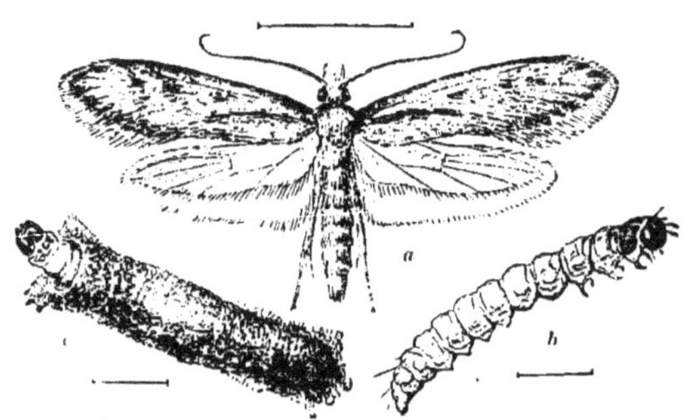

FIG. 40.—*Tinea pellionella*—enlarged—*a*, adult; *b*, larva; *c*, larva in case (after Riley).

or imago state. The statements of habits which are here given are for temperate regions; in more southern regions and in houses kept uninterruptedly warm by furnace or steam heat there is danger of continued injury during winter, and an increased number of generations, where ordinarily in more northern regions there is cessation of injury during the cold season.

The common case-making species is properly called *Tinea pellionella* Linn. The species which makes a gallery of the substance on which it is at work should be known as *Tinea tapetzella* Linn, while the third species, which does not make a case, but in transforming constructs a cocoon by webbing together bits of the substance upon which it feeds should be called *Tineola biselliella* Hummel.

Perhaps the commonest of these in more northern regions is the case-bearing species (*T. pellionella*), shown at Fig. 40. Its habits may thus briefly be stated: The small light-brown moths, distinguished, as shown at Fig. 40 *a*, by the darker spots at intervals on the wings, begin to appear in May and are occasionally seen flitting about as late as August. They pair and the female then searches for suitable places for the deposition of her eggs, working her way into dark corners and deep into the folds of garments, apparently choosing by instinct the least conspicuous places. From these eggs hatch the white, soft-bodied larvæ (see Fig. 40*b*), each of which begins immediately to make a case for itself from the fragments of the cloth upon which it feeds. The case is in the shape of a hollow roll or cylinder and the interior is lined with silk (see Fig. 40*c*). As they grow they enlarge these cases by adding material to either end and by inserting gores down the sides which they slit open for the purpose. The larva reaches its full growth toward winter and then, crawling into some yet more protected spot, remains there torpid through the winter within its case, which is at this time thickened and fastened at either end with silk. I have known these larvæ in autumn to leave the carpet upon which they had fed, drag their heavy cases up a 15-foot wall and fasten them in the angle of the cornice of the ceiling. The transformation to pupa takes place within the case the following spring and the moths soon afterward issue. Such is the life round of the first species. It feeds in all woolen cloths and also in hair cloth, furs, and feathers. Curiously enough a little parasite sometimes enters the house and lays its eggs in the destructive larvæ. The accompanying drawing (Fig. 41) was made from specimens received from Michigan. It may be known as *Hyperacmus tineæ*.

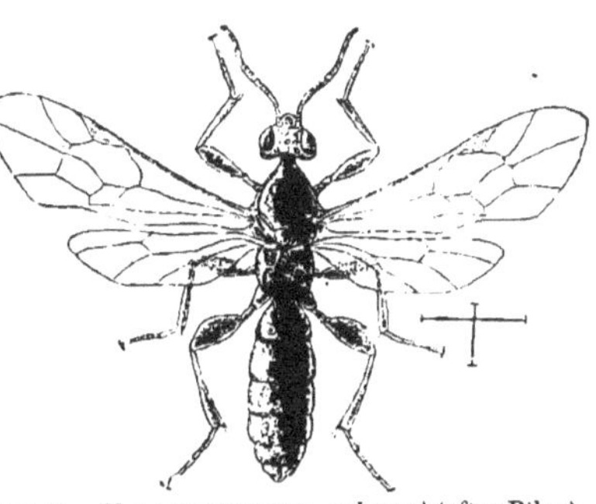

FIG. 41.—*Hyperacmus tineæ*—enlarged (after Riley).

The next species—*Tineola biselliella*—makes no case, but when ready to transform constructs a cocoon mainly from fragments of the material upon which it has been feeding. It spins a certain amount of silk, however, wherever it goes. It is the most common species at Washington, and, so far as my experience goes, in the Southern States. It is generally fond of the same substances upon which the former feeds, and is quite as voracious. A curious instance was brought to my attention in 1884, in which a large stock of feather dusters was completely ruined by this species, while I have often had fine camel's-hair brushes ruined by it when they have been left lying loose in drawers. Its life round is much the same as that of the species just described, but it is commonly believed that

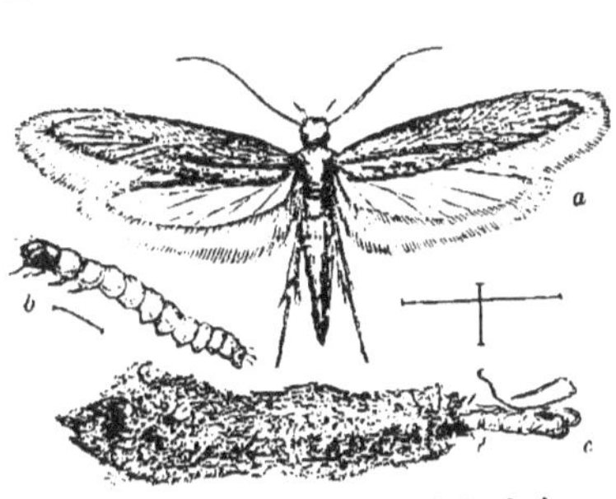

Fig. 42.—*Tineola biselliella*: *a*, adult; *b*, larva; *c*, cocoon and empty pupa-skin—enlarged (after Riley).

there is more than one generation annually in southern latitudes. The parent moth (Fig. 42*a*) is of a delicate straw-color and has no black spots. The larva is shown at Fig. 42*b* and the cocoon at Fig. 42*c*. The latter is often found with the empty pupa-skin protruding from its extremity.

The moth of *Tinea tapetzella*—the last species—is readily distinguished from the others by the fact that the front wings are black from the base to the middle, and white beyond. The white portion is often clouded with dark gray. The habits of this species are much the same as in the others except that the larva forms for itself a silken gallery mixed with fragments of cloth and thus destroys much more material than it needs for food. It remains hidden within some part of the gallery and retreats to another portion when alarmed. It transforms to pupa without other covering than the gallery affords. This is probably the species mentioned by Pliny and referred to in Holy Writ. The moth is shown at Fig. 43.

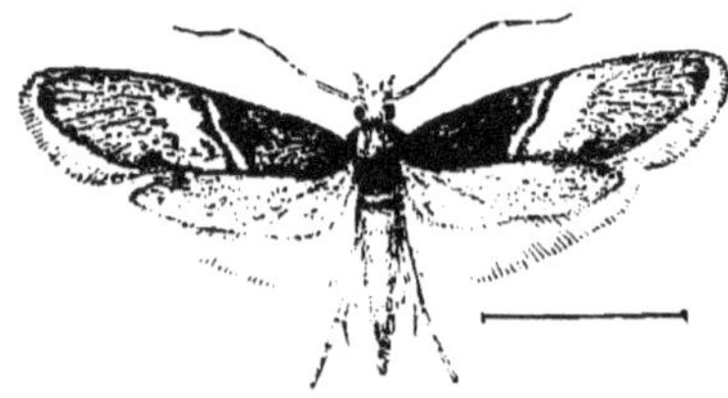

FIG. 43.—*Tinea tapetzella*—enlarged (after Riley).

And now as to the question of remedies: During the latter part of May or early in June a vigorous campaign should be entered upon. All carpets, clothes, cloth-covered furniture, furs, and rugs should be thoroughly shaken and aired, and, if possible, exposed to the sunlight as long as practicable. If the house is badly infested or if any particular article is supposed to be badly infested, a free use of benzine, in the manner mentioned in my last article, will be advisable. All floor cracks and dark closets should be sprayed with this substance. Too much pains can not be taken to destroy every moth and every egg and every newly-hatched larva, for immunity for the rest of the year depends largely—almost entirely—upon the thoroughness with which the work of extermination is carried on at this time. The benzine spray will kill the insect in every stage, and it is one of the few substances which will destroy the egg. I would, however, repeat the caution as to its inflammability. No light should be brought into a room in which it has been used until after a thorough airing and until the odor is almost dissipated.

The proper packing away of furs and winter clothing through the summer is a serious matter. A great deal of unnecessary expenditure in the way of cedar chests and cedar wardrobes and various compounds in the way of powders has been urged by writers on these pests. But experience fully proves that after a thorough treatment in May or June, garments may be safely put away for the rest of the season with no other protection than wrapping them closely in stout paper, to preclude infection through some belated female. My assistant, Mr. L. O. Howard, tells me of an excellent plan which he has adopted. He buys for a small sum from his tailor a number of pasteboard boxes in which they deliver suits, and his wife carefully folds and packs away all clothing, gumming a strip of wrapping paper around the edge of the cover so as to leave no crack. These boxes will last for a life-time with careful use. Others use for the same purpose ordinary paper flour sacks or linen pillow-cases, which answer well. The success of these means depends entirely on the thoroughness of the preliminary work. Camphor, tobacco, napthaline, and other strong odorants are only partial repellants and without the precaution urged are of little avail.

Cloth-covered furniture which is in constant use will not be harmed, and the same may be said of cloth-lined carriages. Where such furniture is stored away or kept unused in a dark room or where the carriages are left in a dark coach-house through the summer, at least two sprayings with benzine, say once in June and once about August 1, will be advisable. Another plan which will act as a protection in such cases is to sponge the cloth linings and covers both sides where possible, with a dilute solution of corrosive sublimate in alcohol made just strong enough not to leave a white mark on a black feather.

IRRIGATION AND INJURIOUS INSECTS.*

The question of the proposed reclamation of the arid lands of the West by irrigation is of great importance from the entomological standpoint, mainly in view of its influence upon the destructive appearances of the Rocky Mountain Locust or Western Grasshopper, which at irregular intervals has greatly damaged the agriculture of certain of our Western States and Territories. The last important invasion of this pest occurred during the years 1875 and 1876, and the devastation which it occasioned at that time is so fresh in the minds of all as to require no elaboration of the importance of the subject. The reports of the U. S. Entomological Commission, an organization founded in March, 1877, and composed of Professors C. V. Riley, A. S. Packard, and Cyrus Thomas, consider the question of the influence of irrigation of a large extent of the arid territory upon the increase of this pest, and from the first report of this Commission, published during the year 1878, and the second report, published in 1880, can be drawn a complete summary of the writings on this subject and the views in full of the Com-

* Reply written by Mr. Howard during Prof. Riley's absence in Europe, in answer to a circular letter from the Assistant Secretary of Agriculture to the heads of certain of the scientific divisions of the Department, asking for the bearings of the proposed Government irrigation of western lands upon the problems comprehended by the work of their respective divisions, for the use of the Senate Committee on Irrigation, of which Senator Stewart is chairman.

mission. Copies of those reports would accompany this statement but they have been long out of print. They may be found, however, in the library of the Geological Survey.

One of the most important results arrived at is the conclusion that an extensive system of irrigation upon a scale of greater magnitude than any which can be undertaken by a pioneer population will be not only necessary to the carrying on of agricultural operations within the belt of territory mapped out as the permanent breeding grounds of the locust, but with the prime result that such an introduction of diversified agriculture into these regions will abolish the conditions necessary to a permanent reproduction of the species, and will consequently reduce the danger of the appearance of destructive migratory swarms to a minimum. The one fact that, according to the careful statistics gathered by the Commission, the loss from this pest during the years 1874 to 1877 amounted to upwards of two hundred million dollars, is a mighty argument for the expenditure of the sums which it is proposed to devote to the purpose which Senator Stewart's committee is now investigating. The words which the Commission have devoted to the discussion of this point are best quoted, and I give in the following pages extracts from the first and second reports above referred to.

It is evident, therefore, that the final and complete solution of the locust problem depends to a certain extent upon the possibility of modifying, to some degree at least, the aridity of the great plains of the Northwest, which undoubtedly form the native breeding grounds of these insects.

By most persons this will be considered equivalent to saying that the locust problem will never be solved. It would scarcely be proper for us here to enter into a discussion of the question of the possibility of modifying the condition of the dry area, but we can not refrain from placing upon record our protest against any such conclusion as this. That man, with a mind that can bring art, science, and mechanics to the perfection now visible on every hand, must be forever unable to convert the desert into fertile fields or to redeem the waste places of earth, we can not believe unless we are shown that the moisture which once supplied these areas has forever taken its departure from our globe.

To what extent these dry areas of the west can be supplied with water and rendered fertile must be determined by those who are proficient in this particular branch of science; but that large sections can be redeemed by proper efforts, if made on a scale of sufficient magnitude, we have no doubt.

By utilizing all the water that flows down from the mountains for the purposes of irrigation; by collecting in reservoirs the winter supply and distributing it in the growing season, a very large section of these plains might be brought under cultivation, and extensive forests grown where now the surface is naked and barren. Every field brought into cultivation, every grove planted, is just so far a step toward the ultimate solution of the locust problem; and the nearer these can be brought to their native home the more effectual will they be in rooting them out. If extensive efforts in this direction were made in British America, north of Montana, also in eastern Montana, western Dakota, and the regions around the Black Hills, it would not only be of immense benefit in supplying new agricultural fields for emigrants from the locust problem; it would also be a most effectual method of settling the Indian question in this region. Just what can be done in the way of redeeming these areas we can not say, but when their settlement depends upon it, and the welfare of a much larger area south and west also depends upon it, certainly the question is worthy of consideration by our national authorities.

The day is not far distant when our National Government will be compelled to meet this important question and to test the ability of man to accomplish the work.

The progress of settlement westward must necessarily be slow when it, as is now beginning to be the case, impinges upon the sterile area; it can only push onward when the front line is backed by a dense population and farms studded with groves. It is possible that if there were no other impediments to overcome than this sterility, formidable as it is, the gradual filling up of the border area with an active population would modify the conditions sufficiently, at least, to allow the pushing into and redemption of a belt of considerable breadth. But when to this difficulty is added the devouring locust the hope of success is greatly diminished. * * *

In the permanent region, which embraces the Rocky Mountain plateau and the bordering plains from the middle of Colorado northward, the rain-fall is insufficient for agricultural purposes, and hence irrigation has to be resorted to; in the temporary region this is unnecessary. The plains and plateaus of the permanent region are to a large extent distinguished by the presence of *Artemisia*, Chenopodiaceous plants, and what is usually termed "bunch-grass;" in short by all the characteristics of a drier climate. One other peculiarity which has not been overlooked appears to mark roughly the southern boundary of the permanent home of the Rocky Mountain locust, and that is the isothermal curve or line of the 50° of mean annual temperature, which also corresponds very nearly with the isothermal curve or line of summer temperature of 70°. But this applies only to that portion of the region which extends upon the plains east of the mountains.

If any practical means of exterminating the locusts in this permanent region could be devised the whole locust problem could be solved, and nothing further would be necessary; but when we take into consideration the vast extent of this area, and the fact that a very large portion of it can not be brought under cultivation without a material change in the climatic conditions, there appears but little hope that such a means of actual extermination will ever be devised, however much we may hope to check the injurious increase of the pest by the means recommended in the concluding chapter of this report. Our discussion of the future prospects of this region in reference to agriculture may as well, therefore, be on this basis.

* * * A careful investigation of this subject for several years and repeated visits in person to this region have served to convince us that, with the advantages afforded the system of irrigation necessarily adopted, there is no reason why the agricultural area lying along the east flank of the range should suffer any more from these pests than portions of the temporary regions.

* * * * * * *

This agricultural belt, extending from Colorado into British America, is partly along the margin of and partly in the very heart of the permanent breeding grounds where the swarms that invade the temporary region originate. It follows, then, as a natural consequence, that just so far as the numbers are lessened by the operations in this section, just so far will the agriculturists of the temporary region be benefited, and, as we will hereafter see, like operations in the latter region will benefit those in the permanent region. We are fully aware of the fact that the part of this vast region which can be irrigated and cultivated is small in comparison with the whole area which forms the native home of the species; but, fortunately, in one respect this cultivated belt occupies, in part at least, the point of departure of the swarms which invade the temporary region. This fact, therefore, renders it more important that it be occupied by an agricultural population.

Although we have admitted that we are unable to present any plan of exterminating the locusts that holds out sufficient promise of success to justify the General Government in undertaking it, it does not necessarily follow that there is no plan of modifying the evil which the Government would be justified in undertaking. On the contrary, if the views we have advanced be correct, they suggest a means by which the General Government might greatly aid in bringing about the desired result; and fortunately the result would be beneficial even should we be mistaken in the opinions advanced.

As will be seen by what has been stated, the great desideratum is to settle the cultivated belt alluded to as rapidly as possible with an agricultural population. Wherever valuable and permanent mines are discovered in the neighboring mountains, the arable areas in the vicinity will be taken up and cultivated to an extent at least sufficient to supply the demand for agricultural products, as in parts of Colorado. But there are large sections where no such influence will be brought to bear, and this is the case along that portion of the belt where the agricultural population is most needed for the purpose mentioned.

An examination of Map No. 1, in our first report, will show that a comparatively limited belt in central Montana, extending from the Big Horn Mountains northwest to the British line, a little west of Cypress Hill, forms the turning point of the locust movements. Without now repeating the data, which may be found in that report, we may summarize it by saying that from this region a large portion of the swarms come which visit Dakota, Minnesota, Nebraska, and Kansas; from this area also proceed a large portion of the swarms that move southwest into Idaho and Utah; this appears to be the point to which most of the returning swarms from the temporary region direct their flight.

That there are other areas in the permanent region which appear to be special breeding grounds, as points of departure, is certainly true, but none to such an extent as this, and none affecting an agricultural area bearing any comparison with the area affected by the locust swarms originating in this belt.

Even should it be shown by subsequent investigations that as a rule the swarms falling on the temporary regions come from intermediate points, as central and southern Dakota and northwestern Nebraska, the facts already ascertained warrant us in asserting that, as a very general rule, they originate in the belt mentioned.

It is evident, therefore, that if any method can be devised by which an agricultural (not pastoral) population can be thrown into this belt it will form one of the best possible means of modifying the evil. If they can be effectually distributed in this area the result will be of immense value to the agricultural interests of Dakota, Minnesota, Nebraska, Iowa, and Kansas, in fact of the entire temporary region. We do not pretend that it will wholly relieve this area from locust invasions, but it will very materially lessen their extent and injury.

In order to carry on agricultural operations to any great extent in this belt, an extensive system of irrigation will be absolutely necessary. It will have to be on a scale of greater magnitude than any that will be undertaken by a pioneer population. We doubt the propriety of the General Government undertaking such a work directly, if it is possible to accomplish it in any other way. This, we think, may possibly be done by giving the land for this purpose. We are fully aware of the opposition at present to the Government's donating any more of the public land, but the circumstances of this case bring it out of the general rule. If donating the entire body of public land in the belt described would suffice to settle it with an agricultural population, not only would the very purpose for which it is held be accomplished, but, if our views are correct, the result would be of immense benefit to the border States.

We therefore suggest the following as probably the most feasible plan of accomplishing the desired end: Let the United States donate a belt of 50 or 60 miles in width, running from the Black Hills west-northwest, so as to strike the Yellowstone River a short distance above the mouth of the Big Horn River; from thence north-northwest by way of Fort Shaw, or the mouth of Sun River, in the direction of Fort Hamilton, in British America—this to be granted on condition that the company to which said land is granted shall, within a given time, construct a railroad from the

Black Hills along the line designated to the international boundary; shall undertake and carry out to an extent to be designated a system of irrigation, and shall equip and keep in operation said road for a certain number of years.

Whether such grant will be sufficient inducement for any competent company to undertake the work specified is probably the chief difficulty in the way of successfully carrying out this plan. On this point we do not feel qualified to express an opinion. That such a road starting from the Black Hills, if once built, would soon be connected southward and eastward with other roads can not be doubted. That it would be the best possible means of bringing an agricultural population into this belt can not be doubted. It would also be an important factor in settling the troublesome Indian problem in this section of the West.

If the plan should be adopted it might be well to colonize, if possible, with Russian peasants who are accustomed to fighting locusts.

The advantage to be derived from this plan consists chiefly in the fact that it is possible to destroy the young to a very large extent by the use of the proper means. If this is done in the very heart of their breeding grounds it greatly lessens the numbers that will migrate. Not only does it prevent the number destroyed from migrating, but of each one killed, so to speak, an entire family brood of the next or migrating generation is destroyed. In other words, the destruction of thousands there would be as effectual as destroying millions of the migrating swarms. The means of destroying the young, as before stated, can be made more effectual in the sections where irrigation is carried on than where it is not.

As shown in our first report, the destruction of the young locusts bred in the temporary region from the invading hordes not only gives immediate relief, but also tends to postpone future invasions by so lessening the numbers in the returning swarms that a longer time is required for development. With an agricultural population in the area designated the work of destruction would then be carried on at each end of their migratory route.

Here we may also remark that the present idea of making that section of our country a peculiarly pastoral area, while doubtless profitable to the present and for two or three generations to come, will in the end entail hardships upon those to follow. It can no longer be doubted that while the destruction of forests was the chief agency, yet the pastoral habit of the people of western Asia and other oriental countries, once so fertile but now barren, was one important factor in producing the present dry and barren condition of those countries. No country in the interior of a continent, unless supplied with numerous lakes or numerous and permanent rivers, can remain permanently fertile and productive if given up largely to pasturage of sheep, goats, and cattle, without cultivation. The rapid destruction of mountain forests, and pasturing their slopes and bordering plains, will most certainly have a tendency to render that portion of our country more dry and barren.

Unless, therefore, our Government adopts some policy by which an agricultural population can be thrown into that area, the day will most assuredly come when it will be as barren and desolate as the plains of Arabia. The development of the locusts is but an incident of the change from a former condition of abundant moisture to the present dry one. But this branch of the subject we propose to omit at present.

It will be seen, therefore, by the foregoing that we think it is possible to modify to a very large extent the operations of the locusts so far as these relate to the area along the east flank of the mountains, and that the General Government may, without any very great expense, very greatly assist in the work.

* * * * * * *

This certainly shows a very moderate climate for this northern latitude. Wheat, oats, rye, and barley grow well, and Indian corn is also raised without difficulty and produces good crops. Such fruits as apples, plums, cherries, currants, raspberries, and gooseberries may be grown and matured here, the climate presenting no serious obstacle.

The amount of land that can be brought under cultivation depends wholly upon the amount of water that can be obtained for irrigation. If the plan for making reservoirs for preserving the winter supply should ever be adopted, the breadth of the agricultural belt would be very largely increased, and this would be doubly beneficial in assisting to destroy the locusts and tending to increase the moisture in the atmosphere by forming a larger evaporating surface. The growth of trees and shrubbery around these reservoirs would also be beneficial in the same direction.

But experience in the settling of these mountain regions and Western Territories shows that no such extensive works will, or in fact can be, undertaken by a pioneer agricultural population. Some efficient aid of some kind must be given if such a scheme is ever carried into effect, and if the land itself will do this, the Government will act wisely in giving it for this purpose.

* * * * * * *

As shown by our first report the region around Salt Lake is subject to repeated locust invasion from the north, apparently the resulting broods of the swarms that originate in that portion of Montana of which we have been speaking, and which, pouring over the mountain-pass at the head of Jefferson River, move down Snake River Valley.

If the scheme we have suggested should be carried out and should prove beneficial in reference to the eastern area, it would have, to some extent at least, a like effect as to this section. If it is possible to establish and maintain an agricultural population in the Upper Snake River Valley, this would have a strong tendency to modify the evil. But the present barren aspect of this region would seem to forbid any hopes of ever accomplishing this desired end. Still there appears to be one possible means of bringing this about, at least to a limited extent. The demand of trade will doubtless complete the railroad already started in that direction, which is one step towards the desired end, but something more is required in this case.

Snake River affords a large body of water which if properly utilized would irrigate a large breadth of land, and notwithstanding the barren appearance of the soil, it is really fertile when irrigated. It is possible, with a moderate expense, to throw dams across this stream at certain favorable spots, and by this means to spread the water over the adjoining plains. A work of this kind would, of course, have to be done by the General Government. The feasibility of this project could easily be ascertained by an officer of the Engineer Corps of the Army; and as this is on the line of the chief inter-montane thoroughfare, and also of the locust invasions of this region, the subject is certainly worthy of the attention of the Government.

As will be seen by what we have presented on this subject, the philosophy of our plan for modifying the evil is to place an agricultural population in the very home of the species, which from necessity would be compelled to wage a constant warfare against them.

By stirring the soil their nests would be disturbed; by fighting the young their numbers would be diminished; and as irrigation would be necessary, the effect of dry seasons on the crops would not be felt as in the temporary region. The possibility of inundating to a considerable extent their egg deposits by the winter supply of water would tend to diminish their numbers. The fact that their breeding-grounds are chiefly in the limited agricultural areas is also another argument in favor of the plan.

That large areas would be left where locusts breed and pour down on the nearest cultivated areas, as in western Colorado, is certainly true, but this does not lessen the value of the plan proposed, nor is it a reason why it should not be put into operation.

The effect of irrigation upon the Rocky Mountain Locust dwarfs into comparative insignificance anything which may be said concerning its influence on other destructive species, yet there are many forms which depend for their existence and multiplication upon a dry climate, and which a thorough system of irrigation would render comparatively harmless.

This has been recognized by the prominent writers upon economic entomology, and I may quote the words of my chief, Professor Riley, as follows:

I have repeatedly laid stress in my writings on the importance of irrigation in combatting several of our worst insect enemies, and, aside from its benefits in this direction, every recurrence of a droughty year convinces me of its guarding against failure of crops from excessive drought. I am glad to know that many farmers, and especially small-fruit growers in the vicinity of New York, are preparing in one way or another for irrigation whenever it becomes necessary, and I was pleased to hear Dr. Hexamer, at the late meeting of the American Pomological Society, urge a general system of irrigation as the most profitable investment the cultivator can make in a climate subject to such periods of drought as ours is known to be.

Perhaps the most striking example among this class of insects is the Chinch Bug—a species which damages certain cereal crops to the extent of upwards of five millions of dollars in years of abundance. This insect is directly influenced by moisture and seldom occurs in numbers in the more eastern States except after two or more successive seasons of drought. After a year of excessive multiplication these insects will often be found to have hibernated in immense numbers, and it is a well-known fact that heavy rain-falls the succeeding spring will destroy them almost completely. This being the case an artificial system of irrigation will enable the agriculturalists to hold this insect completely in check, and such a system as it is proposed to introduce in the West will render the grain-growers of the reclaimed regions independent of the damage which may be done by this insect and will enable them to compete on most advantageous terms with the grain-growers of the more eastern localities, whose crops are occasionally subject to almost total loss by this insect enemy. I may again quote from Professor Riley:

Irrigation where it can be applied—and it can be in much of the territory in the vicinity of the Rocky Mountains, where the insect commits sad havoc, as with a little effort in many regions in the heart of the Mississippi Valley—is the only real available practicable remedy after the bugs have commenced multiplying in the spring. I wish to lay particular stress upon this matter of irrigation, believing, as I do, that it is an effectual remedy against this pest, and that by overflowing a grain field for a couple of days, or by saturating the ground after as many more in the month of May, we may effectually prevent its subsequent injuries.—(Seventh Report Insects of Missouri.)

We may mention also the case of the Grape-vine Phylloxera and may again quote from Professor Riley:

Submersion, where practicable, and where it is total and sufficiently prolonged, is a perfect remedy. This is what even the closest student might expect, as he finds that excessive moisture is very disastrous to the lice. M. Louis Faucon, of Graveson (Bouches-du-Rhône), France, has abundantly proved its efficacy, and has by means of it totally annihilated the insect in his vineyard, which was suffering from it four years ago. From his experience we may draw the following conclusions:

(1) The best season to submerge is in autumn (September and October), when the lice are yet active and the vines have ceased growing. Submergence for 25 to 30 days at this season will generally rout the lice.

(2) A submergence of 40 to 50 days in winter is required, and even where the water is allowed to remain during the whole season the vineyard does not suffer. I should consider this very doubtful.

(3) A vineyard should never be inundated for a longer period than two days in summer or during growth; and, though these brief inundations at that season affect only the few lice near the surface and are by no means essential, they are nevertheless important auxiliaries to the more thorough fall or winter submersion, as they destroy the few lice which are always invading a vineyard in infested districts. These summer inundations will be necessary only after the winged insects begin to appear, and three or four, each lasting less than two days, made between the middle of July and the fall of the leaf, will effect the end desired.

(4) An embankment should be made around the vineyard in order that the water may evaporate and permeate the earth, but not run off and carry away any nutritive properties of the soil.

The varied success which has attended the different attempts to rout the enemy by inundation is owing to the lack of thoroughness in many of them. The ground must be thoroughly soaked for a sufficient length of time. Temporary irrigation does not accomplish the end, for the reason that it does not reach all the lice, and does not break up the numerous air bubbles which form in the soil and prevent the drowning of many of the insects. (Sixth Report Insects of Missouri.)

Too much in fact can not be said of the advantages of a system of irrigation in fighting many insect pests.

A good instance occurred in our experience in the spring of 1879, when the Army Worm appeared in great force upon a large grass plantation near Portsmouth, Va. The plantation was divided into sections by irrigating ditches, and it was only necessary to turn on the water to isolate a badly infested section and to devote it to rolling, fire, or some other means of destruction, preventing ready spread to other sections. In the same way rice planters have a ready means of fighting insect pests at hand.

Other insects might be particularized, but the general statement that from the stand-point of the economic entomologist irrigation in general is a great help in fighting insect pests, and from the marked illustration of the great good accomplished by the reclamation of the arid regions in connection with the damage done by the Rocky Mountain locust it will probably be considered that further elaboration is unnecessary.

Respectfully submitted, May 13, 1889.

NOTE ON THE OVIPOSITION AND EMBRYONIC DEVELOPMENT OF XIPHIDIUM ENSIFERUM, Scud.

By William M. Wheeler, *Milwaukee, Wis.*

Though the Orthoptera have received more attention from students of insect embryology than any other natural order of Hexapoda, there still remain several families which, owing to the difficulty of procuring sufficient material, have not been studied. We possess monographs, more or less complete, on members of the Gryllid, Acridiid, and Blattid groups, but besides a few observations on an European *Mantis* we have no observations on members of the families *Locustidæ*, *Mantidæ*, and *Phasmidæ*. The differences in the details of embryonic development observed in the Orthoptera hitherto investigated are so great that all students of the subject must look forward with considerable interest to any results accruing from the study of representatives of these four families. In my search for insects' eggs of a convenient size, procurable in abundance, and representing families heretofore unstudied, I happened on one of the *Locustidæ*, the eggs of which meet the requirements. The species to which I allude is *Xiphidium ensiferum*, Scud., a very common insect about the meadows and marshlands of Wisconsin and the adjacent states.

Unlike other species of the family whose oviposition has been described, *Xiphidium ensiferum* does not oviposit on or in twigs, but between the scales of a Cecidomyid gall, very common on the willows which grow in the damp situations haunted by the Locustid. Mr. L. O. Howard, who kindly examined a specimen of the scaly turnip-shaped gall for me, pronounces it to be very probably produced by *Cecidomyia salicis-gnaphaloides*, Walsh. On September 8 I observed a female in the act of oviposition. She was perched with her head turned toward the apex of the gall, which contained besides the large white Cecidomyid larva in the center of its base, a number of the smaller orange-red larvæ of an inquiline Cecidomyid between the scale-like leaves. Slowly and sedately she thrust her sword-shaped ovipositor down between the leaves and, after depositing an egg, as slowly withdrew the organ in order to recommence the same operation after taking a few steps to one side of where she had been at work. She soon observed me and slipped away without completing her task.

The subopaque, cream-colored egg is elongate oval, 4 to 4.5^{mm} long and 1^{mm} broad through its middle. One of the poles is somewhat more attenuate than the other and there is a faint curvature in the polar axis which causes one side of the egg to be somewhat more convex than the other. The yolk, very similar in constitution to that of other Orthoptera, is pale yellow. It is inclosed by a delicate vitelline membrane and a thicker, opaque and somewhat leathery chorion which suddenly becomes transparent when immersed in alcohol. The eggs are deposited with their long axes parallel to the long axis of the gall and their attenuate poles upward. They are completely concealed by the leaves, the edges of which close over and very efficiently protect them. The number of eggs found in a gall varies considerably. Sometimes but two or three will be found, more frequently from fifty to one hundred; in one small gall I counted one hundred and seventy and I have opened a few galls which contained more. From these facts I conclude that one female frequently deposits her quantum of eggs in several galls, possibly having some means of selecting the best cradles for her offspring and perhaps trying several till she finds one perfectly adapted to her purposes. Frequently as many as ten eggs will be found under a single scale. When this is the case, the eggs adhere to one another somewhat and are often irregularly placed, as if two or three insects had in succession oviposited in the same place.

Whereas the *Blattidæ* show the greatest fixity in habits of oviposition of any of the Orthopteran families, the *Locustidæ* exhibit the greatest variety. Some species like the American *Anabrus simplex* and the European *Locustæ* oviposit in the ground like the *Acridiidæ*. Others, like many species of *Xiphidium* and *Orchelimum*, oviposit in the pith of easily penetrated twigs. According to Professor Riley *Phaneroptera curvicauda* lays its eggs "singly in the edges of leaves, between the upper and lower cuticles." Other species, approaching *Xiphidium ensiferum* like *Conocephalus ensiger*, lay their eggs between the root-leaves and stems of various plants. The European *Meconema varium*, according to Taschenberg, oviposits under bark scales and occasionally in the galls of the Hymenopteron *Teras terminalis*. Still other forms to which our common Katydid (*Microcentrum retinervis*) belongs, lay their flattened, dark colored eggs in regular rows on twigs, after previously roughening the surface of the bark with their jaws.

The structure of the ovipostor in *Xiphidium ensiferum* would seem to indicate that, like other members of the genus, this species has been in the habit of puncturing the tissues of plants till within comparatively recent times, when it found oviposition in the galls more advantageous. So recent may be the acquisition of this habit that more extended investigation may perhaps show a tendency in some females to puncture twigs, or oviposit, like *Conocephalus*, between the root-leaves and stems of plants.

The Orthoptera present many interesting questions in connection with their habits of oviposition. Most of the species, excepting the aberrant *Phasmidæ*, oviposit in clusters, the eggs of which are arranged in more or less regular rows. This habit is most strenuously adhered to by the *Blattidæ*, though many species of *Acridiidæ*, *Gryllidæ*, *Locustidæ*, and *Mantidæ* are almost equally careful to deposit their eggs in symmetrical series.

During oviposition the two ovaries discharge their eggs alternately in rhythmical sequence, the insect moving a short distance directly forward after the extrusion of each egg or pair of eggs. For what purpose this habit should have been preserved with such tenacity through the long ages during which the Orthoptera have continued to people our earth I am unable to conjecture, unless it be supposed that the primitive species oviposited in portable capsules like those still made by the *Blattidæ*. The method of arranging eggs in two even and alternating series practiced by members of this family is of advantage to the insects, in that it renders the package more compact and more easily carried, just as a box may be made to contain a given number of cigars or similarly shaped objects more easily when they are packed in regular rows than when they are thrown in promiscuously. The *Mantidæ*, which deposit their eggs in cocoons that are no longer carried, may be supposed to represent an intermediate stage as far as the habits of oviposition are concerned between the *Blattidæ* and those numerous forms which either deposit their eggs in exposed situations like *Microcentrum*, or bury them in the earth or the tissues of plants like the *Acridiidæ* and *Gryllidæ*.

The eggs of *Xiphidium ensiferum* begin to develop immediately after their deposition. During the warm days that intervene before the cold of autumn sets in the embryo is formed on the middle of the flat side of the yolk. The head of the embryo points downward towards the insertion of the leaves between which it is placed; consequently the pointed and upward directed pole is the caudal end. The young embryo remains dormant during the winter but continues its development during the warm days of spring. The first larvæ were seen to emerge from the galls on the 17th of May.

I will not here enter into the details of development, many of which I have not yet observed to my own satisfaction. Suffice it to say that the Locustid's ontogeny is strikingly like that of the Gryllid, *Œcanthus niveus* as described by Dr. Howard Ayers. The embryo, as noted above, is developed on the flat ventral face of the egg with its head directed downward. During its growth it gradually moves down the yolk till its head reaches the pole, then it turns and passes up the convex (formerly dorsal) surface of the yolk till its head reaches the pointed (formerly caudal) pole; the body of the embryo meanwhile increases in size and envelops the entire yolk by a very interesting process, the details of which I have not, as yet, been able clearly to elucidate. Considering the position in which the egg is deposited, *i. e.*, with its cephalic pole directed downwards, a revolution like the one described is necessary to bring the embryo's head to the opposite pole, so that in hatching the larva may have no difficulty in crawling out between the scales of the gall.

THE SIX-SPOTTED MITE OF THE ORANGE.

(*Tetranychus 6-maculatus*, n. sp.)

By C. V. Riley.

This mite has done much damage to the orange in Florida since 1886, and we have prepared a preliminary article for the Annual Report of this Department for 1889. As it is deemed wise to exclude purely descriptive matter from the Annual, we give here the diagnosis of the species under the the name of *Tetranychus 6 maculatus* on account of the quite constant markings of its back. In color it is very similar to *T. rosearum* Boisd., *T. tiliarum* Mull. and *T. vitis* Boisd.

Tetranychus 6-maculatus n. sp.—Length of the full-grown specimens 0.3mm. General color, pale greenish-yellow, marked on the abdomen with six or less small dusky spots. General shape oval, somewhat broadest in front of the eyes; laterally slightly constricted just opposite the eyes and at about the middle of the body, at which latter constriction the body is divided by a more or less distinct suture into two parts. There is often, also a distinct, though small tail-like projection at the end of the body. Anterior projection of cephalothorax rather short, somewhat conical, its apex rounded. Terminal joint of legs longest. Eyes, two each side, the anterior one of each pair being blood-red, this pigment extending some distance into the body, giving the appearance of two red eyes on each side; the posterior eyes are colorless and transparent. The spots of the abdomen are arranged in two subdorsal rows, of three spots to each row; they are rounded and quite constant, especially in the smaller and more numerous specimens, though somewhat variable in the larger or full-grown mites.

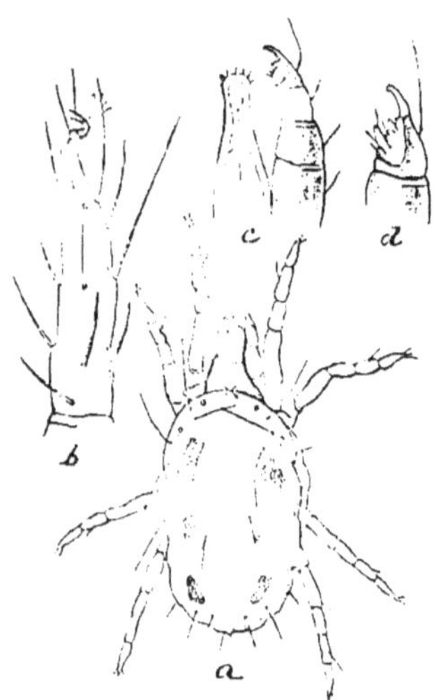

FIG. 44—*Tetranychus 6-maculatus*: *a*, from above—enlarged; *b*, tarsus; *c*, rostrum and palpus—still more enlarged; *d*, tip of palpus—still more enlarged (original).

In the mature specimens the anterior spots, which are arranged close to the dividing suture, are often composed of a collection of eight to twelve, larger or smaller, more or less circular, quite deep black spots, while in others all the spots are single, and with one or the other of the median pair wanting. In the smallest specimens these spots are either wanting or only the anterior or posterior pairs are present, the last pair in this case being generally largest and very distinct. The distribution of the hairs of the body is as follows: Two short, slender hairs medially at anterior margin, directed forward, crossing each other near their tips; each side of these, also close to the margin, at about equal distances from each other and the lateral margin is a pair of transparent, circular pores, resembling those which usually give rise to a bristle. In front of the eyes and removed slightly toward the middle is, on each side, a row of rather long and stout bristles, the anterior pair being directed outward and slightly toward the head, and projecting beyond the lateral margin; the median pair are directed forward and cross each other near their tips. The third pair are longest, situated a little in front of the eyes and directed backward. Besides these stout bristles there is another smaller and slender hair not far from the lateral margin behind the eyes, and another at the margin in front of the eyes. The abdomen is provided on each side with a subdorsal row of three very long bristles, a more slender lateral row, four long dorsal bristles surrounding the end, and four ventral terminal bristles, of which the median pair is smallest.

The eggs are 0.11mm in diameter, globular, either colorless and transparent or very pale greenish-yellow, and are loosely attached to the web.

HARPIPHORUS MACULATUS Norton.

By W. Hague Harrington, *Ottawa, Canada.*

The spotted saw-fly, whose larvæ feed upon the strawberry plant, is widely distributed, and probably well known to all collectors of Hymenoptera, as well as to growers of the delicious fruit which suffers from its ravages. There are, however, one or two points in connection with the species to which attention may be called. Last winter I discussed with Mr. Fletcher the fact that a large proportion of the specimens, which apparently belong to this species, would by the venation of the wings be placed in the genus *Monostegia*, instead of in *Harpiphorus*, and that they agreed closely with the description of *M. obscurata* Cresson.

During the past summer I collected as many specimens of this saw-fly as was possible, in order to further study the species, and to see if there existed sufficient reasons for separating these saw-flies into two species, or on the other hand for including with *H. maculatus* a few specimens which I had previously considered to represent *M. ignota* Norton.

The question has now been made additionally interesting to me by the publication in the November, 1889, number of Insect Life (pp. 137–140) of Mr. F. W. Malley's observations on *M. ignota* as a strawberry pest. The author, after mentioning the similar maculation of the abdomen, states that—

> The most certain method of distinguishing the species is to note the number of submarginal cells in the forewings, *M. ignota* having four, and *H. maculatus* only three.

I found that saw-flies were apparently very scarce last season, but the Strawberry Saw-fly was one of the few species that were moderately abundant. My captures were as follows:

Specimens having three submarginal cells:

Date.	Males.	Females.
May 9	1	1
10	....	1
12	1	....
13	1	....
24	3	3
27	....	1
June 2	....	3
26	....	1
Total	6	10

Specimens having four submarginal cells:

Date.	Males.	Females.
May 5	1	....
9	1	9
11	....	2
12	1	7
24	1	3
June 2	....	1
Total	4	22

This shows the two forms to occur during the same period and in comparatively the same abundance, and the habits of the adults were apparently in all respects similar. With those previously in my collection I have now before me 80 specimens, which appear to belong undoubtedly to the same species. Of these, 16 males and 24 females have *three* submarginals, and 8 males and 29 females have *four*. A connecting link between the two equal groups is formed by the remaining three specimens, which are females, and in each of which the left wing has *four* and the right wing *three* submarginals. Rudiments of the absent—or additional—cross nervure may also be detected in a few of the other specimens.

As might be expected in a series of this length, there is a certain degree of variation in size, coloring, shape of antennæ, etc., but none apparently to warrant a separation into two species, or even varieties. I have, however, two males and one female, collected May 27, which have the abdomen perfectly immaculate, the legs paler and the antennæ shorter, and which appear to be distinct, and to belong to *Monostegia*. The antennæ in these specimens more resemble those of *Monophadnus*, having the second joint as long as the third and fourth united, and the apex blunt; whereas the antennæ in *H. maculatus* (especially in the male) are longer and more tapering, and have the third, fourth, and fifth joints more or less subequal.

Mr. Malley in his excellent plate figures the antennæ of his strawberry pest as of the Monostegia form, and also indicates differences in the larvæ, and possibly the species bred by him may really be a Monostegia and distinct from the specimens with four submarginals which I have taken and consider to be *H. maculatus*. The ornamentation of the abdomen, however, seems so characteristic that one would hardly expect to find insects thus marked feeding upon the same plant and yet belonging to different genera.

ADULTS OF THE AMERICAN CIMBEX INJURING THE WILLOW AND COTTONWOOD IN NEBRASKA.

By F. M. Webster, *Lafayette, Ind.*

Under date of June 11, 1889, Hon. R. M. Pritchard, an old-time friend of the writer in Illinois, but now residing near Pender, Thurston County, Nebr., sent me specimens of both sexes of this species, accompanied by two letters, reading substantially as follows:

> A few days since I was out in my grove of ash, willow, cottonwood, and box-elder, and was not a little startled by finding myself surrounded by what I first thought by their buzzing noise to be great numbers of the large, black hornets; but as the insects were not inclined to attack me, like the hornets of my boyhood days, I began to examine them and watch their movements. There were thousands of them, apparently in the act of mating, but for the most part flying high in the tops of the largest trees, being divided into groups which in their movements seemed to alternately approach and retreat from a central point among the tree-tops, making a noise like a lot of hornets, but moving much slower and more clumsily than hornets. I found a small number settled on the leaves and limbs of the ash and willows, where they seemed to be feeding on the sap. To-day I have been watching them more carefully,

and find that they cut a rough gash almost completely around the limb, seeming to kill the outer bark as far as they cut. This work is done with the jaws. They seem very lively during the middle of the day, and at that time are mostly on the wing, but as the air grows cooler they fasten to the twigs and begin to eat, seemingly being very clumsy and stupid, starting up quickly when approached, but not flying unless forced to do so, and then only a distance of a few feet, often falling to the ground.

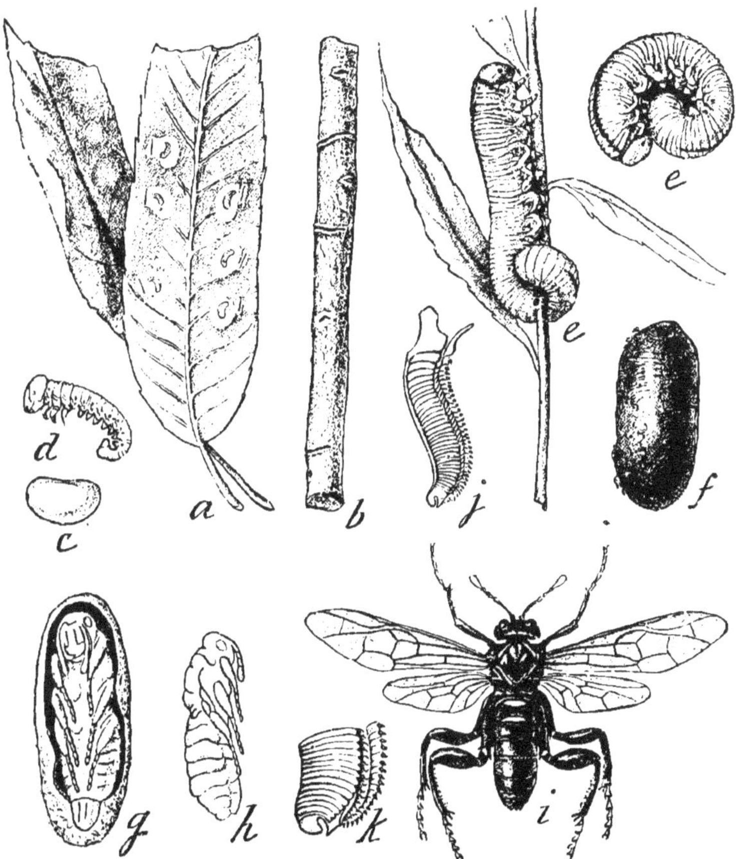

FIG. 45.—*Cimbex americana: a*, willow leaves showing egg-blisters from above and below; *b*, twig showing girdlings; *c*, egg; *d*, newly-hatched larva; *e*, *e*, full-grown larvæ; *f*, cocoon; *g*, cocoon cut open, with pupa; *h*, pupa, side view; *i*, female fly; *j*, her saw detached, side view; *k*, tip of saw—*c*, *d*, *j*, *k*, enlarged, the rest natural size. (After Riley.)

To-day I captured a male and female in the act of pairing, and send them to you for the purpose of learning what they are. I first thought of sending them to Mr. Bruner, an entomologist of high standing residing at West Point, in this State, but decided to send to you on account of "auld lang syne." As you know I have been planting trees all my life, or at least during the last fifty years of it, and I never saw such an insect before. I fear they will damage my grove, but perhaps not. Time will tell.

To my inquiry regarding the varieties of trees attacked, Mr. Pritchard kindly replied under date of July 6:

In only a very few instances do I find that the saw-flies attacked other trees besides the willows. On three or four tender cottonwoods I find they worked as if by mistake. The willows seem to recover and the gash cut by the insects heals over, but the cottonwood breaks off.

The only instance where this cutting habit of the adult saw-flies had been observed, so far as I can learn, is recorded in the Report of the Commissioner of Agriculture for 1884, pp. 334–6, Plate V, Fig. 1. In this case the depredation occurred on the grounds of Admiral Ammen, in the vicinity of Washington, D. C., only the willows suffering, but so great was the injury that the trees were described as looking as if a fire had run over them, or as if they had suffered from a severe frost.

The twigs of willow accompanying Mr. Pritchard's last communication resembled in every respect those figured in the report above referred to, although it would appear that in his case the injury resulting from the attack of the saw-flies was much less than in the case of Admiral Ammen.*

This cutting of the bark of the trees could have had nothing to do with the oviposition of the insect, as the eggs are deposited in slits cut in the leaves by the females. It seems quite possible that they gnaw the bark for the purpose of feeding upon the sap, as intimated by Mr. Pritchard, yet this does not appear to be fully proven. In other words, it would as yet be too much to say that in cutting the incisions the insect has no object in view other than that of obtaining food.

OBSERVATIONS ON MONOMORIUM PHARAONIS Latr.†

By M. A. Bellevoye, *Reims*.

Almost all the old habitations of Reims are infested with a little reddish ant, the *Monomorium pharaonis* Latr. These little insects visit without ceremony our tables; they haunt the side-boards and cupboards which contain eatables; the sugar-box, preserves, and meats are attacked by these small guests which do not ask leave to settle in your house. Their havoc, indeed, does not appear very important, notwithstanding their great numbers; but it is always disagreeable to find animate beings in one's eatables. It is, therefore, necessary to take vigorous measures in this regard; to kill them or use phenic or insecticide powders which drive them away from the places where they abound. The majority of people know only the neuters of this species.

* We have since shown, Insect Life, Vol I, p. 8, that the larger share of the damage at Admiral Ammen's was done by *Phyllœcus integer* and not by the Cimbex.

† Translated and condensed from Annales de la Société Entomologique de France, sixth series, Vol. VIII, 1888, fourth trimestre, Bulletin, pp. clxxvii–clxxxi.

I have just said that the ravages of these ants seem to me very unimportant, and I will show afterwards in what they consist. However, I have read in various authors, among others in the encyclopedia compiled by Dr. Chenu, this note:

The domestic ant of Schenk, a very small species which has of late made great devastation in England, in the houses of parts of London and Brighton, where it has settled and lays waste everything within its reach.

In the remarkable work of Mr. Edm. André, the Species of Hymenoptera of Europe and Algeria, the *Monomorium pharaonis* is indicated as being a native of Algeria, Palestine, and the tropical and sub-tropical regions of the whole world. The following is there given, together with a description of the three kinds or sexes.

This cosmopolitan species, which lives oftenest in houses in the walls or cracks, has acclimated itself in many large cities, such as Paris, Lyons, London, Copenhagen, Hamburg, etc. It causes often great damage by boring holes in furniture to establish its galleries, and by infesting eatables.

Last year on quitting Metz, where I was born and which I did not wish to leave, I came to Reims, and in the apartments which I occupied on Talleyrand street I found in a cupboard, with a quantity of neuters of Monomorium, a half dozen females, of which two had wings, and three males. Happy in discovering the two sexes, which I did not possess, I resolved to search for other specimens, and, if possible, to find the nest itself.

During the winter I saw a few neuters crawling through the dining-room, but nothing revealed to me the presence of any nest, and, until midsummer, although the neuters became more numerous, not a single sexual individual came under my observation. Where, then, was the nest to be found? The sideboards in the dining-room and a new cupboard were particularly frequented by neuter ants, allured by the victuals which were customarily shut up there; but after having several times explored all the corners of these places it became evident that the nest was not to be found there. Ants crawled in numbers upon the floor, where they profited by the falling crumbs from the table; they were going besides in large numbers towards a side of the room where the floor was loosely joined; it was in these clefts of the floor that they disappeared, only to return again to take their food. My neighbor has his pastry oven on that side, and he knows this little ant very well, with its dainty taste for sweetmeats as well as meat. To destroy them he places on the ground, from time to time, ham bones, and the next day he finds them covered with ants, which he destroys by throwing the whole into the fire.

The neighborhood of a pastry shop affords me the advantage of being visited by Blatta (*Kakerlak orientalis*), also *Blatta germanica*, that I kill without mercy; for when I used to allow one to stay on the floor the ants would immediately attack it, and, one hour after, I would see it covered with a hundred ants feeding on the juices contained in its body, which they left whole on the floor.

In the month of August, when flies are numerous in the apartments, I used every day to kill three scores of them which I deposited on a piece of paper in a corner and my boarders would not fail to attend the feast. A big spider was given to them and they liked it so much that by the next day the abdomen had all disappeared; the solid parts, the thorax and feet, remained entire. Sugared fruits and chocolate receive their attention also, but they do not damage them particularly, these substances being too hard for their mandibles. Fallen crumbs answer their purpose better. They do not seem to meet in numbers to carry the least piece of anything away to their nest; they seem to be sure they will always find something to feed on in our houses. Undoubtedly they disgorge to their larvæ the fluid part of the substances they have eaten. No one realizes how little such small animals want.

Up to September 15 I had not perceived either males or females. I then decided to use a more succulent bait, and tried ox liver; I placed a few bits of 5 or 6 centimeters in diameter on a paper, and three or four times a day I shook the paper in a benzine box; thousands of neuters dropped, and at last some males and females. After eight days of search I had taken 20 females, only one of which was winged, and 8 males. From the 16th of September to the 9th of October I captured 131 females, of which two were winged, and 60 males (about 6 females and 3 males per day); from the 10th to the 15th of October I captured 269 females and 90 males (about 54 females and 18 males per day); then the number decreased, and from the 15th to the 25th of October I caught only 159 females, 3 of which were winged, and 74 males (about 16 females and 7 males per day). In all, from the 15th of September to the 25th of October I had therefore captured 577 females, only 14 of which were winged, and 239 males.

In order to know approximately the number of neuters I had taken I counted 1,000 of them, of which the weight was 0.058 gram; 1 gram would therefore contain about 17,000, and as I had gathered 20.56 grams it gives a total of 349,500 neuters secured in six weeks (about 9,000 per day), and this figure is rather below the reality, for I have killed or thrown into the fire a great many of them that were not weighed.

However large these figures may seem, the supply was not exhausted, and every day I saw just as many neuters; the number of the sexual individuals only diminished. I then lifted the wash-board and two boards of the inlaid floor, hoping to find there larvæ and nymphs in their cocoons, but I was disappointed, for clefts in the wall showed me that the progeny of my ants were undoubtedly in the thick wall or in my neighbor's house.

I said at first that the injury by these small beings was almost inappreciable; only the abdomen of a spider had been destroyed, as also the abdomen of a few flies slightly eaten. The bits of raw liver I used as baits did not look damaged after a few days' service, though they were every day covered with ants which fed probably only on blood at first. The pieces which I left to dry up, and which attracted them as well as the fresh liver, were at last furrowed with channels more or less deep. One of these pieces, which served for a score of days, was completely dug through into the center and only the exterior parts remained, which were hardened and bored with holes. In that condition ants were crowding all over them always in as large numbers as at first. How many thousands of ants worked at that piece to reduce it to that condition? Two or three thousand ants working day and night. When I had shaken the piece to gather all the workers, these were replaced an hour after by others; at 11 o'clock at night I found as many as at 7 o'clock in the morning, which proves that the work of the neuters does not stop. The result of these observations, few as they are, seems to determine the time of hatching out of the sexes, which seems to be at the end of September and during the whole month of October. This hatching takes place, of course, successively like the coupling, contrary to what occurs in most species in our country, whose coupling takes place in the air, and of which each female becomes the founder of a new formicary, while the males, becoming useless, die after having wandered aimlessly for a few days. Here, on the contrary, coupling takes place subterraneously, and it appears that the male and female continue to live in the same formicary, which increases indefinitely so long as nothing of an unforeseen character happens to destroy it.

Females lose their wings, of course, immediately after coupling, the superior ones first, for I found several which yet possessed their inferior wings. Their walk is slow, while males, preserving all their wings, run very quickly without my having seen any showing signs of flying away. It may possibly be different in Africa under the influence of a warmer sun than we have in our temperate climate.*

THE DIPTEROUS PARASITE OF DIABROTICA SOROR.

By D. W. COQUILLETT, *Los Angeles.*

Up to the present time but few instances have been recorded of Coleopterous insects being subject to the attacks of Dipterous parasites. In his first report as State Entomologist of Missouri, Professor Riley records having bred the Tachinid, *Exorista* (*Lydella*) *doryphoræ* Riley, from the larvæ of the Colorado Potato-beetle (*Doryphora* 10-*lineata* Say), and in the fifth volume of the *American Naturalist* Dr. Henry Shimer gives an account of the Dexid, *Melanosphora diabroticæ* Shimer,

* Mr. Bellevoye continued to gather these ants during the whole month of November. The neuters were a little less numerous; there was a complete absence of males, but the females were always present, and he captured 203 of them from the first of November to the 6th of December, only there were none with wings, which seems to indicate that there was not another brood of males and females.

which preys on the Striped Squash-beetle (*Diabrotica vittata* Fabr.). In the *Annales de la Societé Entomologique de France* for the month of June, 1888, Mr. M. H. Lucas gives an account of the parasitism of the Tachinid, *Myobia pumila* Macq., on the Asparagus Beetle (*Crioceris asparagi* Linn.), and on page 408 of his well-known *Guide to the Study of Insects* Dr. Packard quotes the French entomologist, Dufour, as authority for the statement that the Tachinids, *Cassidomyia* and *Hyalomyia*, prey respectively on the Tortoise beetle, *Cassida*, and on the Curculionid *Brachyderes*.

The above are the only published references upon this subject that I have been able to find among the limited literature at my command.

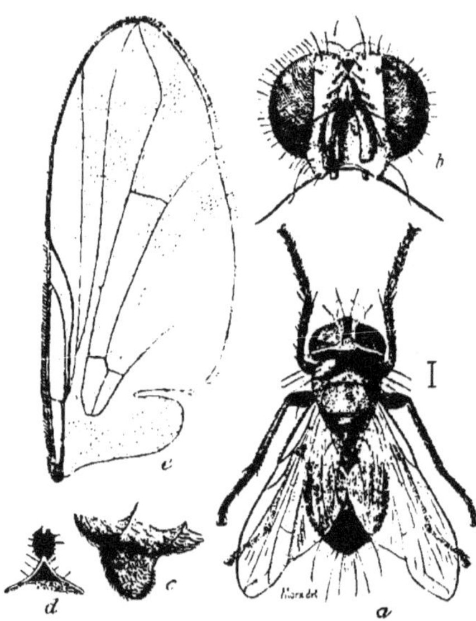

FIG. 46.—*Celatoria crawii*: *a*, adult fly; *b*, head of same from front; *c*, abdominal appendage from side; *d*, abdominal appendage from behind; *e*, wing showing venation—all enlarged (original).

On the 17th of June, 1888, I obtained several Dipterous pupæ from the abdomen of a dead *Calosoma perigrinator* Guér., and three flies issued from these pupæ on the 25th of the same month; they belong to the genus *Masicera* of Macquart, but the species is still undetermined.

On the 21st of June, 1889, I saw a Dipterous larva issue from the abdomen of an adult *Diabrotica soror* Lec.; it soon afterward pupated, the fly issuing on the 5th of the following month. Since this time I have succeeded in obtaining quite a number of the pupæ of this parasite, the flies from which issued at various times during the month of August. The larva in issuing usually breaks away the larger portion of the beetle's abdomen, and pupates wherever it chances to find a resting place—in a depression of a leaf, in the axil, or on the ground, making no attempt at concealing itself. Among a large series of beetles examined fully one-third contained larvæ of this parasite, each infested beetle containing only a single larva.

This parasite is very interesting, not only on account of the fact that it aids the horticulturist in lessening the attacks of the destructive Diabroticas, but also on account of the peculiar structure of the male abdomen, the second segment of which has a large flattened process on the underside—a peculiarity which does not exist in any other Dipteron known to me, nor can I find any reference to it in any work upon this subject to which I have access. I submitted a sketch of it, together with an extensive description of both sexes, to Dr. S. W. Williston, our best authority upon this group of insects, and he writes me that he never saw such a process in any species that he has examined, nor can he find any published reference to it; he further states that the other characters of this species agree quite well with those of the genus *Baumhauria*, a single species of which has heretofore been described, having been bred from a Bombycid belonging to the genus *Arctia*. Our species, however, differs very decidedly from the above genus by characters other than the abdominal process, and therefore I do not think we run any great risk in erecting a new genus for its reception, a description of which I append herewith:

Celatoria, n. gen.—*Head* large, broad as thorax, much broader than high; front in male only slightly wider, in female one-fourth wider than transverse diameter of eye—in both sexes with a single row of bristles each side of frontal stripe extending nearly to insertion of arista, and with two forwardly directed bristles on the crown outside of each of these rows; face much retreating below, bristles bordering median foveæ strong, extending nearly to the lowest in frontal row; vibrissal bristle strong; epistoma but slightly projecting; cheeks small, bristly; palpi well-developed, thickening toward its tip; proboscis soft, wholly retractile, furnished with a large labella; antennæ reaching nearly to oral margin, third joint at least four times as long as the second, rather slender and nearly of an equal width, the upper edge nearly straight; arista sub-basal, very short pubescent, distinctly two-jointed, second joint greatly attenuated on its apical half. *Eyes* bare. *Thorax* nearly as long as the abdomen, furnished with stout bristles. *Scutellum* with three pairs of marginal bristles and a shorter pair of dorsal ones. *Abdomen* oval, thinly depressed pilose, and with several pairs of dorsal bristles besides the usual lateral and anal ones; five abdominal segments, the first nearly as long as the second, the fifth in the male small, in the female concealed in the fourth; venter in the female normal, in the male furnished with a large, longitudinally compressed process on underside of second segment, apex of this process studded with numerous small tubercles; a large cavity in posterior end of venter, inclosing the fifth segment and contracted anteriorly into a narrow groove which extends to the second segment. *Legs* furnished with bristles; posterior tibiæ not ciliated. *Wings* of the usual Muscid type, first posterior cell terminating close to tip of wing, closed in the margin; curvature of the fourth vein in middle of last section of that vein, rounded, and destitute of an appendage; great cross-vein slightly nearer to this curvature than to the small cross-vein, nearly perpendicular; a stout bristle at junction of second and third veins.

Type, *Celatoria crawii* n. sp., which may be further characterized as follows:

Male. Frontal vitta blackish-brown, sides of front white, tinged with yellow; face white; palpi reddish-yellow; antennæ black. Thorax grayish-black, destitute of stripes, the bristles not disposed in rows. Scutellum grayish-black. Abdomen black, mottled with gray, destitute of reddish spots; fifth segment scarcely one-fourth as long as the fourth; a posterior dorsal pair of bristles on the first and second segments, and a posterior transverse row of bristles on the third, fourth, and fifth segments, besides several along the sides of the abdomen; venter concolorous with the dorsum. Legs black, claws and pulvilli much shorter than last tarsal joint. Wings hyaline. Alulæ white. Halteres yellow.

Female. Same as the male except that there is a median pair of bristles on the second, third, and fourth segments. Length $4\frac{1}{2}$ to $5\frac{1}{2}^{mm}$.

Described from three males and two females, bred from adults of *Diabrotica soror* Lec., at Los Angeles, Cal.

PUPARIUM.—Dark brown, cylindrical, the ends rounded; quite thickly covered with black spines of varying lengths, some of the longer ones converging and adhering to each other, forming clusters of from 8 to 14 spines; length $4\frac{1}{2}^{mm}$.

I have dedicated this interesting species to my friend, Mr. Alexander Craw, who first discovered the existence of this parasite and to whom I am indebted for several specimens of the pupa.

SPILOSOMA FULIGINOSA LINN.

By O. LUGGER, *St. Anthony Park, Minn.*

Quite a number of insects are common to northern Europe, Asia, and America. The above insect must be added to these circumpolar species, as it occurs rather abundantly near the experiment station at St. Anthony Park, Minn. Nor is it a recent importation, as I have found it here in some old collections made about twenty years ago.

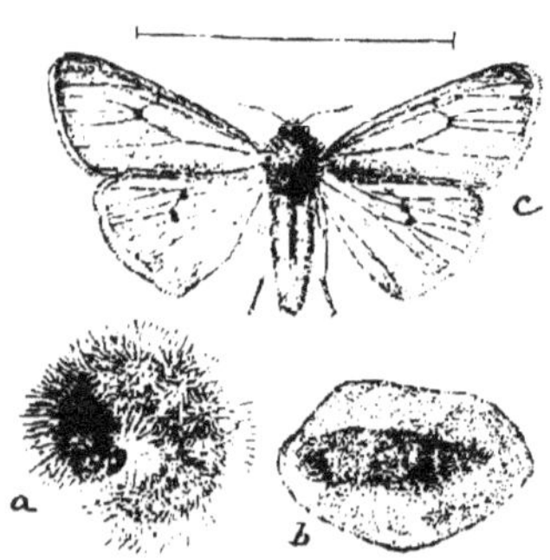

FIG. 47.—*Spilosoma fuliginosa*; *a*, larva; *b*, cocoon; *c*, moth—slightly enlarged (original).

This moth is interesting in many respects. Although I hunted for its larva quite frequently during the summer of 1888 and 1889, I never succeeded in finding it. But late in the autumn, and at a time when the sidewalks are covered every morning with a thick layer of frost, these larvæ are rather abundant. They leave their hiding places and crawl over the sidewalks; at this time they are frequently themselves incrusted with crystals of ice. Some few days ago, with the thermometer ranging from 5° to 3° below zero, I found several of them crawling slowly through the snow. When the sidewalks, made of boards, become warmed up by the rays of the sun, the caterpillars crawl away to the shady and cooler part. The caterpillar has the usual Arctiid shape, is intensely black, and densely covered with hairs, which are pale yellowish near the anterior and posterior ends, but of a dingy pale brown in the central region. The head is polished black.

As I have at present no larvæ, I can not give a closer description, but the illustration will give a good idea of their general appearance. The larvæ are most common wherever the sidewalks are laid in close proximity to clover, yet they are also met with in the vicinity of wild grasses and plants. As soon as such a larva is taken in doors and put into a breeding cage, it will crawl for a few days and soon commence to form a cocoon. This is of a regular oval shape, made of fine threads of dirty white silk, intermixed with a few hairs from the body of the caterpillar. In the course of about ten days the pupal stage is assumed. The pupa is intensely black, highly polished, with rather sparse punctuations. The sutures are reddish brown. If kept in a cool room, the moths commence to issue early in April of the following season, though in a warm room some issued as early as the 3d of February.

The moth, Fig. 47*c*, has rusty black upper wings; the scales are not very close, so that the venation is plainly visible. The under wings are of a similar color, but much lighter, and possess a brick-red, ill-defined space at posterior margin. Both upper and lower wings, with the exception of their anterior margins, are fringed with pale red. Head, thorax, legs, and first two joints of abdomen are rusty brown. The abdomen is blackish, densely covered with rather coarse brick-red hairs; a dorsal and two lateral stripes are blackish. The femora of front legs are bright red. The whole underside of wings is pale reddish brown. Antennæ white, with blackish tips.

This rather handsome moth is very peculiar in its motions. It does not rest like other Arctiids in a more or less perpendicular position upon stems of plants, but prefers some dead leaves, under which it hides. If such a leaf is removed, the moth will rapidly run away to hide again, this time perhaps under a loose lump of soil.

The following extract from my notes illustrates the remarkable vitality of this insect:

December 3, 1889. Found to-day in a little depression of the soil a clear cake of ice, and imbedded in it the larva of the above species. By means of a hot iron I separated a cube of ice with the inclosed larva, and took it to my office. The caterpillar was entirely and solidly inclosed by the ice; no air-spaces could be detected among the hair. How long the caterpillar had been inclosed I could not say. Left the cube of ice in front of my window, where the temperature sunk for two days to 11° below zero. Later the weather moderated, and during the day a little ice would melt near the caterpillar, but never exposing it to the air. After being inclosed for fourteen days, I carefully melted the ice and removed the caterpillar to a piece of blotting paper. In less than thirty minutes the larva was crawling about, not injured in the least. Yet, to escape further experimentation, it has shown good sense and spun up, and transformed into a pupa, healthy to all appearances.

A GRUB SUPPOSED TO HAVE TRAVELED IN THE HUMAN BODY.

During June last we received a communication from Dr. Elizabeth R. Kane, of Kane, McKean County, Pa., from which the following is an extract:

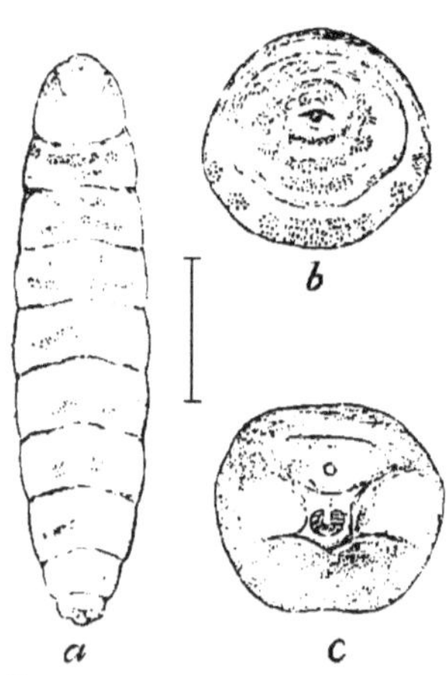

FIG. 48.—Hypoderma larva taken from boy: *a*, ventral surface—enlarged; *b*, anterior end; *c*, anal end—still more enlarged (original).

Numbers 3 and 7 of Vol. 1 of INSECT LIFE contain papers on Larvæ infesting Man and the Squirrel. A case occurring in the practice of Dr. Silvanus D. Freeman, of Smethport, McKean County, Pa, may not be without interest, as touching points alluded to in both papers.

On the 22d of February, 1889, Dr. F. visited a child residing in the country. He had been sent for some days previously, but being unable to go himself had sent his assistant, who reported a threatening of erysipelas. As the child was still suffering, the parents sent again for Dr. Freeman. He found the ear and the tissues around it much swollen, and the swelling plainly erysipelatous. Yet there was no sign of constitutional disturbance, the tongue was clean, breath sweet, and temperature normal. The child, a boy three or four years old, was lively enough to play during the day, but in sufficient pain not to sleep at night. The mother remarked that the cause of suffering was a "pollywog" working under the skin, but no particular attention was paid to the observation.

On February 28, the doctor again visited the child. The swelling under and behind the ear was gone, but a red line of inflammation went up to the under eyelid and then down the cheek. The mother stated that the eye had been closed for twenty-four hours by the swelling, which had traveled about 2 inches since the doctor's first visit, and seemed now about to "point" in the cheek. Placing his finger on the inside, the doctor detected a foreign body in the swelling, which he lanced, and squeezed slightly. A living grub emerged, a little less than half an inch long while living, a little over that when it died a few hours after. Dr. Freeman questioned the child's mother closely and learned that she had first noticed what she called the "pollywog" five months before. It was then under the skin near the sternal end of the right clavicle, and in the five months had traveled (appearing as a tiny lump followed by a red line of inflammation) up and down the chest in front, down one arm as far as the elbow, and over one side of the back, never crossing the median line. Sometimes it had "pointed" and they thought it would come out, but its course had continued on again. Until within a few weeks it had given the child little annoyance, but latterly its nights had been very restless. The mother thought that the "pollywog" traveled at night because she had never detected its movements, and because the child seemed more easy in the day-time. Its increasing suffering was probably caused by the increase of growth of the grub. Taken out February 28, when was the egg deposited? Its movements had been noticed five months before.

A careful examination was made of it under the microscope, and a description written out and sent to an entomologist, who advised the doctor to apply to the Department for Vol. 1 of INSECT LIFE. He found that Dr. Rudolph Matas had figured in No. 3 of that volume a grub found under the skin of certain laborers on the Central American works, who had been stung while bathing, and appeared to be infested with boils. These contained larvæ. Dr. Matas pictured a differently shaped grub from that found by Dr. Freeman, in that the Central American one has a large head and diminishes rapidly towards the caudal end.

In describing Dr. Freeman's grub, he mentions twelve rows of curved black bristles pointing backwards, which he called ciliated epithelia. At the caudal end the three first rings had several pads of these bristles. Dr. Freeman supposed that the maggot propelled itself by their aid. Dr. Matas depicts the same sort of bristles and considers their use to be to keep the grub stationary. He also speaks of the necessity of the maggot's obtaining air. Dr. Freeman's lived five months without it.

Dr. Freeman supposes his grub to be the larva of a gad-fly, as the sting of these insects is very annoying to both horses and human beings in McKean County. It closely resembled, except in being narrower in proportion to the length, the grub figured on page 214 of Vol. 1 of INSECT LIFE as the Emasculating Bot-fly.

I do not suppose that there is anything unusual in finding larvæ living in human flesh, but is not the traveling about unusual?

We immediately wrote Dr. Kane, expressing incredulity regarding the traveling of the grub from the elbow to the eye in the space of five months, and urged strongly that she endeavor to secure the specimen. She wrote to Dr. Freeman, who with great promptness forwarded the specimen in alcohol with the following note:

I am not only willing but anxious the grub be sent to Washington for determination. The evidence of both father and mother, after describing the "pollywog" appearance in its track should be very strong evidence of its being migratory, but putting their statements all aside, I have positive knowledge of its movements, having first seen its track over the scapula, then up the neck to base of ear which was enormously swollen, from there to the outer corner of eye, which was entirely closed, then to middle of cheek where it was plainly felt, and the opening made and expelled. There is no chance for mistake in this case.

We have carefully examined the specimen with the result that it seems without doubt to be a species of *Hypoderma* and closely resembles Brauer's figure of the early stage of *H. diana*, which infests deer in Europe, as also the same stage of the common *H. bovis*.

We have shown this larva at figure 48, *a* representing the entire larva, *b* showing the head, and *c* the anal end of the body. We place the matter on record for what it is worth. The extensive traveling reported we might be inclined to doubt, were it not for the confirmatory evidence in the case of *H. bovis*, published in this issue by Dr. Curtice.

THE DOGWOOD SAW-FLY.

Harpiphorus varianus Norton.

In a recent number of the *Garden and Forest* October 30, 1889, Vol. 2, p. 520), Mr. J. G. Jack presents an interesting article, illustrated by drawings by Mr. C. E. Faxon, under the title "A Destructive Cornel Saw-fly (*Harpiphorus varianus* Norton)," the larvæ of which for two or three years past have been quite destructive to the foliage of various Dogwoods in the Arnold Arboretum.

We have accumulated in the note-books of the Division a number of references to this insect, having first collected the larvæ on *Cornus paniculata* in Missouri in the fall of 1875. Since that time we have collected and received through our correspondents larvæ from various localities, and have succeeded on several occasions in rearing the adult insects. We had purposed to publish the natural history of this insect as soon as opportunity offered, but such publication is now rendered unnecessary in view of the excellent account of its habits and description of its several stages given by Mr. Jack. We will, however, in connection with a brief synopsis of Mr. Jack's paper, put on record our notes relating to the range, date of appearance, and habits of this insect.

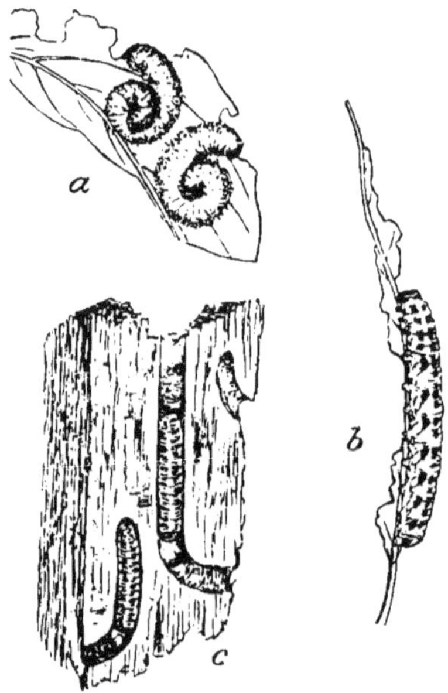

FIG. 49.—*Harpiphorus varianus*: *a*, larvæ before last molt; *b*, larva after last molt; *c*, larvæ in burrows in decaying wood—natural size (after Jack).

As stated above, the larvæ were collected in Missouri in the fall of 1875 on *Cornus paniculata*. These specimens soon entered soil and were found unchanged, excepting being much contracted, on March 23 of the spring following. No adults were obtained. September 27, 1877, larvæ were found on Cornus sp. at Kirkwood, Mo., and these entered soil October 5, but again the adults failed to appear. October 2, 1884, a number of larvæ of this insect were received from Mr. M. S. Crane, of Caldwell, N. J., who seems to have first discovered the peculiar hibernating habits of the larvæ, and who writes of them as follows:

I send you with this a box of saw-fly larvæ found feeding on *Cornus paniculata*. When about to change to a chrysalis the larvæ bore into decayed wood to transform. There are two broods in a season, and the last one remains in the wood until spring. For several years they have been very abundant, but this season they were much scarcer.

October 3 a number of the larvæ received from Mr. Crane bored into rotten wood placed in the breeding-case with them for that purpose. Flies issued from May 27 to June 5 of the year following. Another lot of larvæ was received from Mr. Crane September 2, 1885, concerning which he again writes:

I send you herewith a box of saw-fly larvæ found feeding on panicled dogwood (*C. paniculata*), a few of which I forwarded to you last year, and have not been able to secure any more until a few days ago. Previous to last year they were very abundant, eating the foliage all off of many bushes. For several years I have tried to rear them in a glass jar and succeeded in bringing out one fly by putting a piece of partly decayed wood in the jar, which the larvæ entered to change.

An examination of the larvæ April 8, 1886, showed that they had not then changed to pupæ; the adults appeared from May 28 to June 8.

September 12, 1888, a number of saw-fly larvæ were received from J. G. Barlow, Cadet, Mo., which in every respect were like those previously obtained from Mr. Crane. They were, however, found feeding on the leaves of a wild grapevine, and also on *Polygonum dumetorum*. With us, however, they refused to feed on these plants, but wandered incessantly about in the breeding cage. Flies were obtained August 4 and 9, 1889.

Mr. Charles C. Beach of Hartford, Conn., wrote us in June last describing the habits of a peculiar spotted saw-fly larva found by him burrowing into decayed wood on which he supposed they subsisted. He had in the previous year sent specimens to Dr. A. S. Packard, who was unable to identify them, and who wanted additional material. In the absence of Dr. Packard in Europe, however, he communicated with us, and in compliance with our request forwarded us specimens of the larvæ and adults collected the present year, concerning which he writes under date of August 7, 1889, as follows:

I mailed you yesterday a package containing a number of the live larvæ of the saw-fly of which I wrote you last June; also a small bottle with two of the mature flies. Since the receipt of your letter of June 22 I have searched faithfully for more of the adults, but only succeeded in netting the two which I have sent you in alcohol. The colors have remained practically unchanged. In the box containing the larvæ you will see that most of the specimens are covered with a sort of white bloom, if their journey has not caused it to be rubbed off. This at times, or rather in some instances, is very abundant and continues through all the molts until after the last change, which takes place prior to pupating, when they appear of a black and yellow color and naked. I placed one such in the box with the others. At this stage they are exceedingly restless, ceasing to eat and being found at times a long distance from their food-plant. It is impossible to keep them in a bottle covered simply with gauze as they bite through it, but placed in a bottle with a few bits of dead wood, they make no attempt to escape, but proceed immediately to bore. I have some at present boring, having reached their last molt. When they are well settled in their winter habitat I will mail you some if you desire them.

The imago is a very pretty and active little creature, readily eluding the net, fighting and biting with vigor when captured. I do not know whether the two specimens I inclosed are of one sex or not.

The larvæ of this saw-fly are reported by Mr. C. L. Marlatt to occur not uncommonly at Manhattan, Kans., on Cornus sp.

Mr. Jack was at first unable to rear the adults, but in the spring of 1889, having accidentally found the larvæ burrowing in decaying wood picked up near Cornus bushes, he succeeded about the first of June in obtaining the perfect insects, the larvæ having pupated within the burrows but a short time previous. The life history as given by Mr. Jack is as follows:

On June 10 the first eggs were discovered and within a few days they were quite abundant. The eggs are pale green, oblong, and about four one-hundredths of an inch in length. They are deposited singly within the tissue of the leaf on the upper side. From one or two dozen to three or four hundred eggs may be deposited within a single leaf without any very definite order, although most of them are usually disposed in lines parallel with the midrib, or with the principal veins. Each egg makes a little swelling, noticeable on both the upper and the under sides of the leaf, and, with a little practice, egg-bearing leaves may be readily detected.

On hatching, the larvæ emerge on the under side of the leaf. They are then about six one-hundredths of an inch long and pale green in color, with yellowish head and black eyes. When at rest they keep close together, coiled up on the under side of the leaves; and they appear to feed only in the cooler parts of the day, in cloudy weather, or perhaps at night. Of the first leaves attacked, they eat only the more tender parts, and the leaves are left somewhat skeletonized; but as the larvæ get older they devour every part of the leaf down to the midrib. After the first molt, when they are about twelve one-hundredths of an inch long, the larvæ secrete a peculiar, very white efflorescence, by which the back and sides become covered. This is constant after each molt until the last.

This efflorescence is removed by the slightest touch, and when brushed off the color of the body is a pale greenish white. The head after the first molt is black, and the legs and under sides of the body yellow. At full growth the average length of the larva is about an inch. Some are smaller than others, however, and this possibly indicates the difference of sexes.

When the larva has cast its skin for the last time a complete and surprising change has taken place. All trace of white is gone, and the body is greenish yellow on the back and yellow beneath and along the sides below the spiracles. On each segment along the back are two large and two small black spots, and upon the sides, close above the spiracles, is a row of nearly square, black spots, one for each segment, but so placed as to lap over from one segment to the next. The terminal plate above the ventral segment is black. The legs and prolegs are yellow, the former having a reddish spot on the outer side near the base. The tips of the claws are black. The great change produced by the last molt has led some observers to suppose that there are two distinct species.

Full growth is attained by most of the larvæ early in August, but some may not reach maturity until much later, and this season a few were noticed to pass the last molt about September 20. The season last year was not so advanced, and, in some places, large numbers of larvæ were found in September. The larvæ eat very little after the last molt, and very soon they leave the plant and wander away in search of suitable places in which to hibernate. Stray pieces of decaying wood, fence posts and rails, dead branches and the corky bark of old trees are selected. In the Arboretum, many were found even boring into the soft pith of dead stems of elder bushes. Sometimes two or more occupy one burrow. It is quite possible that some larvæ go into the ground to hibernate, but none have been discovered there.

Figures of the larvæ, showing characteristic position on leaves, and also the nature of their hibernating burrows in decaying wood, are reproduced from Mr. Faxon's figures.

We had identified adults as *H.* (*Emphytus*) *testaceus*, and after again carefully examining our specimens it seems probable that the species just named and *H. varianus* are identical. Those obtained from the larvæ received from Mr. Crane form a very good connecting series between the two species. Those bred from the larvæ sent by Mr. Barlow, together with the adults received from Mr. Beach, agree more closely with *H. varianus.* The variation even in the structural characters of the species is shown in that one female specimen in our collection has in the right anterior wing four perfect submarginal cells and in the left but three, the normal number, and in another female both anterior wings have four submarginal cells; the other specimens are normal. When it is remembered that the number of submarginal cells is used to separate a group of genera, including *Harpiphorus*, *Emphytus*, and *Dolerus*, the confusion likely to result from such variation may be better understood.

The male, of which we have but a single specimen, is much smaller than the smallest female, and the sides of its flattened abdomen are nearly parallel, differing markedly in this respect from the much broader and pointed abdomen of the other sex. In size as well as color there is a wide variation in our specimens, the length ranging from 10^{mm} to 15^{mm}. In color the flies are honey-yellow and reddish, with the thorax and head more or less marked with black; the former in typical specimens being almost entirely black. The four terminal joints of the antennæ, the labrum, tegulæ, scutel, feet, and portions of the legs, white. The two basal joints and more or less of the third joint of the antennæ are reddish. The apical portion of the third joint in all of our specimens and the fourth and fifth joints are brownish black. The basal half of the wings is clear; the outer portion, smoky.

This insect has been recorded from Connecticut, New York, Virginia, Illinois, Massachusetts, and Canada; and *E. testaceus*, which is probably the same insect, from Pennsylvania and Virginia. To these localities we have added New Jersey, Missouri, and Kansas. In Cresson's Catalogue of Described Hymenoptera it is accredited to the United States and Canada.

Of insect enemies, Mr. Jack mentions a number of Hemiptera observed by him to feed on the larva, and he also observed, but failed to secure, a minute fly which was apparently ovipositing upon its eggs.

Certain species of Cornus (*C. florida* and *C. mas*) were found by Mr. Jack not to be attacked by this insect, but the foliage of *C. sericea*, *C. alba*, *C. stolonifera*, *C. paniculata*, *C. sanguinea*, *C. asperifolia*, and one or two others were greedily devoured. Polygonum and Wild Grape may be doubtfully added to the list of its food-plants. Our failure to get the larvæ received from Mr. Barlow to feed on these plants makes additional observations on this point desirable.

PLATYPSYLLUS—EGG AND ULTIMATE LARVA.*

By C. V. Riley.

The egg and the pupa of *Platypsyllus* are yet unknown. I have for some time endeavored to obtain them, and specimens recently received as such gave hope, from the finder's account, that the lacunæ in the life-history of the genus might at last be filled. But examination dispelled the hope; yet not without adding something to our knowledge of the development of this curious beaver parasite. The only reference to the egg is that contained in Dr. Horn's article in the "Transactions of the American Entomological Society" (Vol. XV, p. 25), where it is stated that the eggs were observed, and that "they are minute objects, not fastened to the hair, as is the case with lice, but plastered firmly to the skin among the thickest hair." This, failing in description, might apply to the egg of any other minute creature, and I have, in fact, some reason for concluding that the objects referred to in the observation were not the eggs of *Platypsyllus*, but those of quite a different insect. The eggs, as observed in the oviduct of the female *Platypsyllus*, are sufficiently uncharacteristic, except as to their flattened form; they are 0.4^{mm} long and 0.2^{mm} in broadest diameter, non-sculptured, white, broadly ovoid, but much flattened on two sides. The structure indicates that they may either be thrust under the scales of the skin or fastened thereto.

What was sent as the pupa, proves to be a most interesting larval stage and in keeping with the Mallophagous appearance of the beetle. This larval stage might at first sight be characterized as a Mallophagan by even the most careful zoölogist. The larva, as hitherto described and figured, even in the largest specimens, whether from Dr. Horn's material or my own, has always seemed to me inexplicably small as compared with the imago, and if the form which I now describe is (and I can believe it nothing else) the final larval form of *Platypsyllus*, then the larvæ hitherto described had not yet gone through their final molt. A glance at the accompanying figures suffices to show the remarkable superficial resemblance to the lice in question, and only when the structure, especially of the leg and mouth-parts is studied, does its *Platypsyllus* nature appear. The description will also show how greatly it is modified from the earlier larval stages already described. One is justified from the facilities for grasping which it possesses, as from the position of the head, in inferring this stage quiescent, and in this respect, as well as in the marked deviation from the previous stage, it recalls the pseudo-pupa, or coarctate larva of the Meloids, and of some other parasitic forms. I have but a single specimen and have not been able to clearly make out the spiracles. One can but conjecture as to whether the pupa proper is formed, either partially or wholly, within the skin

* Reprinted from *Entomologica Americana*, February, 1890, p. 27.

of this broadened larva, or whether the skin is completely exuviated in the transformation.

I hope that those who have opportunity to capture beavers will endeavor to obtain the much-desired pupa, and I shall be most glad to communicate with or to receive specimens from any one having such opportunity.

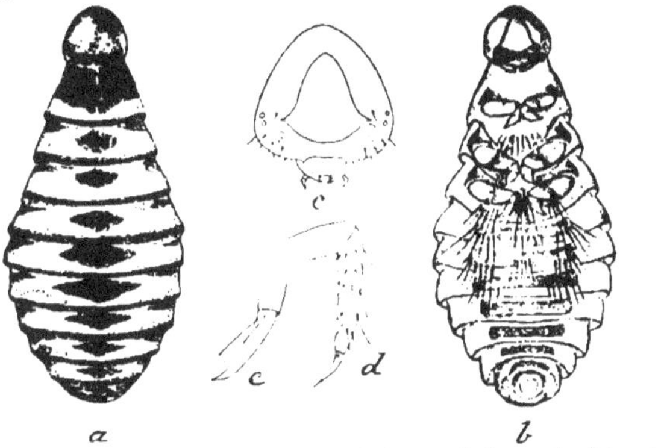

Fig. 50.—*Platypsyllus castoris*, ultimate larva: *a*, dorsal; *b*, ventral view; *c*, head from beneath; *d*, tarsus; *e*, tarsal claw (after Riley).

PLATYPSYLLUS CASTORIS.—*Ultimate Larva*—Length about 2.4^{mm}; greatest diameter about 1.2^{mm}. Nirmiform, flattened, narrowest at thoracic joints and broadest at middle of abdomen. Color grayish white, with brownish, chitinous markings. Head pale brown, peculiar, projecting from joint 1, subtriangular, flattened, occiput without structure, face and vertex completely ventral; the mandibles resting on the prosternum, rather stout and 2 toothed; clypeus very large, triangular; antennæ very small, 3-jointed, inserted in front of the lateral angles of the clypeus, the basal joint rather large, circular, flattened disc-like, the second joint minute, as long as broad; the terminal joint much longer, slender, cylindrical, and bearing a stout bristle at tip; labrum transparent and membranous; palpi apparently 4 jointed (not distinctly made out) the terminal joint cylindrical, about one-half longer than wide and truncated at tip; just outside the antennæ are two black ocelli and several piliferous raised points. Legs rather short, stout, drawn in over the sternum; the tarsi spinose, long, 1-jointed, bearing but a single, long, quite straight claw, with two long, movable spines at base; tibiæ with but a few spines near tip. Dorsally, the prothorax is twice as long as the other joints, which are subequal in width, and the transverse brown markings include the prothorax, except a narrow posterior band, a narrow posterior border across each of the joints (obsolescing on 10, 11 and 12); a median subrhomboidal spot and a subdorsal narrower, somewhat paler spot near the anterior margin of each of joints 2–11. The posterior half of each joint is also beset with numerous pale brown granulations (obsolete on 11 and 12), but without a trace of hair. Ventrally, the thoracic joints are much lengthened, the femora show a transverse shade and the abdominal joints a dusky transverse band, shorter and more conspicuous anally. Patches of long, stout bristles occur on the dusky parts of joints 4, 5, 6, 7 more particularly, and of shorter bristles on the sternum.*

SOME NEW PARASITES OF THE GRAIN PLANT-LOUSE.

By L. O. Howard.

Among the numerous parasites of the Grain Plant-louse reared the past summer and referred to in Insect Life, Vol. II, page 31, are the three following new species. As they belong to groups which I have studied I present the following descriptions at Professor Riley's desire:

There has been considerable doubt concerning the true habits of the species of *Pachyneuron.* It has, beyond question, been bred from Syrphid larvæ in the Division of Entomology and by Mr. Hubbard, in Florida. Professor Cook considered a species reared by him as a Bark-louse parasite, but with the evidence before us at that time I surmised that it might have come from unnoticed Syrphid larvæ. In the same way I was first inclined to discredit Mr. Ashmead's reported rearing of this genus from Aphidids, but Mr. Ashmead tells me that he is quite positive that it does actually feed in plant-lice and the facts concerned in the rearing of the present species seem to indorse his opinion. Our first specimens were reared July 12, 1889, from grain-lice sent from Goshen, Ind., by Mr. Webster and we subsequently reared a rather large series (20 specimens mounted) from lice from different localities in the same State. While it was not observed to actually issue from the lice there seems little chance that Syrphids could have been present in the small mass in such numbers to have harbored such large quantities of the parasites.

The genus *Megaspilus* has been rarely reared in this country. A species has been reared from the Hop Plant-louse in the Division of Entomology and a rather large series from the Grain-louse. I am not familiar with any references to its habits in Europe. The subfamily to which it belongs contains other genera of plant-louse parasites, viz., *Ceraphron* and *Lygocerus.*

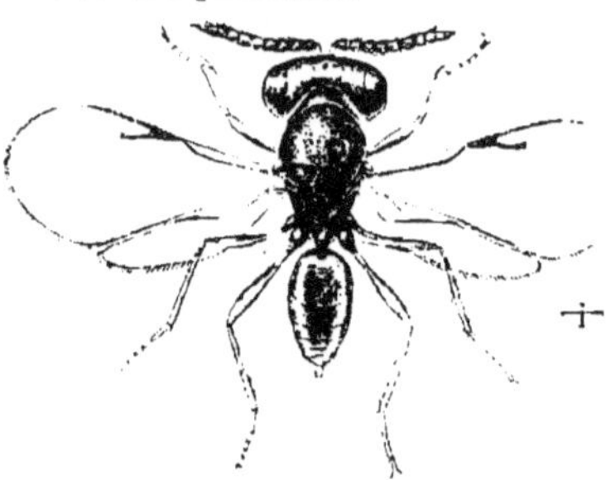

Fig. 51.—*Pachyneuron micans*, female—enlarged (original).

Pachyneuron micans, n. sp.

Female.—Length, 1.28mm; expanse, 2.1mm; greatest width of forewing, 0.46mm. Antennæ short; funicle as long as width of head; first funicle joint succeeding ring-joints as broad as long, not compressed; succeeding joints increasing gradually in

---

*Since this was written, I have ascertained that the spiracles are extremely minute and placed laterally on the posterior border of the joints. The two spots on penultimate joint bordered by short spines correspond to the bases of the cerci.

width, not in length, to club, which is oval, compressed, nearly as long as preceding three joints together; entire funicle with short, appressed hairs. Face and head very delicately shagreened; mesonotum finely punctate; mesoscutum very short and regularly convex, not pointed; metascutum rather strongly punctate near middle, smoother at sides, central carina rounded; abdomen flat, subcampanulate, or oval, nearly as broad as thorax. General color metallic bluish, greenish, or bronzy black; antennæ and all coxæ metallic; all femora metallic on the outside, tipped with dull yellow; tibiæ honey yellow; tarsi somewhat darker, last joint brown.

Male.—Differs as follows: Antennæ longer than in female; pile of funicle longer, more erect, and dirty white instead of silvery white. Abdomen much narrower than thorax, campanulate in shape. The femoral bands are brown instead of metallic, and the hind tibiæ have each a light brown central band.

Described from many male and female specimens reared from *Siphonophora avenæ* from Lafayette and Goshen, Ind.

MEGASPILUS NIGER, n. sp.

Female.—Length, 1.6mm; expanse, 3.33mm; greatest width of fore-wing, 0.62mm. Scape of antennæ very long, somewhat swollen beyond middle; funicle long, curved, all joints increasing gradually in width from pedicel to club; joint 1 of funicle somewhat longer than pedicel, joint 3 shorter, joints 4 to 8 increasing in length very slightly. Head and mesonotum very faintly shagreened, but still glistening; lower portion of mesopleura and all of abdomen perfectly smooth. Abdomen subovoid in shape, acutely pointed at tip. Radial vein only slightly curved, extending a little more than half way from stigma to tip of wing. General color jet black; all trochanters, femora and wing veins dark brown; all tibiæ and tarsi lighter brown.

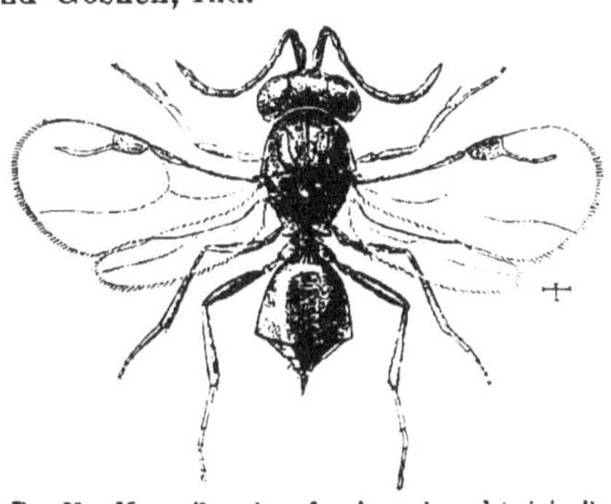

FIG. 52.—*Megaspilus niger*, female—enlarged (original).

Described from five female specimens reared from *Siphonophora avenæ* from Selkirk, Mich., and Lafayette, Ind., July, 1889.

ENCYRTUS WEBSTERI, n. sp.

Female.—Length, 0.93mm; expanse of wings, 2.1mm; greatest width of forewing, 0.35mm. Antennæ short, inserted considerably below the middle of the face; scape cylindrical, not widened below, reaching to vertex; pedicel conical, longer than first funicle joint; all funicle joints as wide as long, the sixth somewhat compressed laterally; club a little longer than last two funicle joints, oval, compressed laterally. Front as broad as one of the eyes, finely shagreened, with sparse, large punctures; ocelli at the angles of a right angle triangle; occipital angle sharp, mesonotum shining, with extremely fine striation; mesoscutellum finely shagreened. Marginal vein wanting; stigmal somewhat longer than postmarginal; wings hyaline; cilia short. Color: Scape of antennæ, all of head, mesoscutum, abdomen and hind thighs, metallic blue-green; funicle of antennæ brown; mes-

FIG. 53.—*Encyrtus Websteri*, male—enlarged (original).

oscutellum bronzy; front and middle femora nearly black with very slight metallic lustre; trochanters and femero-tibial joints yellow; tips of all tibiæ yellow; all tarsi yellow; mesopleura brilliant metallic blue; metapleura shining metallic green.

Male.—Length, 0.8mm, expanse of wings, 1.9mm, greatest width of forewings, 0.35mm; differs from female in its more somber color, the general effect being brown rather than metallic although the mesonotum and head are somewhat lustrous; the antennæ are cylindrical, the segments well separated subcylindrical and furnished with short, finely distributed hair. The general color of the legs is darker; the bands at the joints being narrow and darker; hind tarsi dusky, middle and front tarsi yellow except last joint.

Described from one male and one female reared from *Siphonophora avenæ* by F. M. Webster, at Lafayette, Ind.

This species comes rather close to *Encyrtus clavellatus* Dalman reared in Europe from Cecidomyid galls on willow, but is specifically distinct.

AN AUSTRALIAN HYMENOPTEROUS PARASITE OF THE FLUTED SCALE.

By C. V. Riley.

We have just received from Mr. F. S. Crawford, of Adelaide, the first Hymenopterous parasite of Icerya yet found in Australia. It is

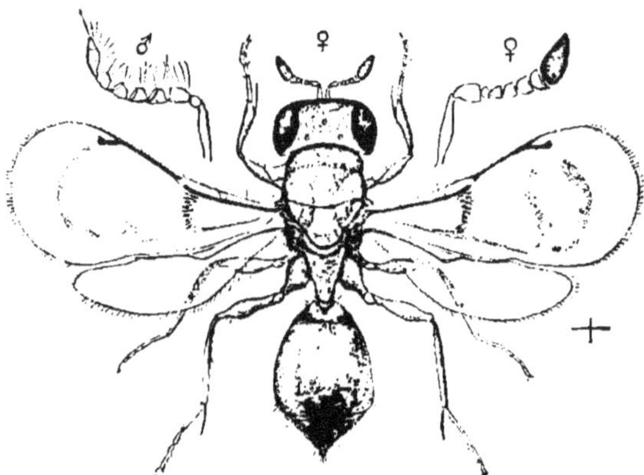

Fig. 54.—*Ophelosia crawfordi*, enlarged (original).

a very interesting form belonging to a new genus, and as it will doubtless become an important factor in the life-chances of Icerya, and it will be convenient to refer to it definitely by name, we take this occasion to characterize it. Its nearest relative is *Dilophogaster californica* Howard, which breeds rapidly in California and is a noted enemy of the Black Scale (*Lecanium oleæ*). So valuable a species is this last that Professor Comstock found that on some trees 75 per cent of the scales were destroyed by it, while in no case was the scale found without its attendant destroyer. Moreover, Mr. Coquillett writes us that in 1889, at Orange, Cal., fully 80 per cent. of the black scales were killed by this parasite.

From these facts it seems probable that the discovery of the new insect will prove important and we have initiated efforts to secure living specimens from Australia. The few facts which Mr. Crawford gives concerning it we quote from his letter of November 24, 1889:

"I received some three months ago some Icerya from a place some 50 miles South of Adelaide, the owner of the orchard not having seen anything of the kind before and wanting to know what they were. These I placed as usual in a bottle loosely stoppered with with cotton wool. With the Icerya was a Chrysopa larva, which for some weeks was feeding on the eggs. One day on examining it I discovered several hymenopters (Proctotrupidæ?), the female yellowish brown, the male almost black. On examination I found that many might have escaped through the cotton stopping being insecure, but I suppose that I have bred about thirty since. It is strange that this is the only instance of a Hymenopterous parasite of the Icerya yet discovered in South Australia. I send you a few of these under separate cover. I presume the small black insect is the male. * * *

Since the following description was drawn up we have received a report* by Mr. Henry Tryon, assistant curator of the Queensland Museum at Brisbane, in which he describes, without name, a Chalcidid parasite of Icerya which he says is very common about Brisbane, and which he believes is responsible for the rarity of Icerya in that vicinity. A careful perusal of his description leads us to believe that he had our insect before him; but as he has proposed no name ours will hold. It is very encouraging to learn that the species is so abundant.

OPHELOSIA, n. g.

Closely resembles in habitus *Dilophogaster* Howard (See Ann. Rept. Dept. Agr., 1880, p. 368, where it is described as *Tomocera*, subsequently changed to *Dilophogaster* on account of the preoccupation of *Tomocera* in *Thysanura*), with which it agrees in many characters, but from which it is sharply defined. The antennal peculiarities are identical in the two forms, viz: The simple, clavate, 10-jointed female antennæ, and the compressed, serrate, hairy, 9-jointed male form. The wings in *Ophelosia* differ markedly, as follows: The sub-marginal vein is not curved downward; the marginal is more than twice as long as stigmal; just below the bend of the sub-marginal in the female is a broad patch of very stout bristles arising from the wing surface. The petiole of the abdomen is nearly as long as the width of the metascutum; the fimbriæ of the callus are very dense, but short. The tufts of hair at base of abdomen are sparse. The hind tibiæ are furnished at tip with a long, slender, slightly-curved spine, nearly as long as first tarsal joint, while in *Dilophogaster* it is entirely unarmed.

O. CRAWFORDI, n. sp.

Female.—Length, 2^{mm}; expanse, 4^{mm}. General color honey-yellow, somewhat darker dorsally than ventrally. Head: face and vertex strongly transverse-rugose; ocelli concolorous; eyes darker; antennæ with club more dusky and with joints 2–6 of flagellum paler than the rest. Thorax: pronotum and mesonotum plainly shagreened, with sparse, appressed concolorous pile; mesoscutellum faintly striate; lateral parts of mesoscutum strongly rugose, the centre faintly so; the four mesoscutellar piliferous tubercles as also the hairs, black, a small spot behind each tegula and the lateral parts of the mesoscutum black or blackish; fimbria of metascutum white;

*This report will be reviewed at length in the next number of INSECT LIFE.

wings with a narrow curved transverse dusky band reaching from the bend of the submarginal vein to hind border of wing including the patch of wing bristles; also with a large nearly circular dusky shade below stigma and reaching nearly across wing; legs uniformly honey-yellow with the coxæ sometimes brownish above. Abdomen with basal joint dark brown, and more or less brown at sides and near tip.

Male.—Slightly smaller; sculpture identical throughout. Pile very inconspicuous, dark. General color black, shining; all legs honey-yellow; the upper sides of the hind femora and tibiæ somewhat darkened; hind coxæ black; front and middle coxæ honey-yellow at tip; antennæ with the scape honey-yellow, and the funicle brownish; wings perfectly hyaline.

Described from four female and two male specimens reared by F. S. Crawford, at Adelaide, from specimens of *Icerya purchasi* received from S. Australia, 50 miles south of Adelaide.

EXTRACTS FROM CORRESPONDENCE.

The Orchid Isosoma in America.

A friend of mine, by occupation a florist, has applied to me for information concerning an insect pest affecting the genus of orchids known as *Cattleya*, more especially *C. trianæ*, *eldorado*, and *gigas*. Said insect belongs to a group I have studied but very little, and as the matter is of general interest I appeal to you.

During the resting season of these plants the pseudo-bulb will sometimes be observed to suddenly start into activity, increasing rapidly in size and becoming swollen spherically. On examination this enlargement is found to contain a cavity in which are several (3 to 8) insects. Those which I have had an opportunity of studying were in the last stages of development; I inclose examples in the light-colored pupa stage, the dark-colored stage preparatory to transformation, and the adult. They make their escape by gnawing a hole from the cavity sufficiently large to allow their egress. The size of the cavity is dependent on the number and state of development of its inhabitants. The larvæ have been described to me as "little white grubs." All the adults that I have seen have had clear wings, but my friend states that is unusual; he has generally found them with dark wings, apparently identical. (I expected to find a Cynips, but is not this a Chalcid?) He thinks the pest is imported with the plants (which mostly come from New Granada), and finds comfortable quarters and a field for activity in greenhouses; he has found them in plants recently imported, together with unmistakable signs of their former presence.

Their depredations are followed by disastrous results. Of course no flowers are to be expected from the bulb attacked, and this abnormal growth taking place during the resting season so saps the vitality of the plant that it behaves as if it were attacked by slow consumption, the leaves lose their vigor and consistence, wither, fade, and gradually die in from one to two years after being seriously attacked.

Any information you can give me concerning this pest, its name, life history, habits, remedies, etc., will be very gratefully received. Will send you sketches of its work if desired.—[Albert P. Morse, South Natick, Mass.

REPLY.—Your letter of November 29, together with specimens of the Chalcidid reared from the pseudo-bulb of *Cattleya*, has been received. These specimens form a very desirable addition to the collection of the National Museum, for the reason that we already possessed the swellings from which they issue, and which were given us a few years since in France. Prof. J. O. Westwood, in the Transactions of the Entomological Society of London for 1882, figures and describes what is probably the same species under the name of *Isosoma orchidearum*. The specimens which you send us differ slightly from those described by Westwood in the coloration of the legs, but this is too unimportant a character to base a new species upon. Prof. Westwood mentions the same insect in *The Gardener's Chronicle*, November 27, 1869, p. 230. I send you by same mail a copy of No. 4 of Vol. 1 of INSECT LIFE, on p. 121 of which you will find a note of our observations on this subject.

If you have plenty of material the most careful observations will be desirable, and if you can send us specimens of the work it will facilitate matters. We shall gladly publish any detailed account of your observations which you may care to write.—[December 3, 1889.]

A Flaxseed Mite.

I send you something by to-day's mail that greatly puzzles me. My father, who has just returned from Kansas, brought with him the bottle which I inclose, containing specimens of a mite, or what appears so to me, which is found in amazing quantities in a warehouse in Paola, Kans., among flaxseed, of which about four thousand bushels was in store there. Crawling masses of these mites several inches deep could be seen on some of the floors and the owner feared they would destroy the entire stock. I have opened a number of the seeds without finding any of the mites on the inside unless the seed happens to be somewhat crushed. I conjecture that the warehouse may be somewhat damp, and that these creatures are feeding on the débris of broken seeds, bits of hulls, stems, etc., which are in a state of partial decay. There was an almost unendurable stench in the compartments where the mites most abounded. What do you think of the matter? The owner of the infested seed is Mr. Z. Hayes, of Paola, Kans.—[Mary E. Murtfeldt, Kirkwood, Mo., December 10, 1889.

REPLY.—Yours of the 10th instant with specimens came safely to hand. The mite which is found in such amazing numbers in flaxseed in Kansas is a species of *Tyroglyphus* differing from any familiar to me; it is quite different from *T. longior*, the common flour and cheese mite. You are probably right in considering that the creatures were feeding upon the débris of the broken seeds, and that they were attracted by the state of partial decay.—[December 13, 1889.]

Abundance of Ægeria acerni.

Your kind favor of November 26, stating that my name had been added to the mailing list duly received with the first four numbers of Vol. 2 of Insect Life. I find them very interesting and instructive, and would ask you to accept my heartfelt thanks for them.

By far the most destructive pest in this city is *Ægeria acerni*. Nine-tenths of Detroit's shade trees are (soft) maple, and the damage done by the borer above named is very considerable. Have seen in June as many as fifteen pupa cases protruding from a single tree.

The inner bark is eaten, often nearly around the trunk, and as the tree grows it leaves either a large hole in the tree or a constriction nearly around it. Dozens of trees thus gouged and girdled are blown down with every high wind.

Trees from two to six inches in diameter seem to suffer the most from their ravages. Have tried several times to remedy the trees infested by painting the holes or rough places in May or June, thinking the moth would not deposit her eggs on painted trees, but the next year noticed pupa-cases protruding through holes in the paint by scores.

Any advice you may care to give me will be gratefully received.—[Charles A. Wiley, Detroit, Mich., December 2, 1889.

REPLY.—Your letter of the 2d instant is just received. We are obliged to you for the note concerning the abundance of *Ægeria acerni* in your city. You will find a full account of this insect with illustrations in my sixth Report on the Insects of Missouri (1874), pages 107 to 110. In my experience these worms are invariably found in such trees as have been injured either by the work of the flat-headed borer, by the rubbing of the tree against a post or board, or in some other way. Where the bark has been kept smooth they do not seem to trouble it. The moth evidently prefers to lay her eggs in cracked or roughened parts. Any application, therefore, which will tend to keep the bark smooth will be of value.—[December 5, 1889.]

Hessian Fly in California.

This insect has been reported as being very abundant during spring (1889) in the central part of the State, destroying most of the wheat around Mt. Eden. Personally it was observed in the Santa Cruz Mountains on May 26. At this time they were found in all stages from young larvæ to pupæ and empty cases (puparia) within barley. Some of these collected the beginning of June produced flies until beginning of July; others collected beginning of July did not hatch up to date, October 30. A few parasites (*Semiotellus destructor*) were also bred; they had issued up to September. Flies are marked 443; parasites 443[c]; and a few Isosoma that had been bred from some straw 443[a]. One small fly was also found in jar; this is marked 443[b]. During September, 1887, a few puparia, evidently of this fly, were found near Alameda on two species of grass, one of these *Elymus americanus* and the second a species of *Agrostis*. Also during the last summer specimens and traces of such were found in the Santa Cruz Mountains upon several species of grass. This is without doubt the Hessian Fly. On October 1, 1889, I found larvæ still remaining within puparia collected in July.—[Albert Koebele, Alameda, Cal., October 30, 1889.

An Ivy Scale-insect.

I have an ivy vine which is badly diseased. Inclosed please find sample of leaves. The ivy is some 30 feet long and runs along the inside of my store. Had one of about the same length destroyed some 2 years ago by the same pest. Kindly inform me what to do to get rid of these pests.—[George Teuchert, Lake View, Ill, Dec. 2, 1889.

REPLY.—Your letter of the 2d instant, inclosing leaves of ivy infested by scale insect, has been received. The insect in question is the common *Aspidiotus nerii*, a cosmopolitan species which infests a great variety of plants, and is by no means confined to the ivy, although occurring commonly upon it. As a remedy I would advise you to spray with a dilute soap emulsion made according to the usual formula. [December 4, 1889.]

Ant Hills and Slugs.

I have resorted to many expedients to get rid of the ant hills that disfigure my lawn and sometimes seriously injure plants and shrubs, and have finally succeeded in conquering them. I first hive them—break up the nest pretty thoroughly and if it is near the roots of a plant draw as much of the débris as possible a little way from it and turn over it a large plant jar. The ants will promptly appropriate the jar, remove their larvæ to it, and fill it with pellets of earth. I then drench this with kerosene emulsion reduced to a strength of 2 to 3 per cent., which will kill every ant thoroughly drenched with it. It is more destructive to them than pure kerosene, which does not adhere to them. In this way I have thoroughly conquered the ants.

The rose slug and the currant worm I keep completely under by use of hellebore, a tablespoonful to a gallon of water, and forcing it violently among the foliage with a hydropult. Commencing in the spring before I can find a slug or a worm, and repeating the drenching once a week for three or four weeks, I can destroy them completely before they do any damage. On one hundred roses I was able this spring to find only two slugs, while the foliage of some common sorts I did not spray was completely destroyed.—[M. C. Read, Hudson, Ohio, September 5, 1889.

A curious Case of insect Litigation.

I recently learned of a case where the good work accomplished by the *Vedalia cardinalis* had been grossly ignored. It appears that a certain adventurer inoculated a number of Icerya-infested orange trees, with the understanding that if by this means he succeeded in destroying all of the Iceryas on these trees he was to receive a certain remuneration for his trouble. A few days after the trees had been inoculated, one of the county inspectors of fruit pests placed a number of the Vedalias in these trees without apprising the experimentor of this fact; at the appointed time the trees were carefully examined and not a living Icerya could be found on them. The experimentor claimed that it was through his inoculating the trees that the Iceryas had been destroyed; the owner of the trees, however, thought that the credit belonged to the Vedalias, and therefore refused to remunerate the quack for his work. Thereupon the latter gentleman brought suit against the owner of the trees and won it, the jury deciding that the fatality among the Iceryas was produced through the inoculation which the infested trees had received, notwithstanding the testimony of the inspector to the contrary, and the fact that the empty pupa cases of the Vedalia were still on the trees! This happened several months ago, at a time when the workings of the Vedalia were not so well known as at the present time.—[D. W. Coquillett, Los Angeles, Cal., December 12, 1889.

Two interesting Parasites.

I send some bottles containing larvæ in alcohol, and a few more slides with specimens for the microscope. Among the latter is an interesting parasite on *Aspidiotus uvæ*, which seems to be doing good work in keeping this pernicious scale-insect in check. More than a dozen of these little flies have emerged from the scales on a bit of cane not 5 inches long. In one of the bottles is a section of a Plusia larva found on Chrysanthemum from which thousands of the minute flies inclosed with it issued. I never saw a more extreme case of parasitism. After spinning up the poor worm lost all semblance to itself. A myriad of the parent flies must have attacked it at once.—[Mary E. Murtfeldt, Kirkwood, Mo., November 23, 1889.

REPLY.—Your parasite on *Aspidiotus uvæ* is a new species of the genus *Centrodora*, and the Plusia larva had evidently been attacked by *Copidosoma truncatellum*, which you will find mentioned as a parasite of *Plusia brassicæ* in my annual report for 1883, p. 121, Plate XI, fig. 6.

Work of White Ants.

I mail you a box to-day containing insects that have done remarkably good work. They bored through paper, then through a full bolt of Conestoga ticking into wood about one-fourth to three-eighths inch deep. The marks in wood were exactly the same as in the ticking I send you a sample of. When alive and killed with naptha (benzine) they drop a brownish fluid from the anus, which I suppose turns into dirt, as it shows on the ticking, lumps being attached to it where eaten, this extending through the whole bolt. The ticking was lying on a shelf (a place not very dark during the day) for about one month. Please let me know their name and habits.—[Eugene R. Fischer, 2707 Winnebago street, St. Louis, Mo., December 21, 1889.

REPLY.—The insect which has done this damage is the commonest of our so-called White Ants, and is known as *Termes flavipes*. This species bores in the woodwork of old buildings, and often does considerable damage. It is a difficult insect to fight, and about the only thing which you can do is to inject steam or hot water or kerosene wherever an opening seems to lead into their burrows in timbers.—[December 27, 1889.]

Importation of Orange Pests from Florida to California.

I am inspector of the Pomona fruit district. There will be a great many orange and lemon trees shipped from Florida this season. I would like to have you inform me of the places that are infested with Red Scale (*Aspidiotus ficus*) or other scales that would be dangerous in this climate, so that I can be on the lookout. Last winter I found Red Scale on trees that came from Orlando, Fla. I treated them with hydrocyanic acid gas and, I believe, killed all of them. Mr. H. G. Hubbard in his report of 1885 speaks of the Red Scale being in Orlando and San Mateo, Fla., but I presume they have spread to other places.—[C. C. Warren, Pomona, Cal., December 10, 1889.

REPLY.—Yours of the 10th instant has just come to hand. I can give you little or no information regarding injurious scale insects of other States which would be likely to be dangerous in California beyond what you will find in Hubbard's Report on Insects Affecting the Orange (1885) and the report on Scale Insects in the Annual Report of this Department for 1880. You are doubtless aware of the fact that the so-called Red Scale of Florida differs from the Red Scale of California.—[December 18, 1889.]

On some Dung Flies.

I send by the same mail that will take this for identification two apparently different species of flies. Those in the smaller bottle, black in color and smaller in size, I have noticed for a month past in great numbers in my poultry house in the barrel which receives the daily droppings. A paper is folded tightly over the barrel on which the cover is placed. On removing these the under side of the paper is often quite black with the minute insects. Mingled with these a few house flies of varying size are also seen, suggesting to me the thought that the minute ones are the early stage of the common house fly. Then comes the idea that most (is it all?) insects having the three stages make all their growth in the larva state and on reaching the imago state are at first of their full size. This is true, may I ask, of the house fly? If this one I send is a species by itself, please give me its name and direct me to its natural history. The other flies, of larger size, lighter color, and with reddish head, I have not noticed till this morning. The pans in which the hen's food is eaten are placed at night in the shed at an open window having a small mosquito screen and the blinds are always shut. On going to the pans this morning these flies arose from them in swarms. There must have been hundreds of them, though not one has been noticed before this year. These, however, are not new to us. We have always noticed them upon fruits, especially when injured, and about cider-mills. Please give its name. I should have said in writing of the others that on the paper where I saw so many of the small, black flies I also noticed crawling about among them other minute creatures of nearly the same size but wingless. Were they in any way related to them? Or can you tell what they probably are without specimens sent.—[S. D. Hunt, South Franklin, Mass., August 31, 1889.

REPLY.—The two flies sent are undoubtedly two different species: The black one can not be recognized without a careful examination and study of the specimens, but it is one of the *Drosophilidæ* and may belong to the genus *Stegana*. It is very distinct from the house fly, and does not belong to the same family. The other larger fly of lighter color is *Drosophila ampelophila* Loew, called by Professor Comstock the "Vine-loving Pomace-fly" An account of its natural history is given by Professor Comstock in the Annual Report of this Department for the years 1881–2, pp. 198–201.—[September 4, 1889.]

Spider Bites.

* * * In the fall of 1847, in southwestern Pennsylvania, I was called to treat a case of spider bite. I saw the young man two or three hours after he was bitten. The puncture was plainly seen on the wrist. The hand and arm were much swollen and the axillary glands swollen and painful. Knowing tincture of lobelia to be a specific in poisoning by poison ivy (*Rhus toxicodendron*), I had his arm enveloped with cloths saturated with the tincture, and gave enough internally to thoroughly empty his stomach. In twelve hours he was well, but the swelling lasted two or three days. That it was a spider bite I never knew, and always doubted. But in the coat-sleeve he had been putting on was a flat circular nest such as spiders often spin in the fall in garments hung in dark places. Those who believe in spider bites ought to show the fangs or other organs with which they can bite, and also the poison-secreting glands

and the poison sacs or cells. Till these are shown or till a spider is seen to bite a person, people will be incredulous.—[Dr. Wm. P. T. Coal, Meadows, Ill., September 3, 1889.

SECOND LETTER.—This morning my sister thought she was bitten by a spider under the sleeve near the wrist and almost immediately in two or three places between that and the shoulder. She crushed the insect with her hand, and on removing the clothes found the fragments which I send inclosed. If you can identify it I would like to know what it is. The bites or stings caused a slight pain and swelling that were gone in a few hours.—[Dr. Wm. P. T. Coal, Meadows, Ill., January 1, 1890.

REPLY.—Your letter of January 1 and the accompany fragments of a spider which is supposed to have bitten your sister have been received. The case is an interesting one and it is extremely unfortunate that the fragments will not enable a definite determination of the species, as the evidence is strong that the bite was made by this creature. Dr. Marx, our authority on spiders, states that the fragments show that the spider belonged to the family *Drassidæ*, and perhaps to the genus *Pythonissa*, the species of which live under stones but may also be found in outhouses. I am very much obliged to you for sending this specimen, and hope that if a similar case ever comes under your observation you will communicate it.—[January 7, 1890.]

GENERAL NOTES.

INSECTS AFFECTING SALSIFY.

Owing, possibly, to the fact that this vegetable is grown only in our gardens, and to a very limited extent, its insect enemies seem to have been but little studied. Mr. John Martin (10th Rep. St. Ent. Ill., p. 139), gives it as one of the food plants of the larvæ of *Prodenia lineatella*; but Mr. Martin seems to have provided it as food for the caterpillars while they were in confinement, they not seeking it from motives of choice; but this is the only species we have noticed on record as depredating upon it.

August 16 of the present year (1889), we found the foliage of these plants being eaten by larvæ, which, as they all fed from within leaves whose edges they had drawn together to form a hollow tube, appeared to belong to the same species. In some of these tubes small chrysalids were also found.

A quantity of infested leaves were gathered and placed in a breeding cage, in which there appeared on August 24 adults of a species of *Pædisca*, followed in a few days by other moths belonging to this species, *Dichelia sulfureana* and *Lophoderus triferana*. A number of larvæ were attacked by parasites, and on September 1 considerable numbers of a species of *Limneria* appeared.

While searching for the larvæ of the preceding a caterpillar of *Spilosoma virginica* was observed leisurely devouring the foliage of this plant, and, also, adults and pupæ of *Lygus pratensis* were noted in abundance among the tender leaves, some of them extracting the juices therefrom.

October 16, plants in this same garden were found to be infested with Aphides, and the top of one of the most thickly populated was removed and placed with living plants in a breeding cage. With the change to a warmer environment, the insects became more active, and instead of of a single species, as at first supposed, there were found to be four, three Aphides, viz, *Siphonophora* near *erigeronensis*, *Aphis* near *plantaginis*, at the time being studied on carrot and *Portulaca*, *Myzus mahaleb*, and a minute *Thrips*, their relative abundance being in the order in which they are here given. All three species of plant lice, and the *Thrips*, developed on the Salsify and remained upon it for several weeks, showing that their occurrence on plants in the garden was not accidental.—[F. M. Webster, December 12, 1889.

AN EGYPTIAN MEALY BUG.

We are indebted to our esteemed correspondent Mr. D. Morris, of the Royal Kew Gardens, for a copy of a letter from Mr. R. W. Blunfield of Alexandria, Egypt. During the past four years the gardens in Alexandria have been infested by a Coccus which destroys all of the trees and is causing the greatest alarm. It first appeared four years ago when Mr. Blunfield noticed it in quantities on the underside of the leaves of the Banyan tree, but it soon spread with extraordinary rapidity and some of the most beautiful gardens of the city full of tropical trees and shrubs have been also destroyed. A breeze sends the cottony pest down in showers in all directions. It seems to attack almost any plant, but the leaves of *Ficus ruginosa* and one or two other kinds of fig seem too tough for it and it will not touch them. He states that it seems almost impossible for a few horticulturists to try to eradicate this pest while their indifferent neighbors are harboring hot-beds of them, and there will have to be some strong measures taken by law to put it down.

Mr. Blunfield sent specimens which were referred to Mr. J. W. Douglas, one of the most prominent British students of Coccidæ, who upon cursory examination decided that it was a species of *Dactylopius*. At the time of this writing Mr. Douglas has not had time to examine it with sufficient care to determine the species. We have written advising the use of one of the resin washes which have proved so effectual against Icerya in California, and have mentioned particularly the one given on page 92 of the current volume of INSECT LIFE.

A CASE OF EXCESSIVE PARASITISM.

The frequency with which the Black Walnut is defoliated by the larvæ of *Datana ministra* has often been a source of regret to admirers of that beautiful and majestic tree. Every autumn, throughout the Western States, September finds many trees as devoid of foliage as in midwinter, the fruit hanging to the naked twigs with the very air of disconsolation. Trees in the forest do not appear to suffer, the caterpillars seeming to prefer isolated individuals or small groups, which are usually planted for ornamentation.

Such a tree stands by the side of the walk midway between my home and the Indiana experiment station, being separated from all others of its kind by nearly a quarter of a mile. During the years 1884 and 1885 this tree was regularly defoliated in August. In 1886, during the usual season, the caterpillars made their appearance and began their work, reaching very near their full growth, when there was a sudden cessation of attack, and the depredators disappeared from the tree with astonishing rapidity, leaving the foliage less than half eaten. This was a change of affairs without a precedent.

An examination of the ground about and beneath the tree at once gave a clue to the mystery, revealing a state of affairs as interesting as unexpected. Everywhere among the short grass and weeds were caterpillars, some of them dead, others dying, while still others were quite active, but all well-nigh covered with eggs of a species of Tachina Fly. The flies were present in myriads, some of them winging their way about, a few inches above the surface of the ground, and others perched on grass, weeds, etc., all evidently watching for caterpillars, while the latter were as evidently hiding from their persecutors, for no sooner would one of them leave its seclusion than perhaps half a dozen flies would give chase, and begin fastening their eggs to various parts of the body, the victim writhing, twisting, and rolling itself about in the dust, in frantic efforts to escape. Even after gaining a place of security, under some leaf or plant, often some portion of the body would be left exposed, and the already half dead caterpillar would be again driven forth from its hiding, like a gored ox. Four caterpillars, fair examples of the whole lot, were forwarded to the Department at Washington, and to their bodies eggs were attached as follows: No. 1, 213; No. 2, 115; No. 3, 131; No. 4, 228. From five others, collected at the same time, we afterwards reared fifty-three adult flies.

During the years 1887 and 1888, not a caterpillar was observed on this tree, though others in the neighborhood were infested, but the present year (1889) they returned again in full force. It would be interesting to know if similar attacks by an allied Tachinid upon the Army Worm were as lasting in effect.—[F. M. Webster, November 28, 1889.

SOME HITHERTO UNRECORDED ENEMIES OF RASPBERRIES AND BLACKBERRIES.

Solenopsis fugax Latr.—These minute ants were observed in great abundance during July, 1886, burrowing into the ripe fruit of the blackberry. The food habits of the species must be exceedingly varied, as we have found them excavating and dragging away the substance of recently planted seed-corn, infesting dead crickets, burrowing into the fatty parts of cured hams, and in attendance upon a species of

Dactylopius infesting the roots of red clover, *Trifolium pratense* L. We have also found them burrowing in ripe apples.

Limonius auripilis Say.—We have observed the adult feeding upon ripe raspberries during July.

Carpophilus brachypterus Say.—These beetles are sometimes quite numerous in the fruit of the raspberry, especially if it be a little over-ripe. Their small size, and the habit of secreting themselves in the cavity of the berry about the receptacle, renders their presence difficult to detect.

Iulus impressus Say.—About the middle of July of the present year (1888) a lady of Lafayette purchased from her grocer a quantity of black raspberries for preserving. The case consisted of 16 quart-boxes, such as are usually employed for holding fruit. On looking the berries over, preparatory to cooking, she began to find these worms intermingled among and devouring the fruit. By the time a small portion of the supply had been inspected, upwards of fifty worms had been found, and the fruit was disposed of in a way rather more summary than that of preserving. Samples of both fruit and worms submitted to me left no doubt as to either the species of *Iulus* engaged, or its appetite for this kind of fruit. Whether the worms infested the fruit in the field, or whether the case was left on the ground and they made their way into the boxes, I was not able to learn, but the latter appears more probable.

Cosmopepla carnifex Fab.—This was reported to me from Livingston County, New York, as injuring the foliage of the black raspberry. See INSECT LIFE, vol. 1, p. 157.—[F. M. Webster, November 30, 1889.

NEBRASKA INSECTS.

We have just received from Prof. Lawrence Bruner his report to the Nebraska State Board of Agriculture for 1888. He considers a number of injurious species, including the Chinch Bug, the Corn Worm, the Box-elder Plant-louse, the Green-striped Maple-worm, the Willow Cimbex, the Apple-tree Flea-beetle, the Apple Twig-borer, the Corn Root-worm, the Army Worm, Cut Worms, the Box-elder Bug, the Imbricated Snout-beetle, the Sculptured Corn Sphenophorus, Tree Crickets, a new enemy to the Colorado Potato-beetle, Ox Warbles, Plum Curculio, Codling Moth, Strawberry Worms. The report is mainly compiled, but contains some account of the author's personal observations in Nebraska of the species mentioned. Among these we may note that the Army Worm is here recorded in injurious numbers for the first time in Nebraska. The damage by the Imbricated Snout-beetle to young corn is also of interest, while the illustrated article on the Box-elder Plant-louse is new. Under the article upon the Plum Curculio he mentions finding a species of *Coccotorus*, which he proposes to name *hirsutus*, feeding upon the Sand Cherry, in Cuming County. This we have since learned is the true *Coccotorus scutellaris* of Leconte (see note in INSECT

LIFE, Vol. I, p. 89) which, by the way, was originally found upon this same plant. Careful comparison of specimens, moreover, shows that the common Plum Gouger (*Anthonomus prunicida* Walsh) is unquestionably a good species, as such go, and not a synonym of *scutellaris* as has been supposed of late years. We illustrate both species at Figs. 55 and 56.

FIG. 55.—*Coccotorus scutellaris*—enlarged (original).

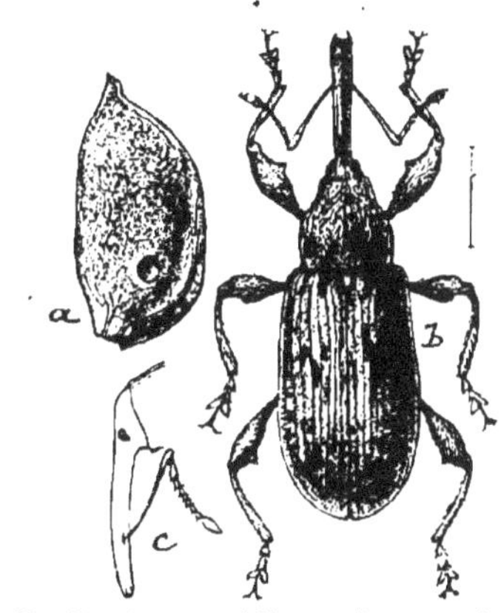

FIG. 56.—*Coccotorus prunicida*; *a*, plum-stone showing exit-hole of larva; *b*, adult; *c*, side view of head of adult—enlarged (original).

A PODURID WHICH DESTROYS THE RED RUST OF WHEAT.

In studying the insect enemies of our cereals during the last five years we have repeatedly come in contact with a small, robust species of *Smynthurus*—species undetermined—both in the field and in breeding cages.

From the fact that we have several times reared the species in cages containing only growing grain and insects preying thereon, and were not able to detect them destroying either one of these, we have been perplexed to understand from what source these little Neuropters obtained their subsistence. During the present year, however, we have twice found individuals feeding upon the Uredo spores of the common wheat-rust, *Puccinia rubigovera*, in both instances on wheat growing in the field and at a time when the rust was first making its appearance on the leaves.

While these observations clear away some of the obscurity surrounding the food habits of these insects, their economic importance is as uncertain as before. We are free to suppose that all rust spores eaten by these insects are destroyed, and to this extent they are benefactors. But their bodies being covered with short bristles, and being such gormandizers in their method of feeding, with every repast they manage to get great numbers of spores caught among the bristles on their bodies, and these spores, it is fair to suppose, are carried away and probably become detached one time and another, more or less of them being left on plants not previously affected by rust. Early in the fall, when rust is only commencing to appear on the young wheat, these *Smynthurus* might destroy many spores, but we have observed them enough to leave no doubt that they may transfer spores from one plant to another in the manner indicated.—[F. M. Webster, November 30, 1889.

INSECTICIDE LITIGATION.

We notice in *The San Francisco Morning Call* of November 22 a statement to the effect that a suit has been commenced in the superior court, by John S. Finch, owner of a ranch at Hayward's, Alameda County, against the Ongerth Grafting Compound Company, to recover $16,500 for losses sustained by reason of the application of the defendant's liquid compound to 106 fruit trees in order to destroy vermin and fungoid growths, whereby the trees were injured and killed. The compound cost Mr. Finch $10. Without any knowledge of the merits of this particular case we would state that we are glad to see the matter brought to trial in order that the responsibilities of the proprietors of patent insecticides may be legally defined.

NORTH EUROPEAN DRAGON FLIES.

We have just received from Dr. Filip Trybom a short paper, entitled "Trollsländor (Odonater) Insamlade under Svenska Expeditionen till Jenisei 1876," in which he describes eight species of Dragon Flies collected mainly in North Sweden, and some as far north as 69° 25′. Four of the species are new.

A CORRECTION.

Professor Forbes calls our attention to the fact that paragraph 6, on page 182, of the December number of INSECT LIFE, should read as follows:

> Mr. Forbes expressed himself as of the opinion that, from our present knowledge of the use of the arsenites as insecticides, they *can not* be recommended for use on the peach.

A PARASITE OF THE MEDITERRANEAN FLOUR-MOTH.

On page 170 of the last number of INSECT LIFE in our article upon this destructive grain pest we mentioned the fact that a small Ichneumon Fly destroyed this insect in the warehouses in the east end of London in the summer of 1887. At the time of writing this article we wrote to Mr. J. B. Bridgman, of Norwich, England, to ascertain whether he knew of this parasite, and have just received a reply in which he states that although he was not familiar with this instance he has since received specimens of *Chremylus rubiginosus* reared from *Ephestia kühniella.*

EFFECTS OF THE OPEN WINTER.

Two interesting effects of the mild weather which we have been having have been brought to our attention recently. December 20 Mr. G. A. Frierson, of Frierson's Mill, La., sent us specimens of the Turkey Gnat

Simulium meridionale), which had issued and were flying around at that date. January 4 he sent us other specimens of the Buffalo Gnat (*S. pecuarum*). January 6 Mr. P. P. Turner, of this city, brought us a living imago of the Fall Web-worm (*Hyphantria cunea*), which had recently issued from the cocoon. If this premature issuing of the latter species is at all general and we have subsequent severe weather the shade trees of Washington will not, in all probability, suffer the coming summer from Web-worms at least.

HONEY BEES AND ARSENICALS USED AS SPRAYS.

Mr. H. O. Kruschke, of Juneau County, Wisconsin, in the *American Garden* for January, 1890, p. 57, warns prospective sprayers that the first man caught applying arsenic to trees in full bloom will be prosecuted—reasoning that the spraying of such trees will result in the storage by the bees of poisoned honey, the consumption of which will be dangerous. In Vol. II, p. 84, of INSECT LIFE, the effect of arsenical insecticides on the honey-bee is briefly discussed, and a well-authenticated case is given which seems to show that such spraying is not attended with ill results either to the bees or the honey. The prevailing belief is, however, the other way, and cases are on record where serious destruction of bees has resulted from spraying. In the case of the Apple, particularly, the application should not be made until the bloom has begun to fall, when no injury will be likely to result. It was because of the possibility of danger that in the beginning we were very slow to recommend the wholesale spraying of orchards with the arsenical mixtures, but experience has shown here, as in other cases, judicious and cautious use is attended only with benefit, and that the possible harm is reduced to such a minimum as to almost justify its being left out of consideration.

ENTOMOLOGICAL SOCIETY OF WASHINGTON.

December 5, 1889. (Fifty-seventh regular meeting.)—The corresponding secretary reported additions to the library.

Professor Riley presented a communication on the ovipositors of Diptera in which he reviewed the general subject of piercing ovipositors in the different orders of insects, stating that in the order Diptera they were very rare, and calling attention to the fact that in *Trypeta* and some allied forms the ovipositor is capable of piercing, and that in *Trypeta pomonella* and in *T. ludens* he had found them to be readily capable of piercing the skins of apples and oranges respectively.

Professor Riley also presented a note upon the genus *Lestophonus*, showing that careful studies which he had made indicated that Mr. Skuse, of Australia, is correct in considering *L. monophlæbi* and *L. iceryæ* as distinct species, and not identical as supposed by Dr. Williston.

Professor Riley further presented a note on dipterous insects passed from the rectum of man, reviewing the older instances, and mentioned particularly the sending of *Eristalis dimidiatus* in the larva state by Dr. J. W. Compton, of Evansville, Ind., who stated that they were passed from the bowels of a young woman. He also mentioned the recent sending of larvæ of *Eristalis tenax* by Dr. J. A. Lintner, to whom they had been sent as having been found under similar circumstances.

Mr. Ashmead read a paper on the Chalcid genus *Halidea*, in which he announced the finding in this country for the first time of a species of this genus, which superficially resembles *Eupelmus*, but is distinguished by the dilated posterior tibiæ and tarsi. The American specimen was captured by Mr. Schwarz at Harper's Ferry, and the species is named by Mr. Ashmead *Halidea schwarzii.*

Mr. Howard read a paper on the Hymenopterous parasites of *Ocneria dispar*, which is incorporated in the article on the Gipsy Moth in this number of INSECT LIFE.

Mr. Townsend presented a communication entitled "Further note on *Dissosteira carolina*," referring to his previous article in the *Canadian Entomologist* for September, 1884, on the peculiar aërial performances of this locust, and giving the results of observations during 1885 and 1886

DIVISION OF ENTOMOLOGY.

PERIODICAL BULLETIN. MARCH, 1890.

Vol. II. No. 9.

INSECT LIFE.

DEVOTED TO THE ECONOMY AND LIFE-HABITS OF INSECTS, ESPECIALLY IN THEIR RELATIONS TO AGRICULTURE.

EDITED BY

C. V. RILEY, Entomologist,

AND

L. O. HOWARD, First Assistant,

WITH THE ASSISTANCE OF OTHER MEMBERS OF THE DIVISIONAL FORCE.

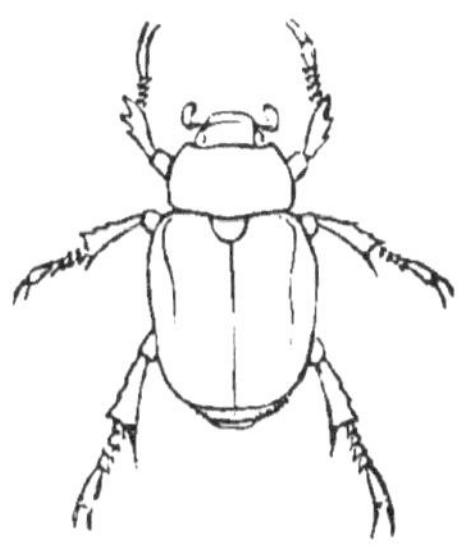

[PUBLISHED BY AUTHORITY OF THE SECRETARY OF AGRICULTURE.]

WASHINGTON:
GOVERNMENT PRINTING OFFICE.
1890.

CONTENTS.

SPECIAL NOTES.

Professor Atkinson's Bulletin on Nematode Root-galls.—We have recently received Bulletin No. 9, new series (Science Contributions, Vol. I, No. 1), of the Alabama Agricultural Experiment Station. It consists of "A preliminary Report on the Life-history and Metamorphoses of a Root-gall Nematode, *Heterodera radicicola* (Graef) Müll., and the injuries produced by it on the roots of various plants." It will be seen that the article deals with the subject of Bulletin No. 20 of this Division, prepared by Dr. Neal. Owing to the fact, stated in the preface to Bulletin No. 20, that Dr. Neal had not access to the literature of the subject, the investigation conducted by him aimed at the discovery of practical remedies rather than scientific accuracy.

The article of Professor Atkinson supplements Dr. Neal's work by giving a careful and accurate account of the life-history and habits of these worms; and as the author is evidently thoroughly familiar with the European writings on Nematodes, little is left to be desired in this direction.

The species is referred with little doubt to *Heterodera radicicola* Müll., which occurs commonly in central Europe in connection with a scarcely distinct species *H. schachtii* Schm. The genus *Heterodera* is shown to be world-wide in distribution. In addition to the species mentioned, one is found in Java in roots of sugar-cane;* in Brazil in roots of the coffee-tree, and one is also recorded from Scotland—all of which are scarcely distinguishable from *H. radicicola*. The structure and histological characteristics of diseased roots of various plants are discussed. The disease of potatoes known as the "potato-scab," the early stages of which are very like the Nematode galls on the potato tubers, the "club-foot" of cabbage, and the functional tubercles on the roots of Leguminoseæ, which have been shown to be of use to the plants in the acquisition of nitrogen, are carefully distinguished from the quite similar root-galls on these plants resulting from Nematode attack.

* See note on page 55 of the present volume.

No experiments were made looking to checking the injuries of this Nematode, and the various recommendations made are in general those already given by Dr. Neal. They consist in the use of various alkaline fertilizers, clean culture, and sterilization of the soil by a system of rotation which introduces crops not subject to their attacks. A German method is given of trapping the worms with catch plants ("Fangenpflanzen"), which are dug up and destroyed after becoming infested and before the worms have escaped.

In the vicinity of Auburn, Ala., some 36 species of plants were found to be affected with Nematode root-galls. A list of the works consulted, 36 in number, is given, most of which are European. The text is admirably supplemented with six plates showing affected roots, entire and in section, and the Nematode in its various stages.

Economic Entomology in India.—We are glad to see that the high standard inaugurated in No. 1 of the "Notes on Indian Entomology," editorially noticed in these pages a short time ago, is maintained in No. 2, which has just been published by the trustees of the Indian Museum, Calcutta.

Mr. E. C. Cotes contributes a translation of an unpublished paper by the late Dr. E. Becker on *Trycolypa bombycis*, a new Tachinid fly, parasitic on Indian silk-worms (*Bombyx fortunatus* and *Attacus ricini*), and figures larva, puparium, and imago.

He follows with original notes on two girdling beetles, *Cœlosterna scabrata* and *Neocerambyx holosericeus*. The former (allied to *Oncideres cingulatus* Say) affect Sal saplings; while *Plocederus pedestris* is found boring in Sal and Jungham, and its larva forms a calcareous egg-like case in which to pupate. A chrysomelid beetle, *Aulacophora abdominalis* G. and H., is destructive to *Cucurbitaceæ*—similar to some of our *Diabroticas* which also attack the Squash family.

Papilio erithonius Cramer produces a caterpillar in appearance like our orange dog, *Papilio cresphontes*, and like it is destructive to the Orange. He says:

> In sending them Mr. Gollan notices that the insect does much damage to young budded oranges, not a plant of which could be raised if boys were not kept to pick off the caterpillars.

A cut-worm, *Agrotis suffusa* (?), often does considerable injury to the young opium poppy, while our well-known Boll Worm, *Heliothis armigera*, is an established pest of the plant. Mr. Cotes says it was described by Mr. John Scott, in his opium report, as *Mamestra papaverorum*.

A brief note is given on *Cecidomyia oryzæ*, a fly allied to our "Hessian-fly," likely to become a serious pest to the rice plant.

Article XI treats of Insecticides, and extracts from some experiments with London purple, made by Mr. Gallan, superintendent of the Government Botanical Gardens, are given. It proved unsuccessful with a beetle on cucumbers, but a complete success in destroying a leaf-hopper, *Idiocerus* sp., on mango trees and a caterpillar on young orange trees. We are pleased to see that the kerosene emulsion, which we have so strongly recommended for the purpose, has been tried on the coffee scale, *Lecanium viride*, and proved eminently successful. Mr. Cotes says:

From Mr. R. H. Morris's experiments, carried out last year in the Nilgiris, there seemed every probability that kerosene emulsion could be effectively employed against the pest, and information has now been received of its having been successfully used in Ceylon over a sufficiently large area to test its practical applicability.

Several pages are then devoted to the life histories of scale insects found on coffee, *Lecanium viride*, *L. coffeæ*, and *L. nigrum*.

The publication terminates with a few notes on Rhynchota by Mr. E. T. Atkinson.

Mr. Tryon's Report on the Insect and Fungus Pests of Queensland.—We have just received from the Under Secretary for Agriculture of Queensland, Australia, a valuable addition to the knowledge of economic entomology and botany of that region in a "Report on Insect and Fungus Pests, No. 1 (1889) by Henry Tryon, Assistant Curator of the Queensland Museum." The work is a pamphlet of 238 pages, and is illustrated with 4 plates showing spraying apparatus. It is to be regretted that no illustrations are given of the pests treated of, and also that the work lacks a good index. It is carefully written, however, and the matter is excellently classified and arranged so that it will be a practical hand-book of the subjects embraced, for orchardists and fruit-growers as well as working entomologists.

The author first treats the subject in a general way—discussing the relation of soil, state of cultivation and drainage to the increase of insect and fungus pests; the introduction and dissemination of pests, and the necessity of discriminating between friends and foes among insects, together with the protection of insectivorous birds, of which a list is given in an appendix.

A classified list of the fruits and cultivated plants of the Toowoomba district follows with a statement in connection with each plant of the principal insects and fungi infesting it. Each plant is afterward taken up in order and its various pests discussed at more length.

Two appendices are added, one relating to insecticide apparatus in which the Riley Nozzle together with certain compound forms is described and figured, and the other being the list of birds already referred to. The author displays a thorough familiarity with the writings of American and European entomologists, and in the discussion of many of the cosmopolitan insect pests, or those that are rapidly becoming so, he has quoted largely from the sources named. The similarity of the insect pests of the Toowoomba district with those of America and Europe enables him frequently to use the writings relating to the closely allied species of older countries. Much of the matter is, however, new, and indicates considerable original investigation on the part of the author.

In this connection we will call attention to the very full account of the Fruit-fly, *Tephritis* sp., an insect closely allied and of similar habits to our *Trypeta pomonella*, but much more injurious and apparently the most serious fruit pest of the district. It infests not only the Apple and allied fruits but also the various stone and citrus fruits.

In connection with the excellent account of the Cottony Cushion-scale, a recognizable description without name is given of a hymenopterous parasite. This is the first published reference to a hymenopterous parasite of Icerya in Australia, and we have no difficulty in connecting the description with a species recently sent us by Mr. Crawford and which we described in the last number of INSECT LIFE as *Ophelosia crawfordi*.

These and other interesting features of the work which might be pointed out will give it a value to all engaged in entomological work.

SOME INSECT PESTS OF THE HOUSEHOLD.

C. V. RILEY.

[*Continued from page* 215.]

IV.—COCKROACHES.*

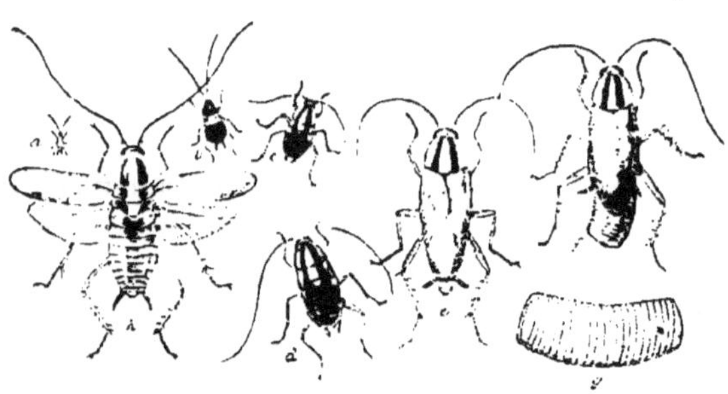

FIG. 57.—The Croton Bug or German Cockroach (*Phyllodromia germanica*): *a*, first stage; *b*, second stage; *c*, third stage; *d*, fourth stage; *e*, adult; *f*, adult female with egg-case; *g*, egg-case—enlarged; *h*, adult with wings spread—all natural size except *g*. (After Riley.)

The cockroaches which commonly annoy the American housekeeper comprise three species, one only of which is indigenous, and this the least harmful of the three. It is the "roach" or "black-beetle" of New England, and is known to science as *Periplaneta americana*. It measures from an inch and a quarter to an inch and three-quarters in length. Its thorax is yellowish with brown mottlings and its antennæ are exceptionally long, reaching considerably beyond the tips of the closed wings,

* Reprinted substantially from *Good Housekeeping*, June 8, 1889.

which themselves are long and powerful and, when closed, reach beyond the tip of the abdomen. The species flies freely in the open air, but when it has once become comfortably domiciled in a kitchen or other favorable location it shows little disposition to use the wings, and, whenever suprised in its nocturnal foraging by sudden light of gas or candle, is content to scramble away on foot—frightened itself, yet too often frightening the overtimid and nervous.

The other two species have been introduced into this country from Europe, and indeed have been carried all over the world in ships, in which they particularly thrive, rendering even large vessels on long tropical journeys almost uninhabitable to fastidious persons. This is particularly true of the larger of the two, which is commonly called "the Oriental cockroach" (*Periplaneta orientalis*). This species is nearly black in color, and is not so large as the American roach, seldom reaching an inch in length. Its wings are also much shorter, not quite reaching to the tip of the abdomen. Its uniform, very dark mahogany color, is unmottled with yellow and its antennæ are relatively shorter than in the former species. It flies well, but not so strongly as *americana*. It swarms in enormous numbers in the holds of vessels, in basement kitchens, and in all dirty, damp places the world over, and is the most noisome and thoroughly disagreeble of all our household pests. A visit at nightfall to a badly infested room is by no means a pleasant experience, even to those not troubled with delicate nerves.

The third species is popularly known all over the country as the "Croton bug," although more properly it might be called the "German cockroach," for its scientific name is *Phyllodromia germanica*. It is also a European species and derives its common name from the fact that its first appearance in force in this country was synchronous with the completion of the Croton system of water-works in New York City. It had in all probability been brought over many years before, but had remained comparatively unnoticed until the extension of the water-works, with their numerous pipes in all residences and places of business, encouraged rapid spread and multiplication; for this species is more fond of water than the other two mentioned, and is often carried by pressure through water-pipes without injury.

The Croton bug is the most prominent cockroach in America to-day, and really does the most damage. It is enormously fecund, and its small size enables it to hide and breed in cracks into which the Oriental or American roaches could hardly push their front feet. When full-grown it never exceeds five-eighths of an inch from the front of the head to the tip of the closed wings, and it is much lighter in color than either of the others. Its color varies considerably, but it is usually of a very light brown with two darker longitudinal stripes on the thorax.

It is this species which I have chosen to figure in detail on account of its greater abundance and powers of destruction and from the fact that it occurs very numerously in northern localities where the other species are seldom seen. Its transformations as shown in the figure will, however, represent in some degree those of the other species. All are closely related and probably pass through the same number of molts, the different stages repeating each other with comparative accuracy in the different species. At Fig. 57 the stages are shown lettered progressively from *a* to *h*. It will be noticed that none of these insects are winged until they cast their skin for the last time and the descriptive remarks which have preceded refer only to the full-grown insects. In point of color, however, they are moderately uniform, except that the newly hatched roaches are very pale—the Croton bug is nearly white—while all are of the same pale hue just after they have cast a skin.

The length of life of none of these species is accurately known, but as with other insects mentioned in this series of articles it doubtless depends largely on food-supply and temperature. They are all nearly omnivorous, but have at the same time preferences in diet. They seem on the whole to prefer animal matter to vegetable, but will eat after all kinds of cooks—good, bad, or indifferent. Almost everything which goes on the table is relished by them.

In the latitude of Washington and further south the Croton bug eats everything which contains paste, and, consequently, wall-paper, photographs, and especially certain kinds of cloth book-bindings suffer severely from their attacks. In a recent number of INSECT LIFE (Vol. I, p. 67) will be found an account of severe injury done to certain of the important files in the Treasury Department in Washington, the bindings of many important public documents being disfigured and destroyed. In the office of the United States Coast and Geodetic Survey they have become an intolerable nuisance by eating off the surface and particularly the blue and red paint from drawings of important maps.

But I need not elaborate further upon the damage which they do. How to kill them and prevent this damage is the question.

Without condemning other useful measures or remedies like borax, I would repeat here what I have already urged in these columns, viz, that in the free and persistent use of California Buhach, or some other fresh and reliable brand of Pyrethrum or Persian Insect Powder, we have the most satisfactory means of dealing with this and the other roaches mentioned.

Just before nightfall go into the infested rooms and puff it into all crevices, under base-boards, into the drawers and cracks of old furniture—in fact wherever there is a crack—and in the morning the floor will be covered with dead and dying or demoralized and paralyzed roaches, which may easily be swept up or otherwise collected and burned. With cleanliness and persistency in these methods the pest may be substantially driven out of a house, and should never be allowed to get full possession by immigrants from without.

For no other insect have so many quack remedies been urged and are so many newspaper remedies published. Many of them have their good points, but the majority are worthless. In fact, rather than put faith in half of those which have been published it were better to rely on the recipe which T. A. Janvier gives in his charming article on "Mexican Superstitions and Folk-lore," published in a recent number of *Scribner's Magazine* (March, 1889, Vol. V, No. 3, p. 350), as current among the Mexicans:

To get rid of cockroaches—Catch three and put them in a bottle, and so carry them to where two roads cross. Here hold the bottle upside down, and as they fall out repeat aloud three *credos*. Then all the cockroaches in the house from which these three came will go away!

TWO SPIDER-EGG PARASITES.

By L. O. Howard.

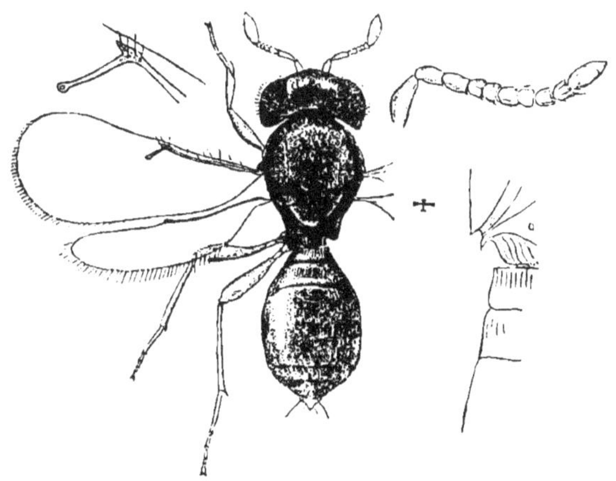

Fig. 58.—*Acoloides saitidis* Howard; female, showing wing veins—greatly enlarged; male antennæ and thorax from side—still more enlarged (original).

Following up the notes published from time to time in these pages on the subject of the hymenopterous parasites of spiders, I present below a description of two interesting new Proctotrupids of the subfamily Scelioninæ, the first of which was reared by Mr. L. Bruner at Lincoln, Nebr., from the eggs of the Araneid *Saitis pulex*. The eggs of this spider are a little more than a millimeter in circumference, and each egg harbors but one parasite, which issues by splitting the egg open rather than by gnawing a regular hole.

ACOLOIDES* n. g. (*Scelioninæ*).

Female antennæ with very large non-jointed club, and 4-jointed funicle. Male antennæ 12-jointed, submoniliform; club small, separable into three joints. Mandibles 3-dentate. Eyes hairy. Lateral ocelli situated on the eye margin. Mesoscutum

* Acolus + εἶδος.

without parapsidal sutures; mesoscutellum distinctly separated. Wings present. Submarginal vein reaching nearly to costa; marginal and postmarginal both exceedingly short; stigmal long, slender. Abdomen short, oval; first and second joints short, abdomen broadening rapidly from first joint; third joint very large; fourth and fifth visible.

It agrees with the points mentioned in the very insufficient characterization of Foerster's genus *Acolus*, except that it is winged. Foerster, however, knew only the female, and only mentions the fact that the antennal club is not jointed, and that the scutellum is developed, while the wings are absent or rudimentary.

Acoloides saitidis, n. sp.

Female.—Length, 1.4^{mm}; expanse, 3.6^{mm}; greatest width of fore-wing, 0.46^{mm}. Antennæ short; pedicel long, nearly one-half the length of scape; joint 1 of funicle one-half as long as pedicel; joints 2, 3, and 4 very short; club very large, oval, and one-third longer than four preceding joints together, but not quite as long as these joints and pedicel together; no articulations can be distinguished, but it is homologically composed of six joints. Eyes hairy; lateral ocelli touching the eye margin. Head, face, and mesonotum densely and finely punctate; parapsidal furrows not present; first and second abdominal segments with fine, close, longitudinal striæ, wanting at smooth posterior border; the very large third segment and short fourth densely and finely punctate, and clothed irregularly with short, whitish pile, which is also present, although sparser, upon the mesonotum, and is quite thick on the vertex; mesopleura finely punctate below; metapleura smooth. The marginal vein is very short and not quite coincident with costa; the post marginal is extremely short; the stigmal is long and slender and terminated by a small rounded knob. General color, deep black; all legs and antennæ honey yellow; all coxæ black, lighter at tips; scape brownish and pedicel darker than club.

Male.—Differs from female only in antennæ which are plainly 12-jointed; joint 1 of funicle as long as pedicel, joints 2 to 7 subequal in length and width, and each as broad as long and well separated; club oval, nearly as long as three preceding joints together. Antennæ uniformly honey yellow.

Described from 9 male, and 1 female specimens.

Genus BÆUS.

Minute wingless *Scelioninæ*, without differentiated scutellum and with non-jointed antennal club.

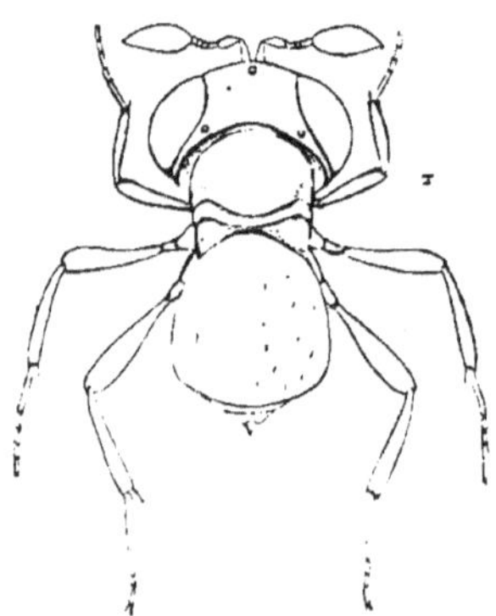

FIG. 59.—*Bæus americanus*. Female—greatly enlarged (original).

Bæus americanus n. sp.

Female.—Length 0.65^{mm}. Length of antennal club .185^{mm}, or in other words the entire body is only three and one-half times as long as the antennal club. Width of antennal club .082^{mm}. General color dark honey-yellow; scape and funicle of antennæ brownish, club lighter, dark at tip; vertex and face light honey-yellow; dorsum of thorax and abdomen dark honey-yellow, almost approaching mahogany; legs throughout concolorous with head; middle and hind tibiæ a little darker near base. Surface of abdomen smooth, shiny; mesonotum very faintly punctate. Thorax and abdomen with extremely fine, sparse, whitish pile; tip of abdomen with a short and contracted fringe of white pile. Antennal club very large, longer than rest of funicle and pedicel together; funicle joints very narrow and short, subequal, pedicel wider and as long as entire funicle except club.

This rather uncharacteristic description is drawn up from three poorly mounted and mutilated female specimens given me ten years ago by Dr. Marx, who I think received them from Col. Nicolas Pike, of Brooklyn, N. Y. They are labeled "Parasites in spider eggs in orange cocoon, collected 1871." After an examination of the eggs, Dr. Marx tells me that nothing can be said with certainty regarding the host except that it belonged to the family *Epeiridæ*.

No species of *Bæus* has yet been described in this country, although Mr. Pergande and myself have collected two or three undescribed species which are deposited in the National Museum collection. But one species is known in Europe—*B. seminulum* Haliday, but as I know of no recognizable description of it the present species is given a new name.

ON THE PARASITIC CASTRATION OF TYPHLOCYBA BY THE LARVA OF A HYMENOPTER (*Aphelopus melaleucus* DALM.), AND THAT OF A DIPTER (*Ateleneura spuria* MEIG.).

By M. A. GIARD.*

The larvæ of the Hymenopterous and Dipterous parasites of *Typhlocyba*, which I have described in a former communication,† belong: the first to *Aphelopus melaleucus* Dalman, the second to *Ateleneura spuria* Meig. (*A. velutina* Macq.; *Chalarus spurius* Schiner).

I have bred in captivity these two insects which have, as also their hosts, *Typhlocyba*, two yearly generations. The first infests the nymphs during the latter half of June, hatching about July 1; the other infests, the second generation of *Typhlocyba*, transforming in the nymphs towards the end of September or in October, and probably passing the winter in that state to yield the perfect insect the following spring.

If one compares these observations with the facts formerly described by Perris (parasitism of *Dryinus pedestris* Dalm. on *Athysanus maritimus* Perris) and by J. Mik (parasitism of *Gonatopus pilosus* Thoms. on *Deltocephalus xanthoneurus* Fieb.), it becomes very probable that Proctotrupids of the family *Dryinidæ* are generally parasites of Homopters of the family *Jassidæ*.

And again, in comparing the results of our investigations with the old ideas of Boheman on the infesting of various leaf-hoppers by Dipterous larvæ, in particular, of *Cicadula virescens* Fall. (*Thamnotettix sulphurella* Zett.) by the larva of *Pipunculus fuscipes* Fall., it becomes equally probable that the Dipters of the family *Pipunculidæ* are in general parasites of Homopters of the family *Jassidæ*.

---

*Translated from *Comptes rendus*, Nov. 4, 1889 (Vol. cix, No. 19, pp. 708–710).

†See *Comptes rendus*, July 8, 1889.

We have been able to procure in abundance and study more completely than has heretofore been done the parasites (Dipterous and Hymenopterous) of *Typhlocyba*, up to the present considered as very rare and captured here and there accidentally.

We have been drawn also to occupy ourselves with some very curious effects of parasitic castration produced by these parasites on their hosts.

Typhlocyba sp., with yellowish or whitish elytra, form a small group of species living often side by side on the same trees and presenting among themselves a mimicry so perfect that it is almost impossible to distinguish them even by a very careful examination of the external characters. To James Edwards, of Norwich, Eng., belongs the credit of having recently attracted the attention of entomologists to the very marked distinctive characters which one can draw from the form of the genital armature of the male to separate these diverse species.

Aided by the work of that acute investigator we have discovered that the *Typhlocyba* of the chestnut, described in our first note under the name of *T. rosæ* L., belongs in reality to two distinct species, viz, *T. hippocastani* J. Edw. and *T. douglasi* J. Edw., which are equally common on the trees of the Luxembourg.

These two species may be parasitized by *Aphelopus* and by *Ateleneura*. But *Aphelopus* infests especially *T. hippocastani* and much less often *T. douglasi*. *Ateleneura* is found, on the contrary, almost always in *T. douglasi* and very rarely in *T. hippocastani*.

The females of *T. hippocastani* and *T. douglasi* are very difficult to distinguish. However, with *T. douglasi*, the ovipositor is more robust and presents only one curvature, while that of *T. hippocastani* is more slender and is doubly curved in the form of a cimeter. With individuals of both species parasitized by *Aphelopus*, the ovipositor is generally considerably reduced and incapable of puncturing. *Ateleneura* seems to have much less influence on the development of that organ.

The genital armature of the male presents some very salient distinctive characters. With *T. douglasi*, the penis is simple and the lateral pieces have the form of legs. The parasitic castration, whether by *Aphelopus* or by *Ateleneura*, induces but very slight modifications.

With *T. hippocastani*, the lateral pieces are slender, simple arcs, but the penis presents a very complex structure, being terminated by a very curious eight-branched fork.

With males parasitized by *Ateleneura*, and especially with those infested by *Aphelopus*, the penis suffers considerable reduction, having but six, four, or even but three branches. The specific characters are thus profoundly changed, and certain of these modified forms would be confounded on superficial examination with *T. roseæ* L. or *T. lethierryi* J. Edw.

Modifications not less great are observed in some singular organs of which the existence in the case of the males of *Typhlocyba* has not yet been noted, so far as I know, and of which the function is altogether enigmatic. They proceed from two invaginations of the exoderm of the ventral side of the first abdominal segment and extend like fingers of a glove to the tip of the fourth segment and sometimes even a little beyond. These organs seem to me homologous to the similar sound organs of male grasshoppers.

With the males of *T. douglasi* and *T. hippocastani* infested either with *Ateleneura* or *Aphelopus*, the ventral invaginations are much reduced, they do not reach in general the second segment of the abdomen and often exist only as two small gussets on the first segment.

Aphelopus melaleucus appears to be rather common; I have found it at Wimereux and in the woods of Meudon infesting *T. hippocastani* and *T. ulmi* L., which live frequently together on the Elms in company with *T. opaca* J. Edw.

In these localities the sac which incloses the larva instead of being yellow, as with the individuals coming from the Luxembourg Garden, is, ordinarily, black. This color is evidently protective to the numerous individuals living on *T. ulmi*, of which the abdomen is black; and it is possible that it is due to heredity in the case of the others. Perhaps, also, *Aphelopus* presents varieties with the various species of *Typhlocyba*, which it infests. It is known, in fact, that Walker has described fifteen different forms of that Hymenopter, and by certain characters the specimens which he has figured differ a little from those which we have studied. Thus it has been impossible for me to find the least trace of the disk cells of the superior wing which, it is true, Walker has represented as very rudimentary. I can affirm further that the palpi possess five joints only, instead of six, which Walker has attributed to them.

It is possible, also, that under the name of *Ateleneura spuria* two allied species of *Ateleneura* have been confused. The rearing of larvæ collected with various Homopters will make the solution of this question easy.

A POISONOUS SPIDER IN MADAGASCAR.

Rev. Paul Camboué, missionary of the Society of Jesus at Tananarive, Madagascar, has recently sent us two papers by himself, the one published in *Les Missions Catholiques*, April 2, 1886, and the other in the *Bulletin Mensuel de la Soc. Nat. d'Acclimatation* on the subject of the beneficial and noxious spiders of Madagascar. What interests us most in these papers is the portion concerning the *Menavody*, a species of *Latrodectus*, a genus which in Madagascar as well as everywhere else is reputed to be very dangerous and to give even fatal bites. He quotes

Dr. Vinson in "de Flacourt's History of the Great Island of Madagascar" concerning the danger of the bite of this spider, and adds his personal experience, which we may freely translate as follows:

I was very desirous of falling in with this terrible spider when, on February 27, 1885, one of our little day scholars of the College of Tamatave brought me a specimen which he had found, so he told me, under a barrel. The child, never doubting the effect attributed to the bite of the spider, had taken it simply in his hand, carried it home and put it in a bottle and had not been injured in the least. I noticed that in this specimen one of the points on the upper surface of the abdomen was red. Having by mistake thrown the spider into alcohol I quickly drew it out again and happily it was still living. Next day it changed its skin and after the molt the spots on the upper side of the abdomen were four in number. The first and third white and the second and fourth red. It died soon after the molt.

The 23d of April following the same scholar brought me two more living females of the *Menarody*. I put them into a jar and was able to continue my observations. On the 24th one of the spiders laid her eggs in a little spherical mass, protected by the white or slightly brownish spheroid cocoon, about a centimeter in diameter, and suspended by a slight web of whitish silk. I had noticed that in this individual the series of spots on the middle of the abdomen did not exist, but were replaced by four depressions, placed in the form of a trapezium, and of the same color as the abdomen. The spider in repose remained below the web in the cocoon. Two little grasshoppers and the other spider were captured in its web and became its prey. It did not devour the substance of its victims, but left their outer skin intact. On the 27th a large living beetle was given to the *Menarody;* it was three times as big as the spider and vigorously defended itself. The *Menarody* displayed all of its means of offense. As it spun its thread it gave out a whitish viscous liquid, which did, it seemed, not a little to help it capture its prey. The beetle died only after a considerable time. On the 29th the spider laid its eggs for the second time. Its cocoon was like the former one. It rested between the two cocoons.

On the 4th of May another cocoon was produced. It then died, and on the 9th I found it at the bottom of the jar.

On the 27th of the same month of May, 1885, on lifting the bark of a large tree, I found several cocoons of the *Menarody*. The eggs from one of these cocoons hatched on June 12. On leaving the eggs the young are of a pale reddish color and the legs are brown. Fifteen days afterwards, on the approach of the first molt, this reddish tint grows darker, particularly on the abdomen. After the first molt, which takes place July 1, the spiders' bodies and the abdomen appear brownish. About the 20th of July a second molt took place. The young spiders killed each other, and there soon remained but two specimens in the jar, the male and the female. Wise disposition of the providence of the Creator and the Ruler of the Universe who thus prevents these venomous insects from multiplying without measure!

August 3, third molt. The red color of the triangular spot becomes more accentuated. The band upon the abdomen is of a slightly reddish white, the six lateral dots are white, those upon the middle of the back are four in number, three reddish white and the fourth white. The cephalothorax, abdomen, and legs have become of a darker color.

Upon the 7th I noticed that the male has become the prey of the female, who has killed him and enveloped him with her web. I continued my observations upon the latter.

August 15, fourth molt; 27, fifth molt. August 31, for the first time, I observed that she cleaned her nest and removed all the bodies of the prey.

September 15, sixth molt; September 26 two of the red spots in the middle of the abdomen, 3 and 4, disappeared.

October 11 she died.

I agree with an ancient writer on the subject of venom of this species: *

"Have spiders venom? Yes, they possess it, but its action is relative to the animal attacked. A fly pierced by a larger spider perishes in a few moments; other insects die more slowly, acording to their size; but a man bitten by a spider, even a large one around Paris, would not be hurt perhaps any more than by the bite of a gnat. In southern climates, however, where these creatures are larger, their wounds can be more serious. They appear to bring about local inflammations which, if the subject is healthy, have no serious consequences, but if the person is predisposed to the action of poison, if he neglects to take care of himself, the heat of the climate will bring more or less grave results, which in certain cases can bring about death."

In this way, upon the shores of Madagascar where the temperature is warm the bite of the *Menavody* is reported as more serious than in the interior of the island where the climate is cooler. In no place, however, does it seem to have more dangerous effects than that of other venomous insects, such as the *Scolopendra*. This opinion is confirmed by information which I have collected from several competent natives. It is related that the Marechal de Saxe was obliged to stop at a tavern where they had only one unoccupied bed, in which all of the travelers who had dared to sleep had died without the cause having ever been ascertained. The Marechal, notwithstanding, took possession of the fatal bed and made his servants sit at the side. Then at the end of some moments they were astonished and frightened to see their master grow pale and appear as if about to die, without seeing anything. In trying to revive him they saw upon his breast a large black spider which was sucking his blood, and which caused the death of the Marechal.

It is, if I do not deceive myself, with our *Fancoho* and *Menavody* as with the black spider of the Marechal de Saxe—the terrible effects of its bite exist only in legendary lore.

* * * * * * *

Since the publication of these notes [M. Camboué writes us], I have heard from the east shore of the island that it is not the bite of the spider, but contact with the crushed body, which produces the inoculation of venom bringing about the gravest symptoms with man and even with the Zebu. I hope later to be able to control the difficulty by inoculations upon different animals, such as poultry, rabbits, and sheep, and I will not fail to inform you of the result of my observations. Even now I believe that my conclusions in my notes are correct and that Latrodectus has without doubt a venom, but a venom whose noxious effect upon man varies with the crowd of circumstances (climate, temperament, etc).

EXTRACTS FROM CORRESPONDENCE.

Injury to Grass from Gastroidea polygoni.

In your reply to inquiry of N. R. Smithson, Winchester, Ill., on page 190, vol. 2, INSECT LIFE, you state that this species injures no crop and feeds solely on weeds of the genus *Polygonum*. While this is true as a rule, there are exceptions. I have observed both larvæ and adults feeding on what seemed to be a species of dock, the specific name of which I do not know, but can ascertain in the future by further observation.

On June 22, 1886, two of the beetles were observed feeding upon heads of timothy, apparently eating both the involucre and incipient seeds. While the species may not be injurious, it will certainly bear watching.—[F. M. Webster, Lafayette, Ind., January 17, 1890.

* Achille Percheron.

Resin Wash against Mealy Bug and Woolly Aphis.

My reason for not answering sooner your letter of January 2 (which was accompanied by report, and duly received) was occasioned through a desire on my part to thoroughly test and report correctly to you the results and effects of my experiments with resin wash upon the foliage of greenhouse plants. I have sprayed several delicate greenhouse plants with it, some of which had a considerable share of the mealy bug on. I have sprayed with from 1 to 12 and 16 per cent. and have seen no bad effects or any injury done to the foliage or plants from its use, while all the mealy bugs were entirely killed.

As to last year's experiments with it on Woolly Aphis and Plum Aphis, I can only say that it killed both, and I consider it a success when properly made and mixed and thoroughly applied with a fine spray.

There is one point to be observed: It should be applied early in the season, that is, as soon as the Woolly Aphis makes its appearance and before the leaves begin to turn yellow, which is caused by the Aphis destroying or checking the vital power that goes to nourish and sustain the leaves and causes them to drop, and which many people believe to be the cause of the wash.—[E. K. McLennan, Berkeley, Cal., February 13, 1888, to Mr. Koebele.

Dryocampa rubicunda.

I send with this some "worms" that are like the locusts of Egypt and "fill the houses." There were a good many last year, but this year they are innumerable. This is the second crop this season, and there was a white miller this spring in great numbers which I suspect to be the "mother of them all." The worms seem to eat nothing but the maples. I have hunted through such reports as I have but can not find out about it. I would like to know what it is and what we can do about it.—[Mrs. Mary T. McCluney, 214 East Sixth street, Sedalia, Mo., September 10, 1888.

REPLY.—The worms belong to the species *Dryocampa* (*Anisota*) *rubicunda*, which is popularly known as the Green-striped Maple-worm. These worms at times are very destructive to the Soft and Silvery Maples. The perfect insect varies somewhat in coloration according to locality. In the west it is nearly all a pale yellow color, with a very faint tinge of rose. The eastern individuals have the rose color quite intense on the front wings and generally a rosy band across the hind wings. In Missouri there are two broods of the insect in a year. In regard to remedies, there is no practical way of destroying them. The worms hold to the tree tenaciously and are not easily jarred down; and before entering the ground they scatter to great distances, so that they could not be found and destroyed while in the chrysalis state. However, this insect is seldom so exceedingly abundant two years in succession. The only directions that can be given to counteract its injuries are to keep close watch for the moths and eggs during the latter part of May, when large numbers of these may be destroyed, and to entrap the worms when they are about to leave the trees by digging a trench around the individual tree or around a grove of trees so affected. This trench should be at least a foot deep, with the outer walls slanting under, in which great numbers of the worms will collect and may easily be killed.—[September 19, 1888.]

Combined Spraying for Bark-lice and Codling Moth.

Having this day sprayed the apple-orchard of Rev. J. S. Fisher, of this place, with an emulsion according to your formula in letter to him of April 16th, I write you to report.

I would say that using one-half common soap, 2 gallons kerosene, and 28 gallons water, I sprayed about 60 trees, and 30 more were sprayed with the same proportion, but using sealed kerosene, such as he had in his can and at his wish to save time, into which was put 12 gallons diluted emulsion and one-fourth pound London purple, thinking to destroy eggs and larvæ of codling moth at same time. The season has been so very late here that apples are only well formed, and I even saw some blossoms on late varieties. We had no apple blossoms for "memorial decorations." Having no microscope at hand I could not tell whether it was just the day to destroy the bark-lice, for the scales seemed to be still fast adhering to twigs. I have other years seen the young lice like yellow dots crawling out on the new wood, but did not see any yesterday.—[J. W. Van Deman, Benzonia, Mich., June 20, 1888.

Greenhouse Pests.

I send you some worms, and one pupa of same (I think), which feed on almost any soft-wooded greenhouse plants; also some flea beetles which feed on Fuschias. I do not think they feed on anything else; at least they do not with us. Both are very destructive, and so far nothing but hand-picking will destroy them. Can you tell me what they are and suggest any remedy for them?—[E. S. Miller, Wading River, Long Island, September 12, 1888.

REPLY.—The larvæ sent are those of *Botis harveyana*. This is a pyralid which has long been known to feed upon various greenhouse plants. The flea-beetles are *Graptodera exapta*. In case these insects are not very abundant, hand-picking will of course be the best remedy. If they should become very numerous an application of an arsenical solution may be made to the plants.—[September 14, 1888.]

Euphoria damaging green Corn.

I send you by to-day's mail three beetles that were found in an ear of sweet corn under the husks, eating the kernels of corn; there were four in the ear, but one got away. They had eaten the ear most all up. I would like to know to what family they belong, and whether they are an old or new enemy to the corn crop.—[Eugene O. Wheelock, Brooklyn, Wis., September 10, 1888.

REPLY.—The beetles belong to a common species, *Euphoria inda*. This species has long been known to attack injured fruit, and is often found congregating in numbers upon injured parts of trees feeding upon the sap. They have not been known to attack sound fruit to our knowledge. It is quite probable that the ear of corn in which you found them had been injured previously by birds or some other agency. We shall be very glad to have you investigate the matter and see whether they attack the corn *before* it has been injured; if so this will prove a new habit. These beetles belong to the same family as the June Beetle and the Rose Chafers.—[September 14, 1888.]

The Indian-meal Moth in Kansas.

I inclose herewith specimens of worms infesting our mill, which, in view of reports in milling journals, have given us some uneasiness. We have not noticed any moth likely to be the parent. The white worm seems to spin for itself a cocoon and pass from that into some other stage. Some of the cocoons have remaining in them a brown shell and we find among the cocoons a brown worm also, some of which are inclosed. We think the pest—whatever it is—came to us in a can of corn purchased in the county east of us (Clark), and as yet is confined to the wareroom containing the corn. The white worm seeks hiding places in folds of sacks and crevices of walls, and there makes its cocoon. The first notice of them was a continuous web spread all over the heap of shelled corn with no worms in this web, but bunches of grains webbed together containing cocoons, and on further search we found them as above mentioned. Please tell me what they are, and if liable to become a serious pest, give remedy if you can.—[J. P. Craig, Memphis, Mo., December 18, 1889.

REPLY.—Your letter of December 18 with specimens came safely. The insect which is infesting your mill is a rather serious pest and is known ordinarily as the

Indian-meal moth (*Ephestia interpunctella*). This is the adult of the white worm which spins the cocoon. The brown worm is the larva of a small beetle known as *Attagenus megatoma* and feeds ordinarily upon dead animal matter. It is probably beneficial in your mill rather than injurious. A larva very closely allied to the one which is troubling you has recently appeared in Canada and is the subject of an article in the last number of INSECT LIFE, the periodical bulletin of this Division, a copy of which I send you by accompanying mail. Your insect is referred to on pages 170 and 171. If the insect appears to be confined to your ware room I would advise energetic treatment to rid your establishment of it. The infested corn should be burnt and the entire room should be thoroughly sprayed with benzine or gasoline, the greatest care being taken to avoid fire, as both of these substances are inflammable and the vapor is explosive. Any further details concerning this matter we shall be glad to receive.—[January 9, 1890.]

A Cocoanut Pest to be guarded against.

Small shipments of cocoanuts leaving this port almost continually for the United States, and the possibility existing that some of these cocoanuts are used as seeds, I have, with much interest, watched the scientific observations made at Havana, Baracoa, and here, with the object of discovering the origin of the mysterious disease which is killing many cocoanut palms and at one time almost threatened to annihilate all the plantations producing cocoanuts for market and export. Opinions of scientists have differed as regards the cause and nature of the disease, Professor Ramos, of Havana, ascribing it to a fungus growth on the base of the leaves, which growth penetrates into the crown of the tree, withering and killing it. This theory was proved to be incorrect, and it is now definitely ascertained that the destroyer of the cocoanut tree is an insect of diminutive size, barely visible to the naked eye, the *Coccus* (*Diaspis*) *vandalicus* Galvez. Professor Gundlach, of Havana, at present here, recommends that all cocoanuts as soon as received in the United States be dipped into boiling water and that the bags they are shipped in be destroyed.—[Otto E. Reimer, Consul, United States Consulate, Santiago de Cuba, December 6, 1889, to Hon. Wm. F. Wharton, Assistant Secretary of State, and referred to this Division.

Food of the Scydmænidæ.

Is it commonly known what the food of the Coleopterous family Scydmænidæ consists of? Both Packard in his "Guide," and LeBaron in his Fourth Ill. Report, are silent on this subject. A few weeks ago I found quite a series of specimens of a *Scydmænus* near *brevicornis*, and eight or ten of them had each a brown mite in its jaws. I found these specimens clinging to the underside of stones lying on the ground near the edge of a small body of water, the ground being very damp. This would indicate that these insects are predaceous, at least in the adult stage.—[D. W. Coquillett, Los Angeles, Cal., January 1, 1890.

Abundance of Bryobia pratensis.

By to-day's mail I send you a vial containing some small insects which I wish to know how to destroy. I first noticed them three years ago last fall, when they were found on windows on the east and south sides of the house. They remained all winter and until May, I think. After that time no signs of them were seen. We thought they had gone for good, but in the fall they came again and remained all winter as before. They are here to-day. They come in at the doors and windows and get on the furniture. I have tried almost everything to drive them away, viz: Carbolic acid, corrosive sublimate dissolved in benzine, insect powder, tobacco, salt, gasoline coal-oil, onion juice. Oil or grease will kill them if it gets on them, but nothing will keep them away that I have tried. They are hatching now. In the vial you will find some of full size, and also some small ones. I wish to know what they are, where they came from, and what they live on. I may add that in the spring the grass is nearly covered with them close to the house. Are they an insect that will disappear bye and bye and stay away? Is there anything that will drive them away? We live in a town of some four thousand inhabitants. I saw one of these insects on a house in town this winter. The first part of May last I saw one on a house 35 miles from here.—[L. H. Ellis, Wilmington, Ohio, December 28, 1889.

REPLY.—Your letter of the 28th ult., addressed to the Smithsonian Institution, has been referred to this Division for reply. The creature which you send is a mite known as *Bryobia pratensis*. It feeds through the summer upon clover and grass and in some places has acquired the habit of migrating to houses in the fall. A number of cases similar to yours have come to our attention within the last two or three years. I know of nothing that will prevent them from entering houses, but after they are in I should say that they could be readily killed with any oily substance. Probably the best thing you can do is to spray the room which is infested with benzine from an atomizer, taking great care with this substance on account of its extreme inflammability. This substance is recommended not only from its insecticide qualities, but on account of the fact that it will evaporate readily and a thorough airing will destroy the odor. It may be well also in the fall, just before the mites begin to appear in the house, to spray the margins of the windows and doors with kerosene, or the grass in the immediate neighborhood of the house may be sprayed.—[January 21, 1890.]

Larval Habits of Xyleborus dispar.

During last autumn the *Xyleborus dispar* appeared very injuriously at Toddington, but since then, to my great regret, I find it has been ravaging unchecked at two or three other localities for a few years—but my present point is the (conjectural) food of the larvæ.

So far as I see I quite agree with Schmidberger that the larvæ feed in the large mother galleries, because in all the specimens I have dissected there are no side galleries, also because I find what I conjecture to be the larva of the *X. dispar* present, and because I find beetles fairly cramming up all the passages, some of these not yet fully colored.

But with regard to food, Schmidberger, in his long account given from minute successive daily examinations, notes that he considers that the larvæ feed on a white material prepared by the mother beetle; other observers have considered that the larvæ of one or more species very nearly allied to the *X. dispar* feed on a mold or fungus that grows in the tunnel.

Now, in my own specimens, I found a white growth which greatly resembled Mycelium of fungus in some of the *dispar* tunnels, and on procuring skilled examination (for I am not a fungoloist), to be made both by microscopic and test examination, it appears likely we shall find that the white material is partly Mycelium and partly white animal matter, thus reconciling the varying observations. At present our observations are quite incomplete for want of specimens, but I have written for some, and then we are going into the subject thoroughly. But meanwhile I thought that the observation, though unfinished, and not proved as yet, might be of some interest, or that what you know of the history in this point of our *dispar*, under your synonym of *pyri* (Peck) might throw some light on the habits of our very destructive pest.—[Eleanor A. Ormerod, St. Albans, England, January 6, 1890.

REPLY.—In regard to the paragraph in your letter of the 6th instant, referring to *Xyleborus dispar*, there is no longer any doubt that in a certain class of Scolytides, to which *X. dispar* belongs, there are no larval galleries, and that, therefore, the food of the larvæ necessarily differs from that of those species whose larvæ excavate galleries of their own. Besides *X. pyri*, which is doubtless a synonym, we have quite a number of allied species in North America, some of them still undescribed, which agree in mode of living, but the real food-habits of the larvæ have not yet been investigated here. In 1844 Th. Hartig had already stated that the "Ambrosia" of Schmidberger is nothing but a fungus which he called *Monila candida*, and that this fungus constitutes exclusively the food of the Xyleborus larva. Eichhoff, on the contrary, believes that the exuding sap, and not the fungus, is the food of the larva. If you can prove that the "Ambrosia" consists of Mycelium and animal matter, Schmidberger's explanation would be partially confirmed. Can you not send us authentic specimens of *dispar* in both sexes ?—[January 25, 1890.]

Since the above was written Miss Ormerod has kindly sent us British specimens of *Xyleborus dispar*, both males and females, and after a careful comparison with North American specimens of *X. pyri*, the males of which we possess through the kindness of Mr. Fletcher, we can only confirm the opinion expressed by other entomologists that the two are specifically identical. In other words, Peck's "Pear Scolytus," described in 1817, is an imported species, which was brought into this country (probably first to Massachusetts) early in the present or late in the past century. Until quite recently only the female beetle was known in this country,* but Dr. Lintner and Mr. Fletcher finally succeeded in finding the male, which in shape of body and other important characters strikingly differs from the female.

In Europe this beetle is known as one of the few really polyphagous Scolytids, since it not only attacks all sorts of deciduous forest trees, but also most of the cultivated fruit trees and even Conifers (see Eichhoff, Europ. Borkask., p. 269). In North America it has hitherto been observed only in various fruit trees (apple, apricot, plum, pear, according to Harris), but it doubtless also infests forest trees, for little attention is paid by our Coleopterists to the life habits of Scolytids, and there is difficulty in finding *in situ* those species which feed within the trunk.

It may now be considered a settled fact that in this and other Scolytids which enter the solid wood of trees, the galleries with all their ramifications are the work of the female parent-beetle, which deposits her eggs irregularly in these galleries. The larvæ are not lignivorous, but their food consists of the peculiar substance already alluded to above.

Insects from Iowa.

I send you in the same mail with this a few insects which I can not determine from the collections here. If you can, through the columns of INSECT LIFE, give me their names and any further information concerning them, I shall be greatly obliged.

Nos. 1 and 2 were reared in considerable numbers from the plum curculio, *Conotrachelus nenuphar*, No. 1 being far more common. I have no specimens of *Sigalphus curculionis* Riley, but these seem to differ from the description of that species in the number of the joints of the antennæ and in the position of the ocelli, at least.

No. 3 is a parasite upon the plum gouger, *Anthonomus scutellatus*. In every case where the work of this parasite has been noticed the larval gouger had prepared its place of exit from the plum pit. Otherwise the parasite could probably never escape. The specimen that I send was cut from a plum where it had eaten its way to the skin.

No. 4 were reared in large numbers early in the spring from the cocoons of *Orgyia leucostigma*.

No. 4^a are secondary parasites reared from No. 4.

No. 5 were reared from the galls of *Rhodites radicum*.

No. 6. This parasite was quite common here this summer on *Meromyza americana*.

No. 7. Several of these flies have appeared in my breeding cages where cut-worms were being reared.

No. 8. This Tachina fly has been reared this summer from cut-worms and from the stalk-borer, *Gortina nitela*.

---

* It is certainly strange that Dr. Harris, who cut quite a number of the beetles from their galleries, never found a male specimen; at least he does not refer to any differences between the specimens found by him.

No. 9. July 5th a cornstalk was noticed to have a number of maggots burrowing down its center. The stalk was brought into the laboratory and twelve of these Dipterons reared from it.

No. 10. A Tineid moth that I have obtained in large numbers from breeding cages containing cut-worms. Can it be that the larvæ of this insect are parasitic upon the cut-worms, or do they live on clover with which the worms are fed?

No. 11. Gall and moth. A small bush of *Amorpha fruticosa* was noticed early in the spring to have one of these galls at the tip of nearly every twig. These galls were brought into the laboratory and the moths began to issue May 22.

No. 12. Three of these Ægerians were reared from a cluster of woody galls on a small limb of *Quercus rubra*. The galls were of last summer's growth and were gathered early in the spring. Aside from the moths nothing but a number of guest gall-flies, *Inquilinæ*, were reared.

No. 13. Dipterons reared from maggots that were mining the leaves of the common "pig-weed," *Chenopodium album*.

Nos. 14 and 15. Reared abundantly from plum twigs that were covered with Aphides.—[C. P. Gillette, Ames, Iowa, August 28, 1889.

REPLY.—List of species referred to in Mr. Gillette's letter of August 28, 1889:

1 and 2. *Sigalphus curculionis* Fitch.
3. *Sigalphus canadensis* Prov.
4. *Pimpla inquisitor* Say.
4a. *Pteromalid*, probably undescribed.
5. *Orthopelma occidentalis* Ashm.
6. *Cœlinius meromyzæ* Forbes.
7. *Anthrax scrobiculata* (?) Loew.
8. *Tachina* sp.
9. *Chætopsis ænea* Wied.
10. *Gelechia* sp.
11. *Walshia amorphella* Clem. and its gall on *Amorpha fruticosa*.
12. *Ægeria nicotiana* H. Edw.
13. *Anthomyia* near *calopteni*.
14. *Scymnus cervicalis* Muls.
15. *Leucopis* n. sp. (?).

There is an immense amount of descriptive work yet to be done in the Pteromalidæ and the Tachinidæ, so that it is impossible at present to identify the majority of the species in these families. It is not at all probable that the little Gelechia, No. 10, is parasitic on the cut-worms, and Mr. Gillette's later surmise is doubtless the correct one.

A Grasshopper Letter from Utah.

I thought a few lines from the Farmers' and Gardeners' Club, of Nephi City, might be interesting to you. The farmers of this place have suffered considerable loss this year by the ravages of the grasshoppers, which came in untold millions and ate every green thing before them. The whole force of the people had to turn out and do their very best to destroy them. The best mode that we found was to dig trenches about 3 feet deep and 2 feet wide, drive the hoppers in, put some straw on them, and then burn them up. It was supposed by this method that we destroyed not less than ten to twelve bushels each day for four or five days. After that there were enough left to do considerable damage to the remaining crops. Some of our farmers did not get as much seed as they put in the ground; some got about half a crop. Then came the very hot weather. The water in our irrigating ditches was not more than one-half as much as we have had in years past, the cause being very little snow in the mountains. Our main dependence, therefore, for crops, agriculture, and horticulture suffered greatly, excepting in some few cases. I have not seen the like in the last twenty-seven years, and I am sorry to say that the farmers have come out this season at the little end of the horn. I sent a specimen of the "hoppers" to Prof. Lawrence Bruner, of the Nebraska Agricultural Experiment Station, at Lincoln. He wrote me that they were of the kind that would stay by us; as they were not the migratory kind we would have to fight them to death. I think that the farmers must have been somewhat neglectful to give them such a start. The trench that I spoke of extended about two miles and a half, so you can judge of the labor that it took to accomplish the work. The apple crop was very light in this part; most of the fruit dropped to the ground before half matured, on the average about one-quarter of a crop. Peaches and plums, however, were in abundance and of the best quality. I never saw finer in these valleys. * * * —[James B. Darton, Nephi City, Utah, November 5, 1889.

Another Insect impressed in Paper.

I have received to-day an interesting pressed specimen of a Neuropterous insect with no other statement in reference to it than that it comes from you. Will you please give me some facts in reference to the specimen and how it came to be so completely pressed? The explanation of this particular example can not be the same as that given upon page 381 of Vol. I, INSECT LIFE, of a species of *Lithobius* that was sent from the Giles Lithographic and Liberty Printing Company, for that was evidently caught up in the surface substance of the paper while it was being manufactured.—[C. V. Riley, December 16, 1889, to Mr. N. O. Wilhelm, 25 Clinton Place, N. Y.

REPLY.—Your letter of December 16 is at hand. The specimen of a Neuropterous insect in a heavy manila paper is an interesting exhibition of the power of the paper machine in incorporating with the paper pulp, into the paper itself, the body, legs, and all except the parchment-like wings of an insect. The wings are quite free from the paper except at the point of union with its owner in life and are yet pressed to the common level. You see all parts of the insect can readily be seen. I think it was curiosity that led to this creature's untimely death. It was evidently facing the crushing rollers, for you see behind the long, tapering discolored band, evidently from the juices of its body. Not only this, but meeting its death through being curious and the numerous empty egg-shells in the surface of the paper persuade me it was a female.—[N. O. Wilhelm, 25 Clinton Place, New York City, December 20, 1889.

The "Katy-did" Call.

By careful observation of several years I have established the fact that the call of "Katydid" is made by the tree cricket. I have captured a number of specimens, and had witnesses who watched them. While making the sound the wings are held upright at right angles to the body, and the sound is made by moving the edges of the wings laterally. * * *—[LeRoy T. Weeks, Osborne, Kans., November 23, 1889, to Smithsonian Institution.

I have observed for several years that the common call of "Katydid" is made by the tree cricket, and that the so-called Katydid makes a continuous "Z" sound.

I have called the attention of many people to the fact. I have caught specimens and kept them in my room. I have reported to Prof. F. H. Snow, K. S. U., and shall report to-day to Harvard, Yale, and Smithsonian Institution.—* * *—[LeRoy T. Weeks, Osborne, Kans., November 23, 1889, to Dr. C. Hart Merriam.

REPLY.—Your letters of the 23d ultimo, addressed to the Smithsonian Institution and to the Ornithologist of the Department of Agriculture have both been referred to me for reply as to the portion referring to tree crickets. You have made a not unnatural mistake in considering that you have found that the insect which makes the Katydid cry is the tree cricket. You probably have not heard the true Katydid. The insect to which you refer which makes the sound not unlike that of the Katydid is *Œcanthus latipennis* Riley. The notes of the Katydids have been carefully studied by several entomologists, and you will find in my sixth Report on the Insects of Missouri, pages 150 to 169, a full account of my own observations, while I have treated of tree crickets in the fifth report of the same series, page 120, and in the general index to the same in Bulletin 6 of the U. S. Entomological Commission, page 163.—[December 4, 1889.]

Notes of the Season from Mississippi.

The cotton worm (*Aletia argillacea*, Hübn.): This worm made his first appearance on bottom land of large plantations in the latter part of July, but its injury was greatly diminished by the use of Paris green. It never appeared on upland farms till August, and in some localities not until September. The percentage of loss averages from 15 to 30 per cent. The late June planting tends to swell the percentage of injury, which was caused by severe drought during the latter part of April and all of May.

The boll or corn worm (*Heliothis armigera*, Hübn.): This worm did but slight damage to the cotton crop in this locality, but has been quite numerous on young corn plants, eating holes in the blades, during June.

The corn-plant louse (*Aphis maidis*): Observed during the summer in large groups on corn and sorghum plants.

The corn-root worm (*Diabrotica 12-punctata*): The larva of the above injured the stand of corn very seriously during April and May.

The cabbage plusia (*Plusia brassicæ*, Riley): Very numerous and destructive in gardens in this locality.

The cabbage pionea (*Pionea rimosalis*, Guenée): This garden pest has been very damaging to the entire cabbage family, generally feeding on the tender leaves surrounding the heart.

The cabbage-plant louse (*Aphis brassicæ*, Linn): Found on a good many plants of the cabbage family in vast groups.

White ants or wood-lice (*Termes flavipes*, K.): Have noticed these insects destroying collard-stalks and turnip-roots by gradually eating out the interior.

Proconia undata: Captured several specimens feeding on cabbage during June.

The bean cut-worm (*Telesilla cinereola*, Guenée): Feeding on bean-pods, doing considerable damage to the bean crop.

The squash-vine borer (*Melittia ceto*, Westw.): Quite numerous, boring the vines of cucumbers, squashes, and cashaws.

The squash bug (*Anasa tristis*, De Geer): One of the most injurious insects known in this locality to most all cucurbitaceous vines, especially squash and pumpkins.

The squash borer (*Endioptis nitidalis*, Cramer): Have noticed this worm boring holes into squashes, cucumbers, melons, and cashaws, feeding on the fleshy pulp, which generally causes rot and decay.

The granulated cut-worm (Larva of *Agrotis annexa*, Treitshke): This larva has been very destructive to most all garden vegetables, also very damaging to young cotton plants.

The shagreened cut-worm (Larva of *Agrotis malepida*, Guen.): Have captured this larva feeding upon cabbage plants and likewise on young cotton plants.

The May-Beetle (*Lachnosterna hirticula*): This beetle has been quite numerous and damaging to the foliage of several forest trees during the past summer.

The tomato worm (*Sphinx carolina*, Linn): Very common on tomato plants, also found them this season on tobacco and pepper plants.—[G. H. Kent, Roxie, Miss.

STEPS TOWARDS A REVISION OF CHAMBERS' INDEX, WITH NOTES AND DESCRIPTIONS OF NEW SPECIES.

By Lord Walsingham.

[*Continued from* p. 155.]

Adela flamensella Chamb.

=*lactimaculella* Wlsm.

This species was originally described from a very bad specimen with antennæ and palpi broken off and therefore presumably with the wing more or less worn.

Imperfect specimens of *lactimaculella*, female, in my own collection agree with the description in having no markings, except a minute whitish spot at the beginning of the costal cilia. The saffron head of the female is also characteristic. Good specimens of the male (which has a black head), show three distinct spots, two costal and one intermediate and dorsal.

Adela simpliciella Wlsm.

A unicolorous species allied to *rufimitrella* Scop. and *violella* Tr. It can not be confused with any North American species, being much smaller than *bella* Chamb.

A very small form apparently undistinguishable from this species occurs in Texas.

Adela punctiferella sp. n.

Antennæ, ♀, 13^{mm} long, whitish tinged with fuscous towards the base.

Palpi, roughly clothed, hoary; the naked apical joint slightly tinged with purple above.

Head and face, roughly clothed, hoary.

Thorax, greenish bronze.

Fore-wings, greenish-bronze, with a small indistinct whitish spot at the end of the cell, a little above the middle of the wing; cilia shading from greenish-bronze to greyish at their tips.

Hind-wings, deep violet, with greenish-brown margins; cilia as in the fore-wings.

Underside of both pairs of wings, violet, sprinkled outwardly with greenish-bronzy scales.

Abdomen, fuscous, hoary beneath.

Exp. al., 10^{mm}.

Hab., Los Angeles, Cal.

Type, ♀, *Mus. Wlsm.*

I am indebted to Dr. Riley for the specimen from which this small but distinct species is described.

Adela bellella Wlk.

= *degeerella* Emmons (*nec* L.).

Walker describes this species as closely allied to *degeerella* L. and I mentioned (P. Z. S., 1880, 78) that it differed from that species "only in the richer coloring and in the darker purple hind wings. The longitudinal stripes before and beyond the central band, as well as the margins of the band itself, are very distinct and of a brilliant shot purple-blue, whereas these and the central band itself are paler in the European species."

Specimens received from Japan are apparently undistinguishable from this species as represented by Walker's type in the British Museum, but without a careful study of the numerous degrees of variation in the many allied Asiatic forms, of which I have a large number of specimens, it would be unsafe to attempt to define its geographical range.

Adela singulella Wlsm.

This species differs from *sulzella* Schiff. in its smaller size, narrower fascia, and in having the antennæ of the female similar to those of the male instead of being thickened to the middle. It has a single narrow fascia on a plain bronzy ground.

Adela septentrionella Wlsm.

This species belongs to the group of which the heads of the male are black and of the female yellowish. It has much the appearance of *trigrapha* Z., in the male sex only, but is smaller, and possesses no third transverse fascia, this being indicated only by a costal spot; moreover the eyes of the male are set much wider apart than in *trigrapha*, and in this respect approaches the genus *Nemotois* Hb. It may be desirable to recognize this genus as occurring in North America, but I prefer to leave this point until the publication of a finally revised index.

Adela purpurea Wlk.

=*biriella* Z.

This very distinct species with its broad post-median white fascia on a bronzy ground, followed by a less conspicuous ante-apical transverse streak, appears to occur only in the northernmost parts of the United States. It is abundantly distinct from all other species.

Adela ridingsella Clem.

= *Dicte corruscifasciella* Chamb.
= *Adela schlægeri* Z.

This species is quite distinct from all European forms, from which the group of black scales and metallic spots at the anal angle at once separate it. It has much the coloration of a *Glyphipteryx*.

Adela bella Chamb.

= *chalybeis* Z.
= *iochroa* Z.

The original description of *bella* Chamb. refers to a "dull brown purple, violaceous, or golden," species (not green) with indistinct *dark-margined* fasciæ near the apex. The antennæ of the female are described as having the basal half dark purple, but it is not recorded that they are thickened at the base with long scales. This agrees in the main with Zeller's description of *chalybeis*, of which the antennæ are four times the length of the body. Zeller's type of *iochroa* in Dr. Staudinger's collection agrees with specimens in my own collection which are not green, but purplish, and have antennæ of the length described. I can find no difference between this and the description of *chalybeis* sufficient to separate them. In my own collection are specimens of a brilliant green *Adela*, from Louisiana, with thickened antennæ in the female and with indistinct transverse lines (scarcely fasciæ), such as described by Chambers in his second notice of *bella* (Can. Ent., IX, 207, and XI, 125), where I think he may have had this undescribed species before him and not *bella*. It seems to require a detailed description and a name.

Adela æruginosella sp. n.

Antennæ, male, with the basal third tinged with purplish-fuscous, the apical two-thirds white, length 22mm, the basal joint enlarged; female, 10–11mm in length, with the basal half thickly clothed with deep purple scales.

Palpi, ferruginous, much mottled with fuscous.

Head, male and female, covered with long ferruginous scales; face purplish-fuscous.

Fore-wings, shining metallic green, deep purplish towards the apex, with a golden tinge along the base of the greenish-purple cilia; on the purple apical portion of the wing are some ill-defined transverse streaks of metallic green, corresponding with the main color of the wing, not dark-margined nor strictly fasciaform.

Hind-wings, deep greenish-purple; cilia tipped with purple, but slightly tinged with golden along their base, especially about their apex.

Thorax and abdomen, dull greenish-fuscous.

Posterior legs, fuscous; tarsal joints with four white spots on the upper side.

Exp. al., 15mm.

Hab., Louisiana (Morrison).

Types, ♂ ♀, *Mus. Wlsm.*

This species differs from *Adela bella* Chamb. and its synonyms in the decidedly green color of the fore-wings, in the absence of golden scales on the apical surface, and in the absence of transverse fasciaform markings on the apical third of the wing, also in the longer antennæ, of which a larger portion towards the base is tinged with purple.

(*To be continued.*)

GENERAL NOTES.

THE WHEAT SAW-FLY.

Mr. W. Hague Harrington, in the February, 1890, number of the Canadian Entomologist, records the collecting of *Cephus pygmæus*, known in England as the "Corn Saw-fly," by sweeping in a meadow, presumably near Ottawa, and also in a collection received from Mr. Van Duzee, collected near Buffalo, N. Y., on the 9th and 11th of June, 1888. Mr. Harrington's specimens were taken in 1887.

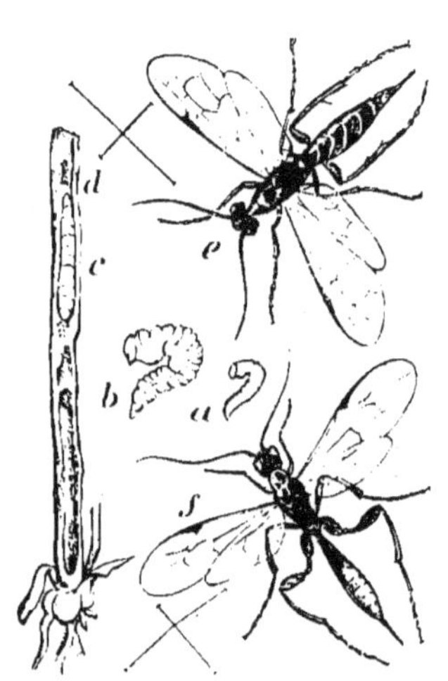

FIG. 60. *Cephus pygmæus*: *a*, outline of larva—nat. size; *b*, larva enlarged; *c*, larva in wheat stalk—nat. size; *d*, frass; *e*, adult female; *f*, female parasite—enlarged (after Curtis).

In this note Mr. Harrington does not refer to Professor Comstock's rearing of this insect from wheat stalks in Ithaca, N. Y., in 1888–'89, which we have noted in a recent number of INSECT LIFE. The figure which we give here is taken from Curtis, and was originally made to show the similarity with the method of work and appearance of *Phylloecus integer*, which bores in the young shoots of willow, and which we treated in No. 1 of Vol. I of INSECT LIFE. A comparison of this figure with the one there given will show the resemblance, and the republication of this figure of *Cephus* will perhaps assist other collectors in recognizing it. The insect figured at *f* is the commonest European parasite—*Pachymerus calcitrator.*

TASMANIAN LADYBIRDS AND THE "AMERICAN BLIGHT."

In reviewing my notes on Australian and Tasmanian insects, published in Vol. I, No. 12, of INSECT LIFE, Mr. Fraser S. Crawford, in The Garden and Field for September, takes exceptions to my statement that the same coccinellid which is so efficient in destroying *Schizoneura lanigera* about Adelaide, South Australia, was found destroying *Rhopalosiphum* on carrot in Tasmania.

When Mr. Koebele and myself parted company in Melbourne, he to go to New Zealand and I to Tasmania, and later to South Australia to secure a supply of the *Schizoneura*-eating coccinellid, I received no description or specimen of the object of my journey to Adelaide, Mr. Koebele stating that Mr. Crawford and myself would have no difficulty in recognizing it.

Of the fruitless search at Heathpool, both Mr. Crawford and myself have written. After rejoining Mr. Koebele at Auckland, New Zealand, on our homeward voyage, and while comparing notes on steamer, I understood Mr. Koebele to say that my Tasmanian species, specimens of which I gave him, was the same as the one I sought to secure at Heathpool. On returning home and preparing the notes for INSECT LIFE, relying on my understanding of Mr. Koebele's statement, I wrote as I did, and not knowing the name of the species, left it blank in the manuscript, and it was supplied in the office of the Division at Washington.

On receipt of the September number of Garden and Field I took pains to have my specimens again determined by the same authority and the species was again pronounced *Leis conformis* Mulsant. The second lady beetle, mentioned as feeding on *Rhopalosiphum*, infesting carrot in Mr. Keen's garden in Kingston, Tasmania, is *Coccinella repanda* Thunberg. Now, Kingston is a small hamlet, surrounded almost entirely by woods and hills, and Mr. Keen's garden is on the outskirts of the village and contains fruits of different kinds, including apples as well as vegetables.

On thinking the matter over again, I remember that the *C. repanda* were much more numerous on the infested carrot tops than *L. conformis*, yet there were a few of the latter present. *Leis conformis* was also very abundant about young bushes of some species of Eucalyptus, infested by *Eriococcus eucalypti* Cr. and, after reading Mr. Crawford's notice, I have no doubt but that they were feeding upon this coccid and some of them had strayed away to Mr. Keen's garden. In reply to Mr. Crawford's objection to the use of the term "little," as applied to *Leis conformis*, I would state that my specimens are from 5^{mm} to 6^{mm} in length. It would not be at all surprising that they were much larger than this in South Australia.

In Tasmania a large number of the pupæ were observed to have been parasitized, and I succeeded in rearing a number of minute Hymenopters from them, but on submitting these to Mr. Howard they were found to be secondary parasites.—[F. M. Webster.

FLIES ON APPLE TWIGS IN NEW ZEALAND.

The New Zealand Farmer for December, 1889, and January, 1890, has contained two articles entitled "Flies on Apple Twigs," which are rather interesting. In the first article an account is given of the occurrence of certain rather large hump-backed flies found sticking upon apple twigs which had apparently "died black" and were covered with a fungus growth. In the second article, however, the fly is determined by Professor Kirk as *Henops brunneus*, and an article is quoted from Mr. Maskell, which states that the black fungus look on the twigs is in reality a mass of eggs laid by the flies. Mr. Maskell saved the eggs until the larva had hatched, but he was unable to keep them alive. He states that the larvæ of none of the *Acroceridæ*, to which this fly belongs, are known, and he is unable to state what these larvæ would have been in the state of nature.

The notes are of considerable interest, especially if the determination should be correct, for upon looking the matter up we find that all of the flies of this family of which the habits are known are parasitic upon spiders. *Acrocera sanguinea* and *A. trigramma* have been reared by C. Koch from the orange-yellow cocoons of *Tegenaria agilis*. *Henops marginatus* or *Ogcodes pallipes* was reared by Menge from *Clubiona putris*, the larva living in the abdomen of the spider. *Astomella lindenii* was reared by Erber, from the abdomen of *Cteniza ariana*. The probabilties are that the discrepancy between the two accounts arises from the wrong determination of the New Zealand insect. The figures are too poor to enable a determination.

NOMENCLATURE OF BLISTER BEETLES.

At the meeting of the French Entomological Society held on November 13, 1889 (*Bull. des Seances*, pp. CCXII–CCXIII), Dr. H. Beauregard proposed some changes in the nomenclature of certain species of Meloidæ, on account of duplicated names. The following apply to our North American fauna:

Nemognatha bicolor Walk. is changed to *N. walkeri*. This change is superfluous as Walker's species has long been known to be a synonym of *N. apicalis* Lec.

Cantharis lugubris Ulke is changed to *C. ulkei* because the specific name conflicts with *Epicauta lugubris* Klug. This change would seem to be unnecessary so long as the genera *Epicauta* and *Cantharis* can be kept apart.

To *Tetraonyx* 4-*maculatus* Fabr. belong as synonyms *T. cruciatus* Cast., described from S. Domingo, and *T. cubensis* Chevr., described from Cuba.—[E. A. Schwarz.

PLANT IMPORTATION INTO ITALY.

We have previously referred in the Bulletins of this Division to the antiphylloxera laws passed at the convention of Berne, and have printed the regulations covering the importation of plants from America into countries represented in the treaty. But as this was some time ago we take occasion to print a letter received by the Italian Minister at Washington from the Italian Department of State, which has reached the Secretary of Agriculture through the Italian Legation in Washington and the Honorable Secretary of State:

ROME, *December* 4, 1889.

Mr. MINISTER: It has happened that certain Royal consular officers in countries which, like the United States of America, do not belong to the International Antiphylloxeric Union, have issued certificates attesting the freedom from phylloxera of plants sent to Italy, or merely the immunity of the countries from which the plants are sent. Now it is well to observe that no plants can be imported from countries that have not adhered to the Antiphylloxeric Convention held at Berne, unless by special previous authorization from the Royal Ministry of Agriculture and Commerce.

Such authorization, in case it is granted, is always dependent upon the presentation of the same documents that are required for plants that are sent from one to another signatory State of the Swiss Convention, and this is because it is expressly provided that States which did not sign that convention can not be treated more favorably than those which did sign it or have subsequently adhered to it.

At the request of the Royal Ministry of Agriculture and Commerce, I inform you of the foregoing, requesting you to give due notice thereof, and to cause such notice to be given to all whom it may concern in the United States, in order that plants sent from that country to Italy may not be refused admission on the Italian frontier.

I will add that, in addition to the aforesaid authorization, the certificate that must accompany shipments of plants must be issued by the local authorities and contain the following declarations:

(1) That the plants shipped are from earth that is at least twenty meters distant from any vine, or that it is separated from any vine by some other obstacle that is deemed sufficient to prevent the extension of the roots of such vine.

(2) That such earth does not contain any vine.

(3) That no vines have been deposited there.

DAMIANI,
Assistant Secretary of State.

The ROYAL LEGATION OF ITALY,
Washington.

TRAPS FOR THE WINTER MOTH USELESS.

Mr. R. McLachlan, in a recent number of the Gardener's Chronicle (Vol. 7, p. 23), calls attention to the fact that the traps which aim at the destruction of the males of the Winter Moth (*Cheimatobia brumata*) will fail of good results, since enough will always escape to fertilize the wingless females, and that it is the latter, rather than the males, that should be guarded against. In this connection is noted the "parthenogenesis" or "agamogenesis" of certain of the wingless female moths, which, of course, would render futile the destruction of the males alone.

A NEW ELM INSECT.

In Garden and Forest for January 15, 1890, p. 30, Prof. J. B. Smith calls attention to a new elm insect (*Zeuzera pyrina* Fabr.) evidently imported from Europe, the moths of which for some time past have occurred in increasing numbers every year in the city of Newark, N. J., particularly about electric lights in the neighborhood of elm trees. Examination failed to show any of the larvæ in the trunks or roots of the elm trees. Recently, however, numbers of the larvæ were found in the small twigs of a felled tree and the pupæ in burrows in the larger branches. The terminal twigs of many of the trees at Newark are reported to be dying as a result, it is supposed, of the attacks of this insect. Recognizable figures of the moths and larvæ are reproduced from drawings by Mr. John Angelmann. The adult insect is a large white moth with blue-black spots, known to English collectors as the leopard moth.

SOOT AS A REMEDY FOR WOOLY APPLE-LOUSE.

The New Zealand Farmer for December, 1889, p. 524, refers to the use of coal soot to destroy the root form of the "American blight" (*Schizoneura lanigera*). The soot is buried 6 or 7 inches below the surface of the affected tree and is said to give very satisfactory results. The use of soot is in the same line as the old remedy of wood ashes which will be found to be equally satisfactory. The alleged efficacy of the soot against all other insect pests of the apple is as is pointed out more than doubtful.

METAMORPHOSES OF FLEAS.

Mr. W. J. Simmons read before the Microscopical Society of Calcutta, March 5, 1888, an interesting paper on "The Metamorphoses of the Dog-flea," which has since appeared in the American Monthly Microscopical Journal, vol. 9, pp. 227–230. He presents some novel phases of flea life, well calculated to excite one's interest in these quite generally anathematized insects. It is stated that there are twenty-five different species of fleas; the dog, cat, fowl, marten, rat, squirrel, hedgehog, mole, pigeon and bat each having its own species, while it is a curious fact that there are also vegetarian species, two of which are mentioned. One of these latter lives in brushwood, while the other is a lover of mushrooms. Besides these, the flea which attacks man has not been mentioned, to which must be added the jigger of tropical America, this being also a true flea. Mr. Simmons makes a considerable point of the order of length of the tarsal joints in the classification of fleas.

Following his notes on the transformations of the dog-flea we find: Eggs were deposited early in the morning of October 17, 1886. These were put in a glass and covered with a pane of the same material. On the morning of October 19, about fifty hours after deposition, most of the nits had hatched out, but a few took twenty-four hours or so longer.

The majority, therefore, required only a little more than two days as their period of incubation. The larvæ were white, eyeless, cylindrical, active grubs; their bodies, exclusive of the head, with thirteen segments. These segments are beset with long hairs, the terminal segment ending in two curved spines, which probably aid the larva in locomotion. They were supplied with no food except blood-pellets (the supposed excreta of the adult flea) that had been left with the nits, etc., on a cloth by a sleeping dog. They were suspected, however, of cannibalism, as their numbers thinned with no other apparent cause. On October 25, the seventh day after leaving the egg-cases, the surviving individuals were found curling up and otherwise acting as though about to pupate. Upon noticing this they were supplied with a fragment of "puttoo," into which, though eyeless, the larvæ quickly swarmed, and there spun little white silken cocoons. November 2, most of them quitted their cocoons as perfect active fleas. They were, therefore, in the eggs for something over two days, as larvæ for six days, and pupæ for eight days, attaining their adult state on the seventeenth day after the deposition of the eggs. This is a much shorter period than given by older writers—Westwood, followed by Packard—who affirm that fleas are larvæ for twelve and pupæ for eleven to sixteen days. However, this may in part be due to the warmer climate of India, where the observations just detailed were made.

THE ENTOMOLOGICAL SOCIETY OF WASHINGTON.

January 9, 1890.—The annual meeting of the Society was held and the following officers were elected for the ensuing year:

President, George Marx; Vice-Presidents, C. V. Riley and L. O. Howard; Recording Secretary, C. L. Marlatt; Corresponding Secretary, Tyler Townsend; Treasurer, B. P. Mann; Executive Committee, E. A. Schwarz, Otto Heidemann, W. H. Fox.

Mr. W. H. Wenzel, of Philadelphia, was elected a corresponding member.

The retiring president, Mr. E. A. Schwarz, then delivered an address upon "North American entomological publications," after which remarks were made upon the address by Messrs. Howard, Riley and Smith.

Mr. Riley expressed the opinion that the recognition of scientific matter, whether descriptive or otherwise, in weekly or monthly periodicals would always depend upon the character of the author of the work and of the periodical; that synonymy should not be affected by the publication of descriptions in newspapers or periodicals which did not have a natural history character, or which did not maintain a regular natural history department.

Mr. J. B. Smith was of the opinion that publications to be recognized in literature should be in accessible journals, or in other words, in works which were put on sale, so that copies could be obtained without favor.

The thanks of the Society were voted to Mr. Schwarz for his address.

B. Pickman Mann,
Acting Recording Secretary.

February 6, 1890.—Mr. Schwarz presented a list of the blind or nearly eyeless Coleoptera, hitherto found in the United States, exhibiting in that connection a very full collection of the blind species. The list of the cave-inhabiting species is the same as published by Dr. Packard; but in that of the non-cavernicolous species, several additions are made and their geographical distribution given. As a preface Mr. Schwarz made some general remarks on blind insects and more especially on their mode of living.

In the remarks on this paper by Messrs. Riley, Howard, and Schwarz, eyeless insects of various orders were discussed, together with the presence or absence of eyes in the different stages of particular insects.

Mr. Riley made some remarks on the larva of Platypsyllus. The discrepancy in size between the larva hitherto described and the mature insect had led him to suspect that the last larval stage as well as the pupa remained to be discovered.

A specimen recently received by him and described and figured (*Entomologica Americana* for February 1890, pp. 27-30) as the "Ultimate Larva," is in general appearance strikingly Mallophagous and a few points may be mentioned as not sufficiently emphasized in the published description. The arrangement of setous hairs on the venter recalled that in the adult, while the raised dorsal points, though unarmed, foreshadowed somewhat the setous points on the dorsal abdominal joints of the adult. Remnants of the anal cerci of the earlier larval stages are noticeable in the two slight swellings on penultimate joint, each surrounded by a series of short spinous hairs. The spiracles are small and lateral, but may be detected with difficulty at the inner angle in the notch between the abdominal joints. The prothoracic spiracle has not been detected.

He had, in the paper already alluded to, raised a parenthetical question as to this being the final form of the Platypsyllus larva, but the position and character of the mouth parts, and particularly the single-jointed tarsi exclude it from the Mallophaga, while its general characteristics, though departing in so many respects from the earlier larva, have caused him to refer it to Platypsyllus. The principal feature that would shake one's faith in this reference is the presence of ocelli, since none occur in the earlier larva nor in the imago, and while such a feature is abnormal under the circumstances, it is no more so than many of the other features of Platypsyllus.

In the discussion, Mr. Schwarz held that if not the ultimate larva of Platypsyllus, it is certainly Coleopterous and can not be referred to the Mallophaga.

In the Coleoptera, the Staphylinid genus *Amblyopinus* is known to be parasitic on terrestrial rodents, two species having been found in the fur of mice and rats in South America and Tasmania. We might reasonably expect to find this genus in North America under similar circumstances, but a glance at Prof. Riley's larva shows that it cannot possibly belong to *Amblyopinus* nor to any other genus of Staphylinidæ.

Dr. Marx discussed a new family of spiders, the species of which are found abundantly in the spring. These spiders come near the family *Dictynidæ*, and belong to the genera *Neophanes* and *Prodalia*. Dr. Marx mentioned also a new remarkable spider, peculiar among other things in having but two spinnerets—a feature which occurs in but three other known genera. These genera differ from all other spiders, and are only related to each other in the number of spinnerets.

Considerable discussion followed relating to the advisability of erecting new families for odd species. The conclusion reached was that generally it would be better to give such species sub-family importance in the nearest related existing family.

Mr. Linell gave some personal observations showing that *Megapenthes limbalis* Hbst. and *M. granulosus* Melsh. were the same species. He had found these two beetles *in coitu*, and as only males of *limbalis* and females of *granulosus* had been previously known, the identity of the two species was fully shown. *M. limbalis* being first described, holds.

C. L. Marlatt,
Recording Secretary.

U. S. DEPARTMENT OF AGRICULTURE.

DIVISION OF ENTOMOLOGY.

PERIODICAL BULLETIN. APRIL, 1890.

Vol. II. No. 10.

INSECT LIFE.

DEVOTED TO THE ECONOMY AND LIFE-HABITS OF INSECTS, ESPECIALLY IN THEIR RELATIONS TO AGRICULTURE.

EDITED BY

C. V. RILEY, Entomologist,

AND

L. O. HOWARD, First Assistant,

WITH THE ASSISTANCE OF OTHER MEMBERS OF THE DIVISIONAL FORCE.

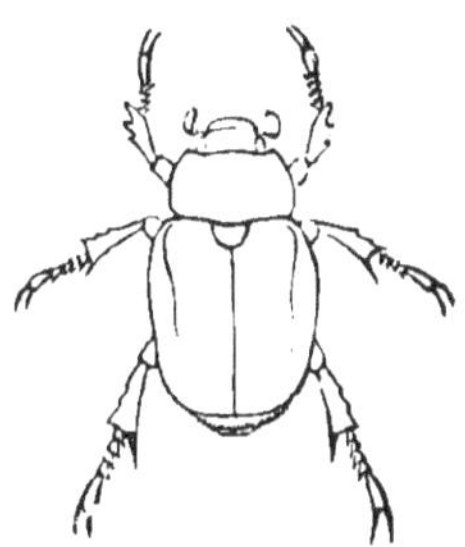

[PUBLISHED BY AUTHORITY OF THE SECRETARY OF AGRICULTURE.]

WASHINGTON:
GOVERNMENT PRINTING OFFICE.
1890.

CONTENTS.

II

SPECIAL NOTES.

On the compound Eyes of Arthropods.—Studies from the Biological Laboratory of Johns Hopkins University, Vol. IV, No. 6, contains a paper "On the Morphology of the Compound Eyes of Arthropods" by Mr. Sho Watase, which is of interest owing to its bearing on the origin of the compound eyes of insects.

The principal subject of the paper is the eye of *Limulus*, but types of the three great groups of Arthropods—Insects, Crustacea, and Arachnids—were studied, and the results are included in the generalizations at the close of the paper.

The primitive type of the *ommatidium*, or visual unit, is traced into a simple open ectodermic pit from which he believes the compound eyes of Arthropods to have developed by a vegetative repetition of similar structures, not unlike what is supposed to have taken place in the formation of certain compound organs in other animals, such as the kidney in vertebrates, or the respiratory organs in Lamellibranchs.

Taking the number of facets as given by Lubbock, the compound eye of the house-fly (*Musca*) would represent about 4,000 invaginations of the skin, and of the dragon-fly (*Æschna*) about 20,000, while an ocellus would represent a single pit.

In an appendix the compound eye of the star-fish is briefly considered and is found to be morphologically strikingly similar to that of an Arthropod. Six lithographic plates accompany the paper and admirably illustrate the author's studies.

More Ohio Notes.—"A Season's Work among the Enemies of the Horticulturist," is the title of a paper by Clarence M. Weed, read December 11, 1889, before the Ohio State Horticultural Society and recently issued in pamphlet form by the author. It treats of both insect and fungus pests and urges the advantage of combining insecticide and fungicide preparations for the simultaneous treatment of both pests whenever possible. The entomological portion of the paper comprises matter for the most part previously published in the bulletins of the Ohio Experiment Station and includes brief accounts of the Striped Cucumber-beetle, the Cherry Tree-slug, a new Strawberry-root Plant-louse (*Aphis forbesi*), described in the August–December, 1889, No. of *Psyche*, and of the "Rhubarb Snout-beetle" (*Lixus concavus*), whose habits are stated (and also in Bulletin Ohio Agricultural Experiment Station, Vol. II, No. 1, second series, No. 8, p. 153), to be for the first time recorded.

In 1872 we studied the habits and reared from the larva found in the stems of *Chenopodium hybridum*, the western representative of this species, *Lixus macer*, while Mr. Webster bred it later from the stems of Helianthus. We briefly recorded these habits and the gall-making habit of *Lixus parcus* from California at the December, 1885, meeting of the Washington Entomological Society (Proc. Ent. Soc. Wash., I, No. 2, 1888, p. 33). That *L. concavus* injures rhubarb in other parts of the country as it does in Ohio and Michigan, was recorded many years ago by Glover, and has been independently observed by Mr. J. G. Barlow and Mr. Wm. B. Alwood. We hope soon to bring our notes on the subject together.

Aquatic Insects of the Mississippi Bottoms.—We have recently received from Prof. S. A. Forbes, Director of the Illinois State Laboratory of Natural History, a paper by H. Garman, entitled "A Preliminary Report on the Animals of the Waters of the Mississippi Bottoms, near Quincy, Illinois."

The report is based on studies and collections made in the summer of 1888, by the State Laboratory of Natural History, the work being aided and facilitated by the Illinois Fish Commission.

After a general description of the peculiar character of the streams and lakes in the locality covered by the investigation, there follows a discussion of the genera and species of the animal life studied, including both the higher forms—mammals, birds, fishes, etc.—and the invertebrates. Among the latter, the Insecta are chiefly considered, and this portion of the work will be of most interest to readers of INSECT LIFE.

The aquatic insects are studied particularly in their relation to fish culture, and those species which are especially important in this connection are chiefly dwelt upon.

Considerable additions are made to our knowledge of food habits in certain cases, and references are given to the published descriptions and accounts of many of the species. Data of importance to the practical ichthyologist are thus brought together.

Insects belonging to the following orders are considered: Diptera, Coleoptera, Trichoptera, Neuroptera, Hemiptera, Ephemeridæ, Plecoptera and Odonata. A single Arachnid is given as occurring near or in the water (*Tetragnatha grallator* Hentz.), and a pale water mite (*Arrenurus* sp.), was frequently taken on the lakes and is believed to be a river species.

Life-histories of some Kansas Moths.—Transactions of the Kansas Academy of Science, Vol. XI, 1887–'88, which we have recently received, contains a paper by Mr. C. L. Marlatt, entitled "Notes on the early stages of three Moths." The species discussed are *Nerica bidentata* Walker, *Anisota stigma* Fabr., and *Callimorpha suffusa* Smith. The life-histories of these moths are quite fully given, together with illustrations of the several stages of each. The species first mentioned breeds on the Elm, the second, as is well known, on the Oak, and the last on Ash.

International Meetings to consider Viticulture and Fungus Diseases.—An International Exposition of apparatus and products for the treatment against mildew, was held at Rome, from the 23d to the 27th of March, 1890, under the auspices of the Italian Œnophile Club. At the same time an International Viticultural Reunion was held, at which various subjects relating to fungus diseases of the vine, investigations on and remedies for the same, were discussed.

THE ROSE CHAFER.

(*Macrodactylus subspinosus*, Fabr.)

By C. V. RILEY.

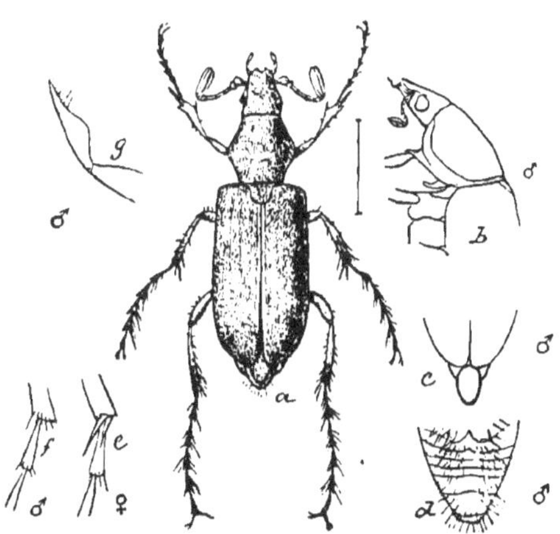

FIG. 61.—*Macrodactylus subspinosus: a*, female; *b*, anterior part of male to show the prosternal metacoxal process; *c*, pygidium of male; *d*, abdomen of male; *e*, tip of hind tibia of female; *f*, ditto of male; *g*, front tibia of male—all enlarged (original).

PAST HISTORY.

Few insects are more often referred to in our horticultural literature than this. The accounts have almost invariably referred to the ravages of the mature beetle, and few persons are familiar with the species in its larval state. In fact, a full life-history with a description of the larva is yet needed, and as we reared it to the imago and made a study of it in the field in 1882 and 1883, and as the beetle attracted more than usual attention the past year we have deemed it advisable at this time to publish the following account.

A native North American insect, there is every reason to believe that this Rose chafer, or Rose bug, as it is more generally called, has increased in number with the progress of horticulture, for the perfect beetle evidently shows a preference for the blossoms and sweeter and more tender fruit of our cultivated plants as compared with those of wild plants. Another reason may be found in the increased area of pasture and meadow lands which form the natural breeding grounds of the species. The first published account of this insect seems to be that given by Dr. Harris in his "Minutes toward a history of some American species of Melolonthæ particularly injurious to vegetation" (Mass. Agric. Report and Journal, X, 1827, pp. 1–12), reported in N. E. Farmer, 1827 (vol. 6, p. 18, ff.). In this account Dr. Harris says that at the time the bugs were first noticed they were confined to the roses, but within forty years they had prodigiously increased in number and had become very injurious to various plants. From this it would appear that as far back as the last century the insect was known as injurious.

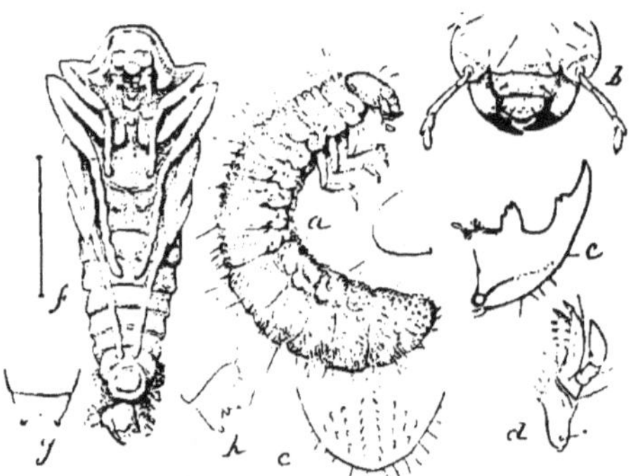

FIG. 62.—*Macrodactylus subspinosus*: *a*, full-grown larva from the side; *b*, head of larva from the front; *c*, left mandible of larva from beneath; *d*, left maxilla of larva from above; *e*, last ventral segment of larva; *f*, pupa from beneath; *g*, tip of last dorsal abdominal segment of pupa; *h*, last segment of pupa from the side—all enlarged (original).

NATURAL HISTORY.

According to Harris the female beetle lays her eggs to the number of about thirty, about the middle of July, at a depth of from 1 to 2 inches beneath the surface of the ground. He does not state the favorite place for oviposition, but in our experience the larvæ are especially abundant in low, open meadow land or in cultivated fields, particularly where the soil is light and sandy. Harris states that the eggs hatch in about twenty days, and, while the period will vary with the temperature, the larva is found fully grown during the autumn months. With the approach of cold weather it works deeper into the ground, but in the spring will frequently be found near the surface or under stones and other similar objects, where it forms a sort of cell in which to pupate. In confinement the pupa state has lasted from two to four weeks. The perfect beetle issues in the New England States about the second week of June, while in the latitude of Washington it is seen about two weeks earlier. It appears suddenly in great numbers, as has often been observed and commented upon, but this is in conformity with the habits of other Lamellicorn beetles, *e. g.*, our common May-beetles (*Lachnosterna*), and this habit is still more marked in certain species of Hoplia and Serica. It remains active a little over a month, and then soon disappears. The species produces, therefore, but one annual generation, the time of the appearing of the beetle in greatest abundance, being coincident with the flowering of the grape-vine.

GEOGRAPHICAL DISTRIBUTION.

The species is recorded by Dr. Horn (Trans. Amer. Ent. Soc., 1876) as occurring from Virginia to Colorado and northward. It is thus not represented in the extreme South and West of the Rocky Mountains. Northward it extends into Maine, and Canada, and Minnesota. It is certainly absent, or at least very scarce in western Kansas, though common and destructive in the eastern and more wooded portions of the State.

Professor Osborn finds the beetle not particularly destructive in Iowa, and our experience shows that as a rule it is less destructive in the Mississippi Valley than in the East. There are, however, numerous specimens marked "Texas" in the collection of the late Mr. Belfrage. Even in the Eastern States the insect is, in certain more or less restricted areas, rare or absent for reasons which are more or less obscure, but which find readiest explanation in the fact that certain moist and open areas or bottom lands, especially of a sandy character, are the preferred breeding places. Thus Dr. Fitch (2d Rep., p. 247) states that in the vicinity of his residence in New York State he took only occasionally a specimen during twenty-five years, and Dr. Lintner mentions (1st New York Rep., pp. 230, 231) a similar case of local exemption. Harris states that *M. subspinosus*, although common in the vicinity of Boston, is, or was a few years ago, unknown in the northern and western parts of Massachusetts, in New Hampshire, and in Maine. Since the species is now common in parts of New Hampshire and very generally over the whole of the State of Massachusetts, it would appear that the species has of late years extended its range.

In the Gulf States it is replaced by a closely allied species, *M. angustatus* Beauv., which has not yet proved to be injurious and is in all probability less abundant. A third species, *M. uniformis* Horn, occurs in the extreme southwest of the country and of this we received in July, 1889, specimens from Judge J. F. Wielandy, of Springer, N. Mex., with the statement that they were injuring apples.

FOOD PLANTS AND RAVAGES.

The food of the larva consists of the roots of grasses and probably also of other low plants. Whether it also feeds on the rootlets of trees and shrubs has not been definitely ascertained, although the larvæ have been found quite numerously around the bases of oak trees near Washington, both by Mr. Koebele and Mr. Schwarz. We found them quite numerous in the sandy low lands of the Merrimac Valley, New Hampshire, on cultivated ground, where they must have fed on the roots of various weeds or on those of meadow grass and cultivated rye and maize. It is probable, however, that they occur yet more numerously in unplowed pasture and meadow land than in cultivated fields.

One peculiarity in the food habits of the larva still remains to be mentioned here, viz, that referred to in our report as U. S. Entomologist for the year 1883 (p. 174): While searching for locust eggs in the infested fields at Boscawen, N. H., we repeatedly found the larva of this Macrodactylus feeding on the egg-pods of *Caloptenus atlanis*. This is certainly a remarkable and exceptional food habit in a plant-feeding larva, but it is paralleled in the common White grub (larva of *Lachnosterna fusca*) which we have shown in the first report of the U. S. Entomological Commission (p. 305) to have a similar habit. The habit is doubtless developed only when the locust eggs are thickly laid in the ground.

The beetle has a partiality for flowers, but also feeds upon leaves of various trees and bushes and attacks certain fruits. It has a predilection for the flowers of roses, wild as well as cultivated,* and, in the experience of many observers, prefers white roses to red ones. Another favorite food is the blossom of the grape-vine, with a decided preference for that of the Clinton. This last fact was first pointed out by Walsh in his first report on the Insects of Illinois (p. 24), and has been confirmed by many other observers and by our own observations. Dr. Lintner, in his first New York Entomological Report (p. 229), contradicts this experience, which only goes to show how the habits of the same species will differ in different sections of the country. Flowers of raspberries and blackberries do not escape its ravages. Mr. E. H. Miller states in the American Agriculturist (see Amer. Nat., v. 17, 1883, p. 1291), that the flowers of *Deutzia scabra* are even preferred by the beetle to the grape-vine. The blossoms of the various species of *Spiræa* are often crowded with the beetles and the same may be said of the blossoms of Sumach, the common Ox-eye Daisy, *Magnolia glauca*, Mock Orange, and some other plants. This list could be greatly extended, but we close it with the state-

* The Cinnamon rose, *Rosa cinnamonica*, is said to enjoy immunity.

ment that the beetles also devour the blossoms of *Pyrethrum cinerariæfolium.*

The foliage of most, if not all, of our cultivated fruit trees and especially Apple, Pear, Peach, Cherry, and Plum at times suffer greatly, the two last-named trees being apparently more attractive than the others. The foliage of cultivated grape-vines is almost as eagerly devoured as the blossoms, and the leaves of Oak, Alder, and other forest trees also serve as food. Of low-growing plants the beetles cut the leaves of strawberries, rhubarb, and of nearly all garden vegetables, as also of sweet potato, corn, wheat, grass, and many wild plants.

Not satisfied with this amount of damage, the beetles attack the fruit of peaches, cherries, apples, and grapes when just forming.

Among ornamental plants the Rose is the greatest sufferer. Harris states that the beetle was first noticed on the Rose (hence its popular name), and that it afterward acquired the habit of feeding on grapevines and fruit trees.

In 1887 a statement went through the daily press and agricultural journals (apparently originating in the Philadelphia *Press* from a communication by E. Williams) that the beetle was poisonous. It is said that a lady who smashed some in her hands had these badly swollen up, and further, that chickens fed with the beetles all died. There is, however, nothing to justify the assumption that the beetle is really poisonous, and if the above reports be true, the affliction was no doubt due to mechanical irritation caused by the long and sharp claws and the spines of the beetle.

NATURAL CHECKS.

As with other insects, there are fluctuations in the numerical abundance of the Rose Chafer; but so far as we yet know they seem to be caused by meteorological conditions, for the species has few natural checks, and no true parasites; while but few enemies of its own Class have been observed. Harris says (Treatise, etc., p. 39):

> Our insect-eating birds undoubtedly devour many of these insects. Rose bugs are also eaten greedily by domesticated fowls; and when they become exhausted and fall to the ground, or when they are about to lay their eggs, they are destroyed by moles, insects, and other animals, which lie in wait to seize them. Dr. Green informs me that a species of Dragon-fly, or devil's needle, devours them.

Toads have been observed to swallow the beetles (see *Mirror and Farmer*, v. 35, July 26, 1883), and it may be inferred that the larvæ are eaten by various ground beetles. While at Boscawen, N. H., in the fall of 1882, we found in the ground in company with the Macrodactylus larvæ a number of an undetermined Elaterid larva. Upon placing both kinds in a tin box filled with earth it was found upon our return to Washington that the Elaterid larvæ had killed and devoured most of those of the Macrodactylus. Since many Elaterid larvæ are either entirely or essentially carnivorous, that observed at Boscawen may thus prove to be one of the natural enemies of the Macrodactylus.

REMEDIES.

It has been assumed by most writers that we can not successfully attack the Rose Chafer in any of its earlier states. To search for the eggs in the ground would be impracticable. It does not, however, follow because of the poor success that has generally resulted from attempts to destroy similar larvæ that they can not be successfully destroyed. In the case of the common European Cock-chafer (larva of *Melolontha vulgaris* and *hippocastani*) and of our own White Grub (*Lachnosterna fusca*) the methods adopted have consisted in plowing and hand-picking. The experiments made, however, on a similar larva with the kerosene-soap emulsion, as narrated in INSECT LIFE (Vol. I, p. 48) clearly show that we have in this insecticide a means of successfully destroying the bulk of the larvæ of the Rose Bug wherever they are known to be sufficiently abundant to justify such treatment. A thorough investigation should be made in the direction of ascertaining the preferred breeding grounds of the species, and it were rash to say here that we have no effectual mode of preventing the insect, nothwithstanding the disfavor in which this mode of warfare has been held in the past.

It is evident, however, that for the present we should concentrate our efforts on the destruction of the beetles especially when they first issue from the ground and congregate in the garden on our roses, grapevines, and fruit trees. A brief statement of the various methods that may be employed for this purpose may prove advantageous. Hand-picking and killing the beetles either by crushing them or throwing them into hot water, or water having a scum of kerosene upon it, has proved useful and satisfactory in a limited way, as also the shaking and knocking down of the beetles into pans or upon sheets saturated or smeared with coal oil. These measures are best carried out and most satisfactorily in the early morning hours and toward evening, as the beetles are then more sluggish and not so quick to take wing as they are during the heat of the day. White roses, Spiræas, or Deutzias, planted on a place, will attract great numbers of the beetles, and thus not only facilitate the destruction of these last, but act as a kind of protection to other plants.

As to other topical applications intended to destroy the beetles, whether directly or by poison taken with the food, the experience with the arsenites is that they are of little avail, and the experience with other materials, like hellebore and pyrethrum, has been so conflicting, that we can not consider either of them reliable or satisfactory. Pyrethrum would seem to have given on the whole the most satisfactory results, and the following experience of Mr. E. S. Carman, editor of the *Rural New Yorker*, would certainly show that it may be used advantageously. It is given in substance from the *Rural New Yorker* of July 7, 1888:

The rose-bugs appeared suddenly on the Rural Grounds in such swarms that their appearance was hardly known until they had half destroyed the grape blossoms. On the morning of the 20th (June, presumably) two hours were spent in spraying rose-bushes, grape-vines, and a *Magnolia macrophylla* about 12 feet high, with Buhach water. The bugs were devouring this latter by hundreds. In fifteen minutes after spraying, thousands of the bugs were found wriggling upon the ground while the tree was virtually cleared of them. Twenty or more of those on the grass were placed in a tomato can and covered with a gauze so as to confine them without excluding air. These soon became paralyzed, and in the evening were apparently dead. Those on the grass crawled about in an aimless way. Towards evening some were found apparently dead. The others had disappeared. Here and there a bug was found on the leaves of the tree. The grape-vines and rose-bushes were also nearly free of the pest during the rest of the day. The next day thousands of rose-bugs were again upon the roses and grape-vines, though few could be seen on the magnolia. All were again sprayed with the same effect as that above recorded, and further spraying has not since been deemed necessary.

Col. A. W. Pearson, of New Jersey, states that the "eau celeste" (solution of sulphate of copper with ammonia) is not only the best remedy for mildew, but also at the same time an effective poison to the Macrodactylus.

The trouble with all these remedies is that the beetles during their brief season continue to issue from the ground and to congregate upon their favored plants in such numbers, under favorable circumstances, that however fatal an application may be it has to be continued, and the most persistent may justly become discouraged in a fight with these beetles when they are abnormally abundant and swarm to the extent we have known them.

As early as 1829 Dr. R. Green, as quoted by Harris, urged as a preventive measure the covering of the grape-vines with millinet, but, however valuable such a method may be for choice vines in limited numbers, it would evidently be too costly for large vineyards or for larger fruit-trees.

Another protective measure (first suggested in the *Rural New Yorker* May 19, 1883) is to dust the plants with air-slaked lime or gypsum, and Prof. C. M. Weed has suggested as an improvement upon it (7th Ann. Rept. Ohio Agr. Exp. St., 1888, p. 151) a liberal spraying of lime water, from one-half to one peck of lime to a barrel of water. Mr. E. A. Dunbar, of Ashtabula, Ohio, who tried this "whitewashing" of his grape-vines and peach trees, reports most satisfactory results.

Many other means that have been tried against this pest are not worthy of serious consideration. Such are the spraying of decoctions of various plants with a view of rendering the leaves unpalatable; methods of hastening the blossoming of grape-vines or other plants by artificial means. These and others that have been urged, even where effective, are hardly likely to be generally employed; and in this case, as with many other insects, success will only follow diligence in the combined application of the insecticides that have been found effective, and the persistent shaking on to sheets or stretchers saturated with coal-oil.

A NEW GENUS AND TWO NEW SPECIES OF AUSTRALIAN LAMELLICORNS.

By Dr. DAVID SHARP, *Wilmington, England.*

The Lamellicorn sent to me by Professor Riley for determination proves to belong to a genus hitherto undefined and is described below, together with another allied species from Adelaide.

Anodontonyx, nov. gen.

Inter genera Haplonycham et Heteronycem locandum. Labium planum. Palpi labiales articulo ultimo dilatato, conico, subtus convexiusculo. Maxillae quinque-dentatæ, palpis simplicibus, articulo ultimo quam penultimo duplo longiore. Labrum sat crassum, angulis parum prominulis. Antennæ brevissimæ, 8-articulatæ, clava perbrevi, tri-articulata. Tarsi elongati, unguiculis simplicibus.

The species of this genus will be readily distinguished by the dilatated joint of the labial taken in conjunction with the simple claws of the feet and the remarkably small club of the antennæ. The maxillæ looked at externally appear to be only three-toothed, but there are two other nearly equally large teeth concealed behind the external teeth.

Although allied in many respects to Scitala, I think it would increase the confusion prevalent in collections if Anodontonyx were merged in that genus. Scitala has a longer club to the antennæ, the male feet not elongated, and in most of the species of the genus the labial palpi have a slender terminal joint; the base of the thorax is sinuate on each side and the hind angles are well marked.

Anodontonyx is probably numerous in species in Australia, as I have five or six others belonging to it in my collection, for none of which can I find names. They are all small and quite unattractive insects, and are apparently of retiring habits, as the specimens obtained are very few in number.

Anodontonyx vigilans, n. sp.

Pallide ferrugineus, crebre punctatus; corpore subtus fere nudo, ad latera parce setoso; elytris inter punctaturam lineis elongatis parum conspicuis. Long., 9–10^{mm}.

Head closely and coarsely punctured, clypeal suture very distinct, margin of clypeus strongly reflexed. Thorax short, moderately, coarsely, and closely punctured; hind angles rounded. Scutellum sparingly punctured. Elytra rather sparingly punctured, each with four longitudinal linear smooth spaces extending nearly the whole length, and with a broader space near the suture which, however, is not free from punctures. Pygidium rather coarsely punctate. Prosternum behind the coxæ armed with a prominent acute lamina. Upper spur of hind tibia elongate, as long as the basal joint of the tarsus.

Australia; Koebele.

No sexual differences are to be seen among the six specimens brought to America by Mr. Koebele, and they are probably all females.

Anodontonyx harti, n. sp.

Oblongus, ferrugineus, vel piceus, convexus, crebre fortiter punctatus; pectore utrinque parum hirsuto; elytris ad latera longuis setosis, inter punctaturam lineis elongatis conspicuis. Long., 12-13mm.

Mas: tarsis omnibus elongatis.

This is not very different in color and punctuation from *A. vigilans*, but is distinguished by some important structural characters. The form is more oblong and elongate. The clypeus is rounded in front, and its margin is very strongly elevated. The sides and hind angles of the thorax are much rounded. The pygidium is somewhat obsoletely punctured at the base, smooth towards the apex. There is only a single carina on the prosternum behind the coxæ. The male has the hind tarsi 5½ millimeters long, whereas in the female they are only 3½. In this latter sex the anterior tibiæ are remarkably broad, the three teeth on it also very broad.

This interesting insect was discovered by the late Mr. Hart during his stay at Adelaide in 1886. Although at that locality only for a short time, and when he was in very weak health, he formed a most interesting collection of Coleoptera. The specimens of *A. harti* described above were given to me by his friend, Mr. W. R. Jeffrey, of Ashford, Kent.

AN INTERESTING TINEID.

(*Menesta melanella* n. sp.)

By Mary E. Murtfeldt, *Kirkwood, Mo.*

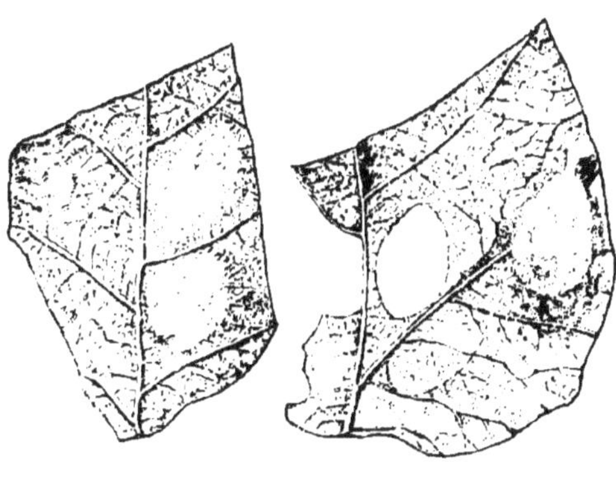

Fig. 63.—*Menesta melanella*: *a*, larval mine; *b*, pupa case, with larval mine cut out (original).

The Tineid genus *Menesta*, to which Professor Riley has kindly referred this species for me, was erected by Dr. Clemens for the reception of a particularly aberrant Gelechiid, which he described from a single captured specimen and named *Menesta tortricella*. The only other species of this genus of which there is any record is a captured unique, obtained by Lord Walsingham from Texas, and described in his lordship's notes on North American *Tineidæ*, in the Proceedings of the London Zoological Society for 1881. This is a small species with dull shaded red fore wings, each of which is ornamented with a minute discoidal ocellus. The hind wings are dark gray.* In all his collecting, personally and by his assistants, Mr. V. T. Chambers, whose studies of American Tineidæ were so extensive, seems never to have met with a species that he could satisfactorily relegate to this genus.

The larval habits of neither of the described species have been observed, but perhaps those of the one which I now propose to characterize may indicate some of their peculiarities.

Menesta melanella n. sp.

Head and thorax above dusky black, face smooth, scales shining-white with golden and iridescent reflections; labial palpi rather short, slender, diverging, slightly curved, second joint scarcely thickened, smooth, tapering to the juncture with the very slender sharply pointed apical joint, inner side white, outer ocherous, dusky at base, maxillary palpi very small, tongue broad, white at base; antennæ brownish-black, with purplish and steely reflections, rough scaled under the lens, but scarcely ciliate, about two-thirds the length of the wings.

Fore wings shining bluish or brownish black, somewhat iridescent, with acute, milk-white, triangular patch on costa midway between base and apex, extending nearly half across the wing, a few white scales near the base; cilia on outer margin pure white, on inner angle dingy black.

Hind wings very dark brown, with rather broad white marginal streak on costal edge, extending from near the base beyond the middle and a patch of white in the cilia near the outer angle, also a few white hairs near the inner angle. Under surface of both fore and hind wings fuscous with leaden reflections, the white costal triangle nearly as well defined beneath as above.

Abdomen above, iridescent, shining black. Thorax beneath and broad ventral abdominal band, white with metallic luster. Front legs white, middle pair white, on femora and tibiæ, with tarsi, dusky, indistinctly annulate with white; hind legs dusky and leaden gray, with broad band of white encircling tibiæ; terminal joint whitish at base, shading to dark gray at tip: upper spurs long, white; lower spurs ocherous. All the white on the under surface has, in certain lights, deep golden and opalescent tints, with a somewhat more stable ocherous shade at the joints. All the legs are coarsely scaled and hairy. The alar expanse is from 10 to 12^{mm}.

This species is pretty and characteristic in its perfect form and interesting in its larval habits and transformations. The larva appears late in summer, on the post oaks (*Q. obtusiloba*) and requires nearly a month to attain maturity. It is at first a miner, but later—probably after first molt—feeds externally on the under surface of the leaf, skeletonizing a large space on one side of the midrib, protecting itself above under a web which is dense in the center and becomes gradually attenuate towards the edges, from under which the frass is ejected. When dis-

* I have since noticed that Lord Walsingham has removed this species from *Menesta* and placed it in the *Tricotaphe* section of *Gelechia*. He also states that it is a synonym of Chambers' *Gelechia refusella*.—M. E. M.

turbed the larva retreats swiftly to the more densely woven part of its cover.

Length of full grown larva from 7 to 8^{mm}, diameter 1.5^{mm}, nearly equal throughout, form depressed, sutures deep, especially laterally; surface smooth; color dingy translucent white, with a broad, smooth pale purple dorsal band extending from the second to the tenth segment. Head small, about one-half the diameter of the first thoracic segment, opaque, yellowish white. Piliferous spots minute, impressed, hairs microscopic. Legs and prolegs yellowish white, almost transparent.

In preparing for its transformations, the larva thickens its tent of white silk, which is externally somewhat disguised by a skillful intermingling of powdery particles of the cuticle of the leaf. It lines the under side also with a mat of silk and then proceeds to cut out a broad oval section around the densest part of the web, about one-half inch in length. This is joined at the edges and forms an *Aspidisca*-like case, which is dragged to some distance from the injured portion of the leaf and firmly attached to the under surface by a broad band of silk from one-eighth to one-fifth of an inch long appearing like a handle to the slightly curled case (Fig. 63*b*). Within this case the larva changes to a somewhat flattened, pale brown pupa, in which state it hibernates.

The imagines usually appear in April, often at long intervals. Of the three specimens bred, one emerged on the 14th of March, one on the 14th of April, and the third on the 24th of the same month. I also captured a single damaged specimen some years ago during the month of May. It is not in this locality at least an abundant species.

EXPERIMENTS WITH THE PLUM CURCULIO.

By F. M. Webster.

These experiments were originally intended as a continuation of those made during 1888 and published in the annual report of the Department for that year, pp. 78, 79. On account of a lack of material, especially of the domestic varieties of plums, the result of previous experiments did not reflect as conclusively upon contested points as desired; and as it would hardly be proper, at the present time, to summarize results based on one set of experiments made during one season, and another the next, under more or less varying conditions, the series this year are also intended to repeat and elaborate some of those made on the wild varieties last season.

The source whence the material was secured is given in the records of each experiment, and I will only add that the larger portion of the first was taken by myself, beneath the trees from which it had fallen, the point being to change the conditions under which it was found only so far as necessary to a change from one locality to another.

The methods employed in carrying on the experiments were the same as last year, except that, in view of the results already obtained, the earth in which the insects had developed was not treated with water, but examined carefully on the dates given, and a record kept of the number of adult beetles found. For vivaria, 8-inch drain tiles, the same as last year, were used.

Experiment No. 1.—June 13, 1889, one hundred and fifty Wild Goose Plums, from Aper's orchard, La Fayette, containing one hundred and eighty-five egg punctures, were placed in vivaria.

Result of examination on September 4: Thirty-eight adults. Dead.

Experiment No. 2.—June 13, fifty Mariana Plums, from Experiment Station orchard, containing eighty-six egg punctures, were placed in vivaria.

Result of examination on September 4: Fourteen adults. Dead.

Experiment No. 3.—June 18, fourteen Kansas Sand Plums, from E. Yenowine, Edwardsville, Ind., containing fourteen egg punctures, placed in vivaria.

Result of examination September 3: Four adults. Dead.

Experiment No. 4.—June 18, six Nectarines, from E. Yenowine, Edwardsville, Ind., containing seven egg punctures, placed in vivaria.

Result of examination September 3: Nothing.

Experiment No. 5.—June 18, twenty-four Chickasaw Plums, from E. Yenowine, Edwardsville, Ind., containing twenty-four egg punctures, placed in vivaria.

Result of examination September 3: Five adults. Dead.

Experiment No. 6.—June 19, two hundred and twenty-five Coe's Golden Drop Plums, from J. G. Kingsbury, Irvington, Ind., containing six hundred and eleven egg punctures, placed in vivaria.

Result of examination September 2: One hundred and nineteen adults. Dead.

Experiment No. 7.—June 19, three hundred and sixty-eight Wild Goose Plums, from orchard of Albertson and Hobbs, Bridgeport, Ind., containing seven hundred egg punctures, placed in vivaria.

Result of examination September 4: One hundred and eighty-one adults. Dead.

Experiment No. 8.—June 19, one hundred and sixty-seven Nectarines, from orchard of Albertson and Hobbs, Bridgeport, Ind., containing five hundred and thirty-three punctures, placed in vivaria.

Result of examination August 28: Fifty-three adults. All living.

Experiment No. 9.—June 20, one hundred and twenty-eight large Damson Plums, from Greencastle, Ind., containing one hundred and thirty-nine egg punctures, placed in vivaria.

Result of examination September 4: Fifty-three adults. Dead.

Experiment No. 10.—June 20, one hundred and sixty-eight Robinson Plums, from Greencastle, Ind., containing two hundred and twenty-three egg punctures, placed in vivaria.

Result of examination September 4: Thirty-five adults. Dead.

Experiment No. 11.—June 20, one hundred and sixty-eight Mariana Plums, from Greencastle, Ind., containing two hundred and nineteen egg punctures, placed in vivaria.

Result of examination September 3: Fifty-three adults. Dead.

Experiment No. 12.—June 20, three hundred and thirteen Lombard Plums, from Greencastle, Ind., containing four hundred and sixty-two egg punctures, placed in vivaria.

Result of examination September 3: Sixty-five adults. Dead.

Experiment No. 13.—June 20, ninety-five Yellow Egg Plums, from Greencastle, Ind., containing one hundred and three egg punctures, placed in vivaria.

Result of examination September 4: Nineteen adults. Dead.*

---

*Six larvæ from this lot were destroyed.

Experiment No. 14.—June 24, one hundred and seventy-six Wild Plums, from woods in Knox County, Ind., containing two hundred and twenty egg punctures, placed in vivaria.

Result of examination September 4: Thirty-five adults. Dead.

Experiment No. 15.—June 24, fifty-nine Lombard Plums, from Knox County, Ind., containing seventy-nine egg punctures, placed in vivaria.

Result of examination September 4: Fifty-one adults. Mostly dead.

Experiment No. 16.—June 24, one hundred and ninety-one Blue Damsom Plums, from Knox County, Ind., containing two hundred and twenty-six egg punctures, placed in vivaria.

Result of examination September 4: Seventy-six adults. Few alive.

Experiment No. 17.*—June 25, fifty-three Apples, from Princeton, Ind., containing sixty-two egg punctures, placed in vivaria.

Result of examination September 6: Five adults. Living.

Experiment No. 18.*—June 25, fifty-four Apples, from same tree as No. 17, containing sixty egg punctures, placed in vivaria.

Result of examination September 6: Nothing.

Experiment No. 19.*—June 25, twenty-nine Apples, from same orchard as No. 17, containing thirty-six egg punctures, but from another tree, placed in vivaria.

Result of examination September 6: Three adults. All living.

Experiment No. 20.—June 26, forty-seven Nectarines, from same tree as Experiment No. 4, and containing forty-eight egg punctures, placed in vivaria.

Result of examination September 4: Six adults. Living.

Experiment No. 21.—June 26, ninety Blue Damson Plums, from E. Yenowine, Edwardsville, Ind., containing one hundred and twenty-five egg punctures, placed in vivaria.

Result of examination September 4: Ten adults. Dead.

Experiment No. 22.—July 12, twenty large Damson Plums, from isolated tree in garden of Hon. E. H. Scott, La Porte, Ind., containing sixty-five egg punctures, placed in vivaria.†

Result of examination August 13, 14, 16: Twenty adults. All living.

Summary of experiments.

Varieties of fruits.	No. of specimens.	No. of punctures.	Adults reared.	Ratio per specimen.	Ratio per puncture.
Large Damson	148	204	73	2.02	2.79
Blue Damson	281	351	86	3.26	4.08
Yellow Egg	95	103	19	5.00	5.42
Lombard	272	541	116	2.34	4.66
Coe's Golden	225	611	119	1.98	5.13
Nectarines	220	588	59	3.72	9.96
Wild Goose	518	885	219	2.36	4.04
Mariana	218	305	67	3.25	4.55
Kansas Sand	14	14	4	3.50	3.50
Chickasaw	24	24	5	4.80	4.80
Wild	176	220	35	5.02	6.27
Robinson	168	223	35	4.80	6.36
Apples	136	158	8	17.00	19.62

As will be observed, the greatest mortality to eggs and larvæ between the time of oviposition and the hatching of the adult occurred in the Wild and Robinson varieties of plums. Also that the apples used this

*A number of plum trees were growing in the immediate vicinity, but I could not get enough fallen plums for experiment.

†The top was so covered with a cone-shaped screen that the adults could be observed as soon as they emerged from the ground.

year were collected on the same day as the latest used last year, and from which nothing was reared. Those used the present year were from a more southern locality, where the season was correspondingly earlier, but the earliest to fall last year were used on experiments of June 20, leaving only the later fallen for the experiment of a few days later, and which gave no adults. Therefore, it would appear that the later punctures either contained fewer eggs, or else a larger portion of the larvæ perished before reaching maturity. If this be true, the variety of plum whose blooming season covers the greatest period of time will best withstand the work of the curculio; the earliest appearing fruit forming a sort of protection for the later.

So far as my experiments have gone the rule seems to hold good among both apples and plums. All of the apples used in both last season's experiments and this were grown among plum trees also fruiting, thereby demonstrating the fact that the planting of plum trees in the apple orchard will not protect the latter and *vice versa*.

From the drift of evidence gained from experiments of both last year and this, it would appear that if anything is to be gained by using another fruit to draw off the curculio and protect the plum, the point is almost as likely to be attained through the Nectarine as the apple. Indeed, this year the apples on the tree from which the fruit for last year's experiment was obtained suffered as bad or worse than the plums on trees growing interjacent. For position of this tree see Diagram.

The apple tree bloomed profusely, and produced a good crop of young apples, but by July 24 there was scarcely a dozen left on the tree, and the condition of these is illustrated by a figure,* drawn from specimens picked on this date, and bearing not only crescent marks in abundance, but also punctures, indicating that the adult beetles had recently been feeding on the pulp.

There seems to be little doubt but that the food punctures were made, in part at least, by the newly emerged adults. I saw an adult puncturing a plum at Greencastle, Ind., on June 22, and Mr. W. O. Fritz, foreman of the experiment farm, on July 23, brought me an adult curculio found that forenoon engaged in the same mischief, and adults were observed in experiment No. 1, July 29, which might have been and doubtless were present some days earlier, as the experiment had not been examined. It seems rather more than probable that the latest appearing individuals of the old brood of beetles may occur simultaneously with the advance individuals of the new brood, both feeding upon the fruit of the plum, apple, etc.

Occasional notices appear in the agricultural papers to the effect that the female curculio will not oviposit in fruit overhanging water. While this seems very doubtful, to say the least, an experiment was made in order to test the matter, but while clearing up the fog in one quarter, the results appear to have still further increased the obscurity in another.

---

* This figure will be published in the next number of INSECT LIFE.

A shallow pan, constructed large enough to cover the ground under one-half of a plum tree, of the Mariana variety, was placed in position on April 24, and kept continually filled with water until August 10. Now, not only were females observed in the act of ovipositing in plums hanging directly over the pan, but the latter contained from time to time quite large numbers of punctured fruit; nevertheless, the only plums on the tree reaching maturity were among those hanging directly over the water.

c	c	c	d	d	d	c	c	d	c	c	c
c	d	c	c	c	c	c	c	c	c	c	c
b	b	c	o	c	c	c	c	c	c	c	c
j	c	c	c	c	c	c	c	i	i	i	i
f	d	g	c	g	g	c	j	c	i	i	c
f	b	d	f	f	f	f	c	f	f	f	f
b	a	a	i	a	f	f	f	f	f	f	f
b	a	a	b	a	a	h	a	a	a	b	b
a	a	e	e	e	e	e	e	e	e	e	a
a	a	a	a	a	a	a	a	a	a	a	a
a	a	a	a	a	a	a	a	a	a	a	a
a	a	a	a	a	a	a	a	a	a	a	a
a	a	a	a	a	a	a	a	a	a	a	a

Explanation.—a=Blackman Plum; b=Wild Plum; c=Wild Goose Plum; d=Boggs' Plum; e=Crab Apple; f=Late Cherry; g=May Cherry; o=Seedling Apple; h=Moore's Arctic Plum; i=Quince; j=Pear.

A single experiment was made to determine the duration of life, and the probability of the female ovipositing, after having partaken of poison. Twelve females taken from the plum tree on May 17, where they were evidently ovipositing, were kept for 24 hours without food, some eggs having in the meantime been deposited in the box where they were confined. At 5 p. m. they were removed and placed separately in receptacles containing a leaf of the plum, thickly dusted with London purple. At 8 p. m., 3 hours later, nearly all seemed to be affected, but were removed and placed separately in clean quarters, and each provided with a fresh plum. At 11 a. m. next day many were dead, the remainder surviving but a few hours longer, but in no case were eggs deposited in the fruit.—[October 1, 1889.]

THE PHYLLOXERA PROBLEM ABROAD AS IT APPEARS TO-DAY.

The report of the Superior Phylloxera Commission has just been published and gives the latest account of Phylloxera matters in France and other foreign countries. Neither law nor effort has prevented the spread of the insect in eleven *arrondissements* in which it made its appearance for the first time the past year, viz: Castellane, Mende, Riom, Joigny, Troyes, Nogent-sur-Seine, Bar-sur-Aube, Vesoul, Gray, Bonneville, and St. Calais. About 240,000 acres have undergone defensive measures, submersion being employed in 72,000, bisulphide of carbon in 145,000, and sulpho-carbonate of potassium in 23,000.

Much good has resulted from the establishment of societies for defense, notably in Haute-Loire. Moreover, it is the small proprietor who derives the largest benefit from the law enacted August 2, 1879. Of twenty-one thousand three hundred and ninety-four proprietors composing a syndicate, each attended to about 4½ acres.

The departments in which vine cultivation is extensive, such as Herault, Gard, and Gironde, contain fewer syndicates for the reason that their Phylloxera work is practically at an end. Each year has shown an increasing acreage of reconstituted vineyards, mostly by means of American stocks, which prove more and more satisfactory and which justify the commission in prophesying the near approach of the time when vine-culture will be as widespread as it was before the era of the Phylloxera. The following approximate tabular statement will be interesting in this connection:

Years.	American vines covered.	Departments.	Years.	American vines covered.	Departments.
	Acres.			*Acres.*	
1881	22,000	17	1886	276,900	37
1882	42,700	22	1887	413,700	38
1883	70,000	28	1888	536,900	43
1884	131,909	34	1889	719,500	44
1885	188,200	34			

If the march of recovery continue at this ratio, in four years vine-planted land in France will reach the unprecedented amount of 6,500,000 acres. Hérault presents 380,000 acres of renewed vineyards; Aude,

68,000; Gard, 60,000; Gironde, 47,000; the western Pyrenees, 75,000; and Var, 47,000.

The efforts to produce by hybridization Phylloxera-proof varieties have so far not proved successful or popular, as most growers still depend on grafting on the American stock. Another noticeable fact is that the Government does not hesitate in its liberal policy of doing all in its power to aid the afflicted vine-grower, and the law of December 1, 1887, by which the land-tax on newly planted or restored vineyards is remitted for four years, is still in force.

Five years ago the Phylloxera first became known in Algeria, and since then it has been kept pretty well in check by the vigorous measures prescribed by the resolution adopted March 21, 1883. The cost has been great, but the results have fully justified the outlay. The vine there covers nearly 250,000 acres, and the vintage of 1889 shows approximately 66,000,000 gallons of wine.

A glance at the viticulture of other vine-growing countries shows that the industry is rapidly developing, especially in Chili, Uruguay, the Argentine Republic, and Australia. The Tunisian vineyards present remarkable developement.

Spain and Italy are yet suffering severely from Phylloxera. In the former the small proprietors are reduced to the necessity of abandoning the cultivation of their fields or selling them at much depreciated prices. The emigration from Malaga to Brazil and the Argentine Republic between April and August, 1889, amounted to eleven thousand persons, and may be taken as an index of the situation.

In Italy about 400,000 acres are affected, and the Government has been forced to forego its first system of defense and resort to American stocks.

Hungary suffers sorely. About one-third of its plantations are attacked and about one-eighth destroyed.

Austria suffers in almost like proportion.

In Switzerland the progress of the Phylloxera has been slow, and in Germany and Russia, owing to the measures taken for its suppression, it makes no progress.

Portugal seems to be in the worst plight of all, for each year the number of invaded districts increases, chiefly in the north, where there are 250,000 acres of infested vines and 90,000 acres of dead ones. The Douro region aggregates 80,000 dead vines out of a possible 125,000.

Nowhere has the combat been carried on more energetically than in France, originally the most sorely stricken country, and nowhere has so much success been achieved against Phylloxera attack.

THE LOS ANGELES COUNTY HORTICULTURAL COMMISSION.

The following copy of the last report of the board of horticultural commissioners of Los Angeles County, Cal,, is taken from the Los Angeles *Evening Express* of March 5, and will not be devoid of interest to our readers. The account of the correspondence between the Secretary and this office is very fair. with the important exception that we insisted upon the necessity of first thoroughly knowing our ground before taking extensive steps for the importation of enemies of the scale insects mentioned. By this we mean ascertaining carefully the range of each species and the probabilities as to its original home.

We respectfully tender herewith the monthly report for February of the county horticultural commission.

The policy adopted by this commission of continued and earnest research for a parasite that will destroy the red and San José scales, or any other pests that are injurious to fruit trees, has been pursued during the past month.

Our secretary was instructed to communicate with United States Entomologist, C. V. Riley, at Washington, requesting him to ask Congress for an appropriation that would enable the Department of Agriculture to seek the world over for parasites that prey on the insect pests that are now threatening the wefare of our great fruit industry. In reply, Professor Riley advises us "that he will not be able to do much with Congress in the way we suggest, but that he hopes and expects that the United States Department of Agriculture will have power to act without such a petition after June next." Professor Riley still further advises us "that the red scale of California (*Aspidiotus aurantii*) has been believed to be of Australian origin, but that it is about as abundant there as it is in California. He says that it does occur in other parts of the world, and much inquiry will have to be made before we can feel sure of its native home; that it has some parasitic enemies in California. and though it doubtless has others in Australia, we know so far only of a fungus and a small beetle that attack it there." Professor Riley also says "that the San José scale (*Aspidiotus perniciosus*) is not as yet known to be an imported species, but that all these scales are amenable to careful treatment by the sprays which we have lately recommended, or by the improved gas treatment."

Notwithstanding the valuable opinion of Professor Riley, this commission feels that in making inquiry for a parasite for the red scale in other countries search should also be made for an enemy for the San José scale insect. This pest, if not speedily destroyed, will utterly ruin the deciduous fruit interests of this coast. It not only checks the growth of the tree, but it covers the tree literally entirely, and the fruit nearly as much so, and if left unchecked, the tree is killed in three years' time.

There is absolutely no parasite at work on the San José scale insect. We find this dangerous pest invading every deciduous fruit district in the county, and have notified owners of such infested orchards to disinfect, giving them the necessary mode of procedure. Unless the San José insect is thoroughly stamped out the deciduous fruit interests of the county will in a few years have dwindled to naught.

In our January report to you we mentioned having been compelled, after exhausting the necessary preliminaries, to place in the hands of the District Attorney for prosecution the case of F. O. Cass of Vernondale. We were led to take this step, not only from our sense of duty to the State law prescribing it, but as a determination of our duty and obligation to the fruit growers of Los Angeles County, wherein we sought to stamp out a dangerous insect pest, the Santa Ana red scale, just obtaining a foot-hold in this county.

The case came up before Justice Rankin and a jury of six, February 14, in San Antonio Township, and was decided against the State on the 17th. The evidence of the defense was simply a line of individual theories, in fact farcical, when compared with the important results of years of study by scientific entomologists and the long and tried experience of the most thorough and intelligent horticulturists of our own county.

It is not and has not been the policy of this commission to enforce indiscriminate spraying without regard to the existence of parisitic insects, but in the case of Mr. Cass it was evident to us that unless prompt measures were taken the Santa Ana red scale would effect a lodgment in this district that would eventuate in its spreading to every citrus fruit orchard in Los Angeles County in another twelve months.

The result of our efforts, when it is considered as the consequence of public opinion, is certainly a sad commentary on Los Angeles County as a citrus or deciduous fruit growing district. This commission endeavors to squelch the most dreaded of all red scale insects in its incipiency, an insect that has no effective parasite, but are prevented from doing so by a jury influenced by the public opinion of Vernon district.

In connection with this deplorable result we hand you herewith a careful compilation of statistics, showing the number of citrus trees now under cultivation in the county. It does not include trees situated in acreage cut up into "town lots," or that have been abandoned or are not worthy of future care.

Description.	Age ten years and over.	Age five years to ten.	Age five years and under.
Number of orange trees	289, 677	119, 530	187, 500
Number of lemon trees	18, 055	29, 345	10, 350
Number of lime trees	4, 575	435	150
Number of citron and pomalo trees	15	15	....

In addition to this there are 395,000 budded orange and lemon trees in nursery form that will be planted the coming season. This does not include seed-bed plants.

Thus there will be 1,054,647 citrus fruit trees, the comparative income from which can be easily computed, that will be threatened with ruin by an insect pest that the commission have been opposed in their endeavors to check.

In our previous reports we have called your attention to the quarantine of other counties against our nursery stock and fruit. The wide publicity given this late obstruction to the law, made to protect and promote the horticultural interests of the State, will still further enact against the county.

We are pleased, however, to report that in some portions of the county producers are alive to the value of our fruit industries, and realize the necessity of vigilant protection. In connection therewith we hand you a communication from the Pomona Board of Trade, inclosing resolutions adopted by that body.

Our instructions to inspectors have been to inforce the law in all cases in reference to infected fruit exposed for sale, and since receipt of Pomona Board of Trade resolution we have renewed said instructions.

It will be apparent to you that if the trees under cultivation and to be set out this season are to return an income to our producers, and if Los Angeles County is to retain its well-earned reputation as a citrus-fruit growing county; and still further, if the thousands of acres in this county so well adapted for fruit-growing are to be settled up and cultivated by fruit-growers, it will be necessary to redouble all previous efforts in a warfare against insect pests.

We regret to report that during the past month a new insect pest, that is, new to this county, has found a lodgment here. We refer to the "Purple scale" that has been introduced on the large number of orange trees now being brought into the State from Florida. Effective means are, however, being used by us in stamping it out, and we are pleased to be able to report that we have been willingly supported in our efforts by all dealers handling the stock.

The State board of horticulture, the State fruit-growers' convention, and the convention of county horticultural commissioners, all meet in Los Angeles March 10 to 15. These bodies will be made up from the leading horticulturists of the State, men of high intelligence and long experience in horticultural pursuits. From their deliberations and determinations we hope for grand results in furthering the fruit industries of our county and in protecting it after such promotion.

We have employed the same number of inspectors this month as during the last. They have inspected 857 acres, containing 49,759 trees, and have served thirty-two notices on owners of infected orchards.

Respectfully,

A. F. KERCHEVAL, President.
F. EDWARD GRAY, Secretary.
County Horticultural Commissioners

EXTRACTS FROM CORRESPONDENCE.

The Pine Lachnus as a Honey-maker.

I send by this mail a box with pine tags, live Aphids, and honey-dew. I put in a section of limb, cut sometime ago, where the insects had sucked the bark dry.

Cutting a limb with the Aphids on, and the leaves covered with honey, I found the next day that they had gone to the cut, where they were fifty deep trying to get at the exuding turpentine. I wish you would send a man, a good chemist and microscopist, to look into this matter. This honey can be seen on the laurel leaves where there are no Aphids. My son, while hunting last week and looking under the pines, noticed that the rays of the sun made visible a fine spray falling from them. Another man told me yesterday he had seen the same. I can show proof that the honey is not a visible exudation from the aphis. I can get you a small vial of this honey gathered drop by drop from the pine leaves. My neighbor has secured 8 small vials full.

We often have honey-dew in summer, sometimes covering the hickory, gum, oak, chestnut, and poplar leaves, but this is the first winter shower of manna we have observed here. It commenced December 20, and ran every day to the 30th; then January 1 to 7, 10, 11, 12, 13, 23, 26, 27, 29 (?), 30 (?), 31; February 1, 3, 4, 12, 13, 15, 16, 17, 18.

My eighteen years' observations have proved to me that the atmosphere is nature's storehouse for honey; my proof and facts I don't think can be overcome. There is so much of this honey on the pines now that my seventy-three colonies of bees can't gather it from them. I estimate 100 pounds of honey on every acre of pines. In the morning it is there like dew in drops as large as peas, but before night it evaporates to thick, ropy honey.—[W. M. Evans, Amherst, Va., February 18, 1890, through the Smithsonian Institution.

REPLY.—The inclosed letter from Mr. W. M. Evans, of Amherst, Va., referring to accession No. 678, is very interesting, and examination of the specimens shows that the plant-louse secreting the honey-dew in such quantities upon the pines is one of the species of the genus *Lachnus* of which several species are known upon coniferous trees. The specimens are dry and can not be determined specifically The facts which Mr. Evans gives us show that the honey-dew is more abundant than I have ever known it before in the Eastern United States, and his letter is well worthy of publication. I shall therefore take the liberty of publishing it in a near number of INSECT LIFE, a copy of which will be sent to him.—[February 21, 1890.]

Root Knot on Apple Trees.

A copy of Bulletin No. 20, Division of Entomology, on the root-knot disease, which was sent to my former address at Glencoe, Nebr., has just reached me. I have been interested in the perusal of Dr. Neal's notes from having had some experience with root-knot myself.

In the spring of the present year I bought several hundred two-year-old trees of willow twig and Ben Davis apple from a local nursery. In planting I found the roots of many were very knotty; those worst affected having few fibrous roots. Not one in ten of some four hundred put out any leaves from the tops, but most of them sent out sprouts from the side of the trunk at or near the ground, which shoots made a weak growth. I had requested the proprietors of the nursery to give me trees of their own raising and supposed they had done as they agreed to do, but some of their employés afterward told me that my trees must have come from Kansas, as they "got all those knotty-rooted trees from that State."

Dr. Neal does not mention apple among the plants affected by Anguillula, and for this reason, and also because he thinks his evidence conclusive that the disease does not exist 150 miles from the coast, I have thought it worth while to bring this matter to your notice. Many of the trees died during the drought of July and August. About seventy-five trees of Ben Davis and Maiden Blush, brought from same nursery same spring, but a few days later, and which had good fibrous roots showing no knots, have grown and done as well as usual in a dry season.—[G. M. Dodge, Louisiana, Pike County, Mo., November 11, 1889.

REPLY.—Your letter of the 11th instant has been received and referred to the Entomologist, who reports that he is obliged to you for your notes on Anguillula, and that he himself has for some time been aware of the fact that many other plants were damaged by these creatures in addition to those mentioned by Dr. Neal: also that the work is by no means confined to the vicinity of the sea-coast. The knots on Apple, however, may have been due to some other cause.—[November 20, 1889.]

A Fuchsia Aleurodes.

The fuchsias in my bay window are infested with scale-like cocoons, on the under side of the leaves, from which emerges a tiny white fly. Please tell me something about it. I send specimens of cocoons and flies in a wax cell mounted on a slide. It is so arranged that you can remove the cover glass if you find it necessary. The flies are alive. What are the peculiar objects like crystals? They polarize prettily, not unlike horn or keratose. Besides the mount, I send leaves infested.—[Samuel Lockwood, Freehold, N. J., January 15, 1890.]

REPLY.—The little insects which you find on fuchsia leaves belong to a species of the genus *Aleurodes*. I have had this form for some time, but it is yet undescribed. The family *Aleurodidæ*, as you know, holds a position between the Aphididæ and the Coccidæ, and has not been studied in this country.—[January 18, 1890.]

The Skein Centipede and Silver Fish.

There are two *creatures* that have the freedom of this town, about which I have heard a great deal of nonsense talked, and now wait for some sensible information. Whether they are *insects* or not, I do not know; I wait for you to tell me, but certainly they must often stroll into the suburbs of your province. I never knew any one who could give a popular name to the first creature, which, for the sake of distinction, I call in the house a *centipede*, which it is not. The first I ever saw was five inches long, at least. I thought it was a skein of brown silk in a tangle, and picked it up from the carpet with thumb and finger. I have never seen another as large, but the wet weather brings them into the bath-room in two sorts, one as I have described it, brown and tangled, the other of the same general shape, but with distinct antennæ at one end, and something similar at the other, black and smoky in color. If you kill either there is no body left, only a ghost, which has no anatomy. I hope you appreciate my scientific knowledge.

The other goes popularly by the name of "*silver fish.*" It also is a creature of the damp. The colored people declare it is the husband of the moth. I killed one in May that looked formidable for it was more than two inches long. When I returned this autumn a bit of flannel that had been carelessly left out was riddled with moths, and as I took it up to throw it in the fire a very large "silver fish" slipped ont to meet a speedy doom. Such is the origin of the myth, I suppose. Now, can you refer me to any Government bulletin which will give me the biography of these unwelcome visitors, or any book? If not, will you give me some of the facts yourself and introduce me to the husbands and wives, if they do not have a family likeness.—[Caroline H. Dall, No. 1603 O Street, Washington, D. C., November 12, 1889.

REPLY.—Your letter of the 12th inst. was duly received, and while it would have been desirable to have received specimens of the "creatures" to which you refer, your interesting description of them leaves little doubt as to their identity. The one which you call a centipede—and it is one—has no definite common name other than "Thousand-legs," or the more inaccurate "ear-wig," but is known to science as *Cermatia forceps.* The two sorts observed by you were only different phases of the same animal, the tangle being either a dead specimen or the exuvium (for, like all Arthropods, it molts). Little is known of the habits or life-history of this widely distributed pest, which of late years, particularly, has frequently occasioned annoyance in houses. It is undoubtedly carnivorous in habit, however, probably feeding on other household pests, which its quick movements enable it to capture. There is current belief, well founded, I think, that it feeds on young roaches. Its bite, while reputed poisonous, is not dangerously so; and I have personally never known of injury so resulting and much doubt if there is foundation for the belief. It may, however, be considered as a friend, but its singular appearance and rapid movement are hardly calculated to inspire confidence.

The "silver fish" of your letter is without doubt the well-known pest of books and clothing (*Lepisma saccharina*), and is entirely distinct from the clothes moths. It feeds particularly on starched clothes and the binding of books, which it eats for the starch, and sometimes injures silks and other fabrics. Pyrethrum will prove effective, also, against this last insect. I am sorry to say that there are no publications of the Department relating to these pests for distribution. The first is fully described by Dr. Lintner in his fourth report on the insects of New York, and the second is briefly described in Packard's "Guide to the Study of Insects," p. 623.—[November 13, 1889.]

A Guava Scale.

I send you a branch of guava tree, the first and only one that I have ever seen infested with any kind of an insect enemy. I suppose this to have come from a large rubber tree near by. The rubber tree is often covered with this black dust and same kind of a scale. Will you please tell me the name of this creature, and whether it will be likely to spread to other guava trees; and if so, how can we best dispose of it.—[E. Gale, Lake Worth, Dade County, Fla., November 13, 1889.]

REPLY.—The specimens which you send from guava are common Florida wax-scales (*Ceroplastes floridensis*). You will find this insect figured and described in Hubbard's Insects affecting the Orange. It commonly affects the gall-berry, but is also found upon quince, apple, and pear, and occasionally upon orange. It can be destroyed by the ordinary kerosene soap emulsion, which, however, should preferably be applied when the insects are young.—[November 20, 1889.]

The Tile-horn Borer.

Last year I sent you specimens of borers which were destroying an ash tree in my yard, and worms also found in a large oak in the same yard. The ash tree died. By this mail I send you another specimen of borer, found in the heart of a large oak which died last fall, the sap timber of which looks like a coarse sponge. I send the oak chip in which the destroyer was found. I very much fear that the fly or moth is depredating upon other oaks in the yard. How can we distinguish it, and is there no protection against it ?—[Carrington Mason, Memphis, Tenn., October, 22, 1889.]

REPLY.—The larva is that of one of the large Tile-horn beetles (genus *Prionus*). The particular species is probably *P. laticollis*. For an illustrated account consult Riley's second report on the insects of Missouri, p. 78.—[October, 1889.]

The Boll Worm.

As you are chief of the entomological division of the Department of Agriculture, I take the liberty of writing you in regard to a pest that is fast destroying the prosperity of the cotton farmer of this section of our State and reducing us to penury, and will ultimately, if help does not come, force us to abandon cotton culture. That pest is the "boll worm." Much discussion has been had in our local press—many things have been advised and tried—but their ravages were greater the past season than ever before, and I feel convinced that something else will have to be done than we have hitherto adopted. Paris green, London purple, burning of lamps to catch the miller, are among the best of the remedies resorted to, but all have proved comparative failures.

I am not an entomologist, but necessity has forced me to give some attention to this matter, and this attention has been followed by the conviction that our most certain method of relief will be found in the line of fostering and caring for those natural enemies, parasites and otherwise, which we know by observation do exist here, or which observation teaches by parallel lines of investigation, may exist elsewhere and can be introduced here. Could you give us any help; and, if so, will you ? I have read carefully reports made by yourself to the Department in 1881 or 1882, and thank you and the Department therefor, but we need something more.

We need, I think, bulletins scattered broadcast throughout this part of Texas (east Texas) containing the information in your reports, and such other practical suggestions as may occur to you or others who are familiar with the life-history of this pest and those parasites, its natural enemies, to be found here; and besides this, a more thorough search for something that will prove of more utility than anything hitherto suggested.—[H. L. Tate, M. D., Lindale, Smith County, Tex., January 12, 1890.

REPLY.—Your letter of January 12 relative to the boll-worm has been received. I have sent you to-day a copy of the Fourth Report of the U. S. Entomological Commission, in which you will find the subject treated from the latest stand-point. If you have read nothing from me upon the subject since 1881–'82, you will find this matter interesting. There is little to be hoped for in the direction which you particularly mention, viz, the assistance of parasites. As it happens, the boll-worm is singularly free from the attacks of parasitic insects, and up to the present time only one or two have been recorded. These, moreover, are very rare and do not seem to breed in any abundance. The best hope is in spraying with Paris green and in worming the neighboring corn-fields, as indicated in the report which I send you. If it seems necessary, we may give some further attention to the matter the coming season.—[January 17, 1890.]

Feather Felting by Dermestids.

I have in my possession a beautiful curiosity, and, as far as I can learn, the only one in existence. I take the liberty of addressing you in regard to it, as you are authority on entomology, and this will probably come under that head.

It is an ordinary feather-pillow tick, which was made of common bed-ticking and filled with the domestic duck feathers about three years ago and the pillow has been in general use about the house since that time. Of late the lady concluded to remove some of the feathers, as the pillow appeared too hard. Upon opening the tick the feathers seemed to be ground up almost into a powder and unfit for further use; therefore they were emptied and the tick turned inside out, and instead of the goods being as when made, it was entirely covered with a fine growth of down as evenly and thickly as the fur on a mole-skin, which it very much resembles; it is firmly attached, the down breaking rather than pull off. Not a piece of the feather is attached to it but as smooth as a piece of velvet, even the seams are covered by the growth. Not an insect can be found in the feathers, but the grinding process was supposed to be done by some insect. The lady made several pillows at the same time and of the same feathers, but when these pillows were opened nothing was found but feathers as when made. This was found about a month ago and the ladies through the country have opened many pillows, some as much as fifty years old, but no such thing can be found. To look at it one would think it the hide of some animal, and would never imagine it to be a pillow-tick except by close inspection. I inclose some of the feathers, which will give you an idea of the color and a description from the local paper which may help to give you an idea of its appearance. Many theories as to its formation are offered, but nothing satisfactory, and the community would like your opinion. * * * —[J. D. Davis, Clarksville, Mo., January 19, 1890.

REPLY.—Yours of the 19th inst., together with the specimens of feathers, duly received. A careful examination of these fragments shows no trace of an insect or of insect remains. The specimens which you have are very interesting, although I have seen the same thing before and several notes have been placed upon record regarding precisely similar cases. In the *American Naturalist* for December, 1882, I mentioned one of these cases and gave an explanation which is as follows:

Pillows in which this felting of the ticking occurs have been infested by one of the Dermestid beetles (in all of the cases with which I am familiar it has been *Attagenus megatoma*) whose work has resulted in the comminution of the feathers, and the felting results from the subsequent mechanical action. The small feather particles are barbed, as you are aware, and, whenever caught in a cotton fabric by their bases, become anchored in such a way that every movement of the pillow anchors them still further. The frequent shaking which pillows receive results ultimately in the formation of this plush-like surface. A similar bit of ticking was exhibited at the Philadelphia Academy of Natural Sciences, April 5, 1883, and elicited the information that one of the members had some years previously examined a similar material known to have been formed from the fragments of gull feathers and that a cloak had been made from it which wore well. * * *—[January 23, 1890.]

Extreme Ravages of Cut-Worms.

As our part of the country has been ravaged this year by the Cut-worm, which I believe is the same one that destroyed the onion crop of Orange County, N. Y., in report for 1885, p. 270, I would like very much to know if you have any subsequent information in regard to the habits of the moth or worm; if so, I would be very grateful for report containing it. Corn has been the crop that has suffered here, and as some fields are totally destroyed, the damage in these parts will amount to many dollars. I will give my case as a sample. I have four pieces of 28, 6, and 5 acres. The first three pieces were planted with Lister, beginning May 10. First planting was a total loss. Replanted all on the 28-acre piece. All destroyed the second time except 3 acres with about one-fourth stand left. Second piece was a total loss, and was sowed with one-half bushel to acre for fodder; at present writing it shows very little corn, as Cut-worms take it not quite as fast as it comes up. Third piece gave less than one-fourth of a stand. Fourth piece was plowed, having been broken last year, and is about one-third stand with Cut-worms still working. I tried cutting, growing rye, and Paris greening to poison them, but the bait was not succulent enough. I think I could have killed a good many if I had used suitable bait. Have found seven worms eating at one stalk of corn under the ground. Last year I also suffered with Cut-worms. Planted 18 acres, replanted 16, and then sowed corn on 3 of it. The sowed corn was all right; balance produced one-fourth stand. If they increase next year over this year, corn planting will be useless.—[Chas. A. Howitt, Neligh, Nebr., July 4, 1888.

REPLY.—In addition to late fall plowing the best remedy which you can use is the poisoned bait with which you are already familiar. The only difficulty is to secure green and succulent vegetation for poisoning; and, of course, being upon the ground, you can more readily decide what will be best and most convenient to use.—[July 10, 1888.]

ANOTHER LETTER.—In looking over the reports of 1884, just received, I notice an article upon Cut-worms, which is of importance to us away up here in northwestern Minnesota, and I desire to ask your advice on how to proceed in my war upon them. I am on the southeast shore of Otter Tail Lake near Otter Tail City: my land is a sandy loam, was timber land in 1850, but now nearly clear of timber by reason of the encroachments of prairie fires. Consequently the soil is a warm productive soil, quick to warm up in the spring, and a good harbor for all sorts of insects.

I have for two years past failed to raise onions, beets, and carrots, and beans too, as well as nearly an acre of sweet corn, on soil only under cultivation three years. Onions, beets, carrots, and beans were sown with a "Planet, Jr.," garden drill; they were cut off as fast as they came to the surface, just below the surface, by very small young Cut-worms. Neither salt nor ashes would stop their work; the crop was an entire loss. Now, what can I do? As I sow two or three acres I can not apply the remedies laid down in the 1884 report, pp. 299–300, as it would lose too much in time and labor. But I do want to raise onions, beets, and carrots as well as beans and sweet corn.—[Washington Muzzy, Balmoral, Otter Tail County, Minn., March 4, 1887.

REPLY.—While late plowing of the fields infested by cut-worms may have a good effect in lessening the numbers the ensuing season, a much better plan will be the adoption of the poisoned ball system recommended in the article to which you refer. This method will not require the expenditure of much time or labor. It simply involves the necessity of a pretty general distribution of poisoned spring grass over the plowed fields a few days before the sowing of your onions, beets, carrots, or beans. There will doubtless be plenty of young grass and weeds up at the proper time, and such should be cut and sprinkled with Paris green solution and little patches placed at intervals about the field. This is absolutely the best remedy known. It works admirably in the South, where there is so much early vegetation, and we shall be glad to learn your opinion of its practicability in Minnesota, or of the success of any experiment you may try with it.—[March 9, 1887.]

Migrations of Plants as affecting those of Insects.

When the writer first came to this part of Kansas, eighteen years ago, two plants which are now very abundant were unknown in this county of Geary, then called Davis. One of these is the *Solanum rostratum*. The region for two or three years suffered from the ravages of the Colorado Potato-beetle, but now, though the beetle is sufficiently abundant every year, the potatoes rarely are damaged. The cause seems to be that *Solanum rostratum*, sometimes called Buffalo Nettle, or Buffalo Thistle, is the native food-plant of this beetle, and where it is scarce *Solanum tuberosum* is accepted as a substitute. The plant belongs to regions farther west, and by some means the beetle traveled in abundance eastward, reaching the other side of the Atlantic years ago, where the plant is still unknown. It is said that the prickly seed-pods of this plant came on the tails of Texas and other cattle from the Southwest, and it is certain that counties remote from the cattle-trails and the through lines of railway were the last to have the plant. The flower is bright yellow, and the whole plant not unhandsome, but its prickles make it a very undesirable weed. Two years ago the writer took particular pains to eradicate it in and around his garden patch, killing every young plant of *S. rostratum* as it came up. The result was a serious attack on the potatoes, which were only saved by twice going over all the plants and collecting and destroying the beetles. That the plant did not migrate easterly at a greater speed—I don't think it has crossed the Mississippi yet—is to be wondered at, as in the region of the one hundred and second meridian, on the wide prairies, it has the tumble-weed habit. The whole plant is subglobose and when ripe snaps off close to the ground and goes bowling along before the wind at a great rate. The winds there, however, are more north and south than from the west, so that probably has delayed the progress of the plant in longitude. The plant is abundant in waste places in towns and by roadsides in all eastern Kansas now, and we rarely hear of the Colorado Beetle damaging potatoes.

Another plant which is traveling eastward is the Mexican Poppy, Prickly poppy, or, as some have called it, California Poppy. It is the *Argemone mexicana*. Many years ago, sixteen or seventeen, the writer first saw it in the region of the one hundredth meridian, and he noticed it more easterly every year since. Several years ago an illustration in *Harper's Magazine* to an article entitled "Ladies' Day at the Rancho," showed it as being a prominent flower in Ellsworth County. It is very abundant in waste lots of Junction City now, and the last season it was seen as far east as Wamego, about the ninety-seventh meridian. It may be further east, but the writer has not seen it. It is a very handsome plant, with a very large white flower, manifestly the variety *albiflora*. It may be that the migration of these plants has elsewhere been recorded, and that it may have proceeded further than is here set down, but it seems that the record is worth preserving if not previously made.—[Robert Hay, Junction City, Kans., February, 1890.

Hymenopterous Parasite of Icerya in Australia.

I have done a little as follows: First, I have bred four hymenopters, which I take to be the parasite which Koebele discovered. All I know about them is I found them alive in a bottle containing some Icerya, and from which some Lestophoni had emerged. The hymenopters had not emerged from any Lestophoni outside the Icerya, but that one would not expect, still I have no proof that they were not parasitic on the Icerya; but I presume they are Koebele's parasite. I found two out of the four.

Second, I received some three months ago some Icerya from a place some 50 miles south of Adelaide, the owner of the orange orchard not having seen anything of the kind before and wanting to know what they were. These I placed as usual in a bottle loosely stoppered with cotton wool. With the Icerya was a Chrysopa larva, which for some weeks was busy feeding on the eggs. One day on examining it I discovered several Hymenopters (Proctotrupidæ?). The female, yellowish-brown (?); male, almost black (?). On examination I found that many might have escaped through the cotton stopping being insecure, but I suppose I have had about thirty since. It is strange that this is the only instance of an hymenopterous parasite of Icerya yet discovered in South Australia. I send you a few of these under separate cover. I presume the small black insect is the male, because I observed them chase the larger brown flies, and then leap on their backs, but so far in front that it would be impossible for any sexual connection to take place (at least in my opinion), and then would commence a rapid movement of the antennæ, as if they were having a bout of fisty-cuffs. I observed this three or four times, but in no case did I observe any act of coition—as the bottle was not very well suited for observation with a coddington lens, it may be that I am mistaken, but such is my impression. This strange proceeding would last a few seconds. Was it a kind of preliminary investigation on the part of the male to discover whether the female had been already impregnated?

I likewise send you some cayenne pepper in which you will find some small beetles which breed in that very hot condiment. A sole diet of cayenne pepper must make them peculiarly hot tempered if beetle life in any way resembles human existence.

Do you know of any such habit in the States?—[Frazer S. Crawford, Adelaide, South Australia, November 24, 1889.

REPLY.— * * * In reference to the specimens which you sent, No. 1 is *Euryischia lestophoni* Riley MS. No. 2 interests me intensely as it is the first primary Hymenopterous parasite on *Icerya* from Australia. I propose to characterize it and name it *Ophelosia crawfordi*,* if you have no objection. It is somewhat near *Dilophogaster*, which is a parasite on some allied Coccids, but will have to form a distinct genus. I shall be very glad to get some additional specimens of this for the Museum collection, but particularly should be obliged to you if you could succeed in sending over a box of living specimens with a few *Icerya* for them to breed on. Better still, if you could get a good supply of *Icerya* from the tree or neighborhood where these were found, the chances would be very good of some of them coming out on the journey, or even after they arrived in California. I should like to have some sent to Mr. Koebele at Alameda, and also some to Mr. Coquillett at Los Angeles.

The beetle in red pepper is the well-known cosmopolitan *Sitodrepa panicea*.—[January 10, 1890.]

Proconia undata Injuring the Vine.

Inclosed please find envelope containing two specimens of an insect; it is of average size and first made its appearance about ten or fifteen days ago. Its mode of operation is to stick its sucker or bill into the young shoot of the vines and commence to *pump*. The water of the vine passes immediately through the bug, which can plainly be seen with the eye by holding your hand under it. When it is at work, your hand in about one minute will be covered with water, just about like the morning dew. Its bill is placed near where its head joins its body. It is very destructive to the vine; the leaves it does not attack, only vines, stems of the leaves, and the stems of the bunches of grapes. Inclosed please find cutting from the vines. The leaves of the vine were also eaten, but by some other insect, as I have failed to find this bug eating the leaves. If there is any remedy for the destruction of them please let me know at once.—[A. B. Daily, San Marcos, Tex., May 10, 1886.

REPLY.—The insect which you send is one of the common leaf-hoppers of the vine and is known as *Proconia undata*. You describe its work very well, and if it appears in sufficient numbers to threaten your vines seriously, it will be well for you to spray them in the heat of the day with an emulsion of kerosene and soap, according to the usual formula. The leaves on being examined showed the appearance of a fungus (*Phyllosticta labruscæ*), which produces the rust colored spots. If this fungus appears extensively you will find a remedy in dusting the vines with sulphur and lime.—[May 17, 1886.]

---

* Since described on p. 249 of the current volume.

STEPS TOWARDS A REVISION OF CHAMBERS' INDEX, WITH NOTES AND DESCRIPTIONS OF NEW SPECIES.

By Lord Walsingham.

[*Continued from page 286 of Vol. II.*]

COPTOTRICHE gen. nov.

Κοπτειν = to cut ; Θριξ = a hair.

Tischeria complanoides F. & B. = *latipennella* Chamb.

Antennæ, ♂, ciliated, a minute projecting hair pencil from the basal joint beneath

Labial palpi, dependent, scarcely longer than the head.

Haustellum, rather long.

Head, clothed with an erect tuft above ; face smooth.

Fore-wings, lanceolate, pointed, clothed with long cilia commencing abruptly at the outer end of a distinct cuticular fold which extends from near the base of the costal margin to three-fourths of the wing-length on the underside, and is of nearly even width throughout; beneath the fold two-thirds of the wing-surface is thickly clothed with long hair-like scales arising most conspicuonsly from the submedian vein. *Neuration*, 9 veins, apical vein forked, the branches ending on opposite sides of the apex ; the remaining veins simple.

Hind-wings, lanceolate, as wide as the fore-wings, the costal margin suddenly depressed at the outer fourth, ending in a sharp point almost in a line with the dorsal margin; the first half of the costal margin is clothed with very long cilia, and the cilia on the dorsal margin are also very long, but at the depression above the apex these are abruptly shortened, giving an excised appearance as if caused by an injury.

Abdomen, anal tuft moderate; terminal segment ending in a pair of well-developed lateral claspers, uncus apparent.

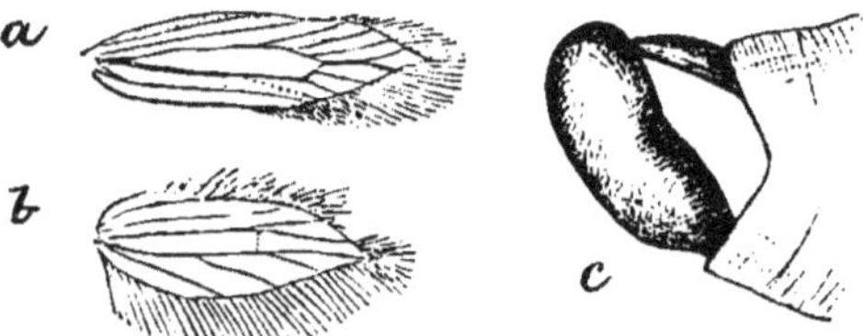

Fig. 64.—*Coptotriche complanoides* : *a, b*, neuration of front and hind wings of male; *c*, genital segments of male—enlarged (Walsingham *del*).

This genus differs from *Tischeria* in its much wider hind-wings, somewhat abruptly pointed downwards at the apex, in the long tufts of hair-like scales on the underside of the fore-wings, and in the conspicuous costal fringe along the basal two-thirds of the hind-wings and in the excised appearance of the costal cilia above the apex.

Coptotriche complanoides F. and B.

Tischeria complanoides F. and B. (*zellerella* F. and B.)
n. syn.=*Tischeria latipennella* Chamb.
(?=*Tischeria zelleriella* Clem.)

Clemens described *zelleriella* [Proc. Ac. Nat. Sci. Phil., 1859, 326. (Stn. Tin. N. Am., 81)] as having bluish-gray hind-wings, the fore-wings yellowish running to reddish saffron towards the tip. He mentions also a supposed female entirely reddish-ferruginous.

Frey and Boll [Stett. Ent. Zeit., XXXIV, 220-1 (1873)] described their specimens of *zellerella* as having the wing-tip of the same color as the base.

Zeller [Ver. Z.-b. Ges. Wien, XXV, 147 (1875)] refers to a specimen sent him by Frey under this name and expresses a doubt whether it is truly Clemens' species ; he also draws attention, for the first time, to the peculiarity of hind-wings which is also the distinguishing character of *latipennella* Chamb.

It is remarkable that neither Frey nor Clemens should have observed this.

The specimen referred to by Zeller is now before me and considering the degree to which the outer portion of the wing is shaded with darker scales it is possible that it may be rightly identified by Frey.

A series of six specimens, all males, received from Miss Murtfeldt and from Monsieur Ragonot (from Boll's collection) show the peculiarity of the hind-wings in a marked degree, sufficiently I think to constitute a separate genus. The difficult question, however, is to decide which of the numerous oak-feeding species described from North America is the female of this form. One specimen regarded by Miss Murtfeldt as *badiella* Chamb., although slightly smaller and lacking the peculiar outline of the hind-wings of the male, appears to me to agree in all necessary particulars; it also differs from *badiella* in the absence of a dorsal spot, agreeing in this respect with *castanella* Chamb. Chambers' remark that *castanella* is larger than *zelleriella* further proves that his idea of that species was not the same as that of Frey and Boll, whose specimen is a large one. It would be rash to presume that *castanella* is merely a synonym of *zelleriella*—this and other allied species require further study. It is, however, quite certain that the *zellerella* of Frey and Boll (for which they suggest the name of *complanoides* if distinct) and of Zeller's writings is equal to *latipennella* Chamb., and it is probable that one of the other species, if described from females only, will turn out to be the same. Frey and Boll refer to the female, but as they overlooked the peculiar form of the male, little, if any, assistance can be derived from their brief notice. Chambers did not mention that he had both sexes of *castanella*.

I shall be greatful to any one who will examine Clemens' type ♂ of *zelleriella* and let me know whether the hind-wings have an excised appearance, caused by the shortening of the cilia above the apex (see Fig. 64 *b*). Until I can assure myself on this point *zelleriella* Clem. must be retained in the index as a distinct species, and Frey and Boll's determination, which was questioned by Zeller, must be regarded as erroneous.

C. complanoides has been received from Texas, Missouri, and North Carolina.

TISCHERIA Z.

Tischeria clemensella Chamb.

=*zelleriella* Chamb. [Cin. Qr. Jr. Sc., II, 109-110 (1877)].

I am quite unable to identify this species from the material in my possession. It may be possibly the true *zelleriella* Clem. as suggested by Chambers [Bull. U. S. G. G. Surv., IV, 98-9 (1878)], in which case Frey and Boll's identification of that species must be incorrect. No reference is given to this name in the Index, but a specimen exists in Cambrige Museum (Mass.), received from Chambers [Hgn. (Frey) Pap., IV, 153 (1884)].

Tischeria castanella Chamb.

I am unacquainted with this species except from the description.

Tischeria citrinipennella Clem.

n. syn.=*badiella* Chamb.

This is a lemon-yellow species. The distinguishing mark noticed by Stainton [Tin. N. Am., 82 (1872)]—a patch of dark scales at the anal angle—was not mentioned in the original description, but exists in a specimen in my own collection compared with the type in 1871. This is characteristic also of *badiella* Chamb., indeed so far as I am aware it occurs only in this species and in the darker *tinctoriella* Chamb. I am unable to trace the patch near the base of the hind-wings mentioned by Clemens. Chambers suggests that his *badiella* may be Clemens' *zelleriella*, but his description agrees in all important points with my example of *citrinipennella*, and I have no hesitation in regarding his name as a synonym.

Tischeria quercitella Clem.

n. syn.—*quercivorella* Chamb.

Chambers in discussing the distinctions between his *quercivorella* and *quercitella* Clem. [Bull. U. S. G. G. Surv., IV, 97 (1878)] regards Frey and Boll's identification of the latter as erroneous. I have a specimen sent by them to Zeller and am certainly disposed to agree with Zeller that it is rightly identified. Despite the minor points relied on by Chambers for its separation, I think the strong fuscous patch at the base of the fore-wings on the under side, and on the base of the hind-wings on the upper side, showing through to the under side, but not actually on that surface as suggested by Chambers, are sufficiently noticeable characters to justify the conclusion that they are the same. I possess also a pair of this species taken at Washington, D. C., April 29, 1871.

Tischeria sulphurea F. and B.

I have specimens of what I can only suppose to be this species received from North Carolina collected by the late H. K. Morrison. I also took it at Washington, D. C. on the 29th of April, 1871, and I am unable to separate from it specimens obtained on Mt. Shasta, Siskiyou County, Cal., in August of the same year—which would prove that there are two broods.

Tischeria fuscomarginella Chamb.

I have received this species from Miss Murtfeldt from Kirkwood, St. Louis, Mo., and from Monsieur Ragonot from Dallas, Tex., from Boll's collection.

Tischeria concolor Z.

I have specimens of this species collected at Kirkwood, St. Louis, Mo., by Miss Murtfeldt, and have received others from Monsieur Ragonot, taken by Boll, at Dallas, Tex.

Tischeria bicolor F. & B.

This species is only known to me from the description.

Tischeria tinctoriella Chamb.

Miss Murtfeldt has kindly sent me specimens of this insect collected at Kirkwood, St. Louis, Mo.

Tischeria helianthi F. & B.

I am indebted to Monsieur Ragonot for four specimens, labelled "Texas, Boll, *Tisch.* von *Helianth, m.*"

Tischeria solidaginifoliella Clem.

I have a single specimen of this species, which was also sent me by Monsieur Ragonot, who received it from Boll from Texas.

Tischeria pruinosella Chamb.

A single specimen of this insect is in my collection. It was received from Belfrage from Texas.

Tischeria pulvella Chamb.—**Tischeria longe-ciliata** F. & B.

These species are only known to me from the descriptions.

Tischeria heliopsiella Chamb.

Tischeria heliopsisella Chamb.
n. syn.—*T. nolckenii* F. & B.

This species is recorded by Chambers as bred from leaves of *Heliopsis lævis* and *Ambrosia trifida* in Kentucky [Cin. Qr. Jr. Sc., II, 113-4 (1875)]. I met with it also on Mount Shasta, Siskiyou County, Cal., in August, 1871, at an elevation of about 6,000 feet, mining the leaves of a species of *Ambrosia;* the mine occupying the whole width of the narrow leaflet. The specimens were bred in the same month. Its general aspect is that of a *Bucculatrix.* Two specimens received from Monsieur Ragonot, collected by Boll in Texas, labeled "*Tischeria nolckenii* F. & B.," agreeing in all respects with the description by Frey & Boll [Stett. Ent. Zeit., XXXVII, 220 (1876)] have been compared with Chambers' description of *heliopsisella*, and also with a specimen, kindly lent me by Miss Murtfeldt, which she received from Chambers, bearing the label "*Tischeria heliopsisella* Chamb. Ky." I have no hesitation in regarding *nolckenii* F. & B. as a synonym of *heliopsisella* Chamb.

Tischeria ambrosiella Chamb.

I have four specimens of this species bred from *Ambrosia trifida* by Miss Murtfeldt, at Kirkwood, St. Louis, Mo. Chambers also records it as bred from *Ambrosia artemisiæfolia* on Miss Murtfeldt's authority [Cin. Qr. Jr. Sc., II, 113 (1875)]. It is apparently a good species and distinct from *heliopsiella.*

Tischeria ceanothi sp. n.

Antennæ, pale grayish-brown, strongly ciliated in the ♂.
Palpi, pale grayish-brown.
Head, roughly clothed with pale grayish-brown scales tending to whitish in front; face whitish.
Fore-wings, grayish-brown, with a faint purplish tinge, in some specimens somewhat paler along the dorsal margin below the fold, a faint indication of a small darker spot about the anal angle; cilia pale grayish-brown. Under side rather shining grayish, slightly darker than the hind-wings.
Hind-wings, pale grayish; cilia scarcely lighter.
Abdomen, the same color as the hind-wings, anal tuft inclining to ochreous.
Legs, luteous; anterior pair darkened with fuscous and having the tarsal joints obscurely spotted.
Exp. al., 6^{mm}.
Habitat, California.
Type, ♂ ♀, *Mus. Wlsm.*

The larva mines the upper side of the leaves of *Ceanothus divaricatus* Nutt., making at first a narrow mine which gradually increases in width, but is apparently never wider than about one-fifth of the leaf; several mines are to be found in a single leaf. I have one before me which contains five. The larva changes to a pupa within the mine. There is no indication whatever of its presence on the under side of the leaf. I met with it at the head of the Noyo, Mendocino County, Cal., on the 8th-11th of June, 1871, in considerable abundance, the whole shrub being covered with mined leaves. I also took it on the wing in Mendocino County, 27th May, 1871. I have received the same species from Dr. Riley, collected at Placer County, Cal., in October, thus showing that the insect is on the wing at three separate times, viz., May, July, and October—possibly three distinct broods.

Tischeria malifoliella Clem.

Two specimens in the Zeller collection, received under this name from Boll from Texas, agree with my specimen compared with Clemens' type in the collection of the American Entomological Society, Philadelphia, in 1871.

Tischeria ænea F. & B.

There is a single specimen of this species in the Zeller collection received from Boll from Texas.

Tischeria roseticola F. & B.

I have specimens of this insect from the Zeller collection, and from Monsieur Ragonot received from Texas from Boll, and am indebted to Dr. Riley for a third example bred from rose.

BEDELLIA Stn.

Bedellia somnulentella Z.

This species, already recorded from North America by Clemens on Stainton's authority [Proc. Ent. Soc. Phil., I, 147-9 (1862)—Stn. Tin. N. Am., 189-91], is very widely distributed, occurring in Australia and New Zealand as well as in Europe. I have received it from Belfrage from Texas, and have myself met with it on McLeod Creek, Siskiyou County, Cal., at the end of July, 1871.

(To be continued.)

GENERAL NOTES.

A RHIZOCOCCUS ON GRASS IN INDIANA.

January 22, of the present year, Director Stockbridge, of the Indiana Experiment Station, placed in my hands a number of egg sacs, seeming to be identical with those mentioned in INSECT LIFE, Vol. I, p. 345, from Dakota, and *loc. cit.*, p. 385, from Nova Scotia. These were given Director Stockbridge by Mr. James Powers, of Lexington, Scott County, Ind., and were attached to blades of dead grass, the dried remains of the female being in most cases still attached to the anterior end of the sac. A week later the sacs were placed on growing plants of timothy and blue-grass, and on February 17 the leaves of these grasses, and also the surface of the soil in the pot containing them, were alive with minute, active, yellowish coccids, having much the color and appearance of young *Thripidæ*, except that they were more robust. The larger portion of these young seemed to forsake the grass and wandered away, while those that remained died in spite of every attempt to rear them.

In a letter to Dr. Stockbridge, written February 18, Mr. Powers gives the following interesting facts: The coccids occurred in spots, comprising the lower portions of about 5 acres of a low meadow, composed of timothy and red top. The meadow was of three years' standing, having been preceded by a crop of wheat. Up to about February 1, the sacs had been observed in great abundance, but a visit to the field on the 17th revealed the fact that all had disappeared—hatched, Mr. Powers supposed. Other meadows in the neighborhood did not appear to be affected.

I have never observed this in Indiana myself, the only coccid found by me being quite different, and affecting blue-grass, where it is not uncommon in August. These occur about the base of the leaves near the surface of the ground; at least this is the only place I have found them. They seem to belong to the genus *Westwoodia*, and I have observed what appears to be the same thing also on blue-grass in Illinois, and understand that Mr. Pergande has also found it on the same plant about Washington.—[F. M. Webster, March 10, 1890.

FURTHER NOTE ON THE EGYPTIAN MEALY BUG.

On page 256 of the current volume we published a note upon this insect, based upon information kindly sent us by Mr. D. Morris, of the Royal Kew Gardens, England. We notice by the March number of the *Entomologists' Monthly Magazine* that Mr. Douglas has found it necessary to erect a new genus for this insect, and that he calls it *Crossotosoma ægyptiacum.*

INDIAN RHYNCHOTA.

Mr. E. T. Atkinson, of Calcutta, has favored us with the second part of his Catalogue of the Insecta of India, which comprises a bibliographic and synonymical list of the family Capsidæ. We have seldom seen a work of this kind which displays such thorough and painstaking work. It is much more extensive than we had anticipated, covering one hundred and eighty odd royal octavo pages of brevier type. There is a full bibliographic list, an index to genera and an index to species.

TWO PARASITES OF THE GARDEN WEB-WORM.

In our article upon *Eurycreon rantalis*, commonly known as the "Garden Web-worm" in our annual report for 1885, the only parasite mentioned was a Tachina fly, reared by Professor Popenoe, at Manhattan, Kans. In 1888 this insect was again abundant in parts of Colorado, Arkansas, and Texas, and we reared an Ichneumonid in early

July from pupæ received from Mr. W. F. Avera, editor of the *Ouachita Herald*, of Camden, Ark., who had noticed the larvæ damaging cotton. This parasite has been described by Mr. Ashmead on p. 437 of the Proceedings of the U. S. National Museum, Vol. XII, 1889, as *Limneria eurycreontis*, and we present herewith a figure of the female sex. The eggs are laid in the larvæ and those specimens which we reared issued from the pupæ. Many Limnerias, it will be remembered, issue from the larvæ of their hosts before the latter have transformed.

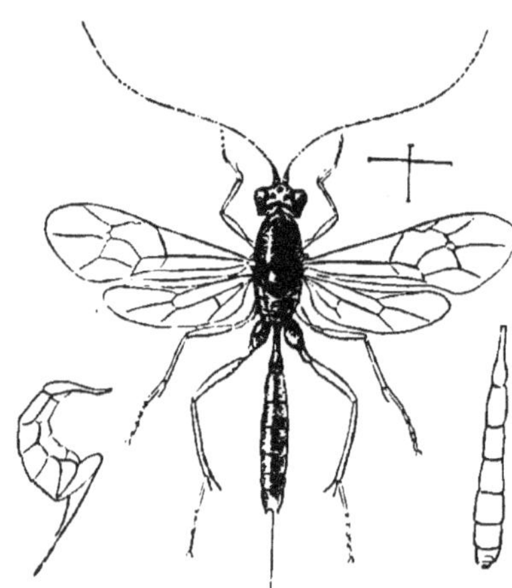

FIG. 64.—*Limneria eurycreontis*—female with abdomen and ovipositor shown detached at left; male abdomen at right—enlarged (original).

We also reared about the same time, from the same lot of web-worm pupæ, specimens of a Braconid parasite, which we have determined as Mr. Cresson's *Agathis exoratus*.

AN APHIS ATTACKING CARROTS.

In his report as State entomologist of New York, for the year 1886, p. 123, Prof. J. A. Lintner records the reported appearance of Aphides on carrots and parsnips, at Oakley Park, Mass., in sufficient numbers to seriously injure the crop. As no further particulars or specimens were furnished the professor, and as this is the only case on record where the carrot in this country has been attacked by Aphides, we are left totally in the dark as to what particular species was engaged in the depredations.

Buckton* states that *Siphocoryne pastinacæ* (Linn.) was found abundantly on carrot, at Haslemere, in July, and Curtis† says that in 1847 a field in Gilford, Surrey, was about one-tenth destroyed by an attack of *Aphis dauci* (Fab.), and another species of Aphis occurs in October about the roots. Miss Ormerod‡ tells us that a serious attack occurred at Newton Farm, near Glasgow, in 1879, and also states that carrots are attacked by several kinds of Aphides, among them *Aphis papaveris* Fab., which infests the leaves, and *A. carrotæ*, which affects the flower stems, and also the below-ground portions of the plant. M. Lichtenstein§ names in his list seven other species which infest the carrot, three of them attacking the parsnip also.

* British Aphides, Vol. II, p. 24.

† Farm Insects, p. 403.

‡ Rep. Obs. Inj. Ins., 1882 (Sixth Report), p. 18.

§ Lintner, Rep. St. Ent., N. Y., 1886, p. 123.

In January, 1889, we observed the seed heads of carrots in a garden near Hobart, Tasmania, thickly populated by a species of *Rhopalosiphum*.*

On October 3, 1888, I found several carrots in a field near La Fayette, Ind., infested with an *Aphis* which Dr. Riley found to closely resemble *A. plantaginis*, and also a species which occurs in the vicinity of Washington, D. C., on the roots of *Portulaca*. At this time those observed by me were clustered on the bases of the *leaf stalks and also on the fibrous rootlets*. A few days later, what appeared to be the same species was found on the roots of *Portulaca oleracea*, and specimens from both this and the carrot were placed in a breeding cage, where both plants were accessible. So far, I had only observed wingless individuals, and these seemed to be all females; at least I saw no males. Soon after the 28th of October the females in the breeding cage began to deposit eggs on both carrot and *Portulaca*. This was continued up to the 5th of November, when I was called away for several weeks, and on my return, November 26, all had disappeared.

The eggs were light-colored immediately after deposition, but soon became shining black, like those of *Aphis mali*, but were rather smaller. As I left home soon after for an absence of several months, no opportunity was offered to watch the development of the eggs.

On August 15, 1889, the same species was found, both on the rootlets of carrot several inches below ground and also on the roots of *Portulaca oleracea*. On the 23d of the same month apterous individuals were observed on the roots of the latter plant, and among them a winged female.

October 16, examples, differently colored, but seemingly belonging to this species, were found on salsify. Although the attempt was made to rear them on this plant, the result proved a failure, and neither eggs nor winged individuals of either sex were obtained. So far I have observed no serious injury to carrots or salsify by reason of the attacks of these insects.—[F. M. Webster, February 15, 1890.

MORE INSECTS INJURING THE TEA-PLANT IN CEYLON.

Mr. E. Ernest Green, of Eton, Punduloya, Ceylon, has sent us the continuation of the articles which he is publishing in the Ceylon Independent upon the above subject. The first nine installments are reviewed upon pages 192–193 of No. 6 of the current volume of INSECT LIFE. The additional insects treated are as follows:

The Tea Aphis (*Aphis* sp.): This insect is a much darker species than the one which occurs upon coffee, and frequently damages seedling plants in the nurseries and the young shoots first thrown out after pruning. The remedies recommended are kerosene emulsion, 1 part to 80 parts of water; phenyle, 1 part to 240 parts of water. The natural enemies mentioned are: Syrphid flies, Chrysopus, Lady-birds, a wasp of the genus *Rhopalum*, and an Aphidiid parasite.

---

* INSECT LIFE, Vol. I, p. 362.

The Dipterous Leaf-miner (*Oscinis* sp.): This insect is so common that it is difficult to find a single tea bush upon which are not a great many leaves marked with the remains of its larva. An internal parasite is mentioned, and it is stated that this miner causes no appreciable damage.

The Black Grub or Cut-Worm (*Agrotis suffusa*): The full-grown larva of this insect shears off a number of young plants at each meal.

The Tineid Leaf-miner (*Gracillaria* sp.): This insect affects the younger leaves only, and has no opportunity of troubling where the bushes are regularly picked.

The Blue striped Nettle-grub (*Parasa lepida*): This is one of the stinging caterpillars, of which we have a number in this country, and it occurs in considerable numbers on the tea plantations, often completely defoliating the trees. The larva is of a brilliant yellow-green color with a rich lilac stripe along the middle of the back and a bright blue stripe on each side. The poisonous spines are pale green and are arranged in tufts along the body. The moth is chocolate brown, with a bright green band obliquing across the fore wings; the hind wings are buff, tinged with chocolate at the margins.

NEW INSECT LEGISLATION.

As exhibiting the lively legislative interest taken in California in regard to insect pests, and as supplementary to the Amended California Horticultural Laws published on pages 81 to 83 of the present volume, we give below a copy of Ordinance No. 26 of San Bernardino County, Cal., which was passed last November.

SECTION 1. No person or persons, either as owner, agent, servant or employé, shall *keep, sell, expose for sale or otherwise distribute* within the limits of San Bernardino, County, California, any fruits, plants, flowers or vegetables infected with live scale or other insects, or their eggs, larvæ or pupæ, detrimental or injurious to fruit-trees or plant-life, or the products thereof, and if any fruits, plants, flowers, or vegetables should, on examination, be found to be infected with scale or other insects, or their eggs, larvæ or pupæ, the said fruits, plants, flowers or vegetables shall be *disinfected or destroyed* under the direction of the county board of horticultural commissioners.

SEC. 2. No person or persons, whether as owner, agent, servant or employé, shall bring, or cause to be brought into the county of San Bernardino, any trees, vines, shrubs, scions, cuttings, grafts, plants, flowers, or vegetables from any district, county, or State declared by the county board of horticultural commissioners of said San Bernardino County to be infested with scale or other insects, d-trimental or injurious to trees, vines, fruits or plant-life or the products thereof.

SEC. 3. No person or persons, as owner, agent, or employé, shall bring, or cause to be brought, into San Bernardino County, California, any trees, vines, shrubs, scions, cuttings, grafts, fruits, plants, flowers, or vegetables, from any district, county or State, *not* declared to be infested, as provided in section two of this ordinance, *without giving notice of their arrival to a member of the county board of horticultural commissioners or the local inspector of the district into which they are brought;* or plant, sell, give away or otherwise distribute them, or cause the same to be done, *until they shall first have been inspected,* and, if necessary, disinfected to the satisfaction of the county board of horticultural commissioners of said San Bernardino County.

[The board respectfully point out to the ladies of the county that the danger of bringing the scale on bouquets and small packages of potted plants, cuttings, etc., from infested districts is as great as from larger packages of trees, shrubs, etc., and ask a hearty compliance on their part with the above.]

SEC. 4. Every owner, or owners, or person or persons, in charge or possession of any orchard, nursery, or other premises in San Bernardino County, on which are growing any trees, vines, shrubs, plants, vegetables, or flowers infected with red or cottony cushion scale, or the eggs, larvæ or pupæ thereof, *shall, when required by the county board of horticultural commissioners*, as in their discretion may seem necessary, *cut back and disinfect* said infested trees, vines, shrubs, plants, vegetables, or flowers to the satisfaction of said board, or dig out and destroy the same as said board may direct.

[From observation and experience so far gained, the board are convinced that the most successful and cheapest method of treatment of the above-mentioned scale is by cutting back and defoliating the tree so that it may be thoroughly scrubbed in every part, subsequently spraying it and the surrounding trees.]

SEC. 5. Any person or persons who shall ship or bring, or cause to be shipped or brought into San Bernardino County, any trees, vines, scions, cuttings, grafts, shrubs, plants, vegetables or flowers, *shall have placed upon or securely attached to each box, package, or separate parcel* of such trees, vines, scions, cuttings, grafts, shrubs, plants, vegetables, or flowers, a distinct mark or label, showing the name of the owner or shipper, and the locality where produced.

[The attention of purchasers and nurserymen is particularly called to this section, and a strict compliance with its provisions will greatly facilitate the work of the board in determining infested districts.]

SEC. 6. The county board of horticultural commissioners shall from time to time, as in their discretion may seem necessary by publication in a newspaper of general circulation published in the county, publish a list of the districts, counties, or States which they declare to be infested for the purpose of this ordinance.

[The board will, as soon as they can obtain the necessary information, publish a list of the districts which they declare to be infested. In the meantime they would urge all persons to refrain from purchasing any trees, etc., from Los Angeles or Orange Counties.]

SEC. 7. Any person violating any of the provisions of this ordinance is punishable by imprisonment in the county jail not less than ten days, and not more than one hundred days, or by a fine not less than ten dollars nor more than one hundred dollars, or both. A judgment that the defendant pay a fine may also direct that he be imprisoned until the fine be satisfied, specifying the extent of imprisonment, which must not exceed one day for every dollar of the fine.

SEC. 8. This ordinance shall take effect and be in force on and after the first day of November, 1889.

A TEST CASE UNDER THE HORTICULTURAL LAW.

Some time during January the Los Angeles county horticultural commission secured the arrest of a fruit-grower who refused to destroy the scale insects upon his trees; and we learn from Mr. Coquillett that the trial has recently taken place, and that it resulted in the acquittal of the individual after the jury had been out but five minutes. The culprit pleaded many extenuating circumstances, and the sympathy of his neighbors was evidently on his side. The prevailing sentiment of the fruit-growers of Los Angeles County is that they are abundantly able to take care of their own trees, and they are strenuously opposed to any dictation as to when they should spray and what they should spray with.

LOCUSTS IN INDIA.

The occurrence in 1889 of swarms of locusts in Northwest India is taken advantage of by Mr. E. C. Cotes, of the Indian Museum, of Calcutta, to elucidate several doubtful points in the history of these destructive insects for a complete report which is being prepared under the direction of the trustees of the Indian Museum. To this end a circular, copies of which we have just received, has been distributed in the regions likely to be overrun, giving, in brief, accounts of the more destructive of the recent locust invasions.

There is some doubt as to the species of locust which invades India, and it is to settle this point and also to determine the distribution and the limits of the permanent breeding grounds that the circulars have been sent out. The locust generally referred to in India is *Acridium peregrinum*, supposed to be the locust of the Bible, but it seems probable that a second species is responsible for the invasion of Madras in 1878 and Deccan in 1882–'83, while the first-named species extends its ravages rather into the dry plains of the Punjab and Rajputana.

The circular gives the life-history and habits of the locusts, together with short accounts of the remedies that have been employed against them. The latter chiefly consist in the destruction of the eggs by plowing, and of the newly hatched locusts by driving them into ditches, where they are covered with earth. The screen system successfully employed against the locusts in Cyprus and Algeria is also described. The winged locusts have been destroyed by driving them into lines of burning straw. We shall look for the full report with considerable interest.

NEW INJURIOUS INSECTS IN COLORADO.

The list of injurious insects of Colorado has recently been augmented by the discovery of three beetles, at Denver, by Mr. H. G. Smith, jr., viz, *Bruchus obsoletus*, var. *fabæ* Riley, *Lachnosterna fusca*, and *Tenebrio obscurus*. Specimens of all of these have been seen by me. The two latter species have been verified by Dr. Horn.—[T. D. A. Cockerell, West Cliffe, Colo., March 3, 1890.

OBITUARY.

The *Entomologists' Monthly Magazine* announces the death, in its February number, of Prof. Heinrich Frey, of Zurich, from apoplexy, on the 17th of January, 1890. The death of Monsieur Lucien Buquet, who was treasurer of the Entomological Society of France for forty-five years (1842 to 1887), is also announced as having occurred the middle of December, 1889. He was appointed honorary treasurer of the French Society on his retirement, in 1887, and published many notes on Coleoptera in the "Annales."

AN ICERYA IN FLORIDA.

Passed Assistant Paymaster H. R. Smith, U. S. N., now stationed at Key West, Fla., sent to this Department on the 24th of March a bark-louse infesting the Rose. April 12th he sent further specimens, including a complete plant and all stages of the insect, except the male. We have recognized in this insect what seems to be a new species of the genus *Icerya*, but which resembles more closely *Icerya sacchari*—the sugar-cane pest of Mauritius—than *I. purchasi*, the citrus pest of California. The young lice are indistinguishable from *I. purchasi*, but the adult females lack the fluted ovisac and the glassy filaments. They are covered with white meal-like wax, and when this is removed they show the contrasting colors of black and red. The black is upon the dorsum of the thorax, and the red is upon the entire ventral surface and the dorsum of the abdomen. The younger stages are entirely red. The antennæ in the different stages are almost indistinguishable from those of *I. purchasi;* the mentum and rostrum are present, and the genito-anal ring lacks bristles. The second stage of the larva possesses not only the six long anal bristles, but has a row of very long bristles on the lateral border of the abdomen. While it is somewhat unsafe to generically refer a Monophlœbid without the male, we hope soon to get this, and will then endeavor to fully characterize and illustrate the species in a near number of INSECT LIFE. Meanwhile, we would propose for it the MS. name *Icerya rosæ.*

ENTOMOLOGICAL SOCIETY OF WASHINGTON.

March 6, 1890.—Mr. Schwarz exhibited and remarked upon the following species of Coleoptera, which are new to the fauna of North America: *Lathridius* (*Coninomus*) *nodifer* Westwood; *Actinopteryx fucicola* Allibert, *Arrhipis lanieri* Guérin, and *Probatius umbratilis* Duval. He also showed specimens of *Temnochila hubbardi* Léveillé, and *Teretriosoma hornii* Lewis, recently described in European journals, from the semi-tropical region of Florida. He finally drew attention to Dr. Horn's recent revision of the North American species of *Ochthebius*, and spoke of the geographical distribution of those aquatic beetles.

Mr. Marlatt presented a note on a dipterous larva infesting the seeds of *Xanthium*. He had found these larvæ at Manhattan, Kans., and, during the past winter, in the District. Drawings were exhibited illustrating the larva and the nature of its work.

He also presented a short note on the food-habits of *Psiloptera drummondi*.

These notes were discussed by Messrs. Schwarz, Townsend, and Howard.

Mr. Townsend read a paper entitled "Notes on Acridiidæ in Michigan," which related more particularly to dates of appearance and habits.

C. L. MARLATT,
Recording Secretary.

o

U. S. DEPARTMENT OF AGRICULTURE.

DIVISION OF ENTOMOLOGY.

PERIODICAL BULLETIN. (Double number.) May and June, 1890.

Vol. II. Nos. 11 and 12.

INSECT LIFE.

DEVOTED TO THE ECONOMY AND LIFE-HABITS OF INSECTS, ESPECIALLY IN THEIR RELATIONS TO AGRICULTURE.

EDITED BY

C. V. RILEY, Entomologist,

AND

L. O. HOWARD, First Assistant,

WITH THE ASSISTANCE OF OTHER MEMBERS OF THE DIVISIONAL FORCE.

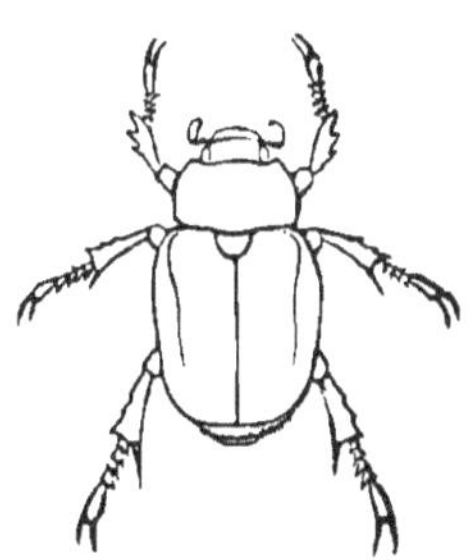

[PUBLISHED BY AUTHORITY OF THE SECRETARY OF AGRICULTURE.]

WASHINGTON:
GOVERNMENT PRINTING OFFICE,
1890.

CONTENTS.

SPECIAL NOTES.

Bibliography of American Economic Entomology.*—We are pleased to be able to announce that Parts I, II, and III of the Bibliography of the more important contributions to American Economic Entomology, by Samuel Henshaw, were published April 7, and are now ready for distribution. The larger share of the edition has been published under four covers, as follows: (1) Part I, the more important writings of Benjamin Dann Walsh, a pamphlet of 49 pages and 385 titles; (2) Part II, the more important writings of B. D. Walsh and C. V. Riley, comprising 46 pages and 478 titles; (3) Part III, the more important writings of Charles Valentine Riley, covering 276 pages and including 1,555 titles; and (4) an index to the first three parts, covering 83 pages and including, besides the general index, systematic indices of the new names proposed by both writers. The remainder of the edition has been published in one volume, cloth bound. We take this occasion to state that although Professor Riley has been greatly interested in the plan and has actively promoted the preparation of the general Bibliography, he is not at all responsible for the present publication, which was decided upon and the proof read during his absence in Paris last summer.

Subsequent parts of the bibliography will include the references to the economic writings of other American entomologists, and its completion is now only a matter of a very few months. Mr. Henshaw has been engaged upon this task for several years and his work has been well and carefully done. We hope that working entomologists will find this volume of assistance in lightening their labors in necessary bibliographical research, and we know from our own experience that the completion of the entire work will result in a great saving of time to investigators.—L. O. H.

*Bibliography of the more important contributions to American Economic Entomology. Prepared, by authority of the Secretary of Agriculture, by Samuel Henshaw. Parts I, II, and III. The more important writings of Benjamin Dann Walsh and Charles Valentine Riley. Washington: 1890.

Bulletin No. 21, Division of Entomology.*—Under this serial number Mr. Koebele's report on his trip to Australia and New Zealand to investigate the natural enemies of *Icerya purchasi* has recently been published. The bulletin is a narrative account and is plain and circumstantial. It is devoted almost entirely to the subject of his quest, but incidentally mentions some of the insect pests to Australian agriculture. Among these are *Otiorhynchus cribricollis*, a common south European Snout-beetle which has been imported into Australia, and is injurious to the Olive; *Aspidiotus rossi* Crawford, a Bark-louse injuring a variety of shrubs, including the olive tree: the Woolly Apple-louse (*Schizoneura lanigera*) probably introduced from this country; *Chortologa australis*, Sauss. MS., a destructive migratory locust which in South Australia takes the place of our *Melanoplus spretus*; the Black Scale (*Lecanium oleæ*), probably introduced direct from Europe on the Olive; several scale-insects of the subfamily *Monophlœbinæ*, injurious to the Eucalyptus; *Mictis profana* Fab. and a new species of *Aspongopus*—two Heteropterous insects injurious to the Orange—and three species of *Melolonthid* beetles injurious in the larval state to wheat crops. These are: *Scitala nigrolineata* Boisd., *S. pruinosa* Dalm., and *Anodontonyx vigilans* Sharp, the latter described in the last number of INSECT LIFE, page 302. The beetles were determined for us by Dr. David Sharp, of England, and the migratory grasshopper by M. Henri de Saussure, of Geneva. We have illustrated the report with 16 figures, 11 of which are new.

Recent important Entomological Reports.—Mr. Fletcher's report as entomologist of the experimental farms of Canada for 1889 reached us April 14† from Canada. He has some 30 pages of interesting matter illustrated with a dozen cuts. The principal articles concern the Hessian Fly, the Grain Aphis, the "Wheat Stem-maggot" (better known as the American Meromyza), Cut-Worms, Mediterranean Flour-moth, Granary Weevils, Spraying with Arsenites, Fuller's Rose-beetle, and a curious account of insects injuring a wooden water pipe. The principal points brought out are the facts that the Meromyza breeds freely in several kinds of grasses, the suggestion that an early sown strip of wheat or barley may be used as a trap for the same insect, and an indorsement of the poison trap remedy for Cut-Worms.

Prof. J. B. Smith has favored us with his bulletin on the insects injuriously affecting Cranberries in New Jersey.‡ He gives full illustra-

---

* U. S. Department of Agriculture. Division of Entomology. Bulletin No 21. (Revised Edition.) Report of a Trip to Australia made under direction of the Entomologist to investigate the Natural Enemies of the Fluted Scale, by Albert Koebele. (Published by authority of the Secretary of Agriculture.) Washington: 1890.

† Experimental Farms. Reports of the Director, Chemist, Entomologist and Botanist, Horticulturist, Poultry Manager, and Superintendents Experimental Farms, for 1889. Ottawa. 1890.

‡ Special Bulletin New Jersey Agricultural College Experiment Station, K, February 28, 1890.

tions and accounts of the Black-headed Cranberry-worm (*Rhophobota vacciniana*), the Cranberry Fruit-worm (*Acrobasis vaccinii*), the Tip Worm (*Cecidomyia vaccinii*), the Cranberry Scale (*Aspidiotus* sp.), Grasshoppers and Locusts and Cranberry Leaf-hoppers. The principal pests are the two first mentioned, and for the first he recommends reflowing, kerosene, and Paris green. For the second he advises an application of Paris green or London purple after all the blossoms are off, or nearly all of them, and the berries are generally set. He follows Professor Fernald in considering *Teras oxycoccana* Pack. as distinct from *T. vacciniana*, though our own conclusions as to the synonymy were based on a specimen of the former determined by Dr. Packard.

Miss Ormerod's report for 1889* reached us during April. The leading article of the report is a consideration of the disease known as clover sickness, produced mainly by an Anguillulid—*Tylenchus devastatrix*. Several measures of prevention and remedy are pointed out, viz, rotation of crops, a dressing of gas lime, avoidance of the use of dung from infested clover or oats, the application of sulphates and deep plowing. The Clover-root Cecidomyia is mentioned and some consideration is given to Millipedes, Clover and Pea Weevils, the Hessian Fly, two species of Oscinidæ, the Wheat Bulb-fly (*Hylemia coarctata*), the Currant Gall mite, the White Currant-scale (*Pulvinaria ribesiæ*), the Mediterranean Flour-moth, the Wheat Fly, and certain orchard insects and a few species injurious to Pine, Plum, and Turnip, together with some further notes on Ox Warbles, repeating her statements regarding "licked beef" and "jelly," reviewed in No. 5 of the current volume of INSECT LIFE and adding further statistics from correspondents. There is an appendix upon *Xyleborus dispar* in which the use of trap wood is recommended and Eichoff's work on Bark-beetles is quoted at length, particularly with reference to the food of the larvæ, a subject which we touched upon on pages 279–280, No. 9 of this volume.

We have just received from Professor Forbes his fourth and fifth reports as State Entomologist of Illinois.† These reports although greatly delayed are none the less welcome. On account of the delay Professor Forbes has been obliged to withdraw several articles already prepared upon subjects which more recent observations will enable him to treat better hereafter.

The fourth report includes articles upon arsenical poisons for the Codling Moth, in which the conclusion is reached that 70 per cent. of the

* Report of Observations on Injurious Insects and Common Farm Pests during the year 1889, with methods of prevention and remedy. 13th Report, by Eleanor A. Ormerod. London, 1890. Price 1s6d.

† Fifteenth Report State Entomologist on the Noxious and Beneficial Insects of the State of Illinois. Fourth Report S. A. Forbes; for the years 1885 and 1886. Springfield, 1889.

Sixteenth Report State Entomologist on the Noxious and Beneficial Insects of the State of Illinois. Fifth Report S. A. Forbes: for the years 1887 and 1888. Springfield, 1890.

crop can be saved by spraying; a second contribution to the knowledge of the life history of the Hessian fly, indicating that the development of a third brood of larvæ may sometimes detract from the effect of late sowing; the life history of the "Wheat Bulb-worm" (the American Meromyza) showing three broods in Illinois; Mr. Weed's article upon an outbreak of injurious locusts in Illinois (the same paper as read before the Society for the Promotion of Agricultural Science in 1888) and an article by the same author upon some common insects affecting the foliage of young Apple trees in the nursery and the orchard. There is an appendix by Professor Forbes on the present condition and prospects of the Chinch Bug in Illinois which was summarized on page 222 of Vol. I of INSECT LIFE. The only illustrated article is that by Mr. Weed upon Apple insects.

The fifth report is more extensive, covering 104 pages and includes three chief articles, viz: Studies on the Chinch Bug, II, the Corn Bill-bugs, and Notes on Cut Worms. The report is illustrated by six beautiful heliotype plates, two and one-half devoted to Bill Bugs, and the others to Cut Worms, the Wheat Thrips, and the Burrowing Web-worm (*Pseudanaphora arcanella*). Professor Forbes gives an excellent account of former observations upon the species of Sphenophorus, ordinarily known as Bill Bugs. As an appendix to the report, an extensive analytic economic bibliography of the Chinch Bug from 1875 to 1888 is given, covering one hundred and twenty-two pages. It seems to be as full and complete as great pains can make it.

Mr. Whitehead's third annual report* has also just been published. It includes a consideration of some thirty topics, several of which are also considered in Miss Ormerod's report. The articles are all short, nearly all are illustrated, and though containing little that is original, the report, as a whole, is well adapted to the use of British farmers and gardeners.

Work at the Cornell Station.—Professor Comstock, in Bulletin 15 of the Cornell Agricultural Experiment Station, December, 1889, entitled "Sundry investigations made during the year," gives an account of the Apple-tree Tent-caterpillar (*Clisiocampa americana*). Though this is a well-known insect, yet, on account of its great increase of late years as a pest, a brief restatement of its habits is not at all out of place.

In the second annual report of the Station, for the year 1889, Professor Comstock presents an outline of his work as entomologist for the year. The Wheat Saw-fly (*Cephus pygmæus*) has been studied and the "clematis disease" has been determined to be due to *Heterodera radicicola*, the same Nematode worm, of the family Anguillulidæ, which

*Third Annual Report on Insects and Fungi injurious to the Crops of the Farm, the Orchard, and the Garden, by Charles Whitehead, esq., F. L. S., F. G. S., 1889, London. 1890. Printed for Her Majesty's stationery office by Eyre & Spottiswood, printers to the Queen's most excellent Majesty.

is the subject of Professor Atkinson's Bulletin No. 9 of the Alabama Agricultural Experiment Station, recently noticed in these pages.

A series of field experiments is proposed by Professor Comstock the present year to determine the best method of combating it. Work has also been prosecuted on wire-worms; a hop-yard has been established for the study of the Hop Aphis; and much attention has been given to a species of Aleurodes (*A. vaporiorum*) which infests in its early stage the under side of the leaves of various plants and has not before been mentioned as occurring in this country, although it is a well-known European pest.

Ohio Station Investigations.—Article XIX in Bulletin 7, volume II (Second Series, No. 14) of the Ohio Agricultural Experiment Station, entitled "Notes on Experiments with Remedies for certain Diseases," by Clarence M. Weed, is interesting as treating of the matter of combining insecticides and fungicides, or applying at the same time a remedy to destroy fungus diseases as well as insects. Mr. Weed states that the practicable application of such a combination originated in the division of entomology and botany of the Ohio Station. The case is well set forth in the article, by an extract from a recent paper by Mr. Weed in *Agricultural Science* (date not given). It is proposed that by combining the copper sulphate solution for blight and the Paris green or London purple solution for the Colorado Potato-beetle, a solution can be made that at the same application will kill both, and lessen the expenditure of time and labor. In the same way a vineyard may be protected from black rot and various leaf-eating beetles by combining such applications as are used for each. A note on the efficacy of "eau celeste" for mildew and the Rose Beetle was published in INSECT LIFE for July, 1888 (Vol. I, p. 32), and we may add that combined applications for insects and fungi have long been made in France.

A résumé of the principal injurious insects noticed by the Ohio Agricultural Experiment Station during the year 1889 is given in the Eighth Annual Report, published in Bulletin 8, Volume II, second series, of the Station. Two insects that gained prominence during the year are the Grain Plant-louse and the White Grub. An original figure of the wingless form of the former is given, and its great abundance in June and sudden decrease from the attacks of Lady-birds and Hymenopterous parasites are noticed.

A new remedy is claimed for the Clover-seed Midge (*Cecidomyia leguminicola*), which consists in mowing the field about the middle of May when the heads are just forming. The new crop of blossoms following matures between the two broods of the midge, thereby escaping. This is but a variation of a remedy originally proposed by Professor Comstock in the Annual Report of this Department for 1879, page 195, and reproposed by Mr. James Fletcher in 1887.

Four clover insects additional to Mr. Weed's recent list are given. These are two butterflies (*Cyaniris pseudargiolus* and *Epargyreus tityrus*) and two plant-lice (*Aphis trifolii* and *Callipterus trifolii*). Successful spraying with arsenites has been carried on against the Plum Curculio and the Codling Moth.

Original figures are given of a Sphinx larva covered with *Apanteles* cocoons, the green Apple Leaf-hopper (*Empoasca albopicta*), the Rose Leaf-hopper (*Typhlocyba rosæ*) and *Belostoma americanum*.

A method is set forth for covering cucumber vines with a gauze-covered frame to protect them from the Striped Beetle (*Diabrotica vittata*). The Bean Weevil (*Bruchus obsoletus*), Pear or Cherry-tree Slug, Imported Cabbage-worm (*Pieris rapæ*), and Strawberry Root-louse (*Aphis forbesi*) are also treated.

Professor Westwood's Revision of the Mantidæ.—The veteran entomologist, Professor Westwood, has just issued a monumental work on the curious insects of this Orthopterous family, which is entitled "Revisio Insectorum Familiæ Mantidarum, Speciebus novis aut minus cognitis descriptis et delineatis." It consists of a synonymical and bibliographical list of the species of the family, full descriptions of one hundred and eight new or little known species, a bibliography of the family, and an alphabetical index of the genera, species, and synonyms. There are fourteen magnificent quarto lithographic plates drawn by the author, comprising figures of one hundred and twenty-seven different forms. Professor Westwood's record has seldom, if ever, been approached among entomological workers. Since 1827 he has constantly been publishing valuable contributions to our science, and now at the age of eighty-four to bring out a work of this character is an event probably beyond precedent.

Of the five hundred and thirty-two species catalogued for the whole world but twelve are found in America north of Mexico. These are the following:

Gonatista grisea Fabr.
Oligonyx uhleri Stål.
Oligonyx scudderi Sauss.
Oligonyx graminis Scudd.
Thesprotia baculina Bates MS.
? *Sphendale infuscata* Sauss.
? *Phasmomantis grandis* Sauss.
Mantis wheeleri Wheeler.
Stagmomantis carolina Johanson.
Stagmomantis dimidiata Burm.
Stagmomantis ? *minor* Scudd.
Pseudovates mexicana Sauss.

Another new entomological Journal.—We have just received the first number of Volume I of the "*Entomological Record and Journal of Variation*," edited by J. W. Tutt, F. E. S., and published by W. H. Allen & Co., of London. This first number relates exclusively to Lepidoptera, but in the prospectus we notice no mention of an intentional restriction to this order. The magazine will be devoted to the wants of English entomologists and restricted to their own fauna and such parts of foreign entomology as they need in the understanding of the British species. The subject of variation will occupy a leading position.

The principal article in the first number is upon the genus "Acronycta and its Allies," by Dr. T. A. Chapman, and is followed by a general consideration of "Melanism and Melanochroism in British Lepidoptera," by Mr. Tutt, who agrees with Mr. Cockerell in considering that melanism depends largely upon humidity for its occurrence. "Notes on Collecting" bring out several interesting points, and the editor contributes some good "Practical Hints" regarding the breeding of rare species.

A necrophagous Dipteron.—In the present number we publish an article by Mr. Webster upon certain flies found infesting a human corpse in Indiana, and under "Extracts from Correspondence" some correspondence upon the general subject as well as upon this particular instance, which will supplement his communication. We had hoped to introduce figures of the species sent by Mr. Webster (*Conicera* sp.), but must defer them for a near number of INSECT LIFE.

Florida Orange Scales in California.—The fruit-growers of California are just at present very much disturbed over the importation of fruit trees from Florida which are infested with several scale-insects which have not before been prevalent in the former State. Among these are the Long Scale (*Mytilaspis gloverii*), the Purple Scale (*M. citricola*), and the Chaff Scale (*Parlatoria pergandii*). The May number of the *Rural Californian* is largely occupied with discussions of the probable damage which will be done by these pests and the necessity for a rigid quarantine. We have received a number of letters also from California asking our opinion and have replied that while there seems reason to believe, and we are inclined to believe, from the evidence at hand, that the scales above mentioned will not flourish in certain parts of southern California like Riverside where the heat and dryness are great, yet it will be unwise to depend too much on the limited experience of the past. We have therefore reiterated our conviction as to the necessity of using every precaution to prevent their introduction.

One pertinent editorial paragraph in the journal referred to strikes us as worthy of quotation:

> There seems to be a feeling awakened that the times of political entomologists are over, and that in the future men who are versed in the science of entomology are only to be appointed to positions requiring some knowledge of that study.

California has taken hold of the subject of insect pests with considerable energy; but, as was to be expected from the number of official positions created, "political entomologists" have been called into existence and the State has suffered from them. Much has been done in the way of county regulations and State laws governing inspection, quarantining, and disinfection, and in many instances these regulations have accomplished a great deal of good. We anticipated this scare about Florida scales and placed ourselves upon record some time ago as to the necessity of the establishment of a quarantine in Florida against infested plants from California and *vice versa*.

THE INSECT COLLECTION OF A LARGE MUSEUM.*

By C. V. Riley.

THE TYPE OR SYSTEMATIC COLLECTION.

The ideal *cabinet* collection of a National Museum should represent, as completely as possible, the insect fauna of the country, properly classified and determined. It can, necessarily, have little interest for the public at large and should be consecrated to the use of the specialist and to the advancement of the science of entomology. For this purpose it should be most carefully guarded and conserved in the best-made drawers and cases and secured alike from light and the too constant handling of the mere curious. It should constitute a study collection to which workers are drawn for unpublished facts and for comparisons and determinations. It should be so well conserved and provided for as to induce describers of new species to add to it their types or authentic duplicates thereof. It will be many years ere such an ideal collection can be gotten together, and none now living may witness it, but the material now on hand forms a good foundation for it.

THE EXHIBIT COLLECTION.

The *exhibit* collection should be something entirely independent and apart from the other, and, on account of the rapid deterioration of insect specimens constantly on exhibition and necessarily much exposed to light, should consist, as far as possible, of duplicates only, or of such commoner species as can be easily replaced. Intended for the instruction and edification of the lay visitor to the Museum, it should illustrate in the boldest possible way the salient characters of the class, the larger classificatory divisions and the structures on which they are based, and the wonderful metamorphoses and economies of the commoner and easily recognized species, particularly in their relations to man either directly or indirectly through injury or benefit.

---

*Extracted, with slight changes, from the Annual Report of the Smithsonian Institution for 1886, Part II, Report of the National Museum, pp. 182-186, Washington, 1890.

The value of such an exhibit collection depends very much on conspicuity, and this can best be obtained by the liberal use of diagrams and enlarged drawings, as the majority of the most interesting species and those which most concern man are almost microscopic in size. Such an exhibit collection will miss its mark and object whenever it exceeds these limits, and by too much detail seeks to interest and instruct the specialist or in other ways trenches on the function of the study collection. As the Museum, in this department, will, in accordance with statute (Revised Statutes, sec. 5586), receive a great deal of its best material through the Department of Agriculture, one of the chief aims of this national collection should be to reciprocate, not only by preserving all systematic material and thus aiding said Department of Agriculture in necessary determinations, but by giving particular attention to the biological side of the collection. This I have endeavored to do, and the collections illustrating the biology of North American insects are the largest in the world.

DRAWERS AND CASES.

The character of the drawers and cabinets employed in such a national collection is important; for upon it the future preservation of specimens very greatly depends. Knowing it to be Professor Goode's desire to adapt, as far as possible, the drawers used in all departments to the unit size which he has adopted for the Museum, some effort was made in this direction; but the adaptation, while possible for the exhibit collection, was found impracticable, or at least very undesirable, for the study collection. Hence, after carefully studying, in person, the different forms and patterns used for entomological collections both in this country and Europe, whether by private individuals or public institutions, I have adopted a drawer and cabinet essentially after the pattern of those used in the British (South Kensington) Museum, but best adapted in size to our own requirements or conception. The drawers are square, with an outside measurement of 18 inches and an outside depth of 3 inches. The sides and back have a thickness of three-eighths of an inch, while the front is five-eighths of an inch thick. The pieces are firmly dove-tailed together, the front being clean and the dove-tailing blind. The bottom is of three-ply cross-grained veneer, run into a groove at the sides, leaving a clear inside depth of $2\frac{1}{16}$ inches to the frame of the cover. The bottoms are lined in all but forty of the drawers with first quality cork one-fourth of an inch thick. At a distance of one-fourth of an inch from the sides and back and three-eighths of an inch from the front there is an inside box of one-eighth inch whitewood closely fitted, and held in place by blocks between it and the outer box. There is thus between the inner and outer box a clear space all round, in which insecticides or disinfectants can be placed to keep out Museum pests, and making it impossible for such to get into the inner box containing the specimens without first passing through this poison chamber.

The entire inside is lined with white paper, or, in the case of the uncorked boxes, painted with zinc-white. The front is furnished with a plain knob. The cover is of glass, set into a frame three-fourths of an inch wide, three-eighths of an inch thick, with a one-fourth-inch tongue fitting closely into the space between the inner lining and outer box, which here serves as a groove. This arrangement furnishes a perfectly tight drawer of convenient size, and not unwieldly for handling when studying the collection.

The material of which these drawers are made is California redwood, except the cover frame, which is mahogany. The cabinets containing these drawers are 36 inches high, 40 inches wide, 21 inches deep (all outside measurements), and are closed by two paneled doors. Each cabinet contains twenty drawers in two rows of ten each, and the drawers slide, by means of a groove on either side, onto hard-wood tongues, and are designed to be interchangeable. * * *

The bulk of the collection is still contained in small folding boxes which are admirably suited for containing a working collection, especially of those orders comprising smaller insects like Coleoptera, Hymenoptera, etc. These folding boxes have the great advantage of being readily re-arranged upon shelves and of being very easily used in study.

The folding boxes finally adopted are of white pine, shellacked and varnished, the bottom and top double, and cross-grained to prevent warping. They are 13 by $8\frac{1}{4}$ inches outside measurement, the top and bottom projecting slightly at the front and sides. The inside measurement is $11\frac{3}{4}$ by 7. The sides, back, and front are five-sixteenths of an inch thick, with a machine joint, which is neat and very secure. The boxes are $2\frac{3}{8}$ inches in outside depth, unequally divided, the lower portion $1\frac{1}{2}$ inches outside depth, lined inside with a thin whitewood strip, projecting three fourths of an inch above the rim of the outside box. Over this projecting lining the lid closes as tightly as practicable and is kept from springing by hooks and eyes. The bottom is cork-lined and covered with a fine white glazed paper.

All the boxes are furnished with neat brass label holders into which a card containing a list of the contents can readily be placed and removed at pleasure. This general form of box has long been used by us and by other collectors, and the chief demerit which I have endeavored to overcome by the above details, is the tendency to warp and crack in the trying steam heat of our Government buildings.

ARRANGEMENT OF BIOLOGIC MATERIAL.

The biologic material is, very much of it, alcoholic; for though many of the immature states of insects may be preserved by dry processes, yet the bulk must needs be kept in liquid. Where the material is in duplicate it is well, when it is not too heavy or cumbersome, to place such biologic material with the systematic collection; yet experience has taught that it is wiser to make a separate *biologic collection*, and this it is proposed to do. This collection will, in fact, be a feature of the Museum collection in the future. Hence it was very desirable to adopt some method of securing the vials in such a manner that they can easily be moved from one place to another, and fastened in the ordinary boxes and drawers employed for pinned insects. The vials in use to preserve such specimens as must be left in alcohol or other liquids are straight glass tubes of varying diameters and lengths with round bottom and smooth, even mouth. The stopples in use are of rubber, which, when tightly put into the vial, the air being nearly all expelled, keep the contents of the vial intact and safe for years.

Various forms of bottles are used in museums for the preservation of minute alcoholic material. I have tried the flattened and the square and have studied various other forms of these vials; but I am satisfied that those just described, which are in use by Dr. Hagen in the Cambridge Museum, are, all things considered, the most convenient and economical. A more difficult problem to solve was a convenient and satisfactory method of holding these vials and of fastening them into drawers or cases held at all angles, from perpendicular to horizontal. Most alcoholic collections are simply kept standing, either in tubes with broad bases or in tubes held in wooden or other receptacles; but for a biologic collection of insects something that could be used in connection with the pinned specimens and that could easily be removed, as above set forth, was desirable. After trying many different contrivances I finally prepared a block, with Mr. Hawley's assistance, which answers every purpose of simplicity, neatness, security, and convenience. It is, so far as I know, unique, and will be of advantage for the same purpose to other museums. Hence I have concluded in this report to give a brief description of it. It has been in use now for the past three years, and has been of great help and satisfaction in the arrangement and preservation of the alcoholic specimens.

The blocks are oblong, one-fourth of an inch thick, the ends (*c c*, fig. 66) beveled, the sides either beveled or straight, the latter preferable They vary in length and breadth, according to the different sizes of the vials, and are painted white. Upon the upper side of these blocks are fastened two curved clamps of music wire (*b b*), forming about two-thirds of a complete circle. The fastening to the block is simple and secure. A bit of the wire of proper length is first doubled and then, by a special contrivance, the two ends are bent around a mandrel so as to form an insertion point or loop. A brad-awl is used to make a slot in the block into which this loop is forced (*e*, fig. 66, 5), a drop of warm water being first put into the slot to soften the wood, which swells and closes so firmly around the wire that considerable force is required to pull it out. Four pointed wire nails (*d d d d*), set into the bottom so as to project about one-fourth inch, serve to hold the block to the cork bottom of the case or drawer in which it is to be placed. The method of use is simple and readily seen from the accompanying figures, which represent the block from all sides.

The advantages of this system are the ease and security with which the block can be placed into or removed from a box; the ease with which a vial can be slipped into or removed from the wire clamps; the security with which it is held, and the fact that practically no part of the contents of the vial is obscured by the holder—the whole being visible from above.

The beveled ends of the block may be used for labeling, or pieces of clean card-board cut so as to project somewhat on all sides may be used for this purpose and will be held secure by the pins between the block and the cork of the drawers.

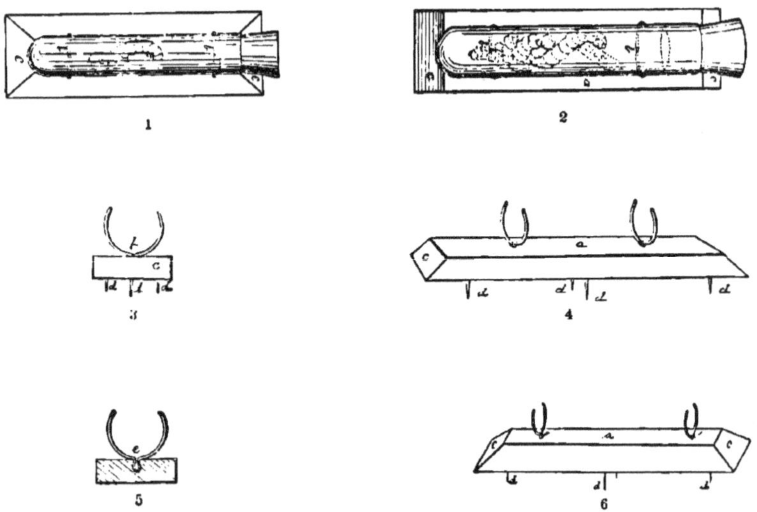

FIG. 66.—Vial-holder; 1, block, with vial beveled on all sides; 2, do. beveled only on ends; 3, block, end view; 5, do. section; 4, 6, do. side views; *a*, block; *b*, spring wire clamps; *c*, beveled ends of block; *d*, pointed wire nails; *e*, point of insertion of clamp (lettering on all figures corresponds.) After Riley.

NOTES ON LANGURIA.

By F. H. CHITTENDEN.

While on a collecting trip during June of last year I observed a specimen of that handsome little Erotylid beetle, *Languria mozardi* Latr. on a Composite plant, the daisy flea-bane—*Erigeron ramosus* Walt. (*strigosus* Muhl.)—the stem of which it was engaged in gnawing, having already cut with its mandibles a fair-sized hole preparatory to the deposition of its eggs.

In the account of the habits of this species published by Prof. J. H. Comstock in the Annual Report of the U. S. Department of Agriculture for 1879 (p. 199) it is stated that "the adult beetles begin to issue in August and on continually, making their exits until late in October. There is probably only one brood in a season, and the insect hibernates in the beetle state. An examination of many stalks (clover) during the winter failed to show the insect in any stage of growth."

When reading this account I remembered having seen during the preceding summer a female *Languria mozardi* ovipositing in a common species of ragweed (*Ambrosia trifida*). A visit in the following November to the locality where this observation was made resulted in the discovery of facts that throw new light on the habits of these beetles. In the ragweed stems were found some half dozen specimens of larvæ, which, with the aid of the description and figure given in the article above referred to, I was enabled to identify as belonging to some species of the genus. Of these larvæ all but one agreed with the published description and were afterward found to be *L. mozardi*. One specimen, however, was larger than the others and differed in other respects from the description. This specimen transformed and proved to be *L. gracilis* Newm. The larvæ did not appear to be feeding on the fresh white pith, but rather in the dead and discolored pith. They have a habit of frequently doubling up, assuming a shape that may be represented by an interrogation point: ⁓. Possibly by thus doubling up they are enabled to crawl up and down in the nearly hollow stems in which they live. Part of the larvæ were kept till April of the ensuing year, proving that they hibernate in this as well as in the adult state.

Can it be said of these beetles as of Cerambycidæ and allied families that they are single-brooded or double-brooded? As is the case with many other Clavicorns they breed the year round and there does not appear to be a well-defined or limited number of broods.

I have frequently observed these species in June and July on the stems of a common nettle (*Urtica dioica*), *L. mozardi* occurring in greater abundance, often *in copula* or busied in gnawing holes in the stems.

To recapitulate, *L. mozardi* is known to breed in the stems of clover, and specimens of larvæ indistinguishable from that of *L. mozardi* as described (*l. c.*) are mentioned by Prof. F. M. Webster (Rept. U. S. Dept. Agr., 1886, p. 674) as infesting the stems of timothy. Both species breed in *Ambrosia*, and their occurrence under the circumstances above recorded on *Urtica* is sufficient evidence that both breed in the stems of this plant as well. The probable oviposition of *mozardi* in *Erigeron ramosus* points to this as a likely food-plant, and the occurrence of the same species on the common ox-eye daisy (*Chrysanthemum*), a near relative of *Erigeron*, would lead to the belief that another Composite plant might be included in the list.

The habits of the two species are very similar, if not identical, and further investigation may show that they breed in the stems of a still greater variety of plants. My observations tend to show that they favor the Compositæ.

SOME OF THE BRED PARASITIC HYMENOPTERA IN THE NATIONAL COLLECTION.

It is our intention, as fast as the material in the National Museum collection can be re-arranged, to record in a series of lists in consecutive numbers of INSECT LIFE the hosts, dates, and localities of those species of Parasitic Hymenoptera which have been reared. New species are indicated in MS names where preliminary descriptions, which we hope to revise and publish, have been drawn up.

The advantages of such lists to working entomologists are too obvious to require elaboration.

Family BRACONIDÆ.

Subfamily Braconinæ.

Parasites.	*Hosts.*
Bracon simplex Cress	*Cerambycid* (unbred) under bark of Oak. Washington, D. C. Collected also at St. Louis, Mo., and in Texas.
Bracon agrili Ashm	*Agrilus fulgens* Lec. under bark of maple. Lafayette, Ind., April 21 to May 4, 1887.
Bracon pectinator? Say	*Saperda vestita* Say, on Elm? Washington, D. C.
Bracon arizonensis Ashm	*Andricus coxii* Bass. Fort Grant, Ariz., March 28 to April 6, 1882.
Bracon solidaginis Riley MS	*Gelechia gallæsolidaginis* Riley. St. Louis, Mo., August, 1867.
Bracon atriceps Riley MS	*Laverna sp.?* on *Epilobium angustifolium.* Cadet, Mo., September 3, 1886.
Bracon cecidomyiæ Ashm	*Cecidomyiid* gall on *Mimulus glutinosus.* Alameda, Cal., Jan. 9, 1886.
Bracon nigripictus Riley MS	*Sannina exitiosa* Say. Washington, D. C., May 10 and June 4, 1879. *Dolba hylæus* Drury. St. Louis, Mo., October, 1870.
Bracon sp.?	*Platynota flavedana* Clem., on Clover. Washington, D. C., September 3, 1879.
Bracon diastatæ Ashm	*Diastata* N. sp. mining leaves of corn Jacksonville, Fla., June 28, 1886. Received also from La Fayette, Ind.
Bracon gastroideæ Ashm	*Gastroidea cyanea* Mels. Columbus, Ohio, June 7, 1886.
Bracon phycidis Riley MS	*Phycis indiginella* Zell. Oxford, Ind., July 9, 1886.
Bracon pissodis Ashm	*Pissodes strobi* Peck. on pine. Wellesley, Mass., August 19, 1886.
Bracon n. sp	*Ægeria exitiosa* Say. Kirkwood, Mo., November, 1872.

Parasites.	*Hosts.*
Bracon xanthostigma Cr.	*Botis penitalis* on Lotus. St. Louis, Mo., September 15, 1875.
	Tortricid gall on Goatweed. Woodburn, Ill., August 6, 1872.
	Gelechia beneficentella Murtf. from bolls of *Solanum carolinense*. Washington, D. C., June 30 to July 4, 1886.
	Gelechia cercidos Murtf. Kirkwood, Mo. Collected also in Texas.
Bracon n. sp.	*Trypeta gibba* Löw. gall on Ambrosia. La Fayette, Ind., May 3, 1889.
Bracon gracilariæ Ashm	*Gracilaria desmodiella* Chamb. Kirkwood, Mo., July 12, 1886.
Bracon bucculatricis Ashm	*Bucculatrix* n. sp. on oak. Kirkwood, Mo., June 10, 1886.
Bracon xanthonotus Ashm	Phalænid pupa on Orange. San Diego, Cal., December 18-20, 1876.
Bracon n. sp	*Clisiocampa constricta* Str. Sacramento, Cal., June 16 and 17, 1882.
Bracon n. sp	*Proteoteras æsculana* Riley. Kirkwood, Mo.
Bracon californicus Riley MS	*Cecidomyiid* gall on *Baccharis pilularis*. Alameda, Cal., February 19, 1886.
Bracon cookii Ashm	Leaf-miner on Basswood. Lansing, Mich.
Bracon notaticeps Ashm	Tineid leaf-skeletonizer on Oak. Washington, D. C., September 30, 1880.
Bracon gelechiæ Ashm	*Gelechia* sp. ? on Oak. Washington, D. C., October 5 and 6, 1880.
	Gelechia cinerella Murtf. Kirkwood, Mo., 1881.
Bracon n. sp.	*Gelechia roseosuffusella* Clem. St. Louis, Mo., May, 1872.
Bracon analcidis Ashm	*Analcis fragariæ* Riley. St. Louis, Mo., September, 1870.
Bracon vernoniæcola Ashm	Dipteron in seeds of *Vernonia*. Kirkwood, Mo., September 14, 1881.
Bracon vernoniæ Ashm	*Platynota sentana* Clem. and *Eudemis botrana* Schiff. in seed capsules of *Vernonia noveboracensis*. Washington, D. C., May 15-18, 1885, and St. Louis, Mo., April 22.
Bracon junci Ashm	*Coleophora* ? on *Juncus balticus*. St. Louis, Mo., September 18, 1876.
Bracon juncicola Ashm	*Coleophora cispiticella* Walsingh. on *Juncus balticus*. St. Louis, Mo., September 11, 1876.
Bracon trifolii Ashm	*Coleophora* sp. ? on white clover. Washington, D. C., June 30, 1879.
Bracon tortricicola Ashm	Tortricid in seeds of *Ambrosia trifida*. Kirkwood, Mo., April 23, 1885.
Bracon euuræ Ashm	Galls of *Euura* on *Salix californica*. Donor P. O., Placer Co., Cal., January, 23, 1886.
Bracon juglandis Ashm	Tineid ? larva in walnuts. Los Angeles, Cal.
Bracon pomifoliellæ Ashm	*Bucculatrix pomifoliella*. St. Louis, Mo.

Parasites.	*Hosts.*
Bracon n. sp.	*Rhyssematus lineaticollis* Say on *Asclepias corymbosa*. La Fayette, Ind., March 29, 1889.
Bracon n. sp.	*Smicronyx tychioides* on *Cuscuta arvensis*. Washington, D. C., July 24, 1879.

Subfamily **Exothecinæ.**

Bathystomus n. sp.	*Tortricid* leaf-roller on Oak. Los Angeles, Cal.
Rhysipolis orchesiæ Ashm.	*Orchesia castanea* Melsh. in woody fungus. Grand Ledge, Mich., July 24, 1881.
Rhysipolis phoxopteridis Riley MS.	*Phoxopteris nubeculana* Clem. on Apple. Kirkwood, Mo., May 5, 1884.

Subfamily **Spathiinæ.**

Spathius abdominalis Riley MS	*Phlœosinus dentatus* Say on Cedar. Salina, Kans., May 23, 1885.
Spathius sequoiæ Ashm.	Coleopterous larva on Red Wood. Alameda, Cal.

Subfamily **Hecabolinæ.**

Cænophanes prodoxi Riley	*Prodoxus decipiens* Riley on Yucca. St. Louis, Mo.
Cænophanes hemiptychi Riley MS.	*Hemiptychus punctatus* Lec., in Grape. Elizabeth, N. J., 1880.
Cænophanes koebelei Riley MS	*Prodoxus æneseens* Riley on Yucca. Los Angeles, Cal., June 5 to 9, 1886.
	Prodoxus n. sp. on *Yucca whipplei*. Los Angeles, Cal., January 6 to February 10, 1887.
	Pronuba n. sp. on Yucca. Los Angeles, Cal., September 15, 1886.
	Prodoxus marginatus Riley on *Yucca whipplei*. Los Angeles, Cal., May 22, 1886.
Cænophanes n. sp.	*Laverna* n. sp. gall-moth on *Trichostomum dichotoma*. Georgiana, Fla., July 11, 1882.
	Gelechia gallæastrella Kellicott on *Aster asteroides*. Bladensburgh, Md., July 5, 1883.

Subfamily **Doryctinæ.**

Doryctes mellipes Ashm.	Borer in rotten Cherry-wood. Kirkwood, Mo., April 27, 1878.

Subfamily **Rhyssalinæ.**

Rhyssalus atriceps Ashm.	*Cacœcia rosaceana* Harr. on Apple. Washington, D. C., July 1, 1882, and August 15, 1886.
Rhyssalus loxotaeniæ Ashm.	*Loxotaenia clemensiana* Fernald on wheat. La Fayette, Ind., June 3, 1885.

Parasites.	*Hosts.*
Rhyssalus n. sp	*Sarrothripa revayana* Dup. on willow. Washington, D. C., July 22, 1886.
Rhyssalus selandriæ Ashm	*Eriocampa cerasi*? Peck. Washington, D. C., July 5, 1879.
Rybssalus antispilæ Ashm	*Antispila ampelopsiella* Cham. on Grape. Kirkwood, Mo.
Rhyssalus trilineatus Ashm	*Coleophora caryæfoliella* Chamb. on Hickory. Washington, D. C., May 5, 1883.
Rhyssalus oscinidis Ashm	*Oscinis* sp.? on *Plantago major*. Washington, D. C., July 6 to 9, 1888.
Rhyssalus californicus Ashm	Gall of *Holcaspis chrysolepsis* Ashm. on *Q. chrysolepis*. Colfax, Cal., December 19, 1885.
Rhyssalus gallicola Ashm	Gall of *Amphibolips trizonata* Ashm. on Oak. Fort Grant, Ariz., June 21 and 23, 1882.
	Gall of *Compsodryoxenus brunneus* Ashm. on Oak. Fort Grant, Ariz., April 27, 1882.
	Gall of *Callirhytes vacciniifolii*? Ashm. on Oak. Fort Grant, Ariz., April 21, 1882.
Colastes microrhopalæ Riley MS.	*Microrhopala vittata* Fab. on different species of Solidago. Washington, D. C., June 21, 1884.
? Colastes	*Gossyparia ulmi* Geoffr. Rye, N. Y., June 23, 1884.
Oncophanes melleus Ashm	Microlepidopterous? larva on Oak. Kirkwood, Mo., August 24, 1884.

Subfamily **Rhogadinæ.**

Heterogamus fumipennis Cr	*Sphinx drupiferarum*? Abb. on Apple. St. Louis, Mo., May 10, 1868.
	Smerinthus juglandis. Abb. St. Louis, Mo., July 15, 1873.
Heterogamus texanus Cr	*Ceratomia amyntor* Hb.? on Elm. Lansing, Mich., June 18, 1887.
Rhogas terminalis Cr	*Leucania unipuncta* Haw. St. Louis, Mo., June 12, 1876.
	Pædisca n. sp. gall moth on *Solidago lanceolata*. Washington. D. C.
	Nephelodes violans Guen.
Rhogas n. sp	*Scopelosoma sidus*? Guen. Washington, D. C., June 17, 1884.
Rhogas lætus Cress	*Acronycta dactylina* Grt. on Alder. Ottawa, Can., August 1888.
	Apparently same species found at West Cliff, Colo., collected also in Texas.
Rhogas harrisinæ Ashm	*Procris (Harrisina) americana* Harr. Jacksonville, Fla., October 9, 1879, and Kirkwood, Mo., October 18, 1881.
Rhogas geometræ Ashm	Geometrid larva. St. Louis, Mo.
Rhogas burrus Cress	*Acronycta hasta* Guen. on Wild Cherry. St. Louis, Mo., July 16, 1872.
	Acronycta lobeliæ Guen. on Oak. St. Louis, Mo., February 1, 1874.

Parasites.	*Hosts.*
Rhogas rileyi Cress	*Acronycta oblinita* S.-A. on Willow. St. Louis, Mo., April, 1868, and La Fayette, Ind.
	Nephelodes violans Guen. Ames, Iowa, June 11, 1887.
Rhogas platypterygis Ashm	*Platypteryx arcuata* Walk. on *Alnus serrulata.* Washington, D. C., October 23, 1883.
Rhogas nolaphanæ Ashm	*Nolophana malana* Grt. St. Louis, Mo., November 3, 1870.
Rhogas simillimus Ashm	Geometrid on Pine. Holderness, N. H., September 8, 1883.
Rhogas desmiæ Ashm	*Desmia maculalis?* Westw. Cadet, Mo., June 17, 1886.
Rhogas canadensis Cr	*Clostera inclusa* Hb. Washington, D. C., September 12, 1882.
Rhogas ceruræ Ashm.	*Cerura* sp. ? on Willow. Napa County, California, August, 1887.
Rhogas melleus Cr	*Aplodes subrifrontaria* Pack. on *Eupatorium.* St. Louis, Mo., 1871.
	Eucrostis zellraria Pack, on *Chrysanthemum.* St. Louis, Mo., October, 1871.
	Aplodes rubivora Riley on *Ageratum.* St. Louis, Mo., April 10, 1881.
	Clostera americana Harr. St. Louis, Mo.
Rhogas intermedius Cr	*Acronycta dactylina* Gr. on *Alnus incana.* Holderness, N. H., September 27 to October 5, 1883.
	Acronycta oblinita S.-A. Oxford, Ind., August 2, 1884. Washington, D. C., September 1, 1880.
	Acronycta sp. ? on Alder. New York, September 25, 1883.
	Acronycta hastulifera A. and S. on Alder. Washington, D. C., June 21, 1883.
	Acronycta americana Harr. on Maple. Kirkwood, Mo., October 7, 1877, and Lincoln, Nebr., November 4, 1889.

Subfamily Cheloninæ.

Phanerotoma tibialis Hald	*Grapholitha caryana* Fitch. Hickory-nuts. Kirkwood, Mo., April 5, 1873.
Sphæropyx bicolor Cr	Arctiid ? Washington, D. C., July 8, 1878.
Chelonus iridescens Cr	Phycid on *Aphyllon tuberosum.* San Diego, Cal., June 27, 1887.
Chelonus lavernæ Ashm	*Laverna cloisella* Clem. Kirkwood, Mo., 1881.
	Laverna sp. on *Epilobium angustifolium,* Cadet, Mo., September 3, 1886.
Chelonus pallidus Ashm	*Gelechia abscondilella* Walk. on Polygonum. Washington, D. C., May 2, 1884.
Chelonus fissus Prov	Lepid. gall on *Ceanothus cureatus.* Colton, Cal., June 18, 1887.

Parasites.	*Hosts.*
Chelonus parvus Say	*Cecidomyia s.-strobiloides* Walsh. Pahreah, Utah.
Chelonus nanus Prov	Nematus gall on Willow. Los Angeles, Cal.
	Subfamily **Sigalphinæ.**
Sigalphus curculionis Fitch	*Conotrachelus nenuphar* Hbst. St. Louis, Mo., June 15 to July 21, 1870.
	Borer in stalk of *Ambrosia.* St. Louis, Mo., May 4, 1873.
Sigalphus copturi Riley MS	*Copturus longulus* Lec. Washington, D. C. June 2, 1883.
Sigalphus nigripes Riley MS	*Andricus coxii* Bass. Fort Grant, Ariz., July 27, 1883.
Schizoprymnus texanus Cr	*Trypeta solidaginis* Fitch on *S. canadensis.* Washington, D. C., May 26, 1880.
	Trypeta gall on *Solidago.* Utah ? 1881.

(To be continued.)

ANTHRAX PARASITIC ON CUT-WORMS.

Four perfect bee-flies (family *Bombyliidæ*) which correspond with the description of *Anthrax hypomelas* Macq., have been sent us by our Indiana agent, Mr. F. M. Webster, and were bred by him last summer from the pupæ of a cut-worm which proved to be that of *Agrotis herilis.* Prof. C. P. Gillette, of the Iowa Agricultural Experiment Station, has also shown us one of three specimens of *Anthrax* (*scrobiculata* Loew) bred by him the past summer from cut-worm larvæ, the species undetermined. More recently Mr. Coquillett sent us a note for publication, covering a similar experience, from which we may quote the following:

Mr. Edwin C. Van Dyke, of this city, who is an enthusiastic young collector of insects, informs me that on one occasion he placed a Lepidopterous chrysalis in a box by itself, and that when next examined this box contained a Dipterous pupæ; the Lepidopterous chrysalis was found to be entirely empty, and in one end of it was a large opening out of which the Dipterous larva had evidently issued and afterward pupated. In due time this pupa produced the perfect fly, and this, together with its cast-off pupa-skin and the chrysalis-skin of its host, was kindly presented to me by Mr. Van Dyke. The chrysalis which it infested closely resembled that of *Taeniocampa rufula* Grote, a Noctuid which is rather common in this locality. The fly proves to be a specimen of *Anthrax molitor* Lœw, one of the commonest *Bombyliids* found in this State and scarcely distinguishable from the common *Anthrax flava* of Europe. The pupa very closely resembles that of *Aphœbantus mus* O. S., figured at 5, 5a and 5b, Plate XVI, of the Second Report of the U. S. Entomological Commission. On either side of the last segment are three short teeth, and on the under side of the head are five black tubercles, the anterior one being the largest, and the remaining four being disposed in two transverse pairs, those comprising the last pair being contiguous at their bases.

Though these are extremely interesting occurrences, and show that some species of *Anthrax* may prove of benefit in destroying cut-worms, they are not without precedent, as the group to which the species belongs is, according to Osten Sacken, known to prey normally on the pupæ of Lepidoptera, especially Noctuæ. In number of species this group is about equally represented in Europe and this country, and we find that this Lepidopterous parasitism, in regard to which both Osten Sacken and Schiner make only a generalized statement, was recorded by Zetterstedt as early as 1842. Meigen in 1820 stated that nothing was known of the early stages of *Anthrax;* Westwood in 1840, in his Introduction, mentions only its Hymenopterous parasitism; but Zetterstedt in the Diptera Scandinaviæ, writing in 1842, states that the eggs of the first section of the genus, which embraces the species with hyaline wings and the tomentum not entirely black (*A. flava* Meig., *A. circumdata* Meig., and *A. cingulata* Meig.), are deposited in the larvæ of Lepidoptera. Walker in 1851 makes the same statement in the Insecta Britannica, that some of the species are parasitic in Lepidopterous larvæ. In the second report of the U. S. Entomological Commission, p. 266, we have referred to Schiner's statement (as quoted by Osten Sacken) that the larvæ of the very nearly allied genus *Argyramœba* were parasitic in Lepidopterous pupæ, which fact has also been referred to by late German writers (*Entomologische Nachrichten*, 1885, p. 306). Osten Sacken refers particularly to this parasitism of *Anthrax* in the Biologia Centrali-Americana, published in 1886, where he states that a certain group of the genus is especially parasitic upon the Noctuæ. Glover in his MS. Notes on the Diptera, and also in Agricultural Report for 1866, states that "an *Anthrax* has been bred from the chrysalis of a moth."

FIG. 67.—*Anthrax hypomelas*: *a*, larva from side; *b*, pupal skin protruding from cut-worm chrysalis; *c*, pupa; *d*, imago—all enlarged (original).

MOUNTAIN SWARMING OF VANESSA CALIFORNICA.

By C. L. Hopkins.

During an ascent of Mount Shasta, made in August, 1889, a most interesting occurrence was noted in the flight of countless myriads of butterflies (*Vanessa californica*) at an altitude far above snow-line.

In our early morning climb of August 29, of the above year, we had left our horses at half past 4 o'clock, at what is known as "Horse Camp," at very near snow-line, where there were many small snow fields close about us. Our progress was very slow and tedious, being all of the time over loose, sliding fragmentary rocks, or the almost smooth, hard-frozen surface of the icy snow, and which latter did not soften till long after the sun had swung high enough to shine full upon it. Some little time after day-light, but long before we could see the sun, as he was hidden from us by the high crest of a sharp ridge on the southwest aspect of the mountain (our ascent being made from Sissons, west of the mountain), a few signs of insect life were seen in the shape of "snow-fleas," two or three large-winged grasshoppers, and, occasionally at first, a butterfly. The last two were stiffened by the cold as if they were there from the day previous. The latter insect increased much in numbers as we ascended, and were many of them found in among and under the loose stones as well as a few upon them.

At perhaps half past 9 we came to a point upon which the sun had long been shining, and here they were flying in the air, the flight being in a southeasterly direction. From here they seemed to increase very rapidly in numbers up the remainder of the ascent to well toward the summit. The latter was reached at 11.20 a. m.; the temperature was noted at 42° Fah. in the open air. We remained here about a half hour, then passed down by way of the Hot Sulphur Springs, and then out on the southerly face of the mountain. We again encountered our beautiful friends at not farther than six or eight hundred feet below the extreme peak, and now in countless numbers, filling the air with their flashing wings, and all passing in the same direction as observed during the ascent—towards the southeast. This strange sight continued until we seemed to pass below them, at an altitude of between 11,000 and 12,000 feet. The fact of its being a continuous flight of these insects across the mountain in one direction during the warm part of the day—a period of nearly five hours—is beyond question. That it was in progress one or more days previous to that upon which I observed it is an easy deduction from the fact of the numbers of the insects found among the rocks and stones while yet stiffened by the cold of the night air. How much longer it may have continued I had no means of knowing.

Where they could have come from, in such vast numbers, and what brought them to such a high altitude, is of course a matter of pure speculation.

I had no means of preserving specimens of these insects except to place them between the leaves of a note-book; in this way some were kept for identification. A gentleman whom I met a few days later pronounced the species to be *Vanessa milberti*, but after presentation of the account of the flight, with the specimens, before the Biological Society of Washington it was determined for me by Mr. Howard as *Vanessa californica*.

March 1, 1890.

NOTES ON A SPECIES OF NECROPHAGOUS DIPTERA.

By F. M. Webster.

The extent to which the mortal part of man is preyed upon by worms and insects, after being consigned to its final resting place, has, no doubt, been greatly exaggerated in the popular mind. Cases of such are doubtless exceptional, the exceptions being by no means common.

The gentleman to whom I am indebted for the specimens and facts upon which this notice is based tells me that within the last five years, and among seven cases of disinterment, this is the only instance which has come under his notice. Of these, four of the bodies had been buried nearly two years or over, and three had been buried about four months. As these disinterments were all made in connection with legal investigations of matters usually of a criminal nature, everything about the graves or on or about the bodies was carefully noted, and, therefore, had anything of the kind occurred in any of the other six cases it would most certainly have not escaped observation.

On February 1 of the present year, Dr. W. H. Peters, physician and analytical chemist, of La Fayette, Ind., placed in my hands, for investigation, a small quantity of light-colored sediment, intermixed in which were quite a number of small flies, later determined by Professor Riley as belonging in or near the genus *Conicera*, numerous pupæ and a single larva, the sediment having been placed temporarily in a vial of water. These insects, in the various stages of development, Dr. Peters stated had been obtained by himself from a corpse which he had examined only two days before.

The body was that of a male, German-American, age sixty-two years, height about 5 feet 9 inches and weight about 165 pounds. The death had been a violent one, and had taken place on January 31, 1888, the body being interred on February 2, two days later. The temperature, according to authentic records, during the time intervening between death and burial ranged from 28° to 37° Fah. The coffin was of wood and of the best modern manufacture, being practically air-tight when closed and the top fastened down, and encased in a box of pine. The grave was of ordinary depth, the soil in which the box and inclosed coffin rested being the upper strata of blue clay—proverbial for its compactness.

The body was exhumed on January 29, 1890, the pine-box being little decayed and the coffin apparently in perfect condition, but on removing the cover of the latter, the body, though exhibiting little indication of putrefaction, presented a very mutilated appearance with every indication that the missing portions had been attacked and destroyed by some element other than natural decay.

The front walls of the abdomen and thorax were gone, except small portions of the ribs and sternum, which were so friable as to be easily broken in the fingers, the ribs being readily severed by a pair of ordinary surgeon's scissors. The thoracic organs were gone, but the back wall of the thorax was only slightly imperfect. The front wall of stomach gone, back wall perfect, as also was the left kidney and spleen, lying beneath, and also portions of the intestines. The liver was unattacked but converted into adipocere, while the right kidney was destroyed. The back wall of abdomen was perfectly preserved; no trace of decomposition being visible. The flesh from the face had entirely disappeared. All of the tissues affected appeared to have been converted into grumous, viscid matter, of small bulk.

A considerable number of the flies were observed by the doctor moving about over the corpse, and living larvæ were noticed in the flesh, while the whole exposed surface of the body was quite thickly covered with pupæ, giving it the appearance of grains of wheat having been strewn over it. Analysis of that portion of the abdominal contents which would have included the contents of the stomach revealed arsenic in small quantities, as did also the liver.

That the larvæ of these flies might subsist upon the flesh of bodies killed by arsenic is by no means surprising, as they are, doubtless, very tenacious of life; yet it will be observed that the best preserved portions of the body and organs were those which would be the most likely to come in contact with the poison contained in the stomach. This, however, must not be taken as proof that the larvæ could not have subsisted upon slightly poisoned flesh, but the following statement found in Woodman and Tidy's "*Forensic Medicine and Toxicology*," p. 303, copied from "*Lancet*," August 23, 1856, p. 231, requires considerable verification before it can be accepted:

> A curious case is recorded, where about one hundred and fifty pheasants were poisoned from eating the maggots generated in some animals destroyed by a strychnia vermin-killer.

These flies, both sexes of which were secured, were entirely new to me, not having before observed anything like them, and while the presence of arsenic in the stomach did not render the presence of these ghoulish feasters more surprising, still, I was and am yet unable to account for their occurrence in the coffin, as observed by Dr. Peters. That adults or larvæ could have made their way to the body through box and coffin, after burial, seems incredible; while that, with the temperature but little above the freezing point, flies should have been attracted to the corpse, while the latter was awaiting interment, and either deposited their eggs upon it, before burial, or have been conveyed within the coffin to the grave and there began reproduction, appears at first thought almost equally impossible. The fact that the man had died suddenly, in the midst of good health, would rather imply the early appearance and rapid progress of decomposition and, thereby lead to the inference that the odors arising from the body would become more generally diffused throughout the house where this body was being kept, and thus attract any flies which might be present in or about the building. On the other hand the condition of the remains on disinterment, together with the well-known preservative effects of arsenic, point directly the other way, and to this feature we must also add the absence of the odors contingent to the sick-room, whatever their influence might be in attracting the flies. Furthermore, the room in which the body reposed was not heated, but the temperature kept as nearly as possible co-equal with that existing outside, viz, 28° to 37° Fah., the single door communicating with other parts of the house being kept closed as continuously as circumstances would permit. The building is of brick, and in the case of this particular room three of the four walls are outside walls.

These details are given thus minutely because if these flies inhabit our dwellings during the winter months, future studies should demonstrate the fact. Besides, Dr. Riley suggests to me that as *Conicera atra* is said by Schiner to breed in decayed radishes in Europe, the present species might have thus originated and been at the time inhabiting the cellar of this house and drawn therefrom by the odors of the corpse. In this case, I am assured that the cellar contained no vegetables except potatoes, which were not decaying, and that the cellar itself was in a cleanly and dry condition, and no portion of it was beneath the room containing the remains, but under an ajoining apartment, and that all of the floors were without holes or cracks. Also, that communication with this cellar was by a stairway leading from a small room, adjoining the one opening into the apartment containing the body, the door of this cellar-way being kept closed except on occasion of the by no frequent visits to the cellar itself. However, while these facts appear to considerably obscure the theory suggested by Professor Riley, I confess my inability to replace it with a more plausible one, and therefore present it as a substitute until some one can, in the future, throw additional light upon the problem.

MARCH 15, 1890.

ADDITIONAL NOTE ON SPIDER-EGG PARASITES.

By L. O. Howard.

Bæus americanus.—The publication of my description of this species on page 270 of the last number of Insect Life, has given me the pleasure of a card from Mr. J. H. Emerton, who informed me that I would find this species among some material sent to the Department by him some months ago, and search has revealed that he is correct. A number of female specimens have been found in a vial labeled in Mr. Emerton's handwriting, "Parasites on spider's eggs in orange cocoon, collected 1871."

In reference to this same species, Mr. W. Hague Harrington has written to Professor Riley as follows:

With reference to the description and excellent figure of *Bæus americanus* in the last number (p. 270) of Insect Life, may I mention that Provancher has described a species of this genus (Additions et Corrections a la Faune Hymenopterologique de la Province de Quebec, p. 209, 25 June, 1887) as a Chalcid, under the name *Trichasius claratus*. After characterizing the new genus formed to receive it, he gives the following (translated) brief description: "Length, .05 inch. Of a uniform reddish brown with the legs yellow. The antennal club black. Thorax densely punctured, metathorax rugose. Legs pale yellow, the last joint of the tarsus brown. Abdomen browner, polished but not metallic. Ottawa. Harrington." Evidently Mr. Howard has not recognized from its position and description the insect described by Provancher. He would hardly look for a *Bæus* among the Chalcididæ. The type, which is in my possession, seems to differ from *B. americanus* in being darker and in having the legs pale. I have not verified the measurement, which would make it about twice the size.

I am greatly obliged to Mr. Harrington, for this note and comparison of Abbé Provancher's description with specimens collected near Washington by Mr. Pergande shows that they are identical. Provancher's species should be known in future as *Bæus clavatus* (Prov.).

Acoloides saitidis.—Mr. F. M. Webster has just sent in twelve specimens of the female of this species which he bred from a spider egg-sac found under the bark of a log at Oxford, Ind., in October, 1884. This indicates that the species is quite wide-spread, as the specimens from which the species was named were reared by Mr. Bruner in Nebraska.

PREPARATORY STAGES OF SYNTOMEIDA EPILAIS Walker AND SCEPSIS EDWARDSII Grote.

By HARRISON G. DYAR, *Buffalo, N. Y.*

SYNTOMEIDA EPILAIS Walk.

Egg.—Hemispherical, the base flat, minutely punctured. Color, shiny pale yellow. Diameter 1mm. Laid in a mass, nearly touching on the under side of the leaf.

First larval stage.—Head brownish, paler down central suture and triangular plate; eyes black; mouth dark brown. Width of head, .5mm. Body pale yellowish white with black spots, arranged much as the warts of the Arctiinæ, each bearing one or more black hairs. Cervical spot brownish, and this as well as the anal plate has a row of small black spots. Feet, all blackish. Length, 2mm. As the stage proceeds, the body becomes pale orange yellow.

Second larval stage.—Head pale yellowish brown, eyes and mouth dark. Width, 8mm. Body, yellowish; spots black, as in mature larva, bearing thin tufts of black hairs, those at the extremities being the longest. Feet, black. Length about 4mm.

Third larval stage.—Head reddish orange; mouth dark. Width 1.1mm. Body reddish orange with black spots bearing pencils of hair as in the last stage, but the hair is only .4mm long. Length of larva about 8mm.

Fourth larval stage.—Head orange red; mouth dark. Width 1.5mm body as in last stage, but the subdorsal and other black marks, not bearing hairs, are absent. Length 15mm.

Fifth larval stage.—Mature larva. Head round, orange red, paler above the mouth. Palpi whitish; eyes and jaws dark brown; a few hairs. Width of head 2mm. Body, orange red with round, elevated, shiny black spots as follows: (1) in subdorsal space, anteriorly on joints 5 to 12 inclusive; (2) subdorsal row; (3) superstigmatal row; (4) stigmatal row of small spots each posterior to a spiracle; (5) and (6) are subventral rows, the lower consisting of large long spots above the base of each leg, while joints 2, 3, and 4 have only one subventral row. Cervical spot and anal plate have a row of small black spots. The subdorsal and stigmatal rows on joints 3 and 4, the superstigmatal on joints 5 to 11, and the subdorsal on joints 12 and 13 bear each a long (10mm) pencil of fine black hair. The others have a thin tuft of short hair. Black marks occur in the subdorsal space joining over the dorsum on the middle segments, situated posteriorly. Another row of spots in stigmatal space also posteriorly, and a fainter row in the subventral space, the latter in some examples nearly forming a band. Thoracic feet black, abdominal, black outwardly. Spiracles small and black. Length of larva about 30mm. Diameter of body 4mm.

Cocoon.—Composed of silk and the larval hairs and constructed in some inclosed place. It is thin and weak.

Pupa.—Depressed behind the thorax; very slightly flat below; abdominal segments without motion and cremaster absent. Color dark orange with black streaks, as follows: A spot on the head; two on the collar; two irregular angulated lines on the thorax; lines on cases of anterior legs and antennæ cases; two large and three or four small streaks on the wing-cases; abdominal segments have a transverse band on each of irregular width, some of them, especially at the anterior and posterior segments, interrupted. Length, 17mm. Width, 5.5mm.

Food-plant.—Oleander, *Nerium odorum.* Larvæ from Dade County, Fla., on the ocean side of Lake Worth.

SCEPSIS EDWARDSII Grote.

Egg.—Probably hemispherical, the base flat; smooth. Diameter, .7mm. The color could not be ascertained, as the egg had hatched and the shell had been nearly devoured by the little larva.

First larval stage.—Head shiny pale straw color, the eyes brown. Width, .4mm. Body, semitransparent whitish; warts arranged as in the mature larva, small and blackish, with scanty, but rather long black and white hairs. Length, 2.5mm.

Second larval stage.—Head shiny light yellow. Width, .5mm. Body whitish, dorsal band purplish obsolete anteriorly, in some examples interrupted by orange spots posteriorly. Warts whitish, some of those on the dorsum black. Hair still rather scanty. Length of larva, 4mm.

Third larval stage.—Head shiny pale yellow, eyes black, mouth whitish. Width, .7mm. The body varies somewhat in its markings, but the design is as follows: Body whitish, a broad dark wine-red dorsal stripe, interrupted by orange spots on joints 4 and 12, the two upper rows of warts on joints 3, 5, 8, 11, and 13 black, the rest whitish; a white subdorsal line. The hairs are long, white, and black. Length of larva, 5mm.

Fourth larval stage.—Head pale yellow, the triangular plate and mouth white; eyes black; width, .9mm. Body pale yellowish white with a white subdorsal line. Joints 3, 5, 8, 11, and 13 are black in the subdorsal space as are the warts. The other segments are tinged with orange, especially joints 4 and 12.

Fifth larval stage.—Head yellowish orange, triangular plate, mouth, and palpi white, the former bordered above by a deep black shade, more or less extensive. Eyes black; width of head, 1.2mm. Body as before; length, 10mm.

Sixth larval stage.—Head as in the mature larva; width, 1.6mm. Body very pale yellow, with a narrow interrupted white stigmatal, rather broad yellowish white subdorsal, and broad black dorsal band, the latter dilated on joints 3, 5, 8, 11, and 13 to inclose and cover the two upper rows of warts, nearly interrupted on joints 4 and 12 by a large orange patch, and on joints 6, 7, 9, and 10 bisecting a fainter orange patch. Hair white, but largely black from the black warts. Length of larva, about 14mm. The pencils of brown hair found on the mature larva on joint 5 are present in some examples, though small.

Seventh larval stage.—As in the previous stage, but the pencils on joint 5 are more prominent and the transverse band on joint 4, found in the mature larva, is present, being yellowish tinged with orange. Width of head, 2.2mm. Length of larva, 20mm.

Eighth larval stage.—Mature larva. Head, brownish red; triangular plate, mouth and palpi white, the former bordered above by a broad black band. Jaws and eyes black. Width of head, 3mm. Cervical spot, blackish, bisected. The warts are arranged as follows: On joint 2, which is much contracted, are two small warts at the spiracle; on joints 5 to 12 is a row of warts in the subdorsal space, situated anteriorly; a subdorsal row; a superstigmatal row; a substigmatal row; two subventral rows on joints 5 to 12, the upper small; only one row on joints 3 and 4. Joint 13 has the upper warts reduced in number and has a row of small ones on the anal plate. Body dirty whitish, a blackish shade on the dorsum, with subdorsal, and traces of stigmatal, yellowish white band; above the former, on joints 5 to 10 and on 12, is a faint orange patch, the brightest being on joint 12. Posteriorly on joint 4, across the subdorsal space, is a pinkish white band with a dark border anteriorly, and on joint 5, from the warts in the subdorsal space (first row) grow two little pencils of brownish red-plumed hairs. Sometimes similar but much smaller pencils appear from the subdorsal warts (second row) of joint 12. The warts all bear yellowish, bristly hairs, some of which overhang the head. Legs concolorous with the body, the claspers of the abdominal tipped with brown. Length of larva, 30mm.

Cocoon.—Spun on any flat surface without covering. It is made of silk and the larval hairs which are laid more roughly at the point at which the imago will emerge. The whole of the under side is fastened to the supporting surface.

Pupa.—Cylindrical, flattened a little in front, the dorsum very slightly depressed behind the thorax. Abdominal segments without motion. Body punctured and wing cases creased, but slightly. Cremaster covered by a bundle of short hooks and surrounded by similar hooks on the last segment, which also extend up the dorsum in little transverse rows. Color, red brown. Length, 14^{mm}.

Throughout the larva is subject to considerable variation. The duration of each stage was three days, except the last two, which were longer. Pupa, 14 days.

Food-plant.—The rubber tree, *Ficus pedunculata*. Larva from Dade County, Fla.

THE TULIP TREE LEAF GALL-FLY.

Diplosis liriodendri O. S.

In the *Garden & Forest* for December 18, 1889, Mr. J. G. Jack again publishes a good account of an insect with which we have long been familiar and about which we have had notes for a long time in the notebooks of the Division which have not seen the light of print.

One of the earliest objects of entomological interest which met our eye when we first came to Washington was a tulip tree, the leaves of which were badly infested by this species and which stood under the window of the Division of Entomology. Attempts were made by Prof. Comstock to rear the adult early in the summer of 1879, but he did not succeed until with a later brood the same season. In October, 1879, however, several adults representing both sexes were reared, and descriptions of these, as well as of the early stages, have since remained unpublished in the notes of the Division.

Mr. Jack, as appears from his article, has recently reared the same insect around Boston, and is the first to record the appearance of the adult. Osten Sacken, in 1862, described the gall and the larva, but did not rear the fly. The appearance of the galls is well described by the latter author in the following words:

> Brown spots with a yellow or greenish aureole on the leaves of the Tulip tree (*Liriodendron tulipifera*). These spots, about two-tenths or three-tenths of an inch in diameter, indicate the presence inside of the leaf of a leaf-mining larva of Cecidomyia. * * *

The effect of the blotches at Boston is described by Mr. Jack and corresponds well with the results of the work of the insect as seen at Washington:

> Many people who have always counted upon their Tulip trees as belonging to one of the few species free from serious insect attacks, have, by midsummer, been disgusted to find the leaves filled with large, brown, and yellow blotches. In some instances the foliage, by the end of August, has become so brown and twisted from the effect of numerous spots in every leaf that it has had the appearance of having been scorched by fire, and many of the leaves having thus become dead and dry fall to the ground.

Each of these spots before maturity contains a single orange-colored maggot which issues, when full-grown, through a slit at the edge of the under side of the blotch and falls to the ground to transform.

Mr. Jack finds three or more annual generations at Boston, the final larvæ dropping to the ground in September and hibernating as pupæ. Our notes indicate that there are also three broods at Washington and, although we have reared the adults in October, we surmise that the species normally hibernates in the larva or pupa state underground.

The figure of the adult accompanying Mr. Jack's article is faulty in regard to the third vein of the wing and in the absence of the cross vein. The female antennæ are also 14-jointed instead of "apparently 13-jointed." His implied criticism of Loew, however, to the effect that the male antennæ are 14 jointed instead of 26-jointed, is probably correct, as in the antennæ of every male Diplosis, with which we are familiar, the true division is at every other bulb instead of at every bulb.

The remedy of late fall or early spring plowing and rolling suggested by Mr. Jack will probably greatly reduce the numbers of the pest.

AN EXPERIMENT WITH COCCINELLIDÆ IN THE CONSERVATORY.

By F. M. Webster.

The extent to which the various species of *Aphididæ* and *Coccidæ* enter into the food of this family of beetles has led to the suggestion that they might be utilized in keeping some of our greenhouse pests in subjection, at least during the winter season. As nothing definite appeared to have been done in this direction, some experiments were begun during the fall of 1889, with a view of learning whether or not the colonization of these beetles, in conservatories, could be made of practical benefit to the florist, and, perhaps, to the market gardener also.

The prospect of realizing any very enthusiastic expectations was somewhat dampened at the start from the fact that the terms "Scale," "Mealy bug," and "Green fly" are far from being specific terms, and might each apply to an indefinite number of species, while considerable evidence has accumulated in this and other countries, going to show that the several species of Coccinellidæ are not indiscriminate feeders, but confine their attention each to some particular species, or, at most, include but a small number on their "bill of fare." Therefore, the results obtained by experimentation with one species might not hold good with another, and, indeed, it might be that, in case one species of beetle proved effective as against its particular favorite among the Aphids, several species might be required to work out beneficial results. From this it will be readily observed that the experiment is one which can not be carried out in a single year, or in a single locality, for the reason that the species of Coccinellidæ are not equally distributed or yearly equally abundant.

Partly because of its great abundance, and partly because it had been observed feeding upon several species of Aphides, among them one infesting the rose, *Coccinella 9-notata* Hbst., was more particularly selected for the purpose of carrying out one portion of the experiment, other species being included in smaller numbers only.

The experiment began July 26, by transferring fifty adults of *C. 9-notata* from the fields to the conservatory. September 24 there were added to these sixty-two, and two days later fifty-six others. These last included also a very few *Megilla maculata*, *Hippodamia convergens* and *H. 13 punctata*. October 1, thirty-four more were placed as the others had been, these being nearly all *9-notata*, and were mating at the time. On October 15, many young larvæ were observed running about over the potted plants, but despite these the Aphides increased so rapidly that it became necessary to fumigate with tobacco smoke to protect the plants, and a very light fumigation was applied. Although the smoke did not appear to affect the larvæ, they continued to decrease in numbers, though only a very few seemed to reach maturity, a single adult, *H. convergens*, being the only evidence that any of the larvæ had developed. At present writing, March 15, of the two hundred and two individuals placed in the conservatory, there remains not a trace, either of themselves or of their progeny, while "green fly" has abounded, as usual.

For the other portion of this colonizing experiment *Chilocorus bivulnerus* Muls. was selected. A couple of white spruce trees *Abies alba*, on the campus of Purdue University, became thickly infested by *Mytilaspis pinifoliæ** which, as is usual in such cases, attracted myriads of the Ladybeetle.

On October 22, several hundred of these beetles were transferred from the spruce to another compartment of the same conservatory, devoted exclusively to tropical and subtropical plants, ferns, etc., upon which were large numbers of *Coccidæ*. For a few weeks after being liberated an occasional beetle would be observed, while dead ones gradually became more numerous until no living beetles could be found. Outside, however, they were present about the spruces in great numbers on warm sunny days, and continued to remain up to date of writing. Not a living individual has been observed in the conservatory for three months, yet the "Scale" and "Mealy bug" have in nowise diminished in numbers. This compartment has not been fumigated, nor has anything been applied to the plants which could in any way affect the Ladybeetles, and therefore both features of the experiment must be set down as yielding information decidedly adverse to the colonization of either of these species of Coccinellidæ in our conservatories.

---

*I may perhaps be pardoned for stepping aside from the tenor of this notice in order to record the fact of this scale being attacked by the Insidious Plant-bug, *Triphleps insidiosus*, and which I several times detected with its beak thrust into the body of the female *Mytilaspis*.

A NORTH AMERICAN AXIMA AND ITS HABITS.

By L. O. Howard.

In the Transactions of the Entomological Society of London for 1862 (p. 373) Mr. Walker described an anomalous genus of Chalcididæ under the name of *Axima*, from specimens collected by Mr. Bates, at St. Paul, Brazil, the sole species receiving the name *Axima spinifrons.* Walker recognized in this genus affinities with the Chalcidinæ, Eurytominæ, and Eucharinæ, and also with certain exotic genera which connect the Pteromalinæ with the Cleonyminæ.

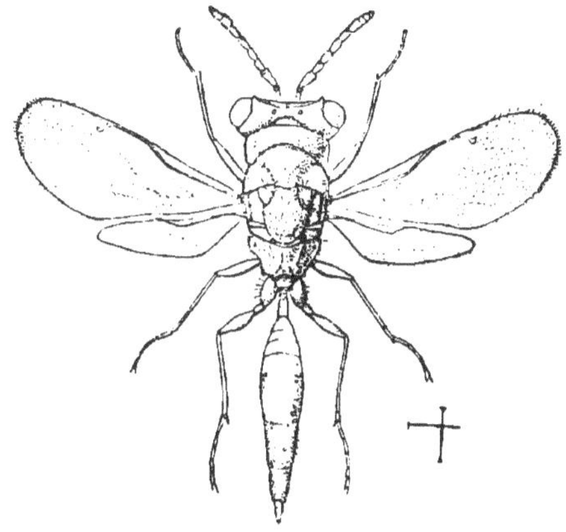

Fig. 68.—*Axima zabriskiei*—Female, from above—enlarged (original).

In July, 1884, Cameron, in the *Biologia Centrali-Americana*, erected upon this genus the subfamily *Aximinæ* and added the Central American genus *Hontalia*. He recognized its relationships with the Chalcidinæ and Eurytominæ. *Hontalia*, however, differs from *Axima* in its thickened and toothed hind femora and in the strongly exserted ovipositor, and Cameron has made a slip in giving as a subfamily character "posterior femora thickened, minutely toothed," which, however well it applies to *Hontalia*, is not applicable to *Axima*.

Mr. Ashmead, in the Proceedings of the Entomological Society of Washington, Vol. I, p. 219, mentions the occurrence of a form closely allied to *Axima* among some South American Chalcididæ collected principally along the Amazon by Mr. Herbert Smith, and which, as a transition form, convinced him that *Axima* really belongs to the *Eurytominæ*.

I had previously reached a nearly similar conclusion from examination of the true species of *Axima*, described in this paper, and also from two transition forms in the collection of the National Museum, the one collected by Branner & Koebele, at Benito, province of Pernambuco, Brazil, in February, 1883, the other occurring in the Belfrage collection from Texas. One of the principal reasons for arriving at this conclusion is the distinctively Eurytoma-like antennæ of the male, as shown in figure 69. Walker did not know the male of his species. Without an examination of the types of *Hontalia*, however, it will be premature to condemn the subfamily *Aximinæ*.

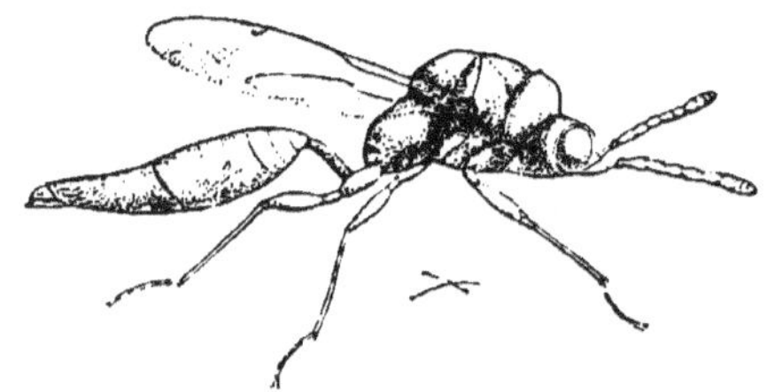

FIG. 69.—*Axima zabriskiei*.—Female, side view—enlarged (original).

To the Rev. J. L. Zabriskie, formerly of Nyack, N. Y. (now of Flatbush, L. I.), is due the credit for first ascertaining the habits of this anomalous group of Chalcidids, although the genus had been found in North America before he reared it, as I recognized in 1887 specimens in the collection of the Cambridge Museum. These were labelled, apparently in the handwriting of Mr. H. G. Hubbard, who left Cambridge in 1874, "Larvæ found in burrows of small blue bee, FreshPond, Mass." I also find in my notes on some of the Chalcids in the Cornell University collection, which I made in 1887, the following entry:

Axima sp. and *Ichneumon* sp. ex. *Ceratina dupla?* Larva of Axima has six or more strong dorsal tubercles and head of pupa is strongly tuberculate.

Mr. Zabriskie on three occasions reared quite a large series of the species about to be described from nests of *Ceratina dupla*, and there can be but slight doubt that *Axima* is a primary parasite of this little bee and probably of allied species. Mr. Zabriskie first reared it in July, 1878, from nests of the Ceratina, in stems of cultivated Black Raspberry, at New Baltimore, Green County, N. Y., and again in April, 1883, and April, 1884, from nests of the same bee, in stems of Sumach (*Rhus typhina*), at Nyack, N. Y. He reared in all twenty-five females and ten males. I briefly mentioned this fact on page 540 of Volume II of the Standard Natural History, but it has not elsewhere been recorded. Eleven specimens were sent by Mr. Zabriskie to Professor Riley, and from them the accompanying figures and descriptions have been made.

Axima zabriskiei n. sp.

Female.—Length 6^{mm}. Expanse, 7^{mm}. Head and thorax coarsely and densely punctate and with faint whitish pile; lateral ocelli just behind ridge extending from one frontal lateral projection to the other; median ocellus just anterior to this ridge, making the ocellar triangle very obtuse and in two different planes; metanotum rugose, with a few irregular longitudinal carinæ; pronotum with a faint median tubercle. Petiole of abdomen as long as metanotum, very finely shagreened and irregularly and faintly carinate. Abdomen smooth, shiny, with patches of fine pubescence; a rounded patch on sides of fourth segment, and fifth and sixth segments almost entirely covered. Fimbria of the metanotal callus quite long and white, and a row of rather long soft white hairs on outer margin of hind coxæ. General color black, with rather indefinite ferruginous markings; all over the thorax the black is so indefinitely blended with ferruginous as to make it impossible to define color areas; the ferruginous is more marked, however, on the sides of the pronotum and mesoscutum; antennæ black, scape reddish at base; all coxæ black and punctate; all trochanters dark honey yellow; all femora and tibiæ black in middle, dark honey yellow at tips; all tarsi honey yellow; abdomen ferruginous at base below. Wings narrow, short, reaching when closed only to middle of fifth abdominal segment, perfectly hyaline, veins very dark brown.

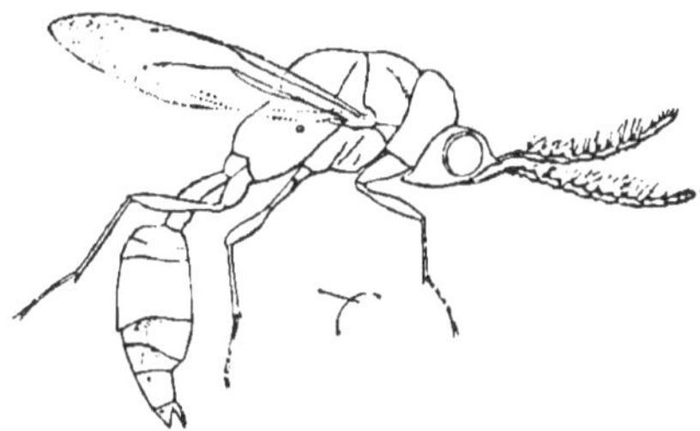

FIG. 70.—*Axima zabriskiei.*—Male, side view—enlarged (original.)

Male.—Differs only in the antennæ and in the shape of the abdomen, as shown in the figure. The frontal projections and the median projection of the pronotum are sharper and more pronounced than in the female.

Described from four female and three male specimens from Rev. J. L. Zabriskie, Nyack, N. Y., reared from nests of *Ceratina dupla.*

EXTRACTS FROM CORRESPONDENCE.

The Scale Question in Florida.

Some time ago a gentleman from Riverside went to Florida for the purpose of examining the orange groves and nurseries in that State to ascertain their condition in regard to being infested with scale insects, and a short time ago he informed me that there was scarcely a single orange grove in Florida over two years old that is not infested with *Mytilaspis citricola.* He further stated that many of the orange groves there were as badly injured by this scale as any orange grove in California has been injured by *Aspidiotus aurantii.* He also stated that next to *M. citricola, Mytilaspis gloverii* is the next most common species, and next to this is *Ceroplastes floridensis.* Yesterday a nurseryman, who is engaged in growing orange trees in Florida and shipping them into this State, called upon me, and informed me that in Florida *Mytilaspis citricola* is harmless; that he has never known it to injure orange trees, during his ten years residence in that State, and that it can not live in Southern California even if imported here. I would like very much to learn from you to what extent *M. citricola* injures orange trees in Florida. It is the commonest species that I receive for identification on trees coming from Florida, and our citrus growers are very anxious to learn to what extent it is injurious. I would also be glad to learn to what extent the Six-spotted Mite, which you recently described as the *Tetranychus* 6-*maculatus*, injures orange trees in Florida. I found it on leaves of orange trees said to have been imported from Florida, but have never found it on trees growing here. It may interest you to know that the Vedalias have survived the winter, unprotected, out of doors. There are at least two places in this city where they are found at the present time. Occasionally a few Iceryas are found, but usually in very small numbers, and as the Vedalias have proved to be able to take care of themselves during the winter season, it is very probable that they will remain with us so long as any Iceryas are to be found.—[D. W. Coquillett, Los Angeles, Cal., April 8, 1890.

REPLY.—Yours of the 8th has come to hand. You have been misinformed as to the state of affairs in Florida. Some sections of that State are naturally more badly damaged by the species of *Mytilaspis* than others, and the Florida Wax-scale is, in my experience, not an especially injurious insect to citrus fruits. The relative importance of the Florida scale-insects is well set forth by Hubbard, and you can learn his opinion by consulting his work. My own experience, in a broad way, from personal observation, may be summarized thus: The three most injurious species in Florida are: *M. citricola*, *M. gloverii*, and *Parlatoria pergandei*. None of these insects are as injurious in Florida as either Icerya or the California Red-scale, or the San José Scale. They are more widely spread throughout the State and do not concentrate so injuriously in given localities. The Florida scales are also more amenable to treatment than the three species mentioned in California. At one time there was considerable alarm from the attacks of *citricola*, and a great many groves have been seriously damaged by it, but the most progressive growers at the present time do not fear it. Men who are ignorant of or fail to apply the best remedies still suffer. What truth is there in the rumor that *citricola* has become established in California? I send you inclosed some galleys from my forthcoming report for 1889, which will give you the latest information as to the damage done by the 6-spotted mite. I am very glad to learn that Vedalia so well survived the winter out of doors.—[April 16, 1890.]

A Palm leaf Scale in Trinidad.

I inclose a piece of palm-leaf of *Prilohardia fibifera*, which is very badly infested by a scale insect of the genus *Mytilaspis*, so far as I am able to make out. The palms were obtained from the botanic gardens in Trinidad, and this insect is only to be found on the species mentioned, while the remainder were absolutely free of them, though they suffered from other pests.

Can you give me any information about the *Mytilaspis?* Unfortunately I have not been able to investigate the life-history of the insect, as the palms are growing in a garden which I can visit only now and then.—[A. Ernst, Caracas, Venezuela, South America, March 9, 1890.

REPLY.—The remarkable Coccid which you send me with your favor of the 9th instant, has only lately been described and figured as a new genus and species, *Ischnaspis filiformis*, by J. W. Douglas, in the *Entomologist's Monthly Magazine*, vol. XXIV, 1887, p. 21. Douglas found it in the conservatories of the Royal Botanic Society, of London, on the leaves of various palms (*Strychnos myriatica*) and other plants. Within the last year or so I find this species under the same conditions in the greenhouses of the U. S. Department of Agriculture at Washington, where it does much damage.—[March 22, 1890.]

The Cigarette Beetle.

My friend, Professor Gill, told me at the Cosmos Club that he had spoken to you about some "troyka" cigarettes that I got at the club, the paper of which had been pierced by a beetle. He told me you said it was the "Death-watch," and gave a latin name, which I did not completely catch, as several people were talking at the same time. To-day I found the inclosed beetle among some of the cigarettes as I was breaking them up. Professor Gill has some of the punctured cigarettes that I gave him; the rest have been destroyed. I inclose the beetle in a vial, and a piece of the punctured cigarette paper with it, and I herewith send the same to you, as the animal may have some interest for your investigation. I do not want it again.—[A. A. Hoehling, M. D., U. S. Naval Hospital, Washington, D. C., April 11, 1890.

REPLY.—I have your favor of the 11th instant and the accompanying specimen of a beetle which you found in cigarettes. This is *Lasioderma serricorne* Fabricius, popularly known as "tobacco beetle," one of the cosmopolitan insects, and known to infest not only dried tobacco leaves but also all sorts of drugs and spices. It is not identical with the so-called "death-watch" (*Anobium pertinax*), but belongs to the same family. Its life-history has often been treated of by various authors but presents no features of especial interest. It is referred to in INSECT LIFE, I, No. 12, pp. 378-9.—[April 14, 1890.]

A Curious Case.

I send you by mail, in a little wooden pen box, marked with my initials, a small black insect for identification. This bug was found in a clothing store here, and had died after cutting through a pair of heavy woolen pantaloons, making eight holes about the size of a buck-shot. It does not seem to be like the moth which usually cuts woolens.—[Thos. C. Harris, curator State museum, Raleigh, N. C., March 27, 1890.

REPLY.—The specimen which accompanies your letter is a wood-boring beetle (*Buprestis striata*), and it is probable that it issued from some of the wood-work within the store, and in endeavoring to make its escape cut through the clothing. The emergence of wood-boring beetles from furniture, which in some cases has been used for years, has been frequently reported. The larvæ in these instances were in the wood before it was used in manufacturing the articles of furniture.—[March 31, 1890.]

Beneficial Beetles infested with Mites.

By to-day's mail I send you a beetle which, with others, has been in a neighbor's cold frame, all of which he says have been covered with the minute ones. Are the small ones the same species, or are they parasites? If parasites, they are fully able to take care of the large ones; he did not say whether the large ones were destructive to his plants. The sleet of last week killed most of the Aphides that were hibernating on the rose bushes, some of which were literally covered from the ground to the very top. They did immense damage in this county (Camden, N. J.) to melons and cucumbers, as well as attacking currants and cherry and apple trees.—[I. W. Nicholson, Camden, N. J., March 13, 1890.

REPLY.—Yours of the 13th, with specimens, duly received. The beetle is one of the ground beetles of predatory habits known as *Harpalus faunus*, and the small creatures upon its back belong to a species of parasitic mite known as *Uropoda americana*. This same species is a common parasite of the Colorado Potato-Beetle, and was first figured and described by me in ninth Report on the Insects of Missouri, page 41.—[March 14, 1890.]

Flea Beetle Injury to Strawberries.

I send you by same mail box containing specimens of small *beetles* which appeared here yesterday. The first I heard of them was in the western part of the county, on Tuesday. They appeared on my strawberries in *thousands*. You can judge of their numbers when I tell you that all sent were taken by holding the box under one leaf and shutting the cover down on it, and I expect you will find at least twenty-five or thirty in the box, and they are numerous all over the patch. All the berry fields in this neighborhood are infested. I have seen them also on weeds of different species and on peach trees. I have tried tobacco dust, wood ashes, and lime dusted over the plants, but these remedies only drive them off for a short time. Will you please inform me if you know the beetle, and if so, how destructive it is and how long it stays. It feeds on the leaves from the *upper side*, eating off all the green part of the leaf and leaving only the skeleton. The beetles are of a very bright shiny dark-green color, and fly about in clouds when disturbed. I am afraid they will destroy all the strawberries, and then I fear for our melons and beans. Please let me hear your opinion of the insect, and if you need more specimens or any further information as to its ravages, I shall be only too glad to give you the results of any thing I can learn of its habits. No one who has seen it here has ever seen it before.—[W. E. Hudson, P. O. box 58, Orlando, Fla., March 27, 1890.

REPLY.—The beetle you send is *Haltica ignita* Illiger. We would recommend dusting with air-slaked lime. It is difficult to treat on plants like the strawberry, on account of risk to fruit from the ordinary insecticides.—[March, 1890.]

Lecanium hesperidum.

In one of your letters, dated May 19, 1887, you incidentally mention that "it has been discovered recently that the male of *L. hesperidum* is often associated with the female scale, an undeveloped, wingless creature." As the *Lecanium viride* of the coffee is considered to be very closely allied to the former species, I have thought that the males may perhaps be found in the same situation. It would assist me greatly in my search if you could give me any further description of the recently discovered male of *L. hesperidum*. Does the male insect differ much in appearance from the female? What are their relative sizes? Does the male undergo any pupal stage, as in the other species of Lecanium? Is it active, and provided with a mouth? This pest is still in activity in legion, although it appears to be slightly decreasing in intensity.

I must thank you again for your extremely interesting periodical INSECT LIFE, which continues to be full of useful information upon all subjects connected with economic entomology.—[E. Ernest Green, Eton, Punduloya, Ceylon, India, February 15, 1890.

REPLY.—Your favor of February 15 came duly to hand. The male of *Lecanium hesperidum* was discovered by Monsieur R. Moniez, who published a description thereof and an account of its development in the *Comptes Rendus des Séances de l'Academie Française*, February 14, 1887, page 449. Various longer and shorter abstracts of this article have been published in several periodicals, *e. g.* in the *Entomologist's Monthly Magazine*, Volume XXIV, 1887, pages 25 to 27, which is probably accessible to you; but I am not aware that any independent investigations on the subject have been made or published subsequent to Moniez's original discovery. The fully developed male is excessively small, with no trace of eyes or wings, but provided with antennæ, legs, and with a short and broad penis. M. Moniez observed three stages of the male: In the first, the body has no appendages whatever, and no visible segmentation; in the second, which represents the pupa stage, the body has a distinct segmentation and contains fully developed spermatozoids and testicles; the third stage is that described above. In no stage has the male been found outside of the body of the parent, and copulation must, therefore, take place within the body of the parent female. I have had no opportunity so far to confirm Mr. Moniez's statements by personal observation, but in past years I have bred the winged males of several of our species of Lecanium.—[March 24, 1890.]

Flies in an exhumed Corpse.

I mail you to-day a species of Diptera in its various stages of development. This matter is of peculiar interest, as the material was taken from the corpse of a man who died two years ago (in midwinter of 1888), and was buried after the usual manner. A few days ago the body was exhumed, the coffin opened, and the front part of the chest and abdomen were found to have been completely eaten away, the mass of flesh and slime being alive with these flies and their larvæ. The material was given me after standing in water for a couple of days, and therefore is in poor condition. When exhumed, both coffin and case containing it were in perfect condition, and the soil was a stiff blue clay. I can not myself account for the presence of these flies, except that the adults were hibernating in the coffin when used, or else the larvæ were in the stomach of the person when death took place. I have never met these flies before. Please let me know what you make out of them, and your idea of the manner of their first securing admission to the corpse.—[F. M. Webster, La Fayette, Ind., February 1, 1890.

REPLY.—Your letters of February 1 and 4 have come to hand, together with the specimens. The fly bred from a corpse belongs in or near the genus *Conicera* of the Phoridæ, although no species of this genus has before been mentioned in this country. The species which you have sent, however, has hairy eyes, while the European species are described as having naked eyes. Little is known of the habits in Europe, although Schiner says that *C. atra* breeds in rotten radishes. The experience which you relate is a most interesting matter and perhaps its publication may bring out further experience.

A number of cases of insects found on or breeding in corpses are on record in Europe. P. Mégnin, in "La faune des tombeaux" (Comptes rendus de l'Ac. des Sciences, v. 105, No. 20, Nov. 14, 1887, pp. 348–351) gives a summary of what is known, from which it appears that on exhumed corpses from two to three years old the following insects have been observed: Diptera, *Calliphora vomitoria*, *Cyrtoneura stabulans*, *Phora* (*Trineura*) *aterrima*, *Anthomyia* sp.; Coleoptera, *Rhizophagus parallelocollis;* Thysanuridæ, *Achorutes armatus*, *Templetonia nitida;* Myriapods, *Julus* sp.

The two first named Diptera cease to work after the lapse of two years, and since they have occurred only on such corpses as have been buried in summer, it is evident that the eggs must have been deposited before burial.

The *Anthomyia*, *Phora*, and *Rhizophagus* are found, on the contrary, on corpses buried whether in summer or winter. Corpses buried two years have been found covered with myriads of the pupæ of the *Trineura aterrima*, and the larvæ of *Rhizophagus* have also been found in large numbers. The eggs of both insects are deposited on the ground, and Mégnin concludes that these larvæ work their way into the coffins through nearly seven feet of ground (2 meters). It is finally stated that the *Phora* prefers lean corpses, whereas the *Rhizophagus* has been found only on *fat* corpses. In your case it would seem the more probable supposition that the eggs were deposited before burial.—[February 10, 1890.]

SECOND LETTER—Glad to hear about the corpse-infesting flies. From the fact that this matter is likely to figure in a supposed murder case, I shall have to ask you to publish nothing for the present. A few additional facts I will, however, give you now, and shall probably get nearer to the bottom later, when I will furnish you with a note for publication.

The person in life weighed about 165 pounds; height, 5 feet 9 inches; age, sixty-two; male. Death accompanied with congestion of lungs, indicating pneumonia, pains in abdomen, and frothing at mouth. Died January 21, 1888. Coffin practically air-tight, constructed of whitewood, and inclosed in ordinary pine case. Undertaker stated at time that he had embalmed body, but now states that it was not embalmed. Substances used in embalming, arsenic and corrosive sublimate.

Body exhumed January 29, 1890. Case and coffin in perfect state of preservation; the latter appearing to be air and water tight. Face, abdominal thorax, front walls of abdomen portions of all abdominal organs, and the less solid part of ribs eaten away. Posterior portion of stomach and body not eaten. Analysis of stomach shows $1\frac{1}{2}$ grains arsenic. Larvæ, pupæ, and adults alive at time of exhumation.

With the criminal and legal features of the case I have nothing to do, but how could these larvæ live in a body containing either arsenic or corrosive sublimate? (The chemist is searching now for the latter and I shall know results in a day or so.) If the man was not poisoned could the larvæ have killed him? It did not at first seem possible that the fly could breed in a body poisoned either before or after death with arsenic, but in "Forensic Medicine and Toxicology," Woodman and Tidy, page 303, is an extract from "Lancet," August 23, 1855, page 231, in which the statement is made that "one hundred and fifty pheasants were poisoned from eating the maggots generated in some animals destroyed by a strychnia vermin-killer." I do not know whether to believe this or not. If we have a fly here in Indiana which can kill a man in mid winter and half devour him within two years, poison and all, it will be worth knowing.

A physician in the city made the analysis and gave me the flies, and has promised me that if it is necessary to exhume the corpse, I shall have the opportunity to inspect it. If you wish more material, or think of any points which can be cleared up in regard to the insect, please write me, and if the chance is offered I will get them.

Please, however, before printing anything on the subject, let me get all the facts possible in the case, when I will put them in shape and send you. Can you figure the different stages with the material you have?—[F. M. Webster, La Fayette, Ind., February 12, 1890.

Reply.—Your letter of the 12th has come to hand, and this further information makes the case of the corpse flies even more interesting. In the first place I do not think there is any possibility that the flies or their larvæ killed the man, and the case which you mention from the *Lancet* is rather improbable. I do not at all doubt that the flies could flourish in the body of the man had he been poisoned by taking a dose of arsenic, but it is less probable that they could live in the body if it had been thoroughly embalmed by injecting the usual arsenic and corrosive sublimate mixtures. Even the latter, however, is not so improbable as it would at first appear, for many of these Dipterous larvæ are very tenacious of life and very little affected by poison. I should by no means say that the fact that they lived in the body and bred in such great numbers is proof positive that the body had not been embalmed. Ptinid larvæ have been known to feed in the corks of bottles containing corrosive sublimate.

I find on examining the specimens here that they were kept in water too long to be in good condition for figuring. If you have other better flies send them on, and if you have an opportunity to secure fresh ones, let me have a set placed directly in alcohol. I will, however, have as good a figure as possible made from those which we have here.—[February 15, 1890.]

The May Beetle and the White Grub.

Have you given any attention to the period of abundance of the May Beetle, *Lachnosterna fusca*, and have you thought it worthy to forecast the year of swarming and attack on their favorite trees for food, as the walnut, hickory, butternut, and ash, invariably stripping off all the June foliage of that year? You are familiar with the life history, indeed your observations are the only ones made by an American entomologist on the common American form of dor-beetle, in reference to its transformations to full development, and I presume you have published the year of great swarming at various times.

I have noted for many years their stages of growth and length of larval and imago life, and by taking the three-year period easily predict their years of swarming, which were for the last decade 1883, 1886, 1889, and will come again in 1892. I have followed this series of broods backward and find it agrees with the swarming in Alabama in 1880, and that in Massachusetts in 1865; also that recorded in eastern New York in 1850. Certainly there is a small number of beetles on the wing every year, and there must be, therefore, two other series of broods, occupying the two intervening years.

I know that the entomologists of Europe predict the year of abundant swarming for their common dor-beetle (a triennial period also, I believe) which led to much preparation for destroying them; but, unfortunately for the reputation of those wise bug-men, something about the weather, fungous diseases, or parasites interrupted, so as to cut off the brood in those districts, and thus the prophesied swarming never came. Has there ever been a break, in any section, respecting their abundance in the swarming year in this country ?

There is considerable usefulness in the record of abundance of the May Beetle for the farmer. He can reckon that those fields in sod in the spring of the swarming year will be the depository for many eggs, the grubs from which will do but little injury that year, but would do much damage to corn or potatoes if planted on the sod ground the following year, but not so much damage the next year, because the grub becomes full-fed and grown to pupa stage by midsummer. The insect really experiences the warmth of four summer suns. The first June, an egg; the second June, a small growing grub; the third June, a nearly full-grown grub; the fourth June, a winged beetle.

I believe the earliest account of this beetle, in respect to a correct exposé of its life history through all the stages and length of time noted, was made in 1852 by David L. Bernard, Clintondale, Ulster County, N. Y., and may be found in Patent Office Report for 1852 (1853), page 219. It is remarkable that he seems not to have known any common or Latin name for the insect. He simply says the grub is the larva of a beetle, and then describes the growth.

It is a matter of constant observation everywhere that skunks feed upon the grubs to the extent of extracting every grub lying anywhere near the surface of the ground, and thus aiding the agriculturists in securing larger and better crops. If they were not trapped off so closely they would rescue the crops from many thousands of dollars damage. Moles feed on them, and I am led to believe the raccoon feeds on grubs in small extent and I presume the hedgehog may have that predilection, but I know of no other American mammal in farming districts so disposed, although some others may be led to acquire melolonthivorous habits; at least, I have found that one class of domesticated animals can be led to acquire a taste for the white grub and very soon exhibit a decided fondness for this grub, literal and pure.

Linne, my little son, without any definite design exactly, began coaxing his dog, a half terrier and spaniel, to eat the grubs. He was quickly successful, and since then this dog and a St. Charles spaniel from an adjoining farm, taking up the habit, both follow the plow all day to eat every freshly exposed grub, and often they scent them underneath the surface and dig them out.

To be sure, if the grubs are very plenty Tony and Ned get a surfeit in an hour, but usually they are in the field nearly the whole time the plowman is there, and they feast on the grubs with as much gusto as at the first, some two years ago. Thus they render a better service than the crows or ravens in those long-ago dreamy rural scenes where troupes of these birds are represented following the plowman to pick up every grub, and indeed some wire-worms, but also crowd in angle-worms and all the beneficial ground beetles and their larvæ.

From trials made with several kinds of domesticated dogs it appears to be easy to induce any variety of this class of quadrupeds to form this habit of eating to a purpose. I am not so sure but wild canines, like the fox, wolf, and coyote, eat grubs and other insects when hard pressed by hunger. The members of the Ursine order are abundantly on record as feeders of the honey, as well as the young grubs, of bees, and the bees too. And bears are known to be fond of the white grubs they dig out from rotten logs, as well as the May Beetle grub they find underneath the logs, besides eating locusts and other insects.—[W. L. Devereaux, Clyde, N. Y., February 8, 1890.

REPLY.—We have established little of a reliably definite nature relative to the life term of the larvæ of this insect, although a large series of notes has accumulated in the endeavor to establish the definite facts. These notes seem to show that at Washington the ordinary length of larval life is three years and that there are no definite broods; that beetles appear and oviposit every summer and that larvæ of all ages can be found in the ground at any given time. We are not prepared to say that these are hard and fast rules for even this one locality, and we should certainly expect a variation with climate. The *Melolontha vulgaris* is said to remain three years in the larval stage in South Europe, and four years in North Europe.—[February, 1890.]

Parorgyia on Cranberry in Wisconsin.

I wish to call your attention again to some insects sent by my brother to you last summer. They were a lot of caterpillars. One species especially had done great damage on a neighboring cranberry marsh. The caterpillar was of a mouse-gray color, 1½ inches long, provided with feelers or horns. On his back there was a tuft of fur or hair, resembling the hump on a camel. You called it a species of *Parorgyia*. I also sent specimens to Professor Henry, at the Madison (Wis.) Agricultural Experiment Station. In his absence Mr. E. S. Goff replied. He called the insect that I speak of *Arctia*, and said that it is an enemy of the cranberry. In the interest of the Wisconsin Cranberry Growers' Association I respectfully ask for a little more light, if you can shed any from the above description or your personal experience. How do they pass the winter? And when does the moth deposit the eggs that furnish the brood which does the damage in July? The vine and fruit worm moths we successfully catch at night by means of lamps set in tin pans containing water, and a little kerosene oil on top. It kills them as soon as they strike the water in the pan. Now, is the moth of the former-described caterpillar of nocturnal flight? If so, they can be caught the same as the fruit moth. I will be thankful for any information that will enable me (not being an entomologist) to study their habits and mode of breeding. I have succeeded in raising a moth from the caterpillar. I wanted it to exhibit to the association last January or I would have sent it to you; perhaps then you could have readily given me the information I now seek.—[H. O. Kruschke, Deuster, Juneau County, Wis., February 24, 1890.

REPLY.—The moth sent by your brother last summer has been reared and proves, as I surmised, to be a species of *Parorgyia*, but the precise species can not be determined at this moment. An allied species lays its eggs late in July and the larvæ attain full growth by fall, hibernate in a web, transform to pupæ in the spring and issue as moths in early summer. The larvæ received from your brother, however, were nearly full-grown August 1, and the solitary moth which we bred issued August 21. This would seem to indicate either two broods or the hibernation of the partly-grown larvæ, moths of which emerge in August. Most of the larvæ which he sent were parasitized. The moths are night-flyers and would probably be captured by the same traps which you use for the vine and fruit worm moths. It is doubtful, however, whether this capturing of the moths will do much good, as careful examination of specimens so captured shows that the vast majority are males, or females which have laid their eggs. The best remedy will be to apply Paris green or London purple, as I suggested in my letter to your brother August 3, last.—[February 27, 1890.]

Helomyza sp. found in Mayfield Cave, Ind.

To-day I send you by mail a number of flies taken in Mayfield Cave on December 28, 1885. They were found under stones on the bottom of the cave, and sticking to the sides of the cave in sheltered places. They were not very torpid, as when I lifted up the stones they would generally commence to move. In the above cave they are abundant. I expect they may be found in other caves around here in equal numbers, although I have not hunted for them. You may keep them or else turn them over to the Smithsonian.—[C. H. Bollman, Bloomington, Ind.

REPLY.—I beg to acknowledge the receipt of your letter and specimens and to state that the flies which you found in Mayfield Cave belong to two different species, both apparently belonging to the genus *Helomyza*. Neither of the species is contained in the collection of the National Museum, and they may be both possibly new, although this is only probable.

A Cave Crustacean in a Well.

A friend of mine has a splendid well of water with a force pump in it. The water is always cool and has been clear until now. Lately it throws forth plenty of the inclosed specimens. Are they not Phylopods, or the Ear-wig, or is this the Lithobius, the crawling fellow we find in our house once in a while? Tell me all about it and how to clear the well.—[J. M. Shaffer, Keokuk, Iowa.

REPLY.—I beg to acknowledge the receipt of yours of the 19th instant, with specimen of an animal found in a well. This is the little crustacean described by Packard as *Cæcidotea stygia* and which has been found in Mammoth and other caves, in the little pools of clear cold water which abound in such locations. You will find a very good figure of this species, with an account of its habits, in the third volume of the *American Entomologist*, pages 35–36 (February, 1880).

Potato Stalk-borer in Corn and Rag-weed.

Mr. O. J. Voorhees brings me this morning samples of growing corn nearly a foot high which are being destroyed by larva unknown to me. I understand that the cornfields are largely infested. Will you please describe fully? Have you knowledge of a remedy? If so state it fully that it may be published for the common good.—[J. M. Shaffer, Keokuk, Iowa.

REPLY.—The worm which you send and which infests corn in your vicinity is the larva of a common Stalk-borer (*Gortyna nitela* Guen). This insect is a very general feeder and ravages not only corn but other cereals and also potatoes, tomatoes and a number of flowering plants which are commonly grown in gardens. By way of compensation it is particularly partial to the stem of the Cockle Bur (*Xanthium strumarium*). On account of its diversity of food plant, and on account of its feeding on the interior of the stalk, it is a most difficult insect to fight. The only remedy which has ever been proposed, and the only one which will result in any practical results, consists in cutting the larvæ out of the stalks which are observed to wilt from its attacks. This of course would be a most tedious operation in large quantities, but it is the only way to lessen the number of worms. The labor of boys could be readily utilized in this work. It has been previously recorded as damaging corn, but I think never to the extent which you describe.

ANOTHER LETTER.—As you request, I to-day send you a box containing a larva of the corn-stalk borer, marked No. 1, and three larvæ of what appears to be the same, which I found in rag-weed stalks, marked No. 2. In the corn they are rare at this season, but are rather common in the rag-weed. On the 29th ultimo I noticed one stalk where the borer had eaten out and left. In large corn they enter the stalk a few inches above the ground, and eat across nearly to the opposite side, and then upward. The first time I ever saw the borer was in 1882, in a piece of ground that had been pastured more than twenty years, and never plowed until that spring. There were quite a good many of them. The next year I had corn on the same ground and there were a few again. These were all near the edge of the field. This year I have corn on the same piece again, and they were all over it. One day about the 1st of June, I killed about fifty worms, and many more at other times. In 1884 I found a few in rag-weed along the edge of this same field, when it was in oats. This spring I found a few in another field over a quarter of a mile from the other infested ground. Others in the neighborhood are not troubled with them to any great extent. The most of their work is done when the corn is from 2 to 10 inches high, and before it begins to joint. Then the heart is eaten out just above the root, leaving the outside of the stalk green. The infested stalks may be known by the central blades being dead. This is the best time to destroy them. They are then from ½ inch to 1 inch long, and are easily killed by pulling the infested stalks up and crushing them.—[Thomas Wikessell, Wauseon, Ohio.

REPLY.—The Corn-stalk Borer (No. 1) and the Rag Weed Borer (No. 2) are both larvæ of *Gortyna nitela* as I supposed from your letter of the 27th ult. *Achatodes zeæ* is quite a different thing. The *Gortyna* is a very common insect and is found on a great many plants in addition to the two which you have mentioned. It first came into prominence as a potato stalk-borer and was described as such by Professor Riley in one of his early Missouri reports, and also in his little book on Potato Pests. It is also found in a number of other cultivated plants and large-stemmed annual weeds. No other remedy has been suggested than cutting them out of the stalks which they infest, by hand, and this of course would be impossible in a field of corn. As they seem to infest the Rag-weed on your place in considerable quantities, numbers can doubtless be killed by cutting and burning the weeds at this time, or while the majority of the larvæ are still within the stalks.

The Melon Worm.

It has come under my observation that the late crop of canteloupes in this section is generally very much injured by a bug or insect which bores a hole in the fruit when about half grown or just ripening, rendering it entirely unfit for use, while early crops are rarely if ever affected by this borer. The canteloupe crop will be much later than usual this year on account of continued excessive rains this spring, and want of warm weather to make the vines grow, and in anticipation of the trouble referred to, I would be very glad to have you give me a remedy if you know of any to avoid it.—[J. H. von Hasseln, Anderson, S. C.

REPLY.—The insect of which you speak is in all probability the Melon Worm of the South (*Phakellura hyalinatalis*). This insect is treated in the annual report of this Department for 1879, pages 218 to 220. The only remedy so far known is to watch for the first brood of the worms, which will probably be found feeding upon the leaves and stems before the young melons have begun to form. They should be killed by hand or by the application of Paris green and flour. At this late date when the second brood of the worms are boring into the melons there is no remedy.

Cut-worms and Carnations.

I send you by this mail some larvæ which I find near the surface of the ground around the roots of our carnations. There is something that eats a hole in the sides of the buds of our carnations and destroys the whole of the flower. Our gardener says that he believes this is the grub that does it, and that it goes up the stem in the night and feeds on the buds, and hides in the soil during the day. As we have not seen it around here very long I send it to you for a name.—[Thomas B. Meehan, Germantown, Philadelphia, Pa.

REPLY.—The insect which you send is the Variegated Cut-worm (larva of *Agrotis saucia*) and it is altogether likely that it is responsible for the damage to carnations which you describe. You will find this larva treated in the annual report of this Department for 1884, pages 297 and 298. The subject of "Remedies for Cut-worms" will be found on pages 298 to 300.

The Plant-feeding Lady-bird and the Potato Stalk-beetle.

I take the liberty of forwarding to your address by mail to-day specimens of a bug (also egg-clusters) which proves to be very destructive to the bean crop in Colorado. It seems to be closely related to the potato-bug. The hairy slug defoliates bean vines in the same way that *D. decemlineata* defoliates potato-vines. I have recommended the Paris green remedy also for this pest. Will you favor us by giving *name* and *history* of the insect? Can you suggest a better remedy than Paris green?

Can anything be done to prevent the ravages of *Baridius trinotatus*, which threatens the destruction of the potato fields in some sections of Pennsylvania?—[Tuisco Greiner, Little Silver, N. J.

REPLY.—This insect belongs to the only genus of the Coccinellidæ or "Lady-birds" which is plant-feeding in its habits. It is *Epilachna corrupta*. I can suggest no better remedy than Paris green.

Baridius trinotatus is an insect which can only be fought by pulling up and burning the infested stalks. It is a tedious remedy but a sure one. As the insect transforms within the stalk this remedy is efficacious at almost any time.

Intrusion of the Elm Leaf-beetle in Houses.

I now take the liberty to intrude upon your time with a few words concerning the habits of this (the Elm-leaf) beetle, with which you may not be so unfortunately acquainted as I am. It was in 1883, in the fall house-cleaning, that my attention was called to these creatures, then unknown to me, massing themselves in close packs behind pictures. In 1884 I noticed the trees for the first time being stripped, and that fall more bugs came in, and in the spring of 1885 they appeared in great quantities about the windows, but soon left the house for the trees, as we suppose. During the summer of 1885 the Elm trees were wretchedly stripped, and last August, as early as the 6th, these beetles came to the house in swarms. The house was thoroughly netted, but nets were of no use. They only disappeared during real winter weather to re-appear this spring, in April and May, in quantities. The old-fashioned garret is full of them; killing off day by day with powder makes no difference except for that day. The shingled roof is full; the window boxes where the cords play are full; the windows are daily covered, especially on the upper part, with quantities. They eat no flannels or woolens of any kind, never bite nor molest the body except liking to sleep in one's bed; they fill drawers, boxes, books, etc., and show no disposition to go out to the trees, and what they subsist on is a puzzle. Please excuse this great liberty; the truth is, that with every effort to bear the plague philosophically, the natural dislike of the housekeeper to be worsted in a battle with any even the most aristocratic insect prevails in my case, and I thought it just possible you might be able to tell me of some one thing that would give me the victory I desire, at the sacrifice of my hospitable instincts.—[H. S. Onderdonk, Great Neck, Long Island, N. Y.

REPLY.—The account which you give of the great numbers in which the Elm Leaf-beetle infests your house is very interesting, but I am sorry to say that I can offer you no encouragement in regard to any remedy beyond what you will find published in Bulletin 6 of this Division (which we have already sent you), and beyond the free use of Persian insect powder in your house.

Re Lestophonus.

Yours of the 21st instant, inclosing duplicates of the articles on the Lestophonus and its parasites, is just received. The facts are so clearly and correctly stated in these articles for INSECT LIFE that I am unable to suggest any change or alteration.

In regard to the manner in which I treated Koebele's second sending of Australian parasites I will say that Mr. Koebele advised me to subject the contents of each box to chloroform, then open each box and destroy all of the Chalcids and transfer the Lestophoni to the tent. However, I was unwilling to expose the Lestophoni to such a risk of life, so I had constructed two sacks of a muslin so thin that I could easily distinguish from the outside the Chalcids from the Lestophoni as they rested on the inside of the sack; the sacks were about 3 feet high by a foot and a half in diameter, and were sufficiently close in texture to prevent the escape of either the Lestophoni or the Chalcids. In these two sacks I emptied the contents of the boxes of parasites, tied up the tops of the sacks, then destroyed the Chalcids by pinching them between the thumb and finger, without opening the sacks, after which the sacks were opened and the Lestophoni liberated into the same tent in which I placed the first consignment of these flies.

The Chalcids are easily distinguished from the Lestophoni as they sit on the inside of the sacks, not only by their more slender form, but especially by their habit of always holding their wings lying flatly upon the back when not in use, instead of holding them partly expanded, as the Lestophoni do. The latter when disturbed usually fly upward, and are thus easily liberated from the sacks, while the Chalcids when disturbed simply leap a short distance and again alight lower down upon the inside of the sack. I have examined these sacks every few days and carefully destroyed the Chalcids and then liberated the Lestophoni. These two muslin sacks I kept inside the tent. The contents of some of the tin boxes which were in worse condition I put in a paper bag, pinned it shut and kept it in my room; nothing but Chalcids have appeared in this bag, and all of these have been carefully destroyed. Altogether there have issued from this second sending up to date twenty-four Lestophoni and one hundred and sixty-one Chalcids.—[D. W. Coquillett, Los Angeles, Cal., January 27, 1889.

GENERAL NOTES.

BOILING WATER FOR PEACH BORER.

Mr. John B. Haas, in the *Pacific Rural Press* for March 22, gives the result of a very conclusive experience in Missouri some years ago. He removed the soil around his infested trees for a depth of 3 or 4 inches, making a trench from 3 to 6 inches in width, and poured a bucketful of water, boiling hot, all around the trunk of the tree, allowing it to remain in the trench. He states that it killed all of the borers present and that his trees, which had been covered at the base with gummy exudation and had been in very bad condition, rapidly improved and bore fine crops.

THE FAMILY PHYLLOXERIDÆ.

Dr. L. Dreyfus, in the "Zoologischer Anzieger," No. 316, 1889, has published a little statement to the effect that his new family which he had erected in his work entitled "Uber Phylloxerinen," Wiesbaden, 1889, should be given the "idæ" termination instead of the "inæ." He therefore gives as the four families of the suborder Phytophthires: (1) Coccidæ; (2) Phylloxeridæ: (3) Aphidæ; (4) Psyllidæ.

THE NEWLY IMPORTED ROSE SAW-FLY.

Mr. J. G. Jack refers in *Garden and Forest* of March 26, 1890, to the introduction of the European *Emphytus cinctus* into this country. He has found it feeding upon the roses in the Arnold arboretum at Cambridge in the summer of 1887 and succeeded in rearing the adult in the autumn of 1888. This species is from two to three times as large as a common Rose Saw-fly, has a white band around the body of the female, and is more active. The eggs are deposited singly on the under side of the leaf and there are two or three annual generations.

TESTIMONIAL TO MR. KOEBELE.

Hon. Ellwood Cooper, the president of the State Board of Horticulture of California, has suggested the raising of funds for the purpose of presenting Mr. Koebele with a testimonial in recognition of his services in importing the insect enemies of the Fluted Scale, and we learn from the *Rural Californian* of April that the sum of $232.50 was raised during the recent convention at Los Angeles. The subsidiary statement which is being quite generally made and which has caused his friends no little anxiety, viz, that Mr. Koebele's health was ruined by his trip to Australia has, we are happy to state, no foundation whatever. Mr. Koebele writes that his health is perfect, and that he is good for three such trips, and it is due him to announce that the statement above-referred to and which has placed him in a false light, was started by secretary of the the State board of horticulture, upon his own confession, "for effect"!

A PARADOX.

It may seem very much like a contradiction in terms to speak of a white black scale, yet this is what we have recently received from Mr. Coquillett. In the midst of a normally colored colony of the Black Scale (*Lecanium oleæ*) on oleander he found a full-grown individual of a uniform perfectly white color. Mr. Coquillett considered this color to have been due to the fact that the specimen had recently molted, but so far as we know the Lecanii have no distinctive molts. It is probably an instance of albiuism, and the first one of the kind which has ever come to our notice among the Coccidæ.

A RARE SPHINGID.

We have just received for the National Museum collection from Mr. W. G. Henry, of the U. S. Coast and Geodetic Survey, a specimen of the female of the rare *Pseudosphinx tetrio.* Mr. Henry gives us an interesting account of its capture, which we may quote:

The insect referred to was captured *at sea*, on January 19, while the *Blake* was at anchor on a current station in the Gulf of Mexico, about 160 miles south of the Mississippi River mouth, and about half way between the Louisiana coast and the Campeche Banks (Yucatan coast), I noticed the insect (I presume it was the same) sitting on the boom. under the awning, and tried to catch it, but it flew away as lightly and easily as a bird and took a straight westerly course across the sea until it was out of sight, and I saw it no more that day. The next day (January 20), I was sitting on deck and saw the insect (presumably the same) come in a straight course from westerly across the sea and alight on board, and, after repeated efforts, it was captured. The *Blake* had been at sea (out of sight of land) for six days, having left the Mississippi on January 13, and the insect was so shy and hard to approach that I think it could not have been on board the ship all that time without being disturbed and seen. For a week previous to its capture there had been no high wind from any direction which could have blown the insect off to sea, and it is therefore natural to suppose that its flight across the sea was entirely voluntary. I sent the insect to you from New Orleans on January 24.

On February 1 (I think) we again left the Mississippi and ran across the Gulf of Mexico to the Campeche Banks, and began to re-occupy the current stations, at intervals of 60 miles, on a line across the Gulf from Campeche Banks to mouth of Mississippi. On February 9 we arrived at and anchored on the same station where the insect sent you was captured, and strange to state, while anchored there another of the same kind of insect came on board. It could be approached near enough to see that it was the same kind of insect, but it eluded every effort to capture it, and finally flew away across the sea. No other insect of that or any other kind had been seen anywhere in the Gulf, and it was rather strange that the only two seen should have been at the same spot, in the center of the Gulf of Mexico, and at an interval of twenty days.

A NEW APPLE PEST.

At a recent meeting of the agricultural bureau of this colony the secretary reported that he had noticed that many of the apples, in a shipment of ten thousand cases from California to Sydney, were perforated and tunnelled as though they had been attacked by the larvæ of the Codlin Moth. He had forwarded some of these to Mr. Frazer S. Crawford, as the matter was urgent, and the following report had been sent on by him to the commissioner of crown lands:

I have received from the secretary of the central agricultural bureau an apple stated to be one of a large importation from California, and which was supposed to be attacked by the codlin moth. On examination I found a number of small channels running through it in various directions, of an average diameter of about one-twentieth of an inch, in some places filled up with fine excreta. From those I extracted seven footless grubs, the largest about one-tenth of an inch long by rather more than half that in width. They are white, or else of a pale rose color, and have a white head. They are evidently the grub of a beetle; but of what species I am unable to say, as no mention of such an insect attacking the apple is made in any English or American work that I have got. I believe it to be a new pest to California, or only one that has only appeared there within the last year or two. If introduced here, I consider it likely to be as destructive as the codlin moth, and one equally as difficult to eradicate I therefore respectfully suggest that every endeavor should be made to trace this shipment of apples, and if possible that all found in the colony should be destroyed, and, furthermore, I wish to point out the advisableness of the other colonies being communicated with in order that the damage of the shipment may be pointed out.

MELBOURNE.

(Melbourne correspondence *Mark Lane Express*, February 17, 1890.)

AMERICAN VINES IN FRANCE AND THE PHYLLOXERA.

The gratifying showing of the rapid increase in the acreage of reconstituted vineyards in France, mostly by the use of American stocks, given in the last number of INSECT LIFE in the article entitled "The Phylloxera Problem Abroad, etc.," hardly leads one to expect the adverse report on the use of American vines given in the *Wine Trade Review* of February 15, 1890, and quoted in the Cape Colony *Agricultural Journal* of February 20.

The quotation is as follows:

An important movement is taking place in the department of Seine-et-Marne, in regard to the introduction of American *cépages* into the vineyards. Many people in France and other countries have been inclined to regard the grafting of French vines on American as one of the most certain methods of arresting the progress of the phylloxera; but it is clear that a different opinion is held in the Champagne country. The prefect of the Marne department last month directed that an inquiry should be opened on the subject, and a few days later the *Syndicat du Commerce des Vins de Champagne de Reims* drew up an important document, in which its views as to American plants were fully stated. In the opinion of the *Syndicat* the introduction of these plants would be infallibly followed by the phylloxera, since they are the conductors and propagators *par excellence* of the pest, and though they may be able to support themselves against it, they rapidly spread it around them. Considering that a great danger is threatened to the vineyards, the *Syndicat* makes an energetic protest against the employment of the American plants, and copies of the document have been sent to the mayors of the seventy-nine communes of the Marne department, as well as to the prefect. The views of the *Syndicat* on such a question as this will doubtless receive the weight they deserve, and then go a long way to indicate the probable result of the inquiry.

A NEW AUSTRALIAN VINE PEST.

We have recently received from the author, through the State Department, advance proof of an article by Charles O. Montrose, editor *Victoria Farmers' Gazette*, relating to a new vine pest which is reported to be seriously ravaging the vineyards, orchards, and gardens of New South Wales.

In this article Mr. F. A. A. Skuse is recorded as stating that the insect in question is a species of plant bug, probably undescribed, belonging to the family Capsidæ, and from the description given, it must be closely allied to our Tarnished Plant-bug.

They are said to attack particularly the fruit-stems of the Grape, Plum, Apple, etc., causing the fruit to dry up instead of ripening. They seem to prefer Plum leaves, and are reported to leave the grape and other plants untouched in the neighborhood of plum trees. They are, however, practically omnivorous, causing great injury to all the common fruits, cereals, and vegetables.

Mr. Montrose has promised to forward specimens, on the receipt of which we may refer to the subject again.

TROUBLE IN CALIFORNIA.

In a recent account of the meeting of the Los Angeles County orange-growers we notice that the board of supervisors has received a petition signed by sixty-seven parties asking for the removal of the board of horticultural commissioners on the ground that spraying is injurious to the trees, and that parasites have been discovered which are effectively cleaning off the White, Red, Black, and San José scales. They claim that spraying kills off the parasites and leaves the scales to "pursue their chosen avocation."

We consider this action short-sighted and unjustified. Proper spraying will not injure the trees, and no effective new parasites of the Red, Black, or San José scales have been discovered. The parasite of the Black scale, discovered by Professor Comstock in 1880 (*Dilophogaster californica* Howard), was at that date considered by him a very effective enemy of this scale, and it is safe to say that, after ten years of uninterrupted work of the parasite, this scale insect is as abundant in California as ever.

Protoparce celeus Hb. (Tomato worm).—Well known as destructive to the foliage of both potato and tomato, but was last autumn observed eating into the fruit of the tomato, six individual tomatoes in one instance being destroyed on a single vine where growing foliage was abundant, but this was scarcely eaten. The trouble was first attributed to fowls and later to sparrows, but both were proven innocent by the worms being surprised in the act.

Daremma catalpæ Bd. (Catalpa Sphinx).—Besides being exceedingly abundant, and the larvæ very destructive to young Catalpa trees in southern Indiana, I have found the larvæ also defoliating trees in the forests of Arkansas in May. Mr. John B. Smith, in his recent monograph of the Sphingidæ,* does not include territory west of the Mississippi River as within the distribution of this species.

Spilosoma virginica Fabr. (Yellow Woolly-bear).—The caterpillar was observed eating holes in ripe muskmelons at La Fayette, Ind., October 15. In one instance an excavation had been made in an otherwise perfect melon, over an inch in diameter, and fully half as deep.

Mamestra legitima Grt.—Adult moth reared during spring of 1889 from larva found feeding within seed pod of *Asclepias incarnata* near La Fayette, Ind., early in November, 1888. The larva appeared to subsist upon the seeds, the pod being attached unopened to the erect plant.

Prodenia lineatella Harv.—Nearly full grown larvæ observed at La Fayette, Ind., October 29, 1888.

Scoliopteryx libatrix L.—Adult moths reared at La Fayette, Ind., September 24. Parasite, *Ophion purgatum* Say, emerged from pupæ of this species October 29.

Aletia xylina Say (Cotton worm).—Adults captured in a large field of red clover near La Fayette, Ind., from about August 20 to October 15, 1889.

Phycis indiginella Zeller (Leaf crumpler).—From a large number of larval cases, collected late in February and placed in warm quarters, there emerged on March 7 two species of parasites, *Hemiteles variegatus* Ashm. and an undescribed species (No. 1092*a*) of *Apanteles*.

Plutella cruciferarum Zeller (Cabbage Plutella).—This pest of the cabbage appeared in some of the market gardens about La Fayette, Ind., during May, 1889, and did serious injury. The moths emerged in great abundance late in May, and about the 10th of June there appeared great numbers of parasites—*Phæogenes discus* Cress.

Wilsonia brevivittella Clem.—Adults of this species were reared from seed pods of Evening Primrose, *Œnothera biennis* L. The larvæ depredate upon the seed pods much as those of *Pronuba yuccasella* Riley do in the seed pods of the Yucca. The larvæ were first observed early in September. The exact date of appearance of moths was not noted, but it must have taken place very late in September, or during October.

* Trans. Am. Ent. Soc., Vol. XV, p. 205.

Callosamia promethea Drury.—The larvæ of this species was very abundant during the season of 1889, and the cocoons were to be found on wild cherry and sassafras in great numbers. Examination of these cocoons in March, 1890, developed the fact that fully two-thirds of them had been parasitized by *Ophion macrurum* Linn.

Agrotis herilis Grote (Western Striped Cut-worm).—In company with other cut-worms, this species is supposed to descend into the earth in the fall for the purpose of hibernating. The winter of 1889-'90, however, proved an exception, and the larvæ, usually about one-fourth to one-third grown, were observed on warm, sunny days during the entire winter feeding above ground upon young wheat in the field, and also upon grass in meadows and other grass lands.

Hadena stipata Morr.—On page 134, Volume II, INSECT LIFE, this species was incidentally mentioned as destroying young corn on newly broken grass lands. Since that notice was written reports of serious depredations have come to me from Clinton, Miami, Madison, and Johnson Counties, Ind., all indicating that this is the most destructive of all our cut-worms in the localities where it occurs; some fields being totally ruined, and that, too, after it is too late in the season for replanting. Both low and high lands, timothy and clover sod, seem alike attacked, even though the ground may have borne but one previous crop of grass or clover.

Lithophana antennata Walk.—Possibly on account of the extreme mild winter just passed these moths made their appearance very early in the season, several being captured at La Fayette, Ind., on the evening of February 24, 1890.—[F. M. Webster, March 29, 1890.]

THE PUNCTURING OF APPLES BY THE PLUM CURCULIO.

In a foot-note to Mr. Webster's article upon "Experiments with Plum Curculio," published on page 308 of the last number, we promised to publish in a future number the figure illustrating the condition of

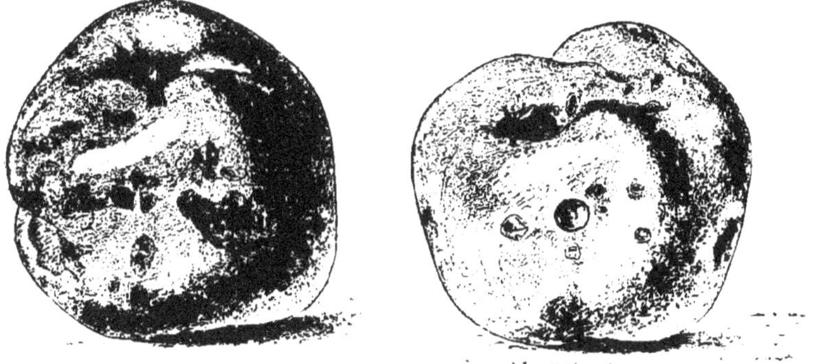

FIG. 71.—Plum Curculio punctures in young apples—natural size (original).

young apples found by Mr. Webster July 24, at La Fayette, Ind., and which illustrated a severe attack of the adult of the Plum Curculio.

The tree from which these apples were picked blossomed profusely and produced a good crop of young apples, but by July 24 all but two dozen had fallen to the condition of this. The figure is drawn from specimens picked on that date and sent in by Mr. Webster, and very well illustrates the work of the Plum Curculio, as we have often witnessed it as much as twenty years ago.

THE VEDALIA IN NEW ZEALAND.—RECENT INCREASE OF ICERYA.

* * * Going back to Vedalia. All parts of New Zealand have been importing plants from Australia for very many years, particularly Citrus and Acacia. Auckland was the first visited by Icerya, which was discovered on an imported plant (*Acacia undulata*), but I am not prepared to say the *individual* plant was imported. It is rather singular that in some districts it appeared first on Australian acacias (plants it seems even more partial to than Citrus, although it is not so rapidly fatal to them), *plants grown from seed* being the first attacked in districts. Auckland was also the first district cleared by Vedalia, and then Takapuna, Wairoa, South Waikomiti, etc., where the Citrus and other plants *were derived from Auckland.* Auckland was cleared so rapidly that no one knew how, till it was over, and it would have also remained a mystery with the other places had I not heard of it and gone there to find the cause. Napier and Nelson are the other two infested parts, and they import direct from Australia *and separately.*

At Napier they received the beetle later than Auckland, but at Nelson they have missed it, and up to the present time the Icerya *is going on unchecked.* I urged them to procure the beetle from Napier while it was yet time, and Mr. Maskell got them the Lestophonus flies (which have done no good). Hamilton sent them the beetles, but Mr. Maskell wrote to me a few posts ago to say that it was doing no good and did not seem to increase. The fact is I feel sure he has sent them the *wrong* insect. It is deplorable to see people making such mistakes and no properly qualified person to set them right. There is one thing I must beg to draw your attention to, and that is, that in my late tour round the North I find (as might be expected) Icerya *returning everywhere* and not a trace of Vedalia; in many places, and around Auckland in particular, it is increasing fast and bids fair to become as bad as ever. This should warn you to take care of Vedalia and conserve a few colonies; the reason is very obvious. * * * —[R. Allan Wight, Paeroa, Auckland, New Zealand, March 15, 1890.

THE PHYLLOXERA IN NEW ZEALAND.

* * * I am ashamed to say that our Government has positively refused to permit me to land any vines from the United States under any circumstances, for fear of importing *Phylloxera vastatrix*, of which a

fine specimen now stands before me in a bottle. I have just come home from a tour round the North, and I have seen it in two vineyards in our principal grape-growing country. I am disgusted. One man dug up the vines and burned them as soon as he was aware of it. The other refused unless his neighbors would pay him £10. What can I do for such a people as this? Maskell is advising the Government to compel all vine owners in infected districts to burn their vines, whether they are infected or not (the insect could do no more). I am advising them to severely punish people who refuse to burn infected vines, when it has once been pointed out to them, and to either compel or encourage others to shift on to proof roots. * * *—[R. Allan Wight, Paeroa, Auckland, New Zealand, March 15, 1890.

PROCEEDINGS OF THE ENTOMOLOGICAL SOCIETY OF WASHINGTON.

Number 4 of Volume I of the Proceedings of the Entomological Society of Washington has just been published. This number is furnished with an index to the whole volume which it completes. It contains about 100 pages and includes, among the shorter notes, papers by Mr. Schwarz, on the Coleoptera common to North America and other countries; notes on the comparative vitality of insects in cold water; stray notes on injurious insects in tropical Florida; notes on the Tobacco Beetle (*Lasioderma serricorne*); notes on *Cicada septendecim* in 1889; food plants and food habits of some North American Coleoptera; Myrmecophilous Coleoptera found in temperate North America, and sudden spread of a new enemy to clover (*Sitones hispidulus*); by Mr. Howard, note on the hairy eyes of some Hymenoptera; note on the mouth-parts of the American Cockroach; authorship of the Family Mymaridæ, and a few additions and corrections to Scudder's Nomenclator Zoölogicus; by Mr. F. V. Coville, notes on Bumble-bees; by Judge L. C. Johnson, the Jigger Flea in Florida; by Mr. Marlatt, swarming of *Lycæna comyntas;* an ingenious method of collecting Bombus and Apathus, and abundance of Oak-feeding Lepidopterous larvæ in the fall of 1889; by Baron Osten Sacken, correction to the monographs of the Diptera of North America, Vol. I, 1862; by Mr. Ashmead, some remarks on South American Chalcids; an anomalous Chalcid (*Hoplocrepis* n. g., *albiclavus* n. sp.), and remarks on the Chalcid genus *Halidea;* by Mr. Townsend, notes on some interesting flies from the vicinity of Washington, D. C.; on the fall occurrence of Bibio and Dilophus, and a further note on *Dissosteira* (*Œdipoda*) *carolina;* by Professor Lugger, on the migrations of the Milkweed Butterfly. In addition to these are many shorter notes by Professor Riley, Dr. Marx, Dr. Fox, Mr. Mann, and others.

The first volume, being now complete, may be obtained from the corresponding secretary of the society, Mr. Tyler Townsend, Department of Agriculture, Washington, for $3.

AN ACKNOWLEDGMENT.

In the March number of INSECT LIFE (p. 290), is a notice of my paper in *Garden and Forest*, on *Zeuzera pyrina*, which makes it necessary to credit the observations to those who gave them to me for use. The figures used were drawn by Mr. C. P. MacChesney, of Arlington, N. J., and were simply put into shape for engraving by me. Mr. Angelman found the larva, and the facts used all came to me from these gentlemen. Mr. MacChesney published his own observations in *Ent. Amer.*, VI, No. 2, and this paper must be credited as the scientific presentation of the matter rather than my popular account to which accident alone gave a date not intended and an apparent priority which it does not deserve.—[John B. Smith, Rutgers College.

THE GENITAL ARMATURE IN MALE HYMENOPTERA.

General Radoszkowski, at the meeting of the French Entomological Society, of September 11, 1889 (see Bulletin Entomologique, p. clxxii), presented a communication on the subject of the use of the male genital armature in Hymenoptera for the separation of species. Following in the line of the investigations of Dufour, Sichel, Fred. Smith, and E. Saunders, and adopting in the main the nomenclature of Dufour, and has found that these parts are of great value in the distinction of species, as they have proven to be with other orders. He has applied this method of diagnosis to more than 40 genera and 500 species. He has expressed himself as of the opinion that of all characters known among the Hymenoptera the form of the genitalia is the surest and most stable for generic and specific characters as well as for varieties. The forms examined seem to belong mainly to the *Anthophila*, *Mutillidæ*, and *Chrysididæ*.

THE MAN INFESTING BOT.

At the 27th of March, 1889, meeting of the French Entomological Society Mr. Émile Gounelle exhibited a larva taken from a man who came from Brazil, and stated that similar cases were not rare, particularly around St. Paul. Mr. Laboulbène added that he had also observed a similar larva taken from a Brazilian woman recently arrived in Paris. It was taken from a painful tumor and recognized as a species of *Dermatobia*. It was placed in a breeding cage, but died before transformation.

THE EGGS OF ATHERIX.

Mr. J. E. Ives, in the March number of *Entomological News* (p. 39), describes a mass of eggs taken from the under surface of a tree overhanging a small stream, which was determined by Dr. Williston as those of the Leptid genus *Atherix*. The same thing has recently been figured and described in England, and certain egg-parasites are also figured. Some thirteen years ago we collected a large number of these eggs upon the piling of Lake Minnetonka, near Minneapolis, and they have formed an interesting part of the Dipterological collection of the National Museum, while more recently we received a bit of piling from the shores of Lake Ontario which were covered with these eggs from which larvæ hatched which we were able to determine as belonging to this genus by comparison with the figures in Dr. Brauer's Monograph of Dipterous larvæ. Our correspondent stated that the wharf piles for hundreds of feet were covered with these eggs.

A MONOGRAPH OF THE EVANIIDÆ.

An important monograph of the family Evaniidæ has been completed by August Schletterer and published in three parts in the *Annalen des K. K. Naturhistorischen Hofmuseums*, Volume IV. Parts I and II bear the date 1889 and part III 1890. The species of the entire world are described by means of analytic tables and lengthy descriptions, and the synonomy is most carefully considered. The monograph is illustrated with 6 lithographic plates of morphological details. He places only the three genera *Evania* Fab. (*Brachygaster* Stephens, *Hyptia* Shuckard), *Gasteruption* Latr. (*Fœnus* Walk., and other authors), and *Aulacus* Jur. (*Aulacostethus* Philippi and *Pammegischia* Prov.) in the family Evaniidæ. The work as a whole is one of the most thorough and complete monographs which we have seen.

COLONEL PEARSON ON THE ROSE CHAFER.

In the article on the Rose Chafer, on page 295 of the last number, we neglected to make mention of an excellent article on this insect by Col. A. W. Pearson in the January 22 number of *Garden and Forest*, in which he states that subsequent experiment with Bordeaux mixture showed that it was not the specific which he formerly considered it to be. Last summer he made a solution of 1 ounce of good Pyrethrum to 2 gallons of water, first wetting the powder to a paste before mixing with all the water. On spraying the vines with this mixture the bugs became paralyzed and fell to the ground. Then he had men pass along both sides of the trellis and jar the vines and kill the bugs with paddles. Insect powder in this strength he finds does not kill them, but only temporarily stupefies them, and they will eventually recover and fly away. Meanwhile they will be quite easily destroyed for some time. As the testimony of a practical man this is of value.

THE COLUMBUS HORTICULTURAL SOCIETY.

We have been favored with No. 1, Volume V, of this Society, which contains some interesting entomological matter. The principal article is by Prof. D. S. Kellicott on "Our Injurious Ægerians." He gives a short account of thirteen species and illustrates upon a well executed plate the Peach Tree-borer, the Pear Tree-borer, the Imported Currant-borer, the Maple Tree-borer, and the Plum Tree-borer.

MR. BUCKTON'S MONOGRAPH OF THE BRITISH CICADÆ AND TETTIGIIDÆ.*

The mere announcement that Mr. G. B. Buckton was about to monograph the British species of the difficult group of insects above mentioned, was a sufficient indication that the work would be well and carefully done, and the two parts which we have before us fully justify our anticipations. The work resembles in character his well-known monograph of the British Aphididæ, although not published as was the former work by the Ray Society. The plates are drawn and lithographed by Mr. Buckton himself, and while a little rough in appearance admirably illustrate the characteristics of the different species. The parts contain 32 pages and 10 plates each, all of the plates being colored except two supplementary ones which indicate details of structure.

EARLY STAGES OF THE ODONATA.†

After many years Mr. Cabot has given us the third part of his monograph, which takes up twenty-three species of Cordulina with a number of forms in the genera Pantala and Tramea. The six lithographic plates are beautifully reproduced from drawings by the author.

INDIAN MUSEUM NOTES No. 3.

The third number of these insect publications has just reached us through the courtesy of Mr. E. C. Cotes. This number is devoted to a description of the "Silk-worms in India," and a surprising number of species actually reared for commercial purposes are treated. Aside from the Mulberry Silk-worm (*Sericaria mori*), the pamphlet considers *Bombyx fortunatus*, the *Desi* or *Chota Polo ; Bombyx crœsi*, the *Nistry* or *Madrassi; Bombyx arracanensis*, the *Nya Paw ; Bombyx sinensis*, the *Sina*, *Cheena*, or *Chota Pat ; Antheræa mylitta*, the *Tusser ; Attacus ricini*, the *Eri ; Antheræa assama*, the *Muga*. Four lithographic plates accompany the treatment of the species.

THE CHINCH BUG DISEASE.

Prof. F. H. Snow, in No. 1 of Volume XII of the Transactions of the Kansas Academy of Sciences (1889), pages 34 to 37, gives the result of his experiments for the artificial dissemination of a contagious disease among the Chinch Bugs. There is little further in this article than that summarized from the *Lawrence* (Kans.) *Daily Journal* on page 126 of the current volume. We repeat our caution as to the too ready acceptance of results of this character.

---

* Monograph of the British Cicadæ and Tettigiidæ, illustrated by more than 400 colored drawings by George Bowdler Buckton, F. R. S. London, Macmillan & Co., and New York, 1890. 8 parts. Price, 8s. per part.

† The Immature State of the Odonata. Part III. Subfamily Cordulina. By Louis Cabot, with 6 plates. Memoirs of the Museum of Comparative Zoölogy, Vol. XVII, No. 1, Cambridge, February, 1890.

STUDY OF THE BIRD LICE.

Mr. Vernon L. Kellogg, in No. 1, Volume XII of the Transactions of the Kansas Academy of Sciences (1889), pages 46 to 48, announces that he has noted and described twenty-four species of Mallophaga representing ten genera taken from Kansas birds. Among these are two new genera. He publishes a figure of *Tetrophthalmus* showing the respiratory system and gives a table of the genera. He has not named his new species, but has given them numbers. We see from this notice, that Mr. Kellogg has gone at this work in the right way, and we hope he will continue his studies.

THE TROPICAL SUGAR-CANE BORER IN LOUISIANA.

Never before have complaints of the tropical cane-borer been so pronounced over so extensive a territory as the present season. It is to be feared that, should the winter prove an open one, they may do very serious damage to the next crop. In 1857 they were so abundant along the lower coast as to have about destroyed the crops on one or two plantations. They again appeared in the same locality, and in Assumption and St. Mary, in large numbers, in 1880, after the open winter of 1879 They attack sorghum and corn in the same manner as cane, and are known near the coast throughout the Gulf States. The moth is of a light, grayish brown color, with about 1¼-inch spread of wings. This lays its eggs upon the leaves of the cane, near the axils, the young borers hatching in a few days. The borer penetrates the stalk at once, usually just above a node, working upward through the soft pith. The full grown borer is about 1 inch long, slender, cylindrical, and cream white in color, with yellow head and black mouth. Several broods are hatched in the course of a season. It is believed to hibernate almost exclusively in the larva or worm state. Those which find shelter in the stubbles, discarded tops and seed cane, alone escape destruction during the harvest of the crop Fortunately, few are found to burrow near the extreme butt of the cane If cut at the surface of the earth very limited numbers will, therefore, be preserved in the ratoons. A speedy burning of the tops, after removal of the crop from the ground, will destroy those which would be carried over to the next season by these. An immediate plowing under of all tops seems the next best alternative, but undesirable. They certainly should not be allowed to remain on the surface of the ground until warm spring weather.

Borers present in seed cane are not so easily dealt with. It is probable that from canes planted in the autumn and rolled the moth is unable to escape. The same is true in less measure of seed put down in windrow, if as heavily dirted as is compatible with the canes' safety. This should be dropped and re-covered as soon after removal from windrow in the spring as possible. Mats, both flat and round, are especially to be avoided for affected canes. It will be safest in all cases to put down as seed such canes as are least attacked. No abandoned forage sorghum should be allowed to go over the winter and corn stalks should also be plowed under, or be otherwise destroyed before winter is past. To neglect these precautions may be ruinous. There are, perhaps, more borers now in your field than you suspect.—[W. J. Thompson in *The Louisiana Planter*, Nov. 2, 1889, Vol. 3, p. 274.—The insect is probably *Chilo saccharalis.*

IMPORTATION OF HESSIAN FLY PARASITES.

With the assistance of Mr. Fred Enock, of London, England, we hope to import during the summer some living specimens of *Semiotellus nigripes*, a Russian parasite of the Hessian fly, in order to endeavor to acclimatize it in this country. Mr. Enock is rearing it extensively and hopes to be able to send us a good supply.

ENTOMOLOGICAL SOCIETY OF WASHINGTON.

April 3, 1890.—Mr B. E. Fernow was elected an active member of the society.

Mr. Fox read a paper on a small group of spiders forming the subgenus *Ceratinella* of the genus *Erigone*. The subgenus includes about seventeen species distinguished by the presence of a shield on the abdomen. All the specimens were collected east of the Alleghanies by Messrs. Marx and Fox, and were found fully developed at all seasons of the year. The paper was illustrated with drawings and a collection of the spiders was shown. Discussion followed by Messrs. Marx and Fox.

Mr. Schwarz read a paper entitled "Labeling Specimens," in which he described the systems of labeling employed in the case of entomological collections, dealing particularly with the systematic collection of the specialist. The various labels employed were described and examples of some of them were shown. The paper called forth considerable discussion which was participated in by Messrs. Riley, Mann, Schwarz, Marx, and Fox.

May 1, 1890.—The committee having in charge the preparation for publication of a list of the insect fauna of the District of Columbia made a partial report, which was discussed at length.

The name of Mr. Townsend was added to the subcommittee on Diptera, and that of Mr. Marlatt to the subcommittee on Hymenoptera.

A revision of the subcommittees will be made at the next meeting.

Mr. Townsend read a list of eighty-seven species of Heteroptera collected by him in southern Michigan, with some brief notes and dates of occurrence. One species, *Corimelæna nitiduloides* Wolff, was taken in a nest of *Formica schaufussi* Mayr.

Mr. Townsend also presented a paper on "Some insects affecting certain forest trees," mostly from Michigan, recording upwards of a hundred Coleoptera and a few of other orders, affecting either the foliage or the sound or decaying trunks of oak, hickory, elm, beech, linden, butternut, iron-wood (*Carpinus*), willow, hazel, etc.

These papers were discussed by Messrs. Schwarz and Riley.

Mr. Dodge read a paper on Artificial Silk, describing the Count de Chardonnet's method, as exhibited at the late Paris Exposition, of making from cellulose a substance closely resembling silk. A detailed account of the process of manufacture was given, illustrated with a figure of the device for producing the thread, and a sample of the silk was exhibited.

Discussion followed by Messrs Philip Walker, Riley, Amory Austin, and others.

Mr. Marx presented some "Arachnological notes" in which he discussed the comparative anatomy of the spinning glands of spiders. The relation of these to the external spinning organs or spinnerets and the importance of both in classification were explained. Careful drawings of the parts discussed were shown.

C. L. Marlatt,
Recording Secretary.

www.ingramcontent.com/pod-product-compliance
Lightning Source LLC
Chambersburg PA
CBHW021350210326
41599CB00011B/819

* 9 7 8 3 7 4 1 1 2 4 8 9 1 *